Handbook of Electrical
and Electronic Insulating Materials

Handbook of Electrical and Electronic Insulating Materials

Second Edition

W. Tillar Shugg
Shugg Enterprises, Inc.

IEEE Dielectrics and Electrical Insulation Society, *Sponsor*

The Institute of Electrical and Electronics Engineers, Inc., New York

This book may be purchased at a discount from the publisher when ordered in bulk quantities. For more information contact:

IEEE PRESS Marketing
Attn: Special Sales
P.O. Box 1331
445 Hoes Lane
Piscataway, NJ 08855-1331
Fax: (908) 981-8062

This book is a revised edition of *Handbook of Electrical and Electronic Insulating Materials*, ©1986 Van Nostrand Reinhold.

Printed in the United States of America

10 9 8 7 6 5 4 3 2 1

ISBN 0-7803-1030-6

IEEE Order Number: PC3780

Library of Congress Cataloging-in-Publication Data

Shugg, W. Tillar.
    Handbook of electrical and electronic insulating materials.—2nd
ed.
      p. cm.
    Includes index.
    ISBN 0-7803-1030-6
    1. Electric insulators and insulation—Handbooks, manuals, etc.
I. Title.
TK3401.S53   1995
621.319'37—dc20   94–44627
                                      CIP

To those developers of insulating materials making possible a century of progress for the electrical and electronics industries.

Contents

CHAPTER 3: Thermoplastic Molding Compounds   **35**

CHAPTER 4: Thermosetting Molding Compounds   109

CHAPTER 10: Dielectric Films   297

CHAPTER 11: Dielectric Papers and Boards   319

CHAPTER 12: Tapes and Coated Fabric   **333**

CHAPTER 19: Government Activities   491

APPENDIX   535

Foreword to the Second Edition

When conceived and created nearly a decade ago, this *Handbook of Electrical and Electronic Insulating Materials* fulfilled a compelling need seen by those involved with the insulation of electrical and electronic machines and devices. Early in this century, in the formative years of electrical insulation, several books on the subject were published and widely used. They thoroughly treated the fundamental properties of materials but were less rigorous in the practical and engineering uses of electrical insulation. New insulating materials such as polymers, plastics, gasses, and liquids appeared frequently in this period of phenomenal growth of electrical machines.

The first edition of this handbook addressed these issues, resulting in an extremely useful reference work which presented properties, uses, standards, and specifications of electrical insulating materials. In the years since the book's inception the pace of the development of electrical machines has somewhat diminished while the breakneck pace of the electronics industry continues unabated. The synthesis of new insulating materials also has slowed but innovative composites that utilize the best properties of their individual components have proliferated.

The Dielectrics and Electrical Insulation Society of the Institute of Electrical and Electronic Engineers leads the electrical and electronic industries worldwide in all matters relating to electrical insulation. It is natural for the DEI Society to have a continuing interest in this handbook. Recently, IEEE Press obtained the rights to this book and has worked closely with the DEI Society to support its revision. Tillar Shugg embarked on the strenuous effort to update and strengthen the considerable body of information found within, culminating in this revised second edition.

Like its predecessor, this authoritative second edition of the handbook is not intended for storage on a library shelf. Rather, it is designed as a desk volume to provide quickly accessible data, design input, comparisons, standards, and suppliers for insulating systems and materials. Every practitioner of electrical and electronics insulation science and technology will find this expanded and updated revision to be indispensable and of immediate and practical value. For those entering the industry, it is a fine educational resource.

RONALD N. SAMPSON
Technology Seminars, Inc.

Foreword to the First Edition

A half century of personal perspective over the 100 years of development in electrical insulation provides understanding of the monumental task needed and undertaken in this handbook. Fifty years ago, it was quite easy to catalog and describe the relatively few insulating materials in use. But even then, the electrical industry was concerned that too many insulating materials were available. For example, General Electric used about 12 drying oil-modified, asphalt, phenol and alkyd varnishes, and wanted to standardize on 8 or less!

Signs of change *in the opposite direction* were already apparent. Glass fiber was being introduced to replace asbestos, but instead ultimately largely replaced cotton, linen, and rayon, Mechanically tough polyvinylformal resin could and would replace bulky cotton-wrapped wire insulation. PVC, plasticized polyvinylchloride, allowed the development of a wholly new concept in extruded cable insulation. Electrical equipment could be made smaller, more efficient, and less expensive. It was an exciting time, and the opportunities seemed and proved to be almost endless.

Twenty-five years later, the number of available insulating materials had grown so rapidly that two comprehensive books, *Die Isolierstoffe der Electrotecknik (Insulating Materials for Electrical Practice)* by W. Oburger and *Insulating Materials for Design and Engineering* by Frank Clark, were written to meet the tremendous need. In the last 25 years, the growth has accelerated so fast that the task of providing an up-to-date handbook for electrical insulation appeared overwhelming. The vast number of insulating materials, which are integrated into complex insulation systems in many kinds of electrical equipment for a host of specialized applications with different environments, indeed makes an overall complete description nearly impossible.

It is fortunate that the enormity of the task did not stop the preparation of this handbook, which provides an important step in the overall evaluation of insulation systems and their application. It is necessary, of course, to recognize that the properties of individual insulating materials vary and also depend upon the way the materials are used. Nevertheless, the ability to compare property values for many insulating materials in a comprehensive fashion provides a base, along with practical application experience, for evaluating insulating materials to be used in the many types of electrical equipment.

KENNETH N. MATHES, P.E.
Consultant on Electrical Insulation
Schenectady, NY

Preface

Prior to this undertaking, there has not been a handbook in print in the United States devoted solely to dielectric materials. This *Handbook of Electrical and Electronic Insulating Materials* is intended to provide a concise but comprehensive reference for all engineers and material specialists in and serving the electrical and electronics industries, as well as those persons with special interest in a particular class of materials. This handbook should fill a wide gap in business, university, and public libraries. It should also be useful as a reference and text in engineering and science courses.

In addition to a word-by-word editing, review, and updating of the first edition, this second edition has been enhanced to include these new and significantly augmented sections:

- Overview of Our Heritage
- Partial Discharge
- Radiation
- Temperature Index of Dielectric Materials
- SI System for Metric Practice
- Definition of Terms Relating to Plastics
- Table showing U.S. consumption of thermoplastics in electrical applications
- Alloys
- Liquid Crystal Polymers
- Conducting Polymers
- Polyacrylate (PAR) Molding Compouneds
- Polyetheretherketone (PEEK) Molding Compounds
- Table showing U.S. consumption of principal thermosetting plastics in electrical applications

- Flooding and Filling Compounds
- Curing Systems
- Polyetherimide (PEI) Film
- Tape Selection
- Hazard Communication Program
- National Institute of Standards and Technology
- Table of Temperature Equivalents

ORGANIZATION OF HANDBOOK

This handbook follows the following format:

- *Preface* (this section), which covers uses for the handbook, its organization, materials covered, sources of information, and acknowledgments.
- *History*, which lists the historical events and personalities of the electrical and electronic industries and presents an overview of our heritage.
- *Fundamentals*, which includes sections on the importance of U.S. and international standards, insulation systems and service life, material aging and breakdown, corona discharge, dipole moment and polarization, material selection steps, temperature index of dielectric materials, standard tests, and the International System of Units (SI).
- Each product group chapter is organized in this way:
  Comprehensive table of contents
  Introduction, including background and overview
  Technology for the group
  Standards and specifications applying to the group
  Manufacturing processes applying to the group
  Development programs
  Market trends
  Summary of properties of materials in the group
- Each insulating material in a product group has this treatment:
  Introduction, including principal producers
  Chemistry (Technology), including formula and manufacturing process
  Standards and specifications applying to this material
  Grades
  Processing methods
  Properties
  End uses
- A chapter on industry activities which lists organizations serving the electrical and electronics industries, summarizing their activities and principal publications. Industry conferences, periodicals, and key reference encyclopedias are also listed.
- A chapter on government activities, including detailed information on Occupational Safety and Health Administration programs relating to worker health and

safety, and Environmental Protection Agency programs aimed at ensuring a clean, safe environment. Also included is information on agencies developing and distributing standards and specifications, and an example of a specification on one of the newer engineering thermoplastics.

- An Appendix, containing conversion factors and formulas, measures and weights, and a sequential listing of key industry and Federal standards and specifications referenced in this handbook. Titles to documents are as they appear on the documents themselves, which may result in some inconsistencies in format.
- Index, comprehensive and cross-referenced for easy use.

MATERIALS COVERED

As shown in the chapter headings listed in the contents, virtually all dielectric materials are discussed according to their broad product group classification, for example, thermoplastic molding compounds, dielectric films, tapes, and coated fabrics. Thus, a certain polymer may be considered in several chapters. For instance, polypropylene appears in these chapters: *Thermoplastic Molding Compounds, Extrusion Compounds,* and *Dielectric Films.* And polyesters are included in these chapters: *Thermoplastic Molding Compounds, Embedding Compounds, Magnet Wire Enamels,* and *Insulating Coatings and Impregnants.* This format is employed since a reader is more often interested in considering all candidates for a specific application, say insulating varnishes, than in researching all possible applications for a specific material, say polyesters. Someone with the latter need will find the index more helpful than the chapter titles.

While I have endeavored to include all commonly used insulating materials in this handbook, there are noteworthy exceptions:

- *Asbestos-base materials*, which are being replaced by other products as a result of the Environmental Protection Agency enforcement of the Toxic Substances Control Act (see Chapter 19, *Government Activities.*)
- *Adhesives*, which are covered in the product group chapters where they apply, for example, in these chapters: *Unclad and Clad Structures* and *Tapes and Coated Fabrics* (pressure-sensitive).

TIMELINESS

The information and data in this second edition are current as of 1994.

SOURCES OF INFORMATION

The principal sources of information for this handbook are the following:

- Manufacturers' catalogues, bulletins, and data sheets
- Attendance at insulation conferences

- Personal discussions and correspondence with authorities
- Papers written by authorities
- Books edited or written by authorities
- Encyclopedias
- ASTM, NEMA, and UL standards
- IEEE publications
- Federal (mostly military) specifications
- Trade magazines
- Patents
- The author's experience

W. TILLAR SHUGG

Acknowledgments

During a lifetime of association with the electrical and electronics industries, there have been thousands of contacts, too numerous to mention, who have helped me acquire an interest in and knowledge of insulating materials. My earliest influence was John Shugg, an uncle, who was among the pioneers of insulating varnishes and varnished cambric at General Electric, and also for a while at Schenectady Varnish (now Schenectady International) and Sterling Varnish (now part of P. D. George). It was his varnished cambric business that I ran through World War II.

Later, at General Electric and Continental-Diamond Fibre, it was Harry Collins who provided opportunities to become involved with most of the materials covered in this handbook. E. O. Hausmann, also formerly of Continental-Diamond, was a great help in product manufacturing and technology. An intensive exposure to mica, tubing, and insulated wire was made available by William Brand of the William Brand Company (now Brand-Rex).

Irving Skeist and Jerry Miron of Skeist Laboratories afforded me opportunities to work on many in-depth studies of insulating materials and their markets. Irving was a consulting editor for the first edition of this handbook.

Special thanks go to Tor Orbeck, Ron Sampson, John Tanaka, Ken Mathes, Phil Alexander, and William Bentley, my colleagues at IEEE, for their guidance and encouragement on this project. This second edition would not have been possible without the enthusiastic support of Greg Stone and Ron Sampson of the IEEE Dielectrics and Electrical Insulation Society and Dudley Kay of IEEE Press.

I am grateful to all who gave their time and expertise in reviewing the chapters:

First Edition
Douglas Bannermann, National Electrical Manufacturers Association
William H. Bentley, Bentley Enterprises

J. Chottiner, Westinghouse

Alan H. Cookson, Westinghouse

Steinar J. Dale, Oak Ridge National Laboratories

E. J. Fisher, Union Carbide

A. N. Hamilton, Du Pont

E. O. Hausmann, Continental-Diamond Fibre

Ralph B. Jackson, Allied Corporation

Lee O. Kaser, Du Pont

John J. Keane, General Electric

Robert H. Lampack, Conap

Lyon Mandlecorn, Westinghouse

Kenneth N. Mathes, General Electric

J. F. Meier, Westinghouse

Jerry Miron, Skeist Laboratories

S. D. Northrup, RTE Corporation

Tor Orbeck, Dow Corning

William Palladino, Conap

Lee J. Payette, Essex Group

R. S. Raghava, Westinghouse

Howard Reymers, Underwriters Laboratories

George L. Richon, John C. Doloph

Ron N. Sampson, Westinghouse

Irving Skeist, Skeist Laboratories

J. D. B. Smith, Westinghouse

Ron Staley, General Electric

William A. Thue, Florida Power & Light

William W. Wareham, Phelps Dodge

Harrison G. Wertz, Schenectady Chemicals

Charles R. Willmore, National Electrical Manufacturers Association

Finally, my gratitude to the IEEE Press staff for their thorough and speedy editing of the second edition, and to Eileen, my wife, for her counsel, her proofreading, and for enabling me to work free of distractions during the more than three years it took me to write this handbook, and the two years to prepare its second edition.

W. TILLAR SHUGG

—————— 1 ——————

History

INTRODUCTION

Electrical and electronic insulating materials, also called dielectric materials, are essential to the proper operation of all electrical and electronic equipment. In fact, equipment size and operating limitations are dictated by the type and amount of material required for insulation. In the early days of the electrical industry, engineers had to adapt wood-finishing varnishes, natural resins, coal tars, and petroleum asphaltic residues as saturants and coatings for tapes used to wrap coils and cables. Today, a virtually unlimited number of insulating materials is available, and the problem becomes one of selecting rather than adapting.

Significant events and personalities causing the growth of the electrical and electronics industries (and with them the insulation industry) are shown in Table 1-1.

No listing of people making major contributions would be complete without including these industry giants:

Charles Proteus Steinmetz (1865–1923) who, while at General Electric, developed the mathematics and the theory of the alternating current.
Marchese Gugliemo Marconi (1874–1937) who produced a practical wireless telegraph system in 1895 and the first transatlantic wireless signal in 1901. He received the Nobel prize in physics in 1909.

An overview is also given of the chronology of insulating materials development with an assessment of future market requirements and trends.

OVERVIEW OF OUR HERITAGE

The beginning of human awareness of electromagnetic phenomena dates back into antiquity. Early humans were mystified by lightning and by the magnetic properties of amber when rubbed with fur, and lodestone, or magnetite as it is now known. Ben Franklin began his electrical experiments in 1747, and later postulated the theory that there are two kinds of electricity, positive and negative. In recognition of his scientific contributions he received honorary degrees from the University of Saint Andrews and the University of Oxford. In addition, he became a Fellow of the Royal Society of London for Improving Natural Knowledge.

It was not until the 1800s, however, that electromagnetic induction was studied and explained by Michael Faraday and Joseph Henry. Faraday also established the principle that different dielectric materials have their own specific inductive capacities, more often referred to now as dielectric constants. In 1855, Faraday completed work on his landmark book, *Experimental Researches in Electricity*.

Henry went on to develop electromagnets of the same form later used in dynamos and motors, and in 1831 built one of the first motors. In later years, he became Director of the newly formed Smithsonian Institution and President of the National Academy of Sciences.

As more electrical equipment was developed to service the telegraph communications industry beginning in 1844 with the first public telegram, and in 1877 with the

**TABLE 1-1   Historical Events and Personalities
of the Electrical and Electronics Industries**

DATA	EVENT	INVENTOR OR ORGANIZER	COUNTRY
1828–1832	Electromagnetic induction	Joseph Henry	U.S.
		Michael Faraday	England
1837	Five-needle telegraph	Charles Wheatstone	England
		William Cooke	England
1838	Simple key telegraph	Samuel Morse	U.S.
		Alfred Vail	U.S.
1873	Treatise on Electricity and Magnetism	James C. Maxwell	Scotland
1876	Telephone	Alexander Bell	U.S.
1879	Incandescent lamp	Thomas Edison	U.S.
		Joseph Swan	England
1883	Transformer	L. Goulard	England
		J. D. Gibbs	England
1884	American Institute of Electrical Engineers founded	Nathanial Keith	U.S.
1884	Trolley car	Frank Sprague	U.S.
		Van De Poele	U.S.
1884	Steam turbine	Charles Parsons	England
1886	Electrical resistance welding	Elihu Thompson	U.S.
1887	Induction motor	Nikola Tesla	U.S.
		Galileo Ferraris	Italy
		Friedrich Haselwander	Germany
1892	Electric vehicle	(several)	France
1906	Triode vacuum tube	Lee De Forest	U.S.
1907	Bakelite	Leo Baekeland	Belgium
1912	Radio signal amplifier	Edwin Armstrong	U.S.
1912	Institute of Radio Engineers founded	John Stone	U.S.
		John Hogan	U.S.
1913	Heterodyne radio receiver	Reginald Fessenden	U.S.
1914	High-frequency alternator	E. F. W. Alexanderson	U.S.
		Reginald Fessenden	U.S.
1921	Oil filled power cable	Emanueli	Italy
1927	Electronic television	P. T. Farnsworth	U.S.
1930	Nylon	Wallace Carothers	U.S.
1930	Neoprene	Wallace Carothers	U.S.
1946	Electronic Numerical Integrator and Computer (ENIAC)	John Mauchly	U.S.
		J. Presper	U.S.
1947	Transistor	William Shockley	U.S.
		Walter Brattain	U.S.
		John Bardeen	U.S.
1958	First Electrical/Electronics Insulation Conference		
1961	First commercial integrated circuit	Fairchild Corp.	U.S.
1963	Institute of Electrical and Electronics Engineers (IEEE) formed from American Institute of Electrical Engineers and Institute of Radio Engineers		U.S.

formation of the Bell Telephone Company, it became necessary to also develop improved insulations over those then available, such as paper, woven fabrics, wood, mica, glass, porcelain, shellac, rosin, copal, rubber and gutta percha, asphalt, linseed and China wood drying oils, and petroleum and coal tar distillates. These materials, of course, are still widely used, but in greatly refined forms and with significantly improved purity.

The first synthetic resins, the reaction products of phenol and formaldehyde, were developed by Belgium chemist Leo Baekeland in 1907 and became available a few years later. Today, phenolic resins are widely used as modifiers to improve temperature resistance of other resins, permitting some formulations to qualify as Class 180 C insulations where they offer low-cost alternatives to silicone varnishes. Molded phenolics continue to be used in a wide variety of heavy duty electrical applications.

Aminos were the next thermoset resins to be widely used for electrical insulations. The two principal types of these resins are urea-formaldehyde and melamine-formaldehyde condensation products of the reaction of urea and melamine with formaldehyde. Melamine compounds are generally superior to urea compounds in resistance to acids, alkalis, heat, and boiling water. Parts made of urea shrink on aging and tend to crack around inserts and sharp corners. Melamine exhibits these characteristics to a lesser extent, and is usually preferred for critical applications.

Another early insulating material, still in use, is vulcanized fibre, used for its high mechanical strength where moisture is not a problem, such as in arc barriers, air circuit breakers, lightning arrestors for heavy duty transformers, switch and appliance parts, and slot liners for rotating equipment. Vulcanized fibre is made by saturating rag-base or woodpulp paper, layered to the desired thickness, in a bath of zinc chloride. This causes the paper to gelatinize and bond into a homogeneous sheet. The zinc chloride is then leached from the sheet with excess water from a nearby stream or river. This process became a problem with the advent of the Clean Water Act of 1977.

The first electrical grade tape, friction tape, was marketed in the late 1920s. It was made by applying to a fabric a soft and extremely tacky mixture of rubber and a softener by means of a three-roll calendar. During World War II, the shortage of natural rubber encouraged the development of synthetic rubber adhesives. Since that time, both products and markets for pressure-sensitive tapes have grown phenomenally as a wide variety of adhesives applied to every conceivable type of backing has become available.

Alkyd saturated polyesters (glyptals) became available in the mid-1920s, and soon became the most widely used coatings and impregnants for application to all types of electrical equipment. Alkyds are produced by partial esterification of a polyhydric alcohol, such as pentaerythitol, with a fatty acid derived from vegetable oils, for example, linseed and China wood (tung) oils, followed by reaction with a dibasic acid, such as phthalic acid. Alkyds may be blended with other varnishes or reacted with other resins, including phenolics for improved water and alkali resistance, epoxies for better adhesion, and silicones for higher thermal and dielectric properties.

Polychlorinated biphenyls (askarels), introduced in the U.S. in 1929 from Europe, soon became the premier liquid dielectric for transformers and certain capacitors. They were, however, phased out in the early 1980s by the Toxic Substances Control Act, although electrical industry leaders in 1994 are asking for a review of the regulations and urging that PCBs again be permitted for use with adequate safeguards. There is as yet no single fluid type suitable for all applications, but rather there is an assortment of fluids, each of which is tailored to a certain end use. Liquids now available include mineral (petroleum) oils, high-molecular-weight paraffinic oils, silicones, and other synthetic oils.

The 1930s brought forth several polymers still widely used. Du Pont's Wallace Carothers discovered nylon in 1930. Nylon, a polyamide polymer, became commercially available in 1938, and was invaluable as a replacement for silk in parachutes and in ladies' stockings when silk became unavailable in World War II. Despite its tendency to absorb moisture, nylon is used extensively as a jacket over primary wire insulation and in molded parts such as coil forms, insulator blocks, and electrical connectors. Nylon 11, a more expensive, low-moisture-absorption grade, is also used as jacketing for aircraft control cable and for hydraulic tubing and hose. The type designation 6/6 shows the number of carbon atoms in each part of a nylon molecule formed by a dibasic acid and a diamine, in this case, adipic acid and hexamethylene diamine. Type 11 indicates there are 11 carbon atoms in the monomer of a nylon formed by polymerization of the ring compound omega-amino undecanoic acid.

Carothers in 1931 also developed Neoprene™ which, chemically, is polymerized chloroprene, useful as a cable jacket for its high resistance to heat, oils, and gasoline.

From pilot plant production in the early 1930s, polyethylene has become by far the most versatile wire and cable insulation. Low-density polyethylene (LDPE), the first polyethylene resin, is produced by polymerizing ethylene gas into long polymer chains in an autoclave, or tublar, reactor. There followed a number of other polyethylene types with designations describing several different synthesis processes. Some examples are LLDPE (linear low-density polyethylene), HDPE (high-density polyethylene), HMW-HDPE (high-molecular-weight high-density polyethylene), and XLPE (crosslinked polyethylene). Polyethylene is often copolymerized with other resins to form useful products such as ethylene propylene diene rubber (EPDM) and ethylene-vinyl acetate (EVA).

The first successor to plain enamel for magnet wire insulation was polyvinyl formal enamel (Formvar™), in use since 1938 and still used extensively where solderability is not required. Monsanto is the developer of the resin made by condensation of formaldehyde and hydrolyzed polyvinyl acetate.

The first fluoropolymer, polytetrafluoroethylene (PTFE), was discovered at Du Pont in 1938, but because early production was devoted to military products before and during World War II, it was not commercially available until 1947. Since that time, the family of fluoropolymers as grown to include several other materials, including perfluoroalkoxy (PFA), fluorinated ethylene-propylene copolymer (FEP), ethylene-tetrafluoroethylene-(ETFE), polyvinylidene fluoride (PVDF), ethylene-chlorotrifluoro ethylene (ECTFE), and polychlorotrifluoroethylene (PCTFE).

A distinction is sometimes made among the terms *fluoropolymers*, *fluorocarbons*, and *fluoroplastics*. *Fluoropolymers* is the generic term which includes fluorocarbons and fluoroplastics. *Fluorocarbons* consist of polymers with only carbon-fluorine bonds. *Fluoroplastics* may have, in addition to carbon-fluorine bonds, carbon-hydrogen and carbon-chlorine bonds. These materials all have in common the widest operating temperature ranges of all plastics, outstanding resistance to attack to chemicals, excellent dielectric properties, and virtually no tendency to absorb water. They are, however, high priced.

Polyvinyl chloride (PVC) was first used in wire and cable insulation and jacketing in the late 1930s. During World War II, its use spread as replacement for rubber, then in short supply. Polyvinyl chloride is one of the most widely used materials in extrusion applications, as producers and fabricators have successfully complied with EPA and OSHA standards and regulations, which in 1974 and 1975 appeared to threaten the survival of the industry. Polyvinyl chloride resins alone are inherently hard and brittle at temperatures up to 180°F (82°C). To be useful, they must be compounded with plasticizers and other additives. A typical electrical grade formulation would contain 50 percent polyvinyl chloride, 25 to 35 percent plasticizer, and the rest other additives. Important considerations in selection of a plasticizer are its compatibility, volatility, tendency to migrate to other materials, toxicity, odor, burning characteristics, dielectric properties, weatherability, and efficiency (concentration required to be effective).

Also in the late 1930s, acrylics, or polymethyl methacrylate (PMMA) resins, were produced for commercial use by Rohm and Haas (Plexiglas™) and Du Pont (Lucite™). A decreasing dielectric constant with increasing frequencies makes acrylics attractive candidates for high-frequency applications. High arc resistance makes them suitable for circuit breakers and other high-voltage applications. They are being used increasingly in fiber optics. They are not suitable where flammability may be a problem.

The first general purpose polystyrene resins were marketed in the U.S. in 1939 by Dow Chemical. The flammability, low impact strength, tendency to craze, and susceptibility to attack by hydrocarbons limit electrical/electronic applications. End uses include computer tape and video cassette reels and compact battery cases. When copolymerized with acrylonitrile, however, styrene-acrylonitrile resin (SAN) is formed with improved toughness, tensile strength, heat distortion, and chemical resistance. When a third monomer is added by grafting SAN on butadiene polymer to make acrylonitrile-butadiene-styrene resin (ABS), a wide range of properties is obtainable, permitting resins to be tailored to specific end uses with emphasis on toughness.

Silicone resins and elastomers, developed by Dow Corning and General Electric, although using different processes, became available in the middle 1940s. The Dow Corning process uses a Grignard reagent, while General Electric employs a direct process. Varnishes and elastomers made from silicones are useful from –50°C (–58°F) to 250°C (486°F), have excellent dielectric properties and moisture resistance, and are among the most resistant of all materials to corona attack. Mechanical properties of silicones are unpretentious.

Epoxies, introduced in 1947, are predominant in embedding compounds and second in usage in insulation varnishes. In combination with phenolics, they make useful novolac thermosetting molding compounds. The basic epoxy resin is the polycondensation product of bisphenol-A and epichlorohydrin, but cycloaliphatic epoxies are also widely used. To be useful, epoxies must be cured by the action of hardeners, including amines, polyamides, anhydrides, and urea-formaldehyde or phenol-formaldehyde resins. The most commonly employed ratio of epoxy resins to hardener is one to one, but may range as high as 100 parts epoxy resin to one part hardener. Epoxies are noted for their excellent dielectric properties, low shrinkage during cure, superior adhesion to most surfaces, good thermal properties, and good chemical and moisture resistance.

About the same time, polyurethanes (PURs) came into general use in magnet wire enamels where they permit soldering without stripping. PURs are also widely used in coatings for their abrasion resistance, and in embedding compounds where, in combination with butadiene, they have overcome their formerly poor hydrolytic stability. The basic building blocks of PUR resins are di- or polyisocynates which react with polyols to form the PUR resin.

Polypropylene (PP), closely related to polyethylene both chemically and in usage, became commercially available in the mid-1950s. The Rural Electrification Administration and cable manufacturers recognize polypropylene and high-density polyethylene as alternatives for each other. The choice for usage depends heavily on cost. The prophylene homopolymer has a relatively high brittleness temperature, but by copolymerizing propylene with 2 to 20 percent ethylene, a resin is formed with significantly more useful properties.

Polycarbonate (PC) resins were discovered in the U.S. at General Electric and were marketed in 1950. The resin, a polyester of carbonic acid, is obtained from bisphenol A and other polyhydric phenols, reacted with phosgene. Polycarbonate resins are noted for their good dielectric properties and corona resistance. Alloys with other polymers are available and are used for a wide range of commercial applications.

The first commercially available acetal homopolymer resin was introduced by Du Pont in 1959 with the trademark Delrin. Two years later, Celanese, now Hoechst Celanese, marketed an acetal copolymer resin Celcon™, which competes for essentially the same end uses, many of which were formerly served by metals.

About 1960, sulfer hexafluoride (SF$_6$) became commercially available from Allied Chemical, now AlliedSignal. It is still the premier dielectric gas for use in circuit breakers for current interruption in high-voltage equipment. SF$_6$ is also the gaseous insulation for high-voltage coaxial power transmission lines with voltage ratings up to 765 kV for use above ground, underground, and underwater where their high cost is justified.

The mid-1960s saw commercialization of several resins with outstanding properties: polysulfone developed by Union Carbide (Udel™), noted for its excellent heat and water resistance; modified polyphenylene ether (PPE), formerly called polyphenylene oxide, discovered at General Electric (Noryl™), featuring good dielectric properties over a wide range of humidity and temperature; and polyimide (PI), produced by Du Pont (Pyre ML™ for wire enamel and Kapton™ for film) with one of the highest thermal ratings.

Polybutylene terephthalate (PBT) resins, introduced in 1969, are thermoplastic polyesters widely used, when reinforced with glass fibers, for industrial motor controls, circuit breakers, terminal blocks, and housings for appliances and hand power tools.

Polyphenylene sulfide (PPS), developed by Phillips Chemical (Ryton™), became commercially available in 1972. Its outstanding property is resistance to thermal degradation, which occurs completely only at temperatures over 1300°F (704°C). It is suitable for applications requiring good dimensional stability, chemical resistance, dielectric properties, and resistance to high temperatures.

Polyethersulfone resins were introduced in the United Kingdom and then in the United States in 1972–1973. They were first produced by ICI Americas (Victrex™). Continuous use temperature index rating for unfilled resin by Underwriters Laboratories, 356°F (180°C), is among the highest for any thermoplastic.

Another high-temperature polymer, polyamideimide (PAI), was brought to market by Amoco in 1973 with the trademark Torlon. Resins have stable dielectric and mechanical properties up to 500°F (260°C).

Specialty resins introduced since 1980 includes polyetherimide (PEI) by General Electric (Ultem™), polyacrylate, with several producers, and polyetherether ketone (PEEK), also with several producers.

So much for chronology, but what about future product development and market trends? No doubt, research and development efforts will continue on polymers, although there is now available a multitude of products from which to choose for any given application. The emphasis will probably be on specialty polymers, such as liquid crystal polymers, alloys, blends, and polymers for cryogenic applications.

The search for improved dielectric liquids will certainly continue, but sulfur hexafluoride will remain the premier dielectric gas.

Efforts will increase to eliminate volatile organic compounds (VOCs) from varnishes and wire enamels to stay in compliance with increasingly stringent environ-regulations.

Insulations for electronics must keep pace with the exciting developments in that technology and in the continuing trend toward miniaturization.

The problem with all this is that insulation technology is an interdisciplinary field involving chemistry, chemical engineering, electrical engineering, mechanical engineering, and quality control. Most practitioners were chemists, electrical engineers, or without formal education, but with a great deal of on-the-job experience, most of which was limited to a few specific applications. With deemphasis on producing electrical insulations by our largest electrical equipment manufacturers, career opportunities in our field are severely limited.

Probably the greatest single opportunity for education and information on new insulating materials and technology is by attending and participating in conferences here and overseas. Only in this way will those involved with dielectric materials have a common forum where ideas and practices can be cross-fertilized with our peers.

2

Fundamentals

INTRODUCTION

This chapter covers subjects which apply generally to all insulating materials and the reader is encouraged to read it next.

U.S. AND INTERNATIONAL STANDARDS

The real test of the suitability of a material for a specific application is how well the part made from the material performs in actual use when subjected to the full range of conditions it is likely to encounter. This method, however, is usually not practical due to the time (often years) it requires. Manufacturers need quicker ways to determine material suitability. To answer this need, there is an ongoing program to develop and refine tests for selected critical properties to make them more indicative, more useful, more precise, and more reproducible. Thus, users of materials have come to rely on standards as the basis for purchase transactions with producers.

All of the industrialized nations of the world have active standards development programs. The U.S. and Russia each have more than 80,000 active specifications and standards, followed by Germany with 24,000, and the Peoples Republic of China with 14,000.

Standards are important documents in many ways:

• An industrial standard comprises definitions, classifications, requirements, test methods, and recommended practices. (The military uses the word *specification* as nonmilitary organizations use the word *standard*.)

• Standards are intended to establish a common understanding between manufacturers and users regarding requirements for materials and the methods to be employed in determining property values.

• Standards were originally developed for products made and used within a country, but because of increasing international trade, there is a growing need for standards acceptable throughout the world. There are now more than 10,000 international standards, of which about 100, developed by the International Electrotechnical Commission (IEC), are concerned with materials for the electrical and electronics industries.

• Purchases for military products and equipment are virtually all made to specification. Because of this practice and the huge volume of expenditures, the Department of Defense (DOD) is by far the largest developer of specifications in the western world. There are over 9,000 specifications for military products and equipment and over 4,700 specifications for materials. The DOD participates in developing nongovernmental standards and adopts these where possible, cancelling the related military specifications.

• In the U.S., the associations most active in the development of standards for insulating materials are the American Society for Testing and Materials (ASTM), the American National Standards Institute (ANSI), the National Electrical Manufacturers Association (NEMA), the Institute of Electrical and Electronics Engineers (IEEE),

Underwriters Laboratories (UL), and the Institute for Interconnecting and Packaging Electronic Circuits (IPC), with ASTM having by far the most standards covering the widest variety of products.

Brief descriptions of principal U.S. and international standards preparing organizations, their publications, and their conferences appear in Chapter 18, *Industry Activities*.

Throughout this handbook, a diligent effort has been made to summarize or reference all current (at the time of publication) military specifications and ASTM, NEMA, UL, and IPC standards relating to the materials covered. A complete military specification for a plastic material is included in Chapter 19, *Government Activities*.

INSULATION SYSTEMS AND SERVICE LIFE

An insulation system is an assembly of insulation materials in a particular type of equipment, as defined by Military Specification MIL-E-917, *Electric Power Equipment, Basic Requirements*. Although there are differing opinions on the subject among some authorities, this specification states the case for determining service life of equipment as follows:

> Experience has shown that the thermal life characteristics of composite insulation systems cannot be reliably inferred solely from information concerning component materials. To assure satisfactory service life, insulation specifications need to be supported by service experience or life tests. Accelerated life tests are being used increasingly to evaluate the many new synthetic insulating materials that are available thus shortening the period of service experience required before they can be used with confidence. Tests on complete insulation systems, representative of each type of equipment, are necessary to confirm the performance of materials for their specific functions in the equipment. The electrical insulation of equipment is made up of many different components selected to withstand the widely different electrical, mechanical, and thermal stresses occurring in different parts of the structure. How long an insulation system will be serviceable depends on the effectiveness of the physical support for the insulation, and the severity of the forces acting on it, as well as on the materials themselves and the service environment. Therefore, the length of useful life of the insulation system will depend on the way that its individual components are arranged, the interactions upon each other, and the contribution of each component to the electrical and mechanical integrity of the system.

UL STANDARDS FOR INSULATION SYSTEMS

Underwriters Laboratories has developed the following standards for insulation systems:

- UL 1097, *Double Insulation Systems for Use in Electrical Equipment*
- UL 1446, *Systems of Insulating Materials—General*

• UL 2097, *Reference Standard for Double Insulation Systems for Use in Electronic Equipment*

MATERIAL AGING AND BREAKDOWN

The principal factors generally recognized as causing the aging and deterioration of an insulation are the following:

• *Thermal stresses* occurring in electrical and electronic equipment. These are caused by internal heating due to current overloads plus ambient temperatures. When exposed for prolonged periods above a temperature specific for each material, chemical and physical breakdown rapidly accelerates, aggravated by mismatches in thermal expansion.

• *Electrical stresses* caused by the voltage gradient in the material. Most equipment is designed with ample safety margin for dielectric strength, so under normal operating conditions, high-voltage gradients below the breakdown voltage do not cause detectable aging. However, at elevated temperatures, electrical stresses may act to further accelerate material degradation.

• *Mechanical stresses* caused by assembly configurations, manufacturing techniques, centrifugal forces, and vibration. These stresses tend to physically damage material.

• *Environmental conditions*, such as exposure to oxidation, ozone, radiation, and chemicals. The destructive effects of oxidation and radiation are discussed in this chapter in the section on "Radiation" (p. 18) and in Chapter 3, sections "Antioxidants" and "Ultraviolet Stabilizers" (pp. 45, 46). The resistance of insulating materials to many chemicals is well documented, so in selecting candidates for a specific application, only those materials should be considered that are resistant to attack by chemicals likely to be encountered.

• *Moisture*, either in the form of water or high humidity, a major cause of lowered dielectric properties or even dielectric failure of an insulation system. Some insulating materials, such as papers and organic fabrics, are highly hygroscopic and are not suitable for use alone as electrical insulation in environments where moisture could be present in appreciable amount. Moisture causes the fibers of these materials to swell, thereby affecting their spacing and the thickness of the material. However, up to 6 percent moisture content is necessary for these materials to exhibit their optimum mechanical properties. Each of these materials has a critical water content below which their usefulness deteriorates rapidly, so they cannot be completely dried before use. It is customary to impregnate and/or coat these materials with insulating liquids or varnishes to impede moisture absorption significantly. When so treated effectively, the paper and fabric perform as substrates and the dielectric properties of the system approach those of the impregnant or coating.

Although not as susceptible to moisture absorption to nearly the same degree as papers and organic fabrics, some plastics will pick up moisture much more readily than other plastics. Examples are polyamides (nylons), polyimides, acrylics, polysulfones, and some polyurethanes. Somewhat lower dielectric properties do not preclude their use for many applications.

Moisture can also form a conductive path on the surfaces of otherwise moisture-resistant materials. This action is exacerbated by the presence of conductive contaminants on a surface, which is often the case with transmission line insulators.

Under certain conditions and at elevated temperatures, moisture may react with some polymers, notably polyurethanes, to cause chemical reversion, thereby destroying the ability to function as a dielectric. This tendency to hydrolyze can be controlled by blending or alloying with less susceptible polymers.

• *Voids and contaminants* within an insulating material providing the location for initiation of destructive action by electrical discharge, moisture, and contaminant reaction with the base material.

Since the aging of insulation systems in service is usually caused by several of the above factors acting simultaneously, testing aging factors independently cannot be expected to portray service life accurately.

Although there is acceptance by most authorities of the avalanche theories on breakdown in gases and liquids (discussed in chapters on *Dielectric Gases* and *Dielectric Liquids*), despite intensive postulation, no theory yet satisfactorily explains all aspects of breakdown in solids. The problem is that dielectric failure in solids is more complex, involving combinations of several breakdown mechanisms. In solids, atoms and molecules are not as free to move or rotate with the application of an electrical field. In functioning electrical equipment, however, impurities, voids, environmental factors, and the configuration of the equipment are more important in determining breakdown than the inherent nature of the material. The probability of breakdown because of these factors is greatly increased by rising temperatures, including localized heating that cannot be effectively dissipated. The *intrinsic (theoretical) dielectric strength* of a material is that value of the electrical field intensity which causes breakdown absent the effect of the above factors.

It is well known that for a given material, the dielectric strength is significantly greater per unit thickness for thin than for thick specimens. The explanation generally given for this phenomenon is that with increased thickness, more dielectrically weaker paths are available, the weakest of which causes breakdown. By similar reasoning, the use of larger area electrodes results in lower dielectric strength values. A rule of thumb states that for homogeneous solids, the dielectric strength varies approximately as the reciprocal of the square root of the thickness of the specimen. However, although this interpretation may satisfy empirical test results, it does not provide an explanation of the true nature of the phenomenon, which awaits the results of further research.

Another factor affecting dielectric strength values is the rate of voltage rise and whether it is continuous or step-by-step. A slow rate of increase usually encourages time-dependent thermal degradation due to local heating, resulting in lower dielectric strength values. To be meaningful, therefore, dielectric breakdown strength values should include a statement of the following:

• Specimen thickness and conditioning
• Method of voltage application

- Type and size of electrodes
- Test temperature
- Any unusual environmental conditions

A phenomenon about which there is little enlightening published information is the significantly higher dielectric breakdown strength perpendicular to laminations of multiple layers of an insulating material compared with the same total thickness of a single layer of the same material. In effect, the boundaries between layers seem to interrupt or lengthen what otherwise would be a breakdown path through a homogeneous material. When tested parallel to laminations, the electrical field has no boundaries to penetrate and the breakdown voltage is significantly lower than for unlaminated material. Here, the boundaries may contain surface discontinuities which present dielectric weaknesses.

PARTIAL DISCHARGE (CORONA)

Although the phenomenon of an electrical discharge which does not completely bridge the insulation between electrodes or conductors is often referred to by both terms, the IEEE prefers *partial discharge*. In general usage, *corona* is accompanied by a faint glow, while *partial discharge* need not be luminous.

Partial discharge may occur in high-voltage transmission lines, resulting in formation of ozone, a highly reactive form of oxygen, and in ionization, often luminous, in the surrounding air. Where conductors are insulated, the insulation may be attacked by ozone and further degraded by electron bombardment.

Partial discharge may also occur in a void within an insulation system where the voltage gradient is sufficiently high. This discharge has a damaging effect on surrounding materials, causing a weakness in the system that can eventually lead to failure. As frequency increases in ac systems, partial discharge damage increases rapidly. Both ac and dc systems are susceptible to ionization attack on insulation surfaces.

Saturated polymers with high resistance to oxidation may have low resistance to ionization attack, as exhibited by polyethylene, polypropylene, and polytetrafluoroethylene. Silicone rubber is rated the best flexible insulation with regard to both ozone and ionization attack for operating temperatures over 130°C (266°F). For lower temperatures, ethylene-propylene copolymer, butyl rubber, nylon, and chlorinated polyolefins are among the least affected polymers. For applications where they are otherwise suitable, paper–oil cables would provide the best protection from both ozone and ionization damage.

SPACE CHARGES

A phenomenon known as *space charges* gained attention and encouraged investigation into their causes and measurement when it was observed that when a positive dc voltage was applied to a cable insulated with crosslinked polyethylene for a significant period of time, the breakdown voltage became perceptibly lower after negative dc voltage was applied. The reason ascribed to this occurrence was that it was caused by a space charge built up in the insulation by free electrons and negative ions. This was of such magnitude that an undersea power transmission cable with XLPE insulation planned between Hokkaido and Honshu islands in Japan was completed instead with oil-filled cable.[1]

Although there are many theories regarding the causes and effects of space charges in dielectrics, the practical consideration is that traditional dielectric strength tests may not always be relied upon as design criteria in certain applications where space charges may be involved.

See also "Breakdown in Gases," p. 435.

RADIATION

When plastics are subjected to ionizing radiation:

- Free radicals are created by breaking chemical bonds.

- Molecules become crosslinked, thereby improving stress-crack, abrasion, and deformation resistance, raising useful service temperature, and increasing resistance to the action of fluids.

Ionizing radiation also promotes rapid room temperature polymerization of solventless organic coatings and vulcanization of natural and synthetic rubbers other than butyl rubber.

However, excessive ionizing radiation may cause serious degradation in these materials. Each material has its own threshold resistance to ionizing radiation beyond which severe damage occurs.

Ionizing radiation may be produced by ultraviolet light (UV) or electron beam (EB) technology.

See also "Ultraviolet Stabilizers," p. 46, and "Radiation Curing," p. 241.

DIPOLE MOMENT AND POLARIZATION

Dipole moment may be defined as follows:

> Molecules in which the atoms and their electrons and nuclei are so arranged that one part of the molecule has a positive electrical charge while the other part is negatively charged. The molecule therefore becomes a small magnet or dipole. Changing electrical or magnetic fields causes the molecule to turn or rotate in one direction or another, depending on the charge of the field. The dipole moment (μ) is the distance between the charges multiplied by the quantity of charge in electrostatic units.[2]

Dipole polarization occurs when normally randomly oriented permanent dipoles of a molecule are aligned by an applied electric field. This phenomenon is facilitated

[1]Y. Li and T. Takada, "Progress in space charge measurement of solid insulating materials in Japan," *IEEE Electrical Insulation Magazine*, vol. 10, no. 5, Sept./Oct. 1994.

[2]Gessner G. Hawley, *Condensed Chemical Dictionary*, 10th ed. New York: Van Nostrand Reinhold, 1981.

at higher temperatures where dipoles are bound less tightly and are more free to align with the field.

The type and arrangement of the atoms in a molecule determine its polarity. In general, the degree of molecular symmetry and the affinity of an atom for its electrons influences polarity. Thus, polyethylene $\left[\begin{array}{cc} H & H \\ | & | \\ -C-C- \\ | & | \\ H & H \end{array}\right]_x$ is nonpolar, polymonofluoroethy-lene $\left[\begin{array}{cc} H & F \\ | & | \\ -C-C- \\ | & | \\ H & H \end{array}\right]_x$ is strongly polar, and polytetrafluoroethylene $\left[\begin{array}{cc} F & F \\ | & | \\ -C-C- \\ | & | \\ F & F \end{array}\right]_x$ exhibits low polarity (because of symmetry and the high affinity of fluorine for its electrons).

The aligned dipole molecules produce a net polarization across the material which has the effect of increasing the dielectric constant. In general, the dielectric constant at 1 megahertz of nonpolar polymers ranges from 2.2 to 2.6, and for polar polymers from 2.6 to over 6. Examples of nonpolar and polar polymers are shown in Table 2-1.

In addition to polarization by permanent dipoles, other forms of polarization include *electronic polarization* and *atomic polarization*. Arthur von Hippel, in his writing on *Molecular Properties of Dielectrics*, describes these phenomena as follows:

> Matter, electrically speaking, consists of positive atomic nuclei surrounded by negative electron clouds. Upon the application of an external electric field the electrons are displaced slightly with respect to the nuclei; induced dipole moments result and cause the so-called electronic polarization of materials. When atoms of different types form molecules, they will normally not share their electrons symmetrically, as the electron clouds will be displaced eccentrically toward the stronger binding atoms. Thus atoms acquire charges of opposite polarity, and an external field acting on these net charges will tend to change

TABLE 2-1   Typical Nonpolar and Polar Polymers

Nonpolar
  Polyethylene
  Polyphenylene oxide
  Polypropylene
  Polystyrene
  Polytetrafluoroethylene
Polar
  ABS resin
  Nylon 6/6
  Polycarbonate
  Polymethyl methacrylate
  Polysulfone
  Polyvinyl chloride

the equilibrium positions of the atoms themselves. By this displacement of charged atoms or groups of atoms with respect to each other, a second type of induced dipole moment is created; it represents the atomic polarization of the dielectric. The asymmetric charge distribution between the unlike partners of a molecule gives rise, in addition, to permanent dipole moments which exist also in the absence of an external field. Such moments experience a torque in an applied field that tends to orient them in the field direction. Consequently, an orientation (or dipole) polarization can arise.

All polymers can be polarized by an electric field, but not necessarily by their permanent dipoles. Where all forms of polarization are present, the highest dielectric constant will be observed in a dc or low-frequency ac field. As frequency increases, a point will be reached where a particular polarization does not have time to form, leading to a decrease in dielectric constant. Permanent dipole polarization takes the longest time, requiring the movement of entire molecular segments, and consequently this form is the first to become inactive. Atomic polarization is next, requiring movement of atoms. The last form to become inactive is electronic polarization, which involves only electrons. All forms, however, persist through power frequencies up to at least 1 megahertz (10^6 Hz).

Following this analysis, the dissipation factor (tan δ) is highest when permanent dipole moments are oscillating in phase at maximum frequency, thereby creating frictional heat, just before the frequency becomes too high for them to oscillate in phase or at all.

For a detailed discussion of polarization, see:

Engineering Dielectrics, Volume IIA, Bartnikas/Eichhorn, American Society for Testing and Materials (includes extensive references).

An Introduction to the Theory of Dielectric Loss in Plastics, A. H. Sharbaugh and J. C. Devins, Institute of Electrical and Electronics Engineers Publication 68C6-EI-33.

MATERIAL SELECTION STEPS

In selecting a material for a specific use, these steps are important:

1. All possible environmental conditions to which the material could be exposed should be determined, including (but not necessarily limited to) temperature, humidity, chemicals, and radiation.

2. The list of candidate materials for the application should be reviewed, eliminating from further consideration all materials known not to be sufficiently resistant to these conditions. As an example, for an application requiring continuous operation at 155°C (311°F), all materials with lower thermal class ratings should be screened out.

3. Considering the equipment or part design and the manufacturing process to be used, only materials which possess the required dielectric and mechanical properties should be selected for further review. For example, unreinforced mica paper tape

would not withstand typical coil winding tensions. Glass fabric reinforced mica tape would, however, withstand the winding process, but at the cost of added bulk and expense. Conformability, deformability, and impact strength are important considerations for many applications, such as housings and control equipment. Arc quenching capability without formation of carbon is required for dielectric gases used in circuit breakers.

4. Compatibility with other compounds in an insulation system is a critical factor and should serve to screen materials. For example, transformer liquids should not affect gaskets (a consideration with silicones), films, or laminates used in the transformer insulation system.

5. Manufacturing ease, time, and cost should be considered, together with material cost to determine final suitability for an end use.

6. Building and testing a prototype should be the final step, with redesign and retesting where necessary.

Property values listed in data sheets and tables should serve only for comparisons among candidate materials and as a screening method where differences are significant. Properties of plastics are time-dependent and cannot be predicted from short-time tests. This is especially so for material degradation by oxidation, radiation, and chemicals.

TEMPERATURE INDEX OF DIELECTRIC MATERIALS

Design engineers require some way of selecting dielectric materials for use in electrical equipment for continuous operation at elevated temperatures. The temperature index has been developed for this purpose to serve as a guide to the relative thermal endurance of commonly used dielectric materials. Since it is impractical to test materials for as long as they are expected to remain effective in service, tests are conducted at selected elevated temperatures, and the time plotted to the end point criteria on a graph with the logarithm of time as the ordinate and the reciprocal of the absolute temperature, K, as the abscissa (Arrhenius plot). The graph is then extrapolated, most commonly to 20,000 hours, to determine the temperature index of the material.

Although this index is of practical use to design engineers, it is to be noted that for many dielectric materials which would have satisfactory endurance for 20,000 hours over the full range of operating temperatures of the electrical equipment they insulate, decomposition occurs at elevated temperatures used to determine temperature indexes, indicating (incorrectly) that these materials would be unsuitable where practical experience has shown them to be effective dielectrics.

Although the terms temperature index, temperature classification, and thermal endurance are often used interchangeably in common parlance, distinctions are made in the technical literature (while still leaving room for some confusion):

• *Temperature Classification* is a term reserved for insulating systems as used in specific equipment, and is no longer recognized as a description of the temperature capability of individual insulating materials (IEEE Standard 98-1984). Note: Individual dielectric materials are, however, generally referred to as Class 90, Class 105, etc., materials.

• *Thermal Endurance* is defined as the relationship between temperature and time spent at that temperature, required to produce such degradation of an electrical insulation that it fails under specified conditions of stress, electric or mechanical, in service or under test (*IEEE Standard Dictionary of Electrical and Electronics Terms*).

• *Temperature Index* is an index that allows relative comparisons of the temperature capability of insulating materials or insulation systems based on specified controlled test conditions (*IEEE Standard Dictionary of Electrical and Electronics Terms*). Temperature ranges and their assigned indexes are shown in Table 2-2.

TABLE 2-2   Preferred Temperature Indexes (°C)

TEMPERATURE RANGE OF FAILURE IN 20,000 H	ASSIGNED TEMPERATURE INDEX
90–104	90
105–129	105
130–154	130
155–179	155
180–199	180
200–219	200
220–249	220
250 and above	none established

STANDARD TESTS FOR DIELECTRIC MATERIALS

Specific tests covering all dielectric materials are described in Chapter 3, *Thermoplastic Molding Materials*, where they are most commonly used for comparing polymeric materials.

SI SYSTEM FOR METRIC PRACTICE (ASTM E 380)

This system is intended as a basis for international standardization of measurement units. The designation SI is an acronym for the French name "Le Système International d'Unités."

SI units and quantity symbols are shown in Table 2-3.

TABLE 2-3   SI Units and Quantity Symbols

UNIT	UNIT SYMBOL	SOMETIMES OCCURS AS: (DO NOT USE)	APPLICATIONS AND NOTES	QUANTITY SYMBOL (FOR USE AS VARIABLES, ETC.)
*ampere	A	amp, a	SI unit of electric current, magnetic (scalar) potential, magnetomotive force.	I U F
ampere-hour	Ah, A·h	amp-hr	Quantity of electricity. 1 Ah = $3.6 \cdot 10^3$ C.	
*ampere-turn	A	At	SI unit of magnetomotive force.	F
*ampere per meter	A/m		SI unit of linear current density, magnetic field strength (note: interpret as ampere turns per meter).	A H
*ampere per square meter	A/m^2		SI unit of current density.	J
*ampere meter squared	A·m^2		SI unit of magnetic (area) moment.	m
angstrom	Å	A°, A	Wavelength. Use not recommended. 1 Å = 10^{-10} m.	
atmosphere, standard	atm		Pressure. Use not recommended. 1 atm =14.7 lb/in^2 = $1.013 \cdot 10^5$ Pa.	
atmosphere, technical	at		Use not recommended. 1 at = 1 kgf/cm^2.	
*atomic mass unit (unified)	u		Atomic mass. The (unified) atomic mass unit is defined as one twelfth of the mass of an atom of the ^{12}C nuclide. Use of the old atomic mass (amu), defined by reference to oxygen, is not recommended.	
*atto	a		SI prefix for 10^{-18}.	
*attoampere	aA		See: ampere.	
*attofarad	aF		See: farad.	
bar	bar	b, barye	Pressure. 1 bar = 10^5 Pa. Use of the bar is strongly discouraged except for limited use in meteorology.	
barn	b		Nuclear capture cross section. In temporary use with SI. 1 b = 10^{-28} m^2.	
barrel	bbl		Volume. 1 bbl = 42 gal$_{us}$ = $1.5899 \cdot 10^{-1}$ m^3.	
barrel per day	bbl/d		Standard barrel used for petroleum, etc. A different standard barrel is used for fruits, vegetables, and dry commodities.	
baud	Bd	baud (w/ prefix)	In telecommunications, a unit of signaling speed equal to one element per second. The signaling speed in bauds is equal to the reciprocal of the signal element length in seconds.	$1/r$

(continued)

TABLE 2-3 *Continued*

UNIT	UNIT SYMBOL	SOMETIMES OCCURS AS: (DO NOT USE)	APPLICATIONS AND NOTES	QUANTITY SYMBOL (FOR USE AS VARIABLES, ETC.)
bel	B	b	Ratio of power. Rarely used. 1 B = 10 dB. See decibel and Appendix B of ANSI-IEEE Std 260-1978 for further guidance concerning notation.	
*becquerel	Bq		SI unit of activity of a radionuclide.	
billion electronvolts	GeV	bev, BeV	Energy of accelerated particles. The name *gigaelectronvolt* is preferred for this unit.	
bit	b	bit	In information theory, the bit is a unit of information content equal to the information content of a message, the *a priori* probability of which is one half.	
			In computer science, the bit is a unit of storage capacity. The capacity, in bits, of a storage device is the logarithm to the base two of the number of possible states of the device.	
bit per second	b/s	bit/s, bits/s, bps, bit/sec		
British thermal unit	Btu		Heat, energy. Use of the joule (SI) is preferred. Conversion factors vary with usage. Consult ANSI/IEEE Std 268-1982.	
byte	byte		A group of bits or adjacent binary digits that a computer processes as a unit.	
calorie (International Table calorie)	cal_{IT}		Heat. Use not recommended. $1\ cal_{IT} = 4.1868 \cdot 10^3$ J.	
calorie (thermo-chemical calorie)	cal		Heat. Use not recommended. $1\ cal = 4.1840 \cdot 10^3$ J.	
*candela	cd		SI unit of luminous intensity.	I
candela per square foot	cd/ft^2		Luminance. Use of the SI unit cd/m^2 is preferred.	L
candela per square inch	cd/in^2		Luminance. Use of the SI unit cd/m^2 is preferred. $1\ cd/in^2 = 1.55 \cdot 10^3\ cd/m^2$.	
*candela per square meter	cd/m^2	nit	SI unit of luminance.	L
candle	cd		The unit of luminous intensity has been given the name *candela*. Use of the name *candle* for this unit is not recommended.	
*centi	c (prefix)		SI prefix for 10^{-2}.	
*centimeter	cm		Length. (Preferred SI unit multiple.)	
*circular mil	cmil		Area (cross section of wire). $1\ cmil = (\pi/4) \cdot 10^{-6}\ in^2 = 5.067 \cdot 10^{-10}\ m^2$.	

TABLE 2-3   *Continued*

UNIT	UNIT SYMBOL	SOMETIMES OCCURS AS: (DO NOT USE)	APPLICATIONS AND NOTES	QUANTITY SYMBOL (FOR USE AS VARIABLES, ETC.)
*coulomb	C	c	SI unit of electric charge, quantity of electricity, electric flux. $1 \text{ C} = 1 \text{ A·s}$.	Q Ψ
*coulomb per meter	C/m		Linear density of charge (SI).	λ
*coulomb meter	C·m		SI unit of electric dipole moment.	p
*coulomb per square meter	C/m^2		SI unit of electrical flux density.	D
*coulomb per cubic meter	C/m^3		SI unit of volume density of charge.	
*cubic centimeter	cm^3	cc	Volume. (Preferred SI unit multiple.)	
cubic foot	ft^3		Volume. $1 \text{ ft}^3 = 2.832 \cdot 10^{-2} \text{ m}^3$.	
cubic foot per minute	ft^3/min	cfm	Flow rate. $1 \text{ ft}^3/\text{min} = 4.719 \cdot 10^{-4} \text{ m}^3/\text{s}$.	
cubic foot per second	ft^3/s		Flow rate. $1 \text{ ft}^3/\text{s} = 2.832 \cdot 10^{-2} \text{ m}^3/\text{s}$.	
cubic inch	in^3		Volume. Section modulus. $1 \text{ in}^3 = 1.639 \cdot 10^{-5} \text{ m}^3$.	
*cubic meter	m^3		SI unit of volume.	
*cubic meter per second	m^3/s		SI unit of flow rate.	
cubic yard	yd^3		$1 \text{ yd}^3 = 0.7646 \text{ m}^3$.	
curie	Ci	C	A unit of activity of radionuclide. Use of the SI unit, the becquerel, is preferred. $1 \text{ Ci} = 3.7 \cdot 10^{10} \text{ Bq}$.	
cycle per second	Hz	c/s, cps, c/sec, cycle	Frequency. See: hertz. The name *hertz* is internationally accepted for this unit; the symbol Hz is used instead of c/s.	
day	d		$1 \text{ day} = 24 \text{ hr} = 86\ 400 \text{ s}$.	
deci	d (prefix)		SI prefix for 10^{-1}.	
decibel	dB	db, DB	Noise intensity, gain, power. See Appendix A of ANSI-IEEE Std 260-1978 for further guidance concerning notation.	
decibel referred to 1 mW	dB (1mW)	dBm	See Appendix A of ANSI/IEEE Std 260-1978 for further guidance concerning notation.	
degree	...°	deg	Plane angle.	
degree Celsius	°C	degree centigrade	SI unit of Celsius temperature. The degree Celsius is a special name for the kelvin, for use in expressing Celsius temperatures or temperature intervals. $T_K = t_{°C} + 273.15$.	t
degree Fahrenheit	°F		Temperature. $T_K = (t_{°F} + 459.67)/1.8$. $t_{°C} = (t_{°F} - 32)/1.8$.	

(continued)

TABLE 2-3 *Continued*

UNIT	UNIT SYMBOL	SOMETIMES OCCURS AS: (DO NOT USE)	APPLICATIONS AND NOTES	QUANTITY SYMBOL (FOR USE AS VARIABLES, ETC.)
degree Kelvin			See: kelvin.	
degree Rankine	°R		Use discouraged. $T_\kappa = T_{°R}/1.8$.	
deka	da (prefix)		SI prefix for 10.	
dyne	dyn	dyne	Force. Use not recommended. 1 dyn $= 10^{-5}$ N.	F
*electronvolt	eV	ev	Energy (nuclear physics). 1 eV = 1.602· 10^{-19} J.	
erg	erg		Work, energy. Use not recommended. 1 erg $= 10^{-7}$ J.	
exa	E (prefix)		SI prefix for 10^{18}.	
*farad	F	f, fd	SI unit of capacitance. 1 F = 1 C/V.	C
*farad per meter	F/m		SI unit of capacity, permittivity.	ϵ
*farad, reciprocal	F^{-1}	daraf	Unit of elastance (SI).	S
*femto	f (prefix)		SI prefix for 10^{-15}.	
foot	ft		Length. 1 ft = 0.3048 m.	
foot per minute	ft/min	fpm	Velocity. 1 ft/min = 5.080·10^{-3} m/s.	
foot per second	ft/s	fps, ft/sec	Velocity. 1 ft/s = 0.3048 m/s.	
foot-pound	ft·lb		A unit of work equal to the work done by a force of one pound acting through a distance of one foot in the direction of the force.	
foot pound-force	ft·lbf		Use joules. 1 ft·lbf = 1.356 J.	
foot poundal (OBSOLETE)	ft·pdl		An absolute unit of work equal to the work done by a force of one poundal acting through a distance of one foot in the direction of the force. Note: No longer used. 1 ft·pdl = 4.214·10^{-2} J.	
footcandle	fc		Illumination. 1 fc = 1 lm/ft². The name *lumen per square foot* is also used for this unit. Use of the SI unit of illuminance, the lux (lumen per square meter), is preferred. 1 fc = 10.764 lx.	
footlambert	fL		Brightness (luminance). 1 fL = $(1/\pi)$ cd/ft². One lumen per square foot leaves a surface whose luminance is one footlambert in all directions within a hemisphere. Use of the SI unit, the candela per square meter, is preferred. 1 fL = 3.426 cd/m².	
gee	g	G	Acceleration of gravity. Standard acceleration of free fall. Use m/s². 1 g = 9.807 m/s².	
gal	Gal		Use strongly discouraged. The gal is used only for the quantity g. 1 Gal = 1 cm/s² $= 10^{-2}$ m/s².	

TABLE 2-3 *Continued*

UNIT	UNIT SYMBOL	SOMETIMES OCCURS AS: (DO NOT USE)	APPLICATIONS AND NOTES	QUANTITY SYMBOL (FOR USE AS VARIABLES, ETC.)
gallon	gal		Volume. Use not recommended. $1\ \text{gal}_{US} = 231\ \text{in}^3 = 3.7854\ \text{L}$. $1\ \text{gal}_{UK} = 4.5461\ \text{L}$.	
gauss	G		Electromagnetic CGS unit of magnetic flux density. Use not recommended. Use tesla. $1\ G = 10^{-4}\ T$.	*B*
*giga	G (prefix)	kM	SI prefix for 10^9.	
*gigabit	Gb		See: bit.	
gigacycle per second	GHz	kMC, Gc/s	Frequency. See: hertz, gigahertz.	f, ν
*gigaelectronvolt	GeV	bev, BeV	Energy. See: electronvolt.	
*gigahertz	GHz	kMHz, KMC, Gc/s	Frequency. (Preferred SI unit multiple.)	f, ν
gilbert	Gb		Electromagnetic CGS unit of magneto-motive force. Use not recommended. $1\ Gb = 0.7958\ A$.	
*gram	g	gm	Mass. (Preferred SI unit multiple.) $1\ g = 10^{-3}\ \text{kg}$.	*m*
*gray	Gy		SI unit of absorbed dose in the field of radiation dosimetry.	
*hecto	h		SI prefix for 10^2.	
*henry (pl. henrys)	H	Hy, hy	SI unit of inductance: (self) inductance, permeance. $1\ H = 1\ \text{Wb/A}$.	L P, P_m
*henry, reciprocal	H^{-1}		Reciprocal inductance (SI). Reluctance (SI).	Γ R, R_m
*henry per meter	H/m		SI unit of (magnetic) permeability, absolute permeability.	μ
*hertz	Hz	cps, c/s, cycle	SI unit of frequency, bandwidth.	f, ν B
horsepower	hp		Power, rate of work. The horsepower is an anachronism in science and technology. Use of the SI unit of power, the watt, is preferred. Conversion factors vary with usage: $1\ \text{hp (electric)} = 7.46 \cdot 10^2\ W$. $1\ \text{hp (metric)} = 7.35 \cdot 10^2\ W$. $1\ \text{hp (U.K.)} = 7.45 \cdot 10^2\ W$.	
*hour	h	hr	Time. $1\ h = 3.6 \cdot 10^2\ s$.	
inch	in	in.	Length. $1\ \text{in} = 2.54 \cdot 10^{-2}\ m$.	
inch per second	in/s	ips	Velocity. $1\ \text{in/s} = 2.54 \cdot 10^{-2}\ \text{m/s}$.	

(continued)

TABLE 2-3   *Continued*

UNIT	UNIT SYMBOL	SOMETIMES OCCURS AS: (DO NOT USE)	APPLICATIONS AND NOTES	QUANTITY SYMBOL (FOR USE AS VARIABLES, ETC.)
*joule	J		SI unit of energy, work, quantity of heat.	E, W W Q
*joule per degree Celsius	J/°C		SI unit of heat capacity, thermal capacitance.	C_θ
*joule per kelvin	J/K		SI unit of entropy.	S
kelvin	K		SI unit of temperature. Previous to 1967 called *degree kelvin*. Note no symbol ° appears with K. $t_{°C} = T_K - 273.15$.	
*kilo	k(prefix)		SI prefix for 10^3.	
*kilobit	kb		See: bit.	
*kilobyte	kilobyte		See: byte.	
kilocycle per second	kHz	kc/s, kc	Frequency. Use kilohertz. See: hertz.	
kilomegacycle per second	GHz	KMC, kMc/s	Frequency. Use gigahertz. See: hertz.	
kilogauss	kG		Use not recommended. See: gauss.	
*kilogram	kg		SI unit of mass.	
kilogram-force	kgf		Use not recommended. Kilogram is SI unit of mass, newton is SI unit of force. 1 kgf = 9.807 N.	
*kilohertz	kHz		Frequency. (Preferred SI unit multiple.)	
*kilohm	kΩ		Resistance. (Preferred SI unit multiple.)	R
*kilojoule	kJ		See: joule.	
*kilometer	km		Length. (Preferred SI unit multiple.)	
*kilometer per hour	km/h		Velocity.	
*kilotesla	kT		See: tesla.	
*kilovar	kvar		Reactive power. (Preferred SI unit multiple.)	Q
*kilovolt	kV		(Preferred SI unit multiple.) See: volt.	
*kilovoltampere	kVA	KVA, kva	Apparent power. (Preferred SI unit multiple.)	
*kilowatt	kW		(Preferred SI unit multiple.) See: watt.	
kilowatthour	kWh		Temporarily in use with SI as a measure of electric energy. Widely used, but should eventually be replaced by the megajoule. 1 kWh = 3.6 MJ.	
knot	kn		1 kn = 1 nmi/h = 0.514 m/s. Use not generally recommended.	
lambert	L		CGS unit of luminance. Use not recommended. 1 L = $(1/\pi) \cdot 10^4$ cd/m² = 3.183 $\cdot 10^3$ cd/m². One lumen per square	L

TABLE 2-3   *Continued*

UNIT	UNIT SYMBOL	SOMETIMES OCCURS AS: (DO NOT USE)	APPLICATIONS AND NOTES	QUANTITY SYMBOL (FOR USE AS VARIABLES, ETC.)
			centimeter leaves a surface whose luminance is one lambert in all directions within a hemisphere.	
*liter	L	l	Volume. $1\ \text{L} = 10^{-3}\ \text{m}^3$. The letter l has been adopted for *liter* by the CGPM, and it is recommended in a number of international standards. In 1978 the CIPM accepted L as an alternative symbol. Because of frequent confusion with the numeral 1, the letter l is no longer recommended for U.S. use. (Script ℓ also not recommended.)	V, v
liter per second	L/s		Flow rate.	
*lumen	lm		SI unit of luminous flux. $1\ \text{lm} = 1\ \text{cd·sr}$.	Φ
lumen per square foot	lm/ft^2		Unit of illuminance and also a unit of luminous exitance. Use of the SI unit, lumen per square meter is preferred. $1\ \text{lm/ft}^2 = 10.764\ \text{lm/m}^2$.	
*lumen per square meter	lm/m^2		SI unit of luminous exitance.	M
*lumen per watt	lm/W		SI unit of spectral luminous efficacy, total luminous efficacy.	$K(\lambda)$ K, K_t
*lumen second	lm·s		SI unit of quantity of light.	Q
*lux	lx		SI unit of illuminance. $1\ \text{lx} = 1\ \text{lm/m}^2$.	E
maxwell	Mx		CGS electromagnetic unit of magnetic flux. Use not recommended. $1\ \text{Mx} = 10^{-8}\ \text{Wb}$.	
*mega	M (prefix)		SI prefix for 10^6.	
*megabyte	megabyte		See: byte.	
*megaelectronvolt	MeV		See: electronvolt.	
*megahertz	MHz		Frequency. (Preferred SI unit multiple.) See: hertz.	
*megohm	$\text{M}\Omega$	M	(Preferred SI unit multiple.) See: ohm.	
*megavolt	MV		(Preferred SI unit multiple.) See: volt.	
*megawatt	MW		(Preferred SI unit multiple.) See: watt.	
*meter	m		SI unit of length, breadth, height, thickness, radius, diameter, length of path, wavelength.	l b h d, δ r d s λ

(continued)

TABLE 2-3 *Continued*

UNIT	UNIT SYMBOL	SOMETIMES OCCURS AS: (DO NOT USE)	APPLICATIONS AND NOTES	QUANTITY SYMBOL (FOR USE AS VARIABLES, ETC.)
*meter, reciprocal	m^{-1}	/m	Wavenumber (SI).	σ (also, $\tilde{\nu}$ in spectroscopy only.)
*square meter	m^2		SI unit of area.	A
*cubic meter	m^3		SI unit of volume.	V, v
*meter per second	m/s		SI unit of velocity.	v
*meter per second squared	m/s^2		SI unit of acceleration.	a, g
*meter to the fourth power	m^4		Second (axial) moment of area (SI). Second (polar) moment of area (SI).	I, I_a J, I_p
*(ion) per cubic meter	m^{-3}	ion/m^3	Ion (number) density (SI).	$n^+; n^-$
*square meter per volt second	$m^2/V{\cdot}s$		Mobility (of a charge carrier in a medium) (SI).	μ
*(electron) per cubic meter second	$m^{-3}{\cdot}s^{-1}$		Rate of production of electrons per unit volume (SI).	q
*cubic meter per second	m^3/s		Recombination coefficient (SI).	α
metric ton	t		Use not recommended. 1 t = 1000 kg.	
mho	mho	Ω^{-1}	Formerly used as the name of the siemens (S). Still in use in the U.S.	
*micro	μ (prefix)		SI prefix for 10^{-6}.	
micromicro	p (prefix)	$\mu\mu$	Prefix for 10^{-12}. Do not use. Use pico.	
*microampere	μA		Electric current. See: ampere.	
*microfarad	μF		Capacitance. (Preferred SI unit multiple.)	
*microgram	μg		See: gram.	
*microhenry	μH		Self inductance. (Preferred SI unit multiple.)	
microinch	μin		1 μin = $2.54{\cdot}10^{-8}$ m.	
*microliter	μL		See: liter.	
*micrometer	μm	μ	Length. (Preferred SI unit multiple.) 1 μm = 10^{-6} m.	
micron	μm	μ	Use not recommended. Micron = micrometer. Use micrometer. Change μ to μm.	
*microsecond	μs		Time. (Preferred SI unit multiple.)	
*microsiemens	μS		Conductance. (Preferred SI unit multiple.)	
*microwatt	μW		Power. (Preferred SI unit multiple.)	
mil	mil		1 mil = 0.001 in = $2.54{\cdot}10^{-5}$ m.	
mile (nautical)	nmi		1 nmi = $1.852{\cdot}10^3$ m.	
mile (statute)	mi (statute)		1 mi = 5280 ft = $1.609{\cdot}10^3$ m.	

TABLE 2-3 *Continued*

UNIT	UNIT SYMBOL	SOMETIMES OCCURS AS: (DO NOT USE)	APPLICATIONS AND NOTES	QUANTITY SYMBOL (FOR USE AS VARIABLES, ETC.)
mile per hour	mi/h	mph	Although use of mph as an abbreviation is common, it should not be used as a unit symbol. 1 mi/h = 0.447 m/s = 1.609 km/h.	
*milli	m (prefix)		SI prefix for 10^{-3}.	
*milliampere	mA		See: ampere.	
millibar	mbar		Use of the bar is strongly discouraged, except for limited use in meteorology. 1 mbar = 100 Pa.	
*milligram	mg		Mass. (Preferred SI unit multiple.)	
*millihenry	mH		Self inductance. (Preferred SI unit multiple.)	
*milliliter	mL		See: liter.	
*millimeter	mm		Length. (Preferred SI unit multiple.)	
millimicron	nm		Use of the name millimicron is not recommended. Use nanometer.	
*millipascal second	mPa·s		Dynamic viscosity. (Preferred SI unit multiple.)	
*millisecond	ms		Time. (Preferred SI unit multiple.)	
*millivolt	mV		(Preferred SI unit multiple.) See: volt.	
*milliwatt	mW		Power. (Preferred SI unit multiple.)	
*minute (plane angle)	. . . '		Used to measure plane angles in surveys, plans, electrical calculations. Radians are used to measure plane angles in scientific and engineering calculations.	
*minute (time)	min		Time. Used to measure work time and in statistics. Note: In engineering measurements and calculations, minutes should be used in time derivative functions only when "second" related quantities (such as "millisecond, microsecond") become impractical. Time may also be designated by means of superscripts as in the following example: $9^h\,46^m\,30^s$.	
*mole	mol		SI unit of amount of a substance.	
month	mo			
*nano	n (prefix)		SI prefix for 10^{-9}.	
*nanoampere	nA		See: ampere.	
*nanofarad	nF		See: farad.	
*nanometer	nm		Length. (Preferred SI unit multiple.)	
*nanosecond	ns		Time. (Preferred SI unit multiple.)	
*nanowatt	nW		Power. (Preferred SI unit multiple.)	

(continued)

TABLE 2-3 *Continued*

UNIT	UNIT SYMBOL	SOMETIMES OCCURS AS: (DO NOT USE)	APPLICATIONS AND NOTES	QUANTITY SYMBOL (FOR USE AS VARIABLES, ETC.)
nat	nat		Natural logarithmic equivalent of the bit.	
nautical mile	nmi		Distance, range. 1 nmi = 1852 m.	
*neper	Np		Natural logarithm of two amounts of power (SI). 1 Np = 8.686 dB.	
*neper per second	Np/s		Damping coefficient (SI).	δ
*neper per meter	Np/m		Attenuation coefficient (SI).	α
*newton	N		SI unit of force. 1 N = 1 kg·m/s².	F
*newton meter	N·m		Moment of force (SI).	M
			Torque (SI).	T
*newton per square meter	N/m²		SI unit of pressure or stress. See: pascal	p, σ, τ
			Young's modulus.	E
			Modulus of elasticity.	E
			Shear modulus.	G
			Bulk modulus.	K
nit			Luminance. The name *nit* is sometimes given to the SI unit of luminance, the candela per square meter. Use of the *nit* is permitted as a name in text but not as a unit symbol. 1 nit = 1 cd/m².	L
oersted	Oe	oe	Electromagnetic CGS unit of magnetic field strength. Use not recommended. 1 Oe = 79.57 A/m.	
*ohm	Ω		SI unit of resistance,	R
			impedance,	Z
			reactance.	X
			1 Ω = 1 V/A.	
*ohm meter	Ω·m		SI unit of resistivity.	
ounce (avoirdupois)	oz		1 oz = 2.835·10⁻² kg.	
*pascal	Pa		SI unit of pressure or stress. 1 Pa = 1 N/m².	
*pascal second	Pa·s		SI unit of dynamic viscosity.	
percent	%			
*peta	P (prefix)		SI prefix for 10¹⁵.	
*pico	p (prefix)		SI prefix for 10⁻¹².	
*picoampere	pA		See: ampere.	
*picofarad	pF		Capacitance. (Preferred SI unit multiple.)	
*picosecond	ps		Time. (Preferred SI unit multiple.)	
*picowatt	pW		Power. See: watt.	
pint	pt		1 pt (U.K.) = 0.5683 L. 1 pt (U.S. dry) = 0.5506 L. 1 pt (U.S. liquid) = 0.4732 L.	
pound (avoirdupois)	lb		1 lb (av) = 0.4536 kg.	
pound per cubic foot	lb/ft³		1 lb/ft³ = 16.018 kg/m³.	

TABLE 2-3   *Continued*

UNIT	UNIT SYMBOL	SOMETIMES OCCURS AS: (DO NOT USE)	APPLICATIONS AND NOTES	QUANTITY SYMBOL (FOR USE AS VARIABLES, ETC.)
pound-force	lbf		1 lbf = 4.448 N.	
pound-force foot	lbf·ft		1 lbf·ft = 1.356 N·m.	
pound-force per square foot	lbf/ft^2		1 lbf/ft^2 = 47.88 Pa.	
pound-force per square inch	lbf/in^2	psi	Although use of the abbreviation psi is common, it should not be used as a unit symbol. 1 lbf/in^2 = $6.895 \cdot 10^3$ Pa.	
quart	qt		1 qt (U.K.) = 1.1365 L. 1 qt (U.S. dry) = 1.1012 L. 1 qt (U.S. liquid) = 0.9464 L.	
rad	rd		Unit of absorbed dose in the field of radiation dosimetry. Use of the SI unit, the gray, is preferred. 1 rd = 0.01 Gy.	
*radian	rad		SI unit of plane angle.	
*radian per second	rad/s		SI unit of angular frequency, angular velocity.	ω
*radian per second squared	rad/s^2		SI unit of angular acceleration.	α
rem	rem		Unit of dose equivalent in the field of radiation dosimetry. Use of the SI unit, the sievert, is preferred. 1 rem = 0.01 Sv.	
revolution per minute	r/min	rpm	Speed of rotation. Although the use of rpm as an abbreviation is common, it should not be used as a unit symbol.	
revolution per second	r/s	rps	Speed of rotation.	
roentgen	R	r	A unit of exposure in the field of radiation dosimetry. 1 R = $2.58 \cdot 10^{-4}$ C/kg.	
*second (plane angle)	...″		1 ″ = $4.848 \cdot 10^{-6}$ rad.	
*second (time)	s	sec	SI unit of time, period.	t T
*second, reciprocal	s^{-1}	/s, /sec	Complex (angular) frequency oscillation constant (SI).	p
*siemens	S		SI unit of conductance. 1 S = 1 Ω^{-1}. The name mho has been used for this unit in the U.S.	
*sievert	Sv		SI unit of dose equivalent in the field of radiation dosimetry.	
slug	slug		FPS system unit of mass. Use not recommended. 1 slug = 14.59 kg.	
square foot	ft^2		Area. 1 ft^2 = 0.0929 m^2.	
square inch	in^2		Area. 1 in^2 = $6.452 \cdot 10^{-4}$ m^2.	

(continued)

TABLE 2-3   *Continued*

UNIT	UNIT SYMBOL	SOMETIMES OCCURS AS: (DO NOT USE)	APPLICATIONS AND NOTES	QUANTITY SYMBOL (FOR USE AS VARIABLES, ETC.)
*square meter	m^2		SI unit of area.	
*square meter per second	m^2/s		SI unit of kinematic viscosity.	
*square millimeter per second	mm^2/s		Kinematic viscosity. (Preferred SI unit multiple.)	
square yard	yd^2		Area. $1 \ yd^2 = 0.8361 \ m^2$.	
*steradian	sr	sterad	SI unit of solid angle.	
*tera	T (prefix)	t	SI prefix for 10^{12}.	
*tesla	T		SI unit of magnetic flux density (magnetic induction). $1 \ T = 1 \ N/(A{\cdot}m) = 1 \ Wb/m^2$.	B
therm	thm		$1 \ thm = 100 \ 000 \ Btu.$ $1 \ thm \ (EEC) = 1.0551{\cdot}10^8 \ J.$ $1 \ thm \ (U.S.) = 1.0548{\cdot}10^8 \ J.$	
ton (short)	ton		$1 \ ton = 2000 \ lb = 907.2 \ kg.$	
ton, metric	t		$1 \ t = 1000 \ kg.$ Use of this name in the U.S. is not recommended.	
torr	torr	Torr	$1 \ torr = 1/760 = 1.333{\cdot}10^2 \ Pa.$ Use not recommended.	
*(unified) atomic mass unit	u		See: atomic mass unit, unified.	
*var	var	VA reactive	SI unit of reactive power.	Q
*volt	V	v	SI unit of voltage, electromotive force. $1 \ V = 1 \ W/A.$	V, E
*voltampere	VA	va	SI unit of apparent power.	S
*volt per meter	V/m		SI unit of electric field strength.	E
*watt	W	w	SI unit of power. $1 \ W = 1 \ J/s.$	P
*watt per meter kelvin	$W/(m{\cdot}K)$		SI unit of thermal conductivity.	λ
*watt per steradian	W/sr		SI unit of radiant intensity.	I
*watt per steradian square meter	$W/sr{\cdot}m^2$		SI unit of radiance.	L
watthour	Wh		$1 \ Wh = 3600 \ J.$	
*watt per square meter	W/m^2		Poynting vector (SI).	S
*weber	Wb		SI unit of magnetic flux, magnetic flux linkage. $1 \ Wb = 1 \ V{\cdot}s.$	Φ Λ
*weber per meter	Wb/m		SI unit of magnetic vector potential.	A
weight percent	wt%		Concentration.	
yard	yd		$1 \ yd = 0.9144 \ m.$	
year	a		In the English language, generally yr. $1 \ yr$ (365 days) $= 3.1536{\cdot}10^7 \ s.$	

Note: Asterisks (*) indicate SI units, preferred multiples of SI units, or other units acceptable for use with SI.

3

Thermoplastic Molding Compounds

INTRODUCTION

Today, the availability and use of thermoplastics is taken for granted. In fact, it would be difficult to imagine life without them, since they are essential to virtually every phase of our daily lives in the home, in commerce, and in industry, as well as being in the forefront of high-technology research and development. Yet, thermoplastics were a novelty as recently as the 1930s, and have since grown to a multibillion dollar industry with a high probability that this astonishing growth rate will continue.

This chapter focuses on the principal engineering thermoplastics used for making parts by injection molding. Thermoplastics used primarily for extrusion are covered in a separate chapter.

As used in this handbook, *thermoplastic* refers to an organic polymer which softens on heating below its decomposition temperature and hardens to its original state on cooling.

DEFINITIONS OF TERMS RELATING TO PLASTICS

Commonly used terms relating to plastics are defined in this section.

A-stage.   An early stage in the preparation of certain thermosetting resins in which the material is still soluble in certain liquids, and may be liquid or capable of becoming liquid upon heating.[1]

Addition Polymerization.   Polymerization in which monomers are linked together without the splitting off of water or other simple molecules.[1]

Aging.   (1) The effect on materials of exposure to an environment for an interval of time. (2) The process of exposing materials to an environment for an interval of time.[1]

Alloy (in plastics).   Two or more immiscible polymers united, usually by another component, to form a resin having enhanced performance properties.[1]

Aromatic Polyester.   A polyester derived from monomers in which all the hydroxyl and carboxyl groups are linked directly to aromatic nuclei.[1]

B-stage.   An intermediate stage in the reaction of certain thermosetting resins in which the material swells when in contact with certain liquids and softens when heated, but may not entirely dissolve or fuse.[1]

Biodegradable Plastic.   A degradable plastic in which the degradation results from the action of naturally-occurring microorganisms such as bacteria, fungi, and algae.[1]

Block Copolymer.   An essentially linear copolymer in which there are repeated sequences of polymeric segments of different chemical structure.[1]

Blocking.   Unintentional adhesion between plastic films or between a film and another surface.[1]

Bloom.   A visible exudation or efflorescence on the surface of a material.[1]

C-stage.   The final stage in the reaction of certain thermosetting materials in which they have become insoluble and infusible.[1]

Chalking.   A powdery residue on the surface of a material resulting from degradation or migration of an ingredient, or both.[1]

Cold Molding.   A process of compression molding in which the molding is formed at room temperature and subsequently baked at elevated temperatures.[1]

Compression Molding.   The method of molding a material already in a confined cavity by applying pressure and (usually) heat.[1]

Condensation Polymerization.   Polymerization in which monomers are linked together with the splitting off of water or other simple compounds.[1]

Crosslinking.   The formation of a three-dimensional polymer by means of interchain reactions resulting in changes in physical properties.[1]

Degradation.   A deleterious change in the chemical structure, physical properties, or appearance of a plastic.[1]

Elastomer.   A macromolecular material that at room temperature returns rapidly to approximately its initial dimensions and shape after substantial deformation by a stress and release of that stress.[1]

Engineering Plastics.   See the section "Engineering Thermoplastics," p. 41 in this chapter.

Extrusion.   A process in which heated or unheated plastic is forced through a shaping orifice (a die) in one continuously formed shape, as a film, sheet, rod, or tubing.[1]

Filler.   A relatively inert material added to a plastic to modify its strength, permanence, working properties, or other qualities, or to lower costs.[1]

Gel.   A colloid in which the dispersed phase has combined with the continuous phase to produce a viscous or jelly-like product.[2]

Glass Transition Temperature (T_g).   The temperature at which an amorphous material changes from or to a brittle, vitreous state to or from a viscous or rubbery condition.[1,2]

Graft Copolymer.   A copolymer in which polymeric side chains have been attached to the main chain of a polymer of different structure.[1]

Homopolymer.   A polymer resulting from polymerization of a single monomer.[1]

Injection Molding.   The process of forming a material by forcing it, in fluid state and under pressure, through a runner system (sprue, gate) into the cavity of a closed mold.[1]

Novolac (Novolak).   A phenol or ortho-cresol-formaldehyde resin which remains thermoplastic until reacted with a source of methylene groups such as hexamethylenetetramine.[1]

Oligomer.   A polymer molecule consisting of only a few monomer units (dimer, trimer, tetramer).[2]

Organosol.   A suspension of a finely divided polymeric resin in a plasticizer together with a volatile organic liquid.[1]

Plastics.   Materials which contain one or more organic polymeric resins of large molecular weight and which are capable of being shaped by flow into finished articles or their component parts.[1]

Plasticizer.   A substance incorporated in a high polymer to improve its workability, flexibility, or distensibility.[1, 2]

Plastisol.   A dispersion of finely divided resin in a plasticizer. It becomes an organisol if a volatile organic solvent is included.[1, 2]

Polymer.   A chemical compound formed by polymerization of a monomer or monomers of different types.

Polymerization.   A chemical reaction in which the molecules of monomers are linked together.[1]

Polyol.   An alcohol having several hydroxyl groups.[1]

Polyolefin.   A group name for thermoplastic polymers such as polyethylene and polypropylene.[2]

Pot Life.   The period of time during which a reacting thermosetting composition remains suitable for its intended processing after mixing with reaction-initiating agents.[1]

Prepolymer.   A polymer of degree of polymerization between that of the monomer and the final polymer.[1]

Prepreg.   In reinforced thermosetting plastics, the admixture of resin, reinforcements, fillers, etc., in a substrate ready for molding.[1]

Resin.   A natural or synthetic polymeric material, usually of high molecular weight. Synthetic resins are classed as either thermoplastic or thermosetting, depending on whether or not they liquefy or soften on heating.

Runner.   The feed channel in an injection or transfer mold that runs from the inner end of the sprue or pot to the cavity gate. Also, the piece formed in this channel.[1]

Sheet Molding Compound.   A fiber-reinforced thermosetting compound in sheet form.[1]

Shrink Mark.   An undesired depression in the surface of a molded piece formed on cooling after molding.[1]

Sprue.   The feed channel that runs from the outer face of an injection or transfer mold to the mold gate in a single cavity mold. Also, the piece formed in this channel.[1]

Stress-cracking.   External or internal cracking, often evidenced by fine lines, caused by tensile stresses or environmental factors acting on a plastic material.

Thermoplastic.   A high-molecular-weight polymer that may be softened repeatedly when heated after returning to its original condition when cooled to room temperature.[2]

Thermosetting Plastic.   A plastic capable of being changed into an infusible or insoluble substance when cured by heat or other means.[1]

Transfer Molding.   The process for making thermosetting parts in which a measured charge of suitably preheated material is placed in a chamber from which it is forced through channels into a closed cavity (mold) where it is cured at a controlled temperature.

Vacuum Forming.   A forming process in which a heated plastic sheet is drawn against the mold surface by evacuating the air between it and the mold.[1]

Viscosity.   The internal resistance to flow exhibited by a fluid.[2]

Vulcanization.   The irreversible process resulting from crosslinking of the unsaturated hydrocarbon chain in rubber with sulfur, usually with the application of heat.[2]

---

[1]ASTM D 883, *Standard Terminology Relating to Plastics.*

[2]Gessner G. Hawley, *Condensed Chemical Dictionary*, 10th ed. New York: Van Nostrand Reinhold, 1981.

PLASTICS INDUSTRY

Since the early 1930s, thermoplastic technology has experienced steady, strong growth, resulting in one of the world's major industries comprised of companies ranging in size from among the world's largest to the smallest. The plastics industry is comprised of the following:

• *Monomer and polymer manufacturers*, mostly large chemical companies that convert basic raw materials from coal and petroleum into intermediate chemicals, and from these make monomers and polymers.

• *Additive, reinforcement, and filler manufacturers*, a diversified group making catalysts, stabilizers, antioxidants, flame retardants, fiber reinforcements, and fine particles which, when mixed with polymers, make them more suitable for specific end uses.

• *Compounders*, including most polymer manufacturers, independent specialists, and in-house operations that formulate resins from polymers, additives, reinforcements, and fillers.

TABLE 3-1   1993 U.S. Consumption for Thermoplastics in Electrical and Electronic Applications

MATERIAL	MILLION POUNDS
Acrylonitrile-butadiene-styrene	126
Polyamide (nylon)	108
Polyacetal	4
Polycarbonate	39
Polyester	58
Polyethylene	
High density	140
Low density	389
Polyphenylene alloys	29
Polypropylene	28
Polystyrene	300
Polyvinyl chloride	530
Styrene acrylonitrile	4
Total	1,755

Source: Modern Plastics Magazine.

• *Processors*, including companies engaged in molding, extruding, casting, calendering, and laminating plastics.

• *Finishers*, companies that assemble, decorate, and otherwise make plastic products suitable for marketing.

Most of these companies employ specialists whose job it is to help prospective customers in the area of their expertise.

Table 3-1 shows the estimated electrical and electronic industries' consumption of widely used thermoplastics. These figures include materials used for dielectric films and extruded wire and cable insulations.

ENGINEERING THERMOPLASTICS

Most of the polymers discussed in this chapter are classed as *engineering thermoplastics*, a designation that at best is loosely defined. Generally speaking, these polymers can be molded with close tolerances into complex precision parts. The implication here is that other (commodity) polymers lack this capability, which is not necessarily so. Under certain conditions, some engineering thermoplastics also do not satisfy this generalization. For example, polyamide parts may change dimensions significantly as they absorb moisture, which they do more readily than other engineering or commodity polymers.

A broader definition of engineering thermoplastics states that these materials have a good balance of properties which is retained over a wide range of environmental conditions. Yet, as shown in the section of this chapter entitled "Summary of Rankings of Thermoplastic Polymers," no polymer has a monopoly on high or low rankings for all key properties, and there is little consistency in the rankings of any polymer vis-à-vis any other polymer.

However defined, there is general agreement on the polymers classed as engineering thermoplastics. These polymers are identified in the section of this chapter entitled "Principal Thermoplastic Polymers for Electrical/Electronic Uses" (p. 68).

POLYMERIZATION

Homopolymerization is a chemical reaction in which molecules of a simple structure (monomer) link together to form large, often complex, molecules (macromolecules) in a chainlike configuration. When two or more different monomers are involved, the process is termed *copolymerization* or *heteropolymerization*.

Linear or *straight-chain* polymers, typified by polyethylene, form long, virtually continuous chains, theoretically without crosslinking between chains. Chains, however, are not one-dimensional, but rather three-dimensional and always changing in configuration because of molecular motion. Chain components may have side groups attached to a main chain, thereby modifying polymer properties. Polymers of this type are called *branched* polymers. When side groups (R) are arranged randomly along the backbone chain, the polymer is termed *atatic*:

```
       H   R   H   R   H   H
       |   |   |   |   |   |
    —C — C — C — C — C — C —
       |   |   |   |   |   |
       H   H   H   H   H   R
```

An *isotatic* polymer has side groups located on one side of the backbone chain:

```
       H   R   H   R   H   R
       |   |   |   |   |   |
    —C — C — C — C — C — C —
       |   |   |   |   |   |
       H   H   H   H   H   H
```

A *syndotatic* polymer has side groups arranged in a symmetrical and recurring pattern on both sides of the backbone chain:

```
       H   H   R   H   H   H   R
       |   |   |   |   |   |   |
    —C — C — C — C — C — C — C —
       |   |   |   |   |   |   |
       R   H   H   H   R   H   H
```

Polymerization may also proceed by *step growth*, in which linear polymers with intermediate degree of polymerization interact to form macromolecules:

$$—A—A—A—A—\qquad\qquad —(A—A—A)—(A—A—A)—$$

chain growth   step growth

Crosslinked, or *network* polymers are formed when linear chains are interconnected as chain growth proceeds as a result of interaction of reactive sites on separate linear chains.

Addition polymers are polymers in which free radicals are the initiating agents that react with the double bond of the monomer by adding to it on one side while producing a new free electron on the other side, thereby making the chain self-propagating.

Polymers may also be formed by *polycondensation*, a chemical reaction in which two or more monomers combine to form large molecules giving off water, salt, or other simple compounds. This process is typified by the reaction between phenol and formaldehyde, as shown in "Phenolics" (p. 128) of Chapter 4, *Thermosetting Molding Compounds*.

When two or more polymers are arranged in alternating sections, or blocks, along the backbone chain, the polymer is referred to as a *block* copolymer:

$$——(A—A—A—A)——(B—B—B)——(A—A—A—A)——(B—B—B)——$$

The term *graft* copolymer is used to describe a backbone chain formed from monomers (A) to which side chains formed from other monomers (B) are attached (grafted) at various points. Thus, a graft copolymer is similar in structure to a branched polymer where the side group is replaced with a polymer:

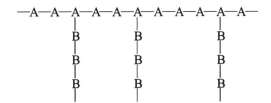

Polymer *alloys* are formed by physically blending two or more polymers of different properties into a homogeneous mixture through addition of special wetting agents and compounding under high shear conditions to prevent phase separation.

Polymer properties are significantly affected by polymer structure and molecular weight. Branched molecules are usually less crystalline with lower melting points. Chains that can crystallize and which have polar groups have high melting points. Longer chains have high melt viscosities. Increasing molecular weight imparts greater toughness characteristics. Polymer structures may be determined by infrared, X-ray diffraction, and nuclear magnetic resonance techniques.

The five common polymerization techniques are:

1. *Gas phase* polymerization requires high pressures and elevated temperatures. The process is usually initiated by a free radical source. Some grades of polyethylene and polypropylene are made by this method employing pressures up to 45,000 psi.

2. *Bulk, mass,* or *batch* polymerization, usually in liquid phase, is carried out at normal pressure and at elevated temperatures, with or without catalysts. Polyvinyl chloride may be made by this method. The reaction is exothermic, and removal of heat becomes a significant problem as viscosity increases. Control of molecular weight is difficult using this process, and the resins produced usually have a broad molecular weight distribution.

3. *Solution* polymerization employs a suitable solvent for the monomer and polymer, thereby facilitating removal of reaction heat. Reaction temperature must remain below the boiling point of the solvent, which may lengthen reaction time significantly. Other possible disadvantages are solvent flammability, toxicity, and the problems associated with solvent recovery. High-molecular-weight polymers are difficult to produce using this method. Acrylic resins may be made by this process.

4. *Emulsion* polymerization requires the monomer to be dispersed in water containing an emulsifying agent and a water-soluble initiator. This method obviates solvent problems and permits faster reaction rates than bulk or solution processes. Emulsion polymerization is suitable for making high-molecular-weight copolymers with narrow molecular weight distribution. Resins require a drying step, necessary to remove all moisture before use.

5. *Suspension* polymerization is a method in which a monomer is mechanically dispersed in a nonsolvent without use of an emulsifying agent, and linking is catalyst-initiated or by free-radical addition. This process facilitates control over the reaction variables and the final particle size of the polymer. This method is used to make high-molecular-weight polymers with relatively narrow molecular weight distributions. Resin particles must be washed and dried before use.

REINFORCEMENTS AND FILLERS

Thermoplastic resin properties are often improved significantly by compounding polymers with reinforcements and fillers.

Glass is the most common reinforcing fiber. It is low in cost and contributes to improve mechanical strength and electrical insulation properties with reduced flammability. Chopped strands of "E" glass in lengths of $1/8$–$1/4$ inch comprise 10–40 percent of compound weight for optimum results. Glass-fiber-reinforced grades may be purchased from resin suppliers or compounded by the molder at the injection molding machine. Other fibers often used in combination with glass for special applications include *aramid, boron, sisal, polyester*, and *carbon*.

Since all polymers are not affected to the same degree by the addition of fibers, reference should be made to manufacturers' data sheets for a particular polymer to determine advantages and limitations of reinforced versus general purpose grades.

Fillers, sometimes called *extenders, fine particles*, and *nonfibrous property enhancers*, also are important ingredients in thermoplastic resin compounds. Proper use of fillers has the following effects:

- Increased flexural modulus
- Increased dimensional stability
- Increased heat deflection temperature
- Reduced tensile strength
- Reduced creep
- Improved dielectric properties
- Reduced flammability
- Higher thermal conductivity
- Increased melt viscosity
- Improved processing
- Higher part weight

Commonly used fillers for electrical grades include:

- *Calcium carbonate* (also contains small amounts of magnesium carbonate and oxides of aluminum, iron, silicon, and manganese) which lowers cost, modifies flow characteristics, and improves chemical resistance and some mechanical properties
- *Alumina trihydrate* which reduces flammability
- *Kaolin*, or clay (mostly oxides of aluminum and silicon), which controls flow and improves chemical resistance and some mechanical properties
- *Mica flake* which has a much lower cost than glass fiber, contributes to arc resistance, tensile and flexural strength, and to modulus, but reduces impact strength
- *Talc* (hydrous magnesium silicate) which imparts impact strength and modulus

To achieve maximum effectiveness, filler particles, often submicron in size, must be uniformly dispersed throughout the compound and wet by the polymer. With judicious use, fillers permit resin cost to be reduced without sacrifice of essential properties in many applications. Filler concentrates, purchased from custom compounders, provide a convenient way for molders to incorporate fillers.

FLAME RETARDANTS

Each thermoplastic polymer is unique in its flammability characteristics. Some polymers, such as polyamide-imide, polyarylate, polyetheretherketone, polyetherimide, polyethersulfone, and polyphenylene sulfide are inherently resistant to burning and require no further treatment. Other polymers, such as polyamides, polycarbonate, and polysulfone, are somewhat less resistant, and several polymers burn readily. However, the flammability of these polymers may be significantly reduced by compounding with reactive and nonreactive halogenated compounds, phosphate esters, and antimony oxide.

Examples of reactive halogenated compounds include tetrabromo- and tetrachlorobisphonol A for polycarbonates.

Nonreactive halogenated compounds include chloroparaffins and halogenated biphenyls used in styrenics and polyamides (nylons).

Phosphate esters with flame-retardant properties include triaryl phosphate, tricresyl phosphate, isodecyl diphenyl phosphate, and trioctyl phosphate. Styrenics and modified polyphenylene oxide use these compounds.

Antimony oxide (antimony trioxide) is used in conjunction with halogenated compounds forming antimony oxychloride which greatly enhances flame-retardant properties.

A principal objective in developing flame-retardant grades is to minimize the degradation in polymer properties which invariably occurs.

Hazards other than the flammability of a polymer or its flame-retardant resin grade are smoke generation and its toxicity, and tendency to melt drip.

Two of the most widely used tests for flammability with reported results are UL 94 and Oxygen Index, described under "Standard Tests" (p. 47) in this chapter.

ANTIOXIDANTS

Oxidation is the principal degrading reaction that limits the useful life of a plastic. This process is inherent with plastics, and is often initiated by the presence of impurities with exposure to oxygen or other oxidizing agent. Oxidation is accelerated by heat, light, humidity, and radiation.

Oxidative degradation causes polymer chain scission and/or crosslinking. In the process hydroxyl, carbonyl, acid, and ester groups are formed and polar groups may develop. Unstable peroxide radicals resulting from the oxidative process themselves decompose, initiating further degradation and significant polymer breakdown.

Effective antioxidants include hindered phenolics and secondary aryl amines with reactive OH or NH functional groups, respectively. These groups act to transfer hydrogen to free radicals, principally peroxy radicals, thereby limiting further degradation. Amines, additionally, act as chain terminators, but may cause discoloration of the plastic. Phenolic antioxidants include butylated hydroxytoluene, polyphenolics, and thio-bisphenolics. Substituted para-phenylene-diamine and diphenylamines are examples of amine antioxidants.

Another type of antioxidant includes phosphites and thioesters which act to decompose hydroperoxides before they form free radicals. Phosphites reduce hydroperoxides to alcohols and are themselves oxided to phosphates. Thioesters act to make phenolic and amine antioxidants more effective. Examples of this type of antioxidant include tris(mononylphenyl) phosphite, disteryl pentaerythritol diphosphite, and dialkyl thiodiprorionates.

Antioxidants are effective in the range of 0.1–0.5 percent of polymer weight. However, since other ingredients in a recipe may affect antioxidant action, the precise amount of antioxidant for optimum effectiveness for a given plastic should be determined for that plastic's specified formulation.

ULTRAVIOLET STABILIZERS

As noted above, the degradation process in plastics is accelerated by radiation. Ultraviolet (UV) light from the sun and fluorescent lamps is a common type of radiation to which plastics are exposed. Most plastics contain light-absorbing chemical groups (often formed in processing) such as hydroperoxides and carbonyl groups (chromophores). On absorbing UV energy, these groups form free radicals, which then react with other base polymer and additive polymer groups, initiating further degradation, much as in oxidative degradation.

Certain additives will retard this UV action because they have higher UV absorptivity than chromophores, thereby preventing most of the UV energy from affecting these groups. Examples of UV absorbers include 2-hydroxybenzophenones, 2-hydroxybenzotriazoles, acrylic esters, salicylates, and oxanilides. Hindered amines and phosphites, described previously in the "Antioxidants" section, also act as UV stabilizers in much the same way as they retard oxidative degradation.

Polyethylene is a plastic whose UV resistance (which is inherently poor) can be dramatically improved by addition of UV stabilizers along with 2–3 percent carbon black.

UV stabilizers are effective in the range of 0.25–1.0 percent of polymer weight.

INJECTION MOLDING

The process for forming thermoplastic resins into parts is injection molding. Compression molding, used for thermosetting resins, could also be used, but this method is not as efficient, requiring more energy and slower cycles.

Injection molding consists of these basic steps:

1. Dried thermoplastic compound is fed from a hopper into a heating chamber.

2. After reaching a viscous state, the plastic is forced (injected) by a plunger or screw into a water-cooled mold of the desired shape.

3. After cooling to the point where it can be handled without deformation, the part is ejected from the mold cavity, and sprues and runners (used to convey the plastic from the chamber to the mold) are trimmed.

Early injection molding machines used a reciprocating plunger, or ram, to inject plastic into the mold. While satisfactory parts may be made with this type of machine, there are several inherent disadvantages:

- There is virtually no mixing action to homogenize the plastic.
- Uniform heating throughout the charge, or "shot," is difficult to achieve in thick sections.
- Nozzle pressures tend to be erratic.
- Metering of material is not precise.

Newer injection molding machines now in common use employ a screw instead of a plunger. Compound is mixed and fed from the hopper through the screw tip backflow stop valve by the rotating screw. Accumulation of melt in front of the screw tip forces the screw to back up in the barrel. After the desired amount of melt has accumulated, the screw is powered forward hydraulically as an injection piston to shoot the charge into the mold. This type of machine, shown in Fig. 3-1, overcomes the disadvantages of machines using plunger feeds. A more complex machine employs two intermeshing screws, which provide even better mixing with reduced backflow along the screw flight.

The screw is divided into three zones: feed, compression, and metering. The feed section transports the resin from the hopper to the heated portion of the barrel. Flight volume is constant past adjacent flights. The compression section compacts the material from its powder or granular state into a melt. The volumes of the flight in this section decrease in the direction of the nozzle, compensating for the increasing density of the material as it melts. The melting characteristics of the resin determine the geometry of the screw, the sharper the melting point of the resin, the more abrupt the compression. In the metering zone, the plastic is mixed and heated into a homogeneous melt. A large amount of heat is generated by the action of the screw, and external heating must be adjusted accordingly. This zone has constant flight volume.

The objective is to determine the optimum balance of temperatures and pressures throughout the molding process. Mold and screw design are critical. Inadequate processing techniques may cause molecular weight degradation and buildup of low molecular weight ends to polymer chains resulting in significant property loss.

STANDARD TESTS

Selected standard tests described below are used to compare plastics and indicate their suitability for proposed applications. The accompanying tables rank polymers in descending order of their desirability. Values are obtained from the tables of key properties included in the polymer sections of this chapter.

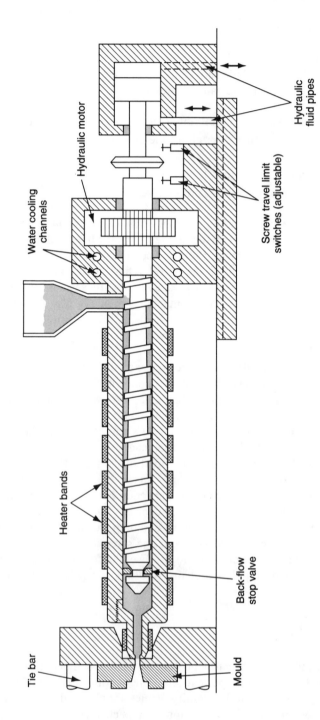

Fig. 3-1. Single-screw injection molding machine. (*Courtesy ICI Americas Inc.*)

48

Since these values are typical of standard grades without additives, it is note-worthy that higher or lower values would apply to other grades and, within limits, could change the rankings. The values shown indicate the inherent properties of the polymers and are significant when differences in values are appreciable.

A designation of na in the tables means data are not available.

Useful and convenient manuals on testing are published by Underwriters Laboratories:

- UL 746A, *Polymeric Materials—Short Term Property Evaluations*
- UL 746B, *Polymeric Materials—Long Term Property Evaluations*
- UL 746C, *Polymeric Materials—Use in Electrical Equipment Evaluations*

Practice for Conditioning Plastics and Electrical Insulating Materials for Testing (ASTM D 618)

Since moisture and temperature affect the properties of electrical insulating materials, it is essential that test specimens be uniformly conditioned before they are tested to ensure comparability of test results. This test calls for conditioning test specimens in standard laboratory atmosphere of 50% ± 2% relative humidity, 73.4° ± 1.8°F (23° ± 1°C), with adequate air circulation:

SPECIMEN THICKNESS	HOURS
0.25 in or under	40
over 0.25 in	88

Specific Gravity and Density (ASTM D 792)

Weight per unit volume is the value needed to calculate how many parts can be molded from a given weight of resin. Thus, it is as important as price in determining the cost of a molded part. Production and purchase specifications for a resin include an acceptable value or range for specific gravity.

The terms *specific gravity* and *density* both refer to weight/volume relationships and are often used interchangeably, although in property tables for electrical insulating materials, *specific gravity* is more often encountered.

Specific gravity is the ratio of the weight of a given volume of material at 73.4°F (23°C) to that of an equal volume of water at the same temperature. It is properly expressed as *Specific Gravity, 23/23°C.*

Density is the weight per unit volume of a material at 23°C and is expressed as *D23C, g/cm³*. Specific gravity and density values for the same material differ slightly because water at 23°C has a density slightly less than one: D23C, g/cm^3 = Specific Gravity 23/23°C × 0.99756.

To determine specific gravity, a piece of molded or extruded part is held by a fine wire, weighed, and submerged in water at 73.4°F (23°C). While it is in the water, it is weighed again. The difference in weight equals the weight of an equal volume of water.

TABLE 3-2   Specific Gravity of Thermoplastic Polymers
(23/23°C)
(ASTM D 792)

RANK	POLYMER	VALUE
1	Polypropylene	0.90
2	Acrylonitrile-butadiene-styrene	1.03
2	Polystyrene-butadiene	1.03
4	Polystyrene	1.04
4	Nylon 11	1.04
6	Modified polyphenylene oxide	1.06
7	Styrene-acrylonitrile	1.07
8	Nylon 6/6	1.14
9	Acrylic	1.19
10	Polycarbonate	1.20
11	Polysulfone	1.25
12	Polyetherimide	1.27
13	Polybutylene terephthalate	1.31
14	Polyphenylene sulfide	1.34
15	Polyethersulfone	1.37
16	Polyamide-imide	1.40
17	Acetal copolymer	1.41
18	Acetal homopolymer	1.42

Specific gravity equals the weight of the piece in air divided by the weight of an equal volume of water.

Specific gravities of commonly used thermoplastic polymers are listed and ranked in Table 3-2.

Dielectric Strength (ASTM D 149)

Dielectric strength, expressed as volts per mil, is the voltage gradient at which a continuous arc is formed through the specimen. To be comparable, specimens of different materials must be identical in geometry and conditioning, since values decrease as specimen thicknesses increase and are adversely affected with increasing moisture content and test environment temperature.

Specimens are thin sheets having parallel plane surfaces with sufficient area to prevent arc flashover. Published tables often show values for $1/8$ inch thick specimens for comparability. Specimens are placed between electrodes of prescribed size. There are two ways to conduct the test:

1. *Short-Time Test* (often published in comparison tables). The voltage is increased from zero to breakdown at a uniform specified rate from 0.5 to 1.0 kilovolt per second.

2. *Step-by-Step Test*. The initial voltage applied is 50 percent of the breakdown voltage determined by the short-time test. Voltage is increased in stated increments and at prescribed rates until breakdown occurs.

Dielectric strengths of commonly used thermoplastic polymers are listed and ranked in Table 3-3.

**TABLE 3-3   Dielectric Strength of Thermoplastic
Polymers
(Short Time, 125 mils, volts/mil)
(ASTM D 149)**

RANK	POLYMER	VALUE*
1	Polypropylene	600
1	Nylon 6/6	600 (dry)
3	Polyamide-imide	580
4	Modified polyphenylene oxide	550
5	Polystyrene	500
5	Acrylic	500
7	Polyetherimide	480
8	Nylon 11	425 (dry)
8	Polysulfone	425
8	Styrene-acrylonitrile	425
8	Acrylonitrile-butadiene-styrene	425
11	Polybutylene terephthalate	420
12	Polyethersulfone	400
12	Polystyrene-butadiene	400
15	Acetal homopolymer	380
15	Acetal copolymer	380
15	Polyphenylene sulfide	380
15	Polycarbonate	380

*At room temperature and 60 Hz.

Dielectric Constant, Permittivity, Specific Inductive Capacity (ASTM D 150)

Dielectric constant, permittivity, and specific inductive capacity (SIC) are different names for the same property. Since values for this property of an insulating material change with temperature and frequency, the term *constant* is really not appropriate, and *permittivity* or *relative permittivity* have come to be the preferred terms and are so designated by ASTM. SIC is no longer a widely used term. Because *dielectric constant* values and references still appear more commonly in the literature and in data tables, this term is used throughout this handbook.

Dielectric constant is a critical property in the selection of an insulating material for a specific end use. Where the primary function of an insulation is to minimize electric power losses, a low dielectric constant is required. For a capacitor dielectric, however, a high value is desirable to enable the capacitor to be designed as small as possible. For most insulating materials, dielectric constant values increase with temperature. The effect of increasing frequency is unique with each insulating material. Dry air has a dielectric constant of one (lowest) under normal conditions. The value for another material is determined by calculating the ratio of the capacitance of a capacitor with the material as dielectric to the capacitance of the same electrode system with air as dielectric. Frequencies commonly used are 60 Hertz and 10^6 Hertz.

Dielectric constants of commonly used thermoplastic polymers are listed and ranked in Table 3-4.

TABLE 3-4   Dielectric Constant of Thermoplastic Polymers
(ASTM D 150)

RANK	POLYMER	DIELECTRIC CONSTANT*	
		(60 Hz)	(10^6 Hz)
1	Polypropylene	2.2	2.2
2	Polystyrene	2.5	2.4
2	Polystyrene-butadiene	2.5	2.4
4	Styrene-acrylonitrile	2.6	2.6
4	Acrylonitrile-butadiene-styrene	2.6	2.6
4	Modified polyphenylene oxide	2.6	2.6
7	Polysulfone	3.1	3.0
8	Polyphenylene sulfide	3.1	3.2
9	Polycarbonate	3.2	3.0
10	Polyetherimide	3.2	3.1
11	Polybutylene terephthalate	3.3	3.1
12	Polyethersulfone	3.5	3.5
13	Acrylic	3.7	2.2
14	Acetal homopolymer	3.7	3.7
14	Acetal copolymer	3.7	3.7
16	Nylon 11	3.9	3.1
17	Polyamide-imide	4.3	3.9
18	Nylon 6/6	8.0	4.6

*At room temperature.

Dissipation Factor, Loss Tangent, Tan Delta (δ)
(ASTM D 150)

Dissipation factor, loss tangent, and tan (δ) are commonly used terms for the same property. For consistency, *dissipation factor* is the term used throughout this handbook.

Numerically, this property is the tangent of the loss angle. The charging current of a perfect dielectric will lead the applied voltage by 90 degrees. The angle by which the current vector of a less-than-perfect dielectric is less than 90 degrees is called the loss angle.

Power factor, a similar property, is determined by dividing the number of watts indicated by a wattmeter in a circuit, by the number of watts calculated by multiplying the reading on an ammeter times the reading on a voltmeter in the same circuit. The apparent wattage must be multiplied by that factor to give the power available for use. When the dissipation factor is less than 0.1, the power factor differs from the dissipation factor by less than 0.5 percent.

Electrical Resistance (ASTM D 257)

Electrical resistance is expressed in ohms: 1 ohm equals the resistance of a circuit in which a potential difference of 1 volt produces a current of 1 ampere. A high resistance is a desirable property in an electrical insulation. This property is adversely affected by increasing moisture content in the specimen and by higher humidity and higher temperature in the test environment. Surface condition of the specimen is another impor-

TABLE 3-5   Dissipation Factor of Thermoplastic Polymers (ASTM D 150)

RANK	POLYMER	DISSIPATION FACTOR* (60 HZ)	DISSIPATION FACTOR* (10^6 HZ)
1	Polystyrene	0.0001	0.0001
2	Polyphenylene sulfide	0.0003	0.0007
3	Modified polyphenylene oxide	0.0004	0.0009
4	Polypropylene	<0.0005	<0.0005
5	Polysulfone	0.0008	0.0034
6	Polyethersulfone	0.001	0.004
7	Acetal homopolymer	na	0.0048
8	Acetal copolymer	0.001	0.006
9	Styrene-butadiene	0.002	0.002
10	Polybutylene terephthalate	0.002	0.02
11	Polyetherimide	0.004	0.006
12	Acrylonitrile-butadiene-styrene	0.004	0.007
13	Styrene-acrylonitrile	0.007	0.007
14	Polycarbonate	0.009	0.010
15	Polyamide-imide	0.025	0.031
16	Nylon 11	0.04	0.05
17	Acrylic	0.05	0.3
18	Nylon 6/6	0.2	0.1

*At room temperature.

tant factor affecting resistance values. For test results to be comparable and meaningful, it is essential that test conditions be defined and controlled.

There are several indicators of electrical resistance.

Insulation resistance is the ratio of direct voltage applied to electrodes embedded in the test material to the total current between them. Both volume and surface resistance are factors.

Volume resistivity, a more widely published property, measures the electrical resistance between opposite faces of a unit cube of material.

Surface resistivity is the resistance between two opposite edges of a unit square of material.

ASTM D 257 describes test methods for determining these and other electrical resistance properties.

Electrical resistances of commonly used thermoplastic polymers are listed and ranked in Table 3-6.

Arc Resistance (ASTM D 495)

Higher values of arc resistance indicate greater resistance to surface breakdown caused by arcing or tracking. Values are determined by the time in seconds for breakdown along the surface of the test material. Thus, the condition of the surface and the humidity present in the test environment significantly affect values.

TABLE 3-6 Volume Resistivity of Thermoplastic Polymers (ASTM D-257)

RANK	POLYMER	$\Omega \cdot cm$*
1	Acrylic	1×10^{18}
2	Polyetherimide	6.7×10^{17}
3	Polyethersulfone	$>1 \times 10^{17}$
4	Polycarbonate	8.2×10^{16}
5	Polyamide-imide	8×10^{16}
6	Polysulfone	5×10^{16}
7	Polybutylene terephthalate	4×10^{16}
8	Acrylonitrile-butadiene-styrene	2×10^{16}
9	Polystyrene	$>1 \times 10^{16}$
9	Styrene-acrylonitrile	$>1 \times 10^{16}$
9	Polypropylene	$>1 \times 10^{16}$
12	Styrene-butadiene	1×10^{16}
12	Polyphenylene sulfide	1×10^{16}
12	Modified polyphenylene oxide	1×10^{16}
15	Acetal homopolymer	1×10^{15}
16	Acetal copolymer	1×10^{14}
16	Nylon 11	1×10^{14}
18	Nylon 6/6	1×10^{13}

*At room temperature.

Test specimens are 0.125 inches thick and have a flat surface. Two electrodes are placed on the surface of the specimen so that an arc occurs when sufficient voltage is applied between them. The arc is caused to occur intermittently with decreasing intervals between arcing. In later stages of the test, the current is also increased. The time of arcing remains constant. There are several distinct types of failure:

- A conductive fine line, or track is formed between electrodes
- The surface of the specimen carbonizes to the point where it carries the current
- The specimen bursts into flame
- Some inorganic dielectrics become incandescent and conduct current, but regain their dielectric properties on cooling.

Arc resistances of commonly used thermoplastic polymers are listed and ranked in Table 3-7.

Water Absorption (ASTM D 570)

The amount of water which a plastic absorbs affects its properties, lowering electrical values and dimensional stability, and is a key indicator of the suitability of a plastic for electrical/electronic applications.

For molding materials, specimens are disks of 2 inches in diameter by $^1/_8$ inch thick. For sheet materials, specimens are 3 inches long by 1 inch wide by the thickness of the sheet. Specimens are conditioned for 24 hours at 122.0°F (50°C), cooled in a desiccator,

**TABLE 3-7   Arc Resistance of Thermoplastic Polymers
(ASTM D 495)**

RANK	POLYMER	SECONDS
1	Acrylic	no tracking
2	Acetal copolymer	240 (burns)
3	Polyamide-imide	230
4	Acetal homopolymer	220
5	Polybutylene terephthalate	184
6	Polypropylene	150
7	Nylon 6/6	130
8	Polyetherimide	128
9	Nylon 11	123
10	Polysulfone	122
11	Polycarbonate	120
12	Styrene-acrylonitrile	115
13	Acrylonitrile-butadiene-styrene	100
14	Modified polyphenylene oxide	75
15	Polyethersulfone	70
16	Polystyrene	65
17	Polystyrene-butadiene	50
18	Polyphenylene sulfide*	34

*Reinforced and filled grades have much higher values.

and immediately weighed. The specimen is immersed in water at 73.4°C (23°C) for 24 hours. Upon removal, the specimen is wiped dry and immediately weighed. The increase in weight is reported as percentage gained. For materials which lose soluble matter during immersion, the specimen must be redried and reweighed, from which percent the soluble matter is calculated. The percent gain in weight plus the percent soluble matter lost equals the percent water absorption. Specimens may be tested at other temperatures, such as 212°F (100°C), and for other periods of immersion, including equilibrium.

Water absorption of commonly used thermoplastic polymers are listed and ranked in Table 3-8.

Deflection Temperature (ASTM D 648)

Deflection temperature is the temperature at which a specified deflection occurs in a specimen under a selected load. This test is useful in indicating the temperature at which a plastic distorts under test conditions, although it does not necessarily indicate its high temperature usage limit.

A specimen 5 inches long by $^1/_2$ inch wide by thickness in the $^1/_8$–$^1/_2$ range is placed on supports 4 inches apart. A load of 66 or 264 pounds per square inch is applied at the center of the specimen. The temperature in the chamber of the testing device is raised at the rate of 3.6° ± 0.36°F (2° ± 0.2°C) per minute. The temperature at which the specimen deflects 0.010 inch is the deflection temperature at the selected load. (See Fig. 3-2.)

Deflection temperatures of commonly used thermoplastic polymers are listed and ranked in Table 3-9.

TABLE 3-8   Water Absorption of Thermoplastic Polymers
($\frac{1}{8}$ in Thick Specimen)
(ASTM D 570)

RANK	POLYMER	% GAIN IN WEIGHT AFTER 24 h @ 73°F (23°C)
1	Polypropylene	0.01
2	Polystyrene	0.02
2	Polyphenylene sulfide	0.02
4	Polystyrene-butadiene	0.06
4	Modified polyphenylene oxide	0.06
6	Polybutylene terephthalate	0.08
7	Polycarbonate	0.15
8	Styrene-acrylonitrile	0.20
9	Acrylonitrile-butadiene-styrene	0.22
9	Acetal copolymer	0.22
11	Acetal homopolymer	0.25
11	Polyetherimide	0.25
13	Polyamide-imide	0.30
13	Polysulfone	0.30
13	Nylon 11	0.30
13	Acrylic	0.30
17	Polyethersulfone	0.43
18	Nylon 6/6	1.20

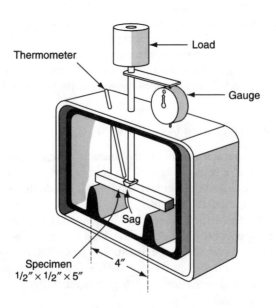

Fig. 3-2   Deflection temperature device. (*Courtesy Celanese.*)

**TABLE 3-9   Deflection Temperature of Thermoplastic
Polymers
(ASTM D 648)**

RANK	POLYMER	FLEXURAL LOAD 264 psi	
		°F	°C
1	Polyamide-imide	532	278
2	Polyether sulfone	397	203
3	Polyetherimide	392	200
4	Polysulfone	345	174
5	Polyphenylene sulfide	275	135
6	Polycarbonate	265	129
6	Modified polyphenylene oxide	265	129
8	Acetal homopolymer	255	124
9	Acetal copolymer	230	110
10	Acrylonitrile-butadiene-styrene	210*	99*
11	Styrene-acrylonitrile	200	93
11	Polystyrene	200	93
11	Polystyrene-butadiene	200	93
11	Acrylic	200	93
15	Nylon 6/6	194	90
16	Polypropylene	140	60
17	Nylon 11	131	55
18	Polybutylene terephthalate	130	54

*Annealed.

Tensile Properties (ASTM D 638)

Tensile strength is one of the key mechanical properties, measuring in pounds per square inch the force necessary to elongate and break the specimen and in percent how much the material stretches before breaking. The tensile (elastic) modulus is the ratio of the applied stress to the strain it produces in the region in which strain is linearly proportional to stress. The modulus is usually expressed in terms of pounds per square inch values. Parts with little elongation and high tensile strength would tend to be brittle. Hence, moderate elongation is necessary for toughness. High elongation, however, indicates parts would be rubbery.

Specimens with dimensions shown below are prepared by injection molding or by machining from compression-molded plaques. Specimens may also be rod-shaped or tubular, depending on the shape of the end product. An Inston tensile testing machine is employed in which both ends of the specimen are tightly clamped in the machine heads. One jaw is fixed and the other is movable at rates of 0.05, 0.2, 2, or 20 inches/minute. The machine automatically plots stress against strain on graph paper. (See Fig. 3-3.)

Tensile strengths of commonly used thermoplastic polymers are listed and ranked in Table 3-10.

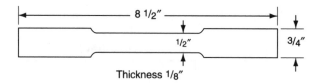

Fig. 3-3   Tensile test specimen.

Izod Impact Resistance (ASTM D 256)

The degree to which a plastic is notch-sensitive may be determined by measuring the energy required to break notched specimens under standard conditions. In designing parts made with plastics with low impact resistance, care should be taken in their design to avoid sharp corners and grooves. It is noteworthy that some otherwise tough plastics (nylon and acetals are examples) are notch-sensitive and exhibit relatively low Izod impact resistance.

Specimens are usually $2^{1}/_{2}$ inches long by $^{1}/_{2}$ inch wide by $^{1}/_{8}$ inch thick, although thicknesses up to $^{1}/_{2}$ inch may be used. A notch is cut in the middle of the narrow face of the specimen, which is then clamped in the base of a pendulum testing machine with the notch facing the direction of impact. The pendulum is released and the force required to break the specimen is calculated from the height the pendulum reaches on

TABLE 3-10   Tensile Strength of Thermoplastic Polymers (ASTM D 638)

RANK	POLYMER	psi AT BREAK*
1	Polyamide-imide	27,800
2	Polyetherimide	15,200
3	Polyethersulfone	12,200
4	Nylon 6/6	11,200
5	Polysulfone	10,200
6	Acrylic	10,000
6	Acetal homopolymer	10,000
8	Modified polyphenylene oxide	9,600
9	Polyphenylene sulfide**	9,500
9	Styrene-acrylonitrile	9,500
11	Polycarbonate	9,000
12	Acetal copolymer	8,800
13	Nylon 11	8,200
14	Polybutylene terephthalate	8,200
15	Acrylonitrile-butadiene-styrene	7,000
16	Polystyrene	6,000
16	Polystyrene-butadiene	6,000
18	Polypropylene	5,400

*At room temperature

**Reinforced and filled grades have much higher values.

the follow-through. Izod impact strength is expressed as foot pounds per inch of notch. (See Figs. 3-4 and 3-5.)

Izod impact strengths of commonly used thermoplastic polymers are listed and ranked in Table 3-11.

Flammability (UL 94)

Flammability of a plastic material depends on part size and shape, as well as the tendency of the plastic to burn. Important characteristics are ease of ignition, burning rate, time to extinguish, and products of combustion. UL 94 is intended to indicate the suitability of a plastic for applications where flammability is a factor.

Specimens are 5 inches long by $^1/_2$ inch wide by thickness usually in the 0.030–0.250 inch range. In the UL 94HB test procedure, specimens are conditioned for 48 hours at 73.4° ± 3.6°F (23° ± 2°C) in relative humidity of 50 ± 5 percent prior

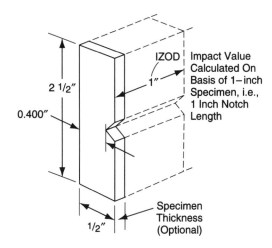

Fig. 3-4   Izod specimen. (*Courtesy Celanese.*)

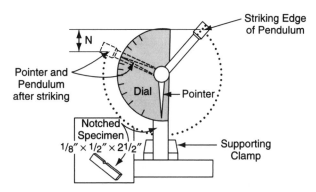

Fig. 3-5   Impact testing machine. (*Courtesy Celanese.*)

TABLE 3-11   Izod Impact Strength of Thermoplastic Polymers
(Foot pounds per inch notched $^1/_8$ in specimen)
(ASTM D 256)

RANK	POLYMER	VALUE*
1	Polycarbonate	12
2	Acrylonitrile-butadiene-styrene	6.0
3	Modified polyphenylene oxide	5.0
4	Polyamide-imide	2.7
5	Polystyrene-butadiene	2.5
6	Nylon 6/6	2.1
7	Acetal copolymer	1.4
8	Acetal homopolymer	1.3
8	Polysulfone	1.3
10	Polyethersulfone	1.2
11	Polybutylene terephthalate	1.0
11	Polyetherimide	1.0
13	Nylon 11	0.75
14	Polypropylene	0.50
15	Styrene acrylonitrile	0.40
15	Acrylic	0.40
15	Polystyrene	0.40
18	Polyphenylene sulfide**	0.30

*At room temperature.

**Reinforced and filled grades have much higher values.

to testing. A specimen is clamped horizontally and canted at a 45° angle to horizontal. A wire gauze is positioned beneath the specimen at a distance of $^3/_8$ inch. The free end of the specimen is ignited with a Bunsen burner and the burning rate is measured. The material is classed 94HB if:

• Burning rate does not exceed 1.5 inches/minute over a 3.0 inch span for specimens having a thickness of 0.120–0.500 inch

• Burning rate does not exceed 3.0 inches/minute over a 3 inch span for specimens having a thickness less than 0.12 inch

• Burning ceases before the flame reaches a reference mark 4.0 inches along the specimen.

For a 94V classification, a set of specimens is conditioned as above and another set is conditioned in an air-circulating oven for 168 hours at 158° ± 1.8°F (70° ± 1°C), after which the set is immediately placed in a desiccator for at least 4 hours prior to testing. The burning test consists of clamping a specimen from either set vertically with its lower end 12 inches above a layer of dry cotton. A Bunsen burner is placed under the lower end of the specimen for 10 seconds. After the flame is removed, the duration of flaming of the specimen is noted. The procedure is repeated on the same specimen.

Table 3-12 shows the test criteria for 94V-0, V-1, and V-2 classifications.

TABLE 3-12   UL 94 Flammability Test Criteria

	94 CLASSIFICATIONS		
	V-0	V-1	V-2
Maximum time of flaming after either test (seconds)	10	30	30
Maximum total flaming time for each set of 5 specimens (2 flame applications each specimen) (seconds)	30	250	250
Number of specimens that burn to clamp	0	0	0
Number of specimens that drip flaming particles igniting cotton 12 in below specimen	0	0	some
Maximum number of seconds with glowing combustion for any specimen after second removal of flame	30	60	60

For a 94-5V classification, a set of specimens (or plaques 6 inches × 6 inches × 0.5 inch) is conditioned for at least 48 hours at $73.4° ± 3.6°F$ ($23° ± 2°C$) and relative humidity of $50 ± 5$ percent prior to testing. Another set is conditioned in a circulating air oven for 60 days at $249.8° ± 1.8°F$ ($121° ± 1°C$), then cooled in a desiccator for at least 4 hours prior to testing. A specimen from either set is clamped vertically. Then a flame from a Bunsen burner is applied for 5 seconds to a lower corner of the specimen and removed for 5 seconds. The procedure is repeated five times. After the fifth removal of the flame, no specimen shall burn for more than 60 seconds or drip any particles.

Flammability ratings of commonly used thermoplastic polymers are listed and ranked in Table 3-13.

TABLE 3-13   Typical Flammability Ratings of Thermoplastic Polymers. Based on UL Standard 94 ($^1/_{16}$ in Specimen)

	POLYMER	STANDARD GRADE	FLAME-RETARDANT GRADE
1	Polyamide-imide	V-0	not needed
1	Polyetherimide	V-0	not needed
1	Polyethersulfone	V-0	not needed
1	Polyphenylene sulfide	V-0	not needed
5	Nylon 6/6	V-2	V-0
5	Nylon 11	V-2	none
5	Polycarbonate	V-2	V-0
5	Polysulfone	V-2	V-0
9	Acrylonitrile-butadiene-styrene	HB	V-0/5V
9	Polybutylene terephthalate	HB	V-0
9	Modified polyphenylene oxide	HB	V-0
9	Polystyrene	HB	V-0
9	Polystyrene-butadiene	HB	V-0
9	Styrene-acrylonitrile	HB	V-0
9	Polypropylene	HB	V-0
16	Acetal homopolymer	HB	none
16	Acetal copolymer	HB	none
16	Acrylic	HB	none

The Oxygen Index (OI) (ASTM D 2863)

This test is designed to provide a graduated, numerical measure of flammability, rather than a classification (UL 94), and to improve the reproducibility of tests.

The oxygen index is the percent oxygen in an oxygen–nitrogen mixture that will just sustain combustion of a vertically mounted bar specimen that has been ignited by application of a gas flame to its *upper* end (as a candle). The oxygen content of air at sea level, approximately 20 percent, is considered the reference point in determining specimen susceptibility to burning. An index of 100 means the specimen could not be ignited or would not continue to burn after ignition in 100 percent oxygen. Materials with OI greater than 28 are usually classed UL 94V-0.

Oxygen indexes of commonly used thermoplastic polymers are listed and ranked in Table 3-14.

TABLE 3-14   Oxygen Index of Thermoplastic Polymers (ASTM D 2863)

RANK	POLYMER	STANDARD GRADE	FLAME-RETARDANT GRADE
1	Polyetherimide	47	not needed
2	Polyphenylene sulfide	46	not needed
3	Polyamide-imide	43	not needed
4	Polyethersulfone	38	not needed
5	Nylon 6/6	31	na
6	Polysulfone	30	32
7	Polycarbonate	25	35
8	Nylon 11	24.5	none
9	Polybutylene terephthalate	22	29
10	Acrylonitrile-butadiene-styrene	19	na
11	Polystyrene-butadiene	18	28
12	Polypropylene	18	27
13	Polystyrene	18	none
14	Acetal homopolymer	15	none
15	Acetal copolymer	14.9	none
16	Modified polyphenylene oxide	na	30
17	Acrylic	na	none
17	Styrene-acrylonitrile	na	na

MAXIMUM SERVICE TEMPERATURE OF THERMOPLASTIC POLYMERS

All properties of thermoplastic resins are affected by temperature, and each resin has a useful thermal range beyond which it will not perform satisfactorily. Each resin is unique in its reaction to temperature. For some resins, some property values hold up well throughout this range, but beyond it, deteriorate rapidly. For other resins, some property values steadily worsen with increasing temperature, making it more difficult to define the point beyond which the resin is no longer useful. In selecting a resin for a specific application, it is important not only to determine the expected operating temperature range of a part, but also to consider how resin characteristics change within that range.

Table 3-15 lists maximum service temperatures below which manufacturers predict their resins will retain effective dielectric and mechanical properties.

TABLE 3-15   Maximum Recommended Service Temperatures of Thermoplastic Polymers

RANK	POLYMER	°F	°C
1	Polyamide-imide	500	260
2	Polyphenylene sulfide	400	205
3	Polyethersulfone	355	180
4	Polyetherimide	340	170
5	Polysulfone	300	150
5	Polybutylene terephthalate*	300	150
7	Nylon 6/6	265	130
8	Polycarbonate	240	115
9	Modified polyphenylene oxide	220	105
9	Acetal copolymer	220	105
9	Polypropylene	220	105
12	Acrylonitrile-butadiene-styrene (heat stabilized)	200	95
12	Acrylic	200	95
14	Acetal homopolymer	195	90
15	Styrene-acrylonitrile	180	80
16	Polystyrene	160	70
17	Polystyrene-butadiene	150	65
18	Nylon 11*	150	65

*Reinforced grades have much higher values.

RELATIVE COSTS OF THERMOPLASTIC POLYMERS

There are many factors to be considered in determining part cost, among which are the following:

- Resin price per pound
- Specific gravity (determines the number of cubic inches per pound)
- Part design (can make use of superior resin properties to reduce wall thickness)
- Molding cycle and machine capabilities
- Number of mold cavities

Thus, it is possible for a high-priced resin with relatively low specific gravity to compete successfully with a low-priced resin having relatively high specific gravity, especially if the high-priced resin has a significantly shorter molding cycle and superior properties permitting design economies.

In selecting a resin for a specific application, these steps should be taken:

1. Candidates with appropriate properties should first be determined.

2. Resin cost per cubic inch should be obtained from suppliers or calculated by dividing the resin price per pound by cubic inches per pound for the resin (see Table 3-16).

3. Part design and resin properties should be reviewed to see which resin permits the most economical design.

4. Molding cycles should be estimated in consultation with resin suppliers and machine manufacturers.

5. Calculate finishing costs when comparing thermoplastic and thermosetting resins (which typically cost more to finish).

6. Decide on the number of mold cavities, considering part volume needed to supply forecast market demand, and production constraints such as machine capabilities and availability.

Table 3-16 shows cubic inches per pound for selected thermoplastic polymers, from which are calculated price advantages of lighter polymers. For example, if the resin price of nylon 6/6 were 25 percent higher than acetal, part cost would be equal due to the higher specific volume of nylon 6/6 (24.3 versus 19.5).

TABLE 3-16   Specific Volume and Cost Factor of Thermoplastic Polymers

	CUBIC* INCHES PER LB	% PRICE/LB DIFFERENCE FOR EQUAL PART COST
1. Polypropylene	30.8	58
2. Acrylonitrile-butadiene-styrene	27.4	41
3. Polystyrene-butadiene	26.9	38
4. Polystyrene	26.6	36
5. Nylon 11	26.6	36
6. Modified polyphenylene oxide	26.1	34
7. Styrene-acrylonitrile	25.9	33
8. Nylon 6/6	24.3	25
9. Acrylic	23.7	22
10. Polycarbonate	23.1	18
11. Polysulfone	22.1	13
12. Polyetherimide	21.8	12
13. Polybutylene terephthalate	21.1	8
14. Polyphenylene sulfide	20.6	6
15. Polyethersulfone	20.2	4
16. Polyamide-imide	19.8	2
17. Acetal copolymer	19.6	—
18. Acetal homopolymer	19.5	—

ALLOYS

Alloys are formed by interfacial adhesion between selected polymers during extrusion to form homogeneous melt. Alloys differ from blends in that, in blends, distinct polymer phases remain within the resin matrix. Although alloys exhibit many of the properties of a macromolecule, the close affinity of component polymers is physical and not chemical. Producers include Dexter, Du Pont, General Electric, Miles, Rohm & Haas, and Union Carbide.

Also described as alloys are elastomeric products consisting of a continuous phase of a polyolefin in which is dispersed a finely divided highly vulcanized elastomer, for example, ethylene propylene diene rubber (EPDM). Monsanto (Santoprene™, Geolast™) and Du Pont (Alcrym™) are producers of this type of elastomeric alloy.

The key to successful alloying is the role of compatibilizers which must be miscible with all the ingredient polymers acting to bond them together. Compatibilizing agents are highly proprietary with alloy manufacturers. Examples of successful agents include maleic anhydride, isocyanate, stearic acid, ethylene-vinyl acetate, styrene-ethylene/butylene-styrene block copolymer (Kraton™, Shell Chemical), and nitrile-butadiene elastomer (Chemigum P-83™, Goodyear).

Possible combinations of polymers to produce alloys are virtually limitless. Properties of highly miscible polymer alloys may be better than those of component polymers.

See also the "Polymerization" section, p. 41 in this chapter.

LIQUID CRYSTAL POLYMERS (LCPS)

Among the newest and most intriguing of materials, liquid crystal polymer resins have been commercially available since 1984 from Dartco (Xydar™), now Amoco Performance Products. Other producers now include Hoechst Celanese (Vectra™) and Montedison (Granlar™).

Liquid crystal polymers comprise a family of thermoplastic aromatic copolyesters with outstanding high-temperature performance and excellent resistance to chemicals, radiation, and burning.

Chemistry

Vectra resins are made from p-dihydroxybenzoic acid and 6,2 hydroxynaphthoic acid monomers. Xydar resins are polyesters based on terephthalic acid, isophthalic acid, dihydroxybiphenyl, and p-dihydroxybenzoic acid. Granlar resins are reported to be made from substituted hydroquinones and terephthalic acid monomers.

Liquid crystal polymers are classed as engineering thermoplastics.

Grades

Available grades for electrical/electronic applications include unreinforced, up to 40 percent glass-reinforced, and mineral-filled grades. Low-viscosity grades permit very thin sections.

Processing

Liquid crystal polymers may be injection molded on modified conventional equipment. Molding conditions vary widely with the grade and type of resin used. Resins must be thoroughly dried before use.

Properties

The polymer is comprised of densely packed fibrous chains that account for exceptional mechanical strength. Continuous use temperature is in the range of –60° to 465°F (–51°–240°C). Dimensional stability is excellent and creep resistance is superior to most other plastics.

Parts made from liquid crystal polymers are resistant to attack by virtually all chemicals, except sodium hydroxide in concentrations of 30 percent or higher. They are unaffected by exposure to most solvents and radiation. Polymers possess Oxygen Index ratings in the 35 to 47 range and do not burn in air.

Electrical/Electronic Uses

Where property requirements justify their relatively high price, liquid crystal polymer resins are suitable for connectors, chip carriers, sockets, automotive underhood electrical parts, and air conditioning electrical parts.

CONDUCTING POLYMERS

Although still in the early stages of commercial development, conducting polymers are of intense interest for high-technology applications such as batteries for watches, timers for video-cassette recorders, and TV remote controls. A possible high-volume application and one which would greatly advance the evolution of electric vehicles is the development of a practical conducting polymer type battery which would have high power density, long service life, adequate mechanical strength, and rapid rechargeability. The organization spearheading this program is the U.S. Advanced Battery Consortium (USABC), Detroit, Michigan.

Other applications with commercial potential include solder replacements, biomedical devices, electrolytic capacitors, and conducting skins for military equipment such as tanks and aircraft.

Conducting polymers are prepared at Westinghouse from conventional polymers which have considerable delocalization of electrons along polymer chains. Next, the polymers are reacted with an oxidizing or reducing agent which withdraws or adds electrons. These dopants include halogens (iodine, bromine), organic oxidizing agents (chloramil, dicholorocyanoquinone), and alkali metals (sodium, potassium).

At the outset, conductivity rises sharply with increasing dopant content to a point where additional dopant causes only incremental conductivity increase. A major problem of considerable concern is that conductivity decreases on exposure to elevated temperatures and on normal aging. Some conducting polymers (most are proprietary) may be blended with polymers such as polyethylene or polyvinyl chloride, forming a readily processible partially conducting material.

An earlier and still used method for making conducting polymers is to add to conventional polymers conducting fillers such as carbon black, or copper or aluminum flakes.

DEVELOPMENT PROGRAMS

Polymers with a wide range of properties and prices are available to design and materials engineers. While research and development for new polymers continues, the efforts of many key resin producers and compounders are directed toward modifying existing polymers in these ways:

- Physically blending, or "alloying," different polymers
- Copolymerization or polycondensation of two or more monomers by addition reaction (see "Polymerization" (p. 41) in this chapter)
- Graft copolymerization in which side chains of one polymer are attached to another polymer serving as backbone
- Block copolymerization in which a row, or "block," of units of one polymer are joined with a row of units of another polymer (—AAAA—BBBB—)
- Compounding polymers with reinforcements and fillers (see "Reinforcements and Fillers" (p. 44) in this chapter)

The development of resins formulated with styrene, butadiene, and acrylonitrile and their blends, or alloys, with polycarbonate, polysulfone, and polyvinyl chloride are successful examples of these programs. Blending ratios may be designed to achieve certain properties required for a specific end use.

Underway at several companies are intensive development programs for new alloys, liquid crystal polymers, and conducting polymers, all covered separately in this chapter.

The focus of reinforcement and filler technology is on improving wetting characteristics by surface treatment of fibers or particles with stearates, resinates, silanes and titanates.

For those resins not inherently flame retardant, efforts continue to develop improved FR grades with minimal sacrifice of other properties.

MARKET TRENDS

In high-volume electrical/electronic applications for thermoplastic molding resins (such as transportation, appliances, and computers) where minimal dielectric properties are acceptable, cost effectiveness and design ingenuity are primary concerns. A goal of design engineers is to make the most practical use of materials so that the lowest possible manufacturing cost is attained consistent with specified performance standards. In this regard, where there are no overriding cost factors, plastics with low specific gravity, such as styrenics and nylons, are favored.

In applications requiring flame-retardant properties, likely candidates are compounds with UL 94V-0 or 5V ratings having combustion products with minimal toxicity. These would include imide, polyarylate, polycarbonate, polyetheretherketone, and modified polyphenylene oxide compounds.

Where optimum dielectric properties are required, styrenics and modified polyphenylene oxide have the lowest dielectric constant, nylon 6/6 and polyamide-imide have the highest dielectric strength, and acrylics, acetals, and polyamide-imides are best for arc resistance.

Usage of conducting polymers is growing rapidly.

SUMMARY OF RANKINGS OF THERMOPLASTIC POLYMERS

Table 3-17 ranks thermoplastic polymers according to key properties for each polymer. Rankings range from 1, most favorable from an electrical insulation viewpoint, to 18, least favorable. Note that no polymer ranks either highest or lowest in all properties. This table is intended as an aid to design engineers desiring to investigate polymers potentially suitable to meet critical criteria. Values are for basically unmodified polymers. Polymers with reinforcements (usually glass fibers), fillers (often mineral), and additives (flame retardants, antioxidants, UV inhibitors, etc.) may have significantly higher or lower values for each property.

PRINCIPAL THERMOPLASTIC POLYMERS FOR ELECTRICAL/ ELECTRONIC USES

The chemistry, grades, processing, properties, and electrical/electronic uses of principal thermoplastic polymers are described in the sections which follow. The values shown in tables are for general purpose grades, unreinforced, except as noted.

The following polymers are covered:

- Acetals: Homopolymer
- Acetals: Copolymer
- Acrylic (PMMA)
- Amide Polymers: Nylon 6/6
- Amide Polymers: Nylon 11
- Imide Polymers: Polyamide-Imide (PAI)
- Imide Polymers: Polyetherimide
- Polyarylate (PAR)
- Polybutylene Terephthalate (BPT)
- Polycarbonate (PC)
- Polyetheretherketone (PEEK)
- Modified Polyphenylene Oxide

TABLE 3-17 Ranking of Thermoplastic Polymers by Key Property Values

POLYMER	SPECIFIC GRAVITY	DIELECTRIC STRENGTH	DIELECTRIC CONSTANT	DISSIPATION FACTOR	VOLUME RESISTIVITY	ARC RESISTANCE	WATER ABSORPTION	DEFLECTION TEMPERATURE	MAXIMUM SERVICE TEMPERATURE	TENSILE STRENGTH	IZOD IMPACT STRENGTH
Acetal homopolymer	18	15	14	7	15	4	11	8	14	6	8
Acetal copolymer	17	15	14	8	16	2	9	9	9	12	7
Acrylic	9	5	13	17	1	1	13	11	12	6	15
Nylon 6/6	8	1	18	18	18	7	18	15	7	4	6
Nylon 11	4	8	16	16	16	9	13	17	17	13	13
Polyamide-imide	16	3	17	15	5	3	13	1	1	1	4
Polyetherimide	12	7	10	11	2	8	11	3	4	2	11
Polybutylene terephthalate	13	11	11	10	7	5	6	18	5	14	11
Polycarbonate	10	15	9	14	4	11	7	6	8	11	1
Modified polyphenylene oxide	6	4	4	3	12	14	4	6	9	8	3
Polyphenylene sulfide	14	15	8	2	12	18	2	5	2	9	18
Polypropylene	1	1	1	4	9	6	1	16	9	18	14
Polystyrene	4	5	2	1	9	16	2	11	16	16	15
Polystyrene-butadiene	2	12	2	9	12	17	4	11	17	16	5
Styrene-acrylonitrile	7	8	4	13	9	12	8	11	15	9	15
Acrylonitrile-butadiene-styrene	2	8	4	12	8	13	9	10	12	15	2
Polysulfone	11	8	7	5	6	10	13	4	5	5	8
Polyethersulfone	15	12	12	6	3	15	17	2	3	3	10

- Polyphenylene Sulfide (PPS)
- Polypropylene (PP)
- Styrenics: Polystyrene (PS)
- Styrenics: Polystyrene-Butadiene
- Styrenics: Styrene-Acrylonitrile (SAN)
- Styrenics: Acrylonitrile-Butadiene-Styrene (ABS)
- Sulfone Polymers: Polysulfone
- Sulfone Polymers: Polyethersulfone

Note: A designation of "na" in the tables means data are not available.

Acetals: Homopolymer (Molding Resin)

The first commercially available acetal homopolymer resin was introduced in 1959 by Du Pont with the trademark Delrin. Two years later, Celanese (now Hoechst Celanese) marketed an acetal copolymer resin, Celcon™, which competes for essentially the same end uses, many of which were formerly served by metals.

Chemistry

Acetal homopolymers are linear polymers formed by polymerizing formaldehyde in the presence of a catalyst to polyoxymethylene:

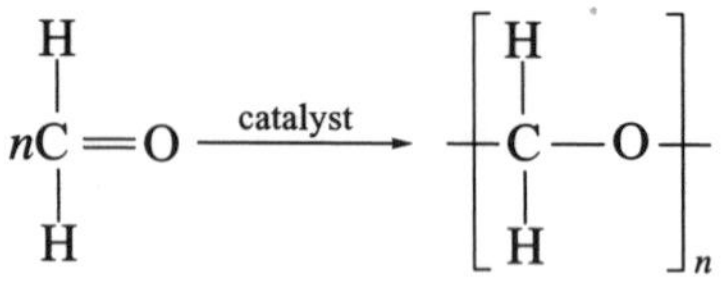

$$n \overset{\overset{\textstyle H}{|}}{\underset{\underset{\textstyle H}{|}}{C}} = O \xrightarrow{\text{catalyst}} \left[\overset{\overset{\textstyle H}{|}}{\underset{\underset{\textstyle H}{|}}{C}} - O \right]_n$$

    An acid anhydride may be used to terminate the ends of the molecule, giving the polymer greater stability.

    Acetal homopolymer is classed as an engineering thermoplastic.

Grades

Several injection molding grades of Delrin resin are produced:

- General purpose molding grade
- Higher melt flow grade with lower impact strength and elongation
- High-productivity, high-flow, fast-molding grades
- 20% glass fiber-filled general purpose grade
- Low-wear/low-friction, lubricated grade
- Lowest-wear/low friction, 21% Teflon™ TFE fluoroplastic fibers-filled grade

No flame-retardant grades are available.

The related standard for both military and commercial purposes is ASTM D 4181, *Specification for Acetal (POM) Molding and Extrusion Materials.*

Processing

Acetal homopolymer resins may be injection molded in conventional equipment. Resins do not require drying prior to molding.

Properties

Acetal homopolymer resin is superior to the copolymer resin in mechanical properties, but has lower processability and resistance to heat, hot water, and strong alkalis. Resistance is good to aliphatics, but poor to aromatics. Sonic welding may be used in finishing operations.

Table 3-18 shows key property values for both acetal homopolymer and copolymer resins.

TABLE 3-18   Key Properties of Acetal Homopolymer and Copolymer Molding Resins

	ASTM TEST METHOD	HOMOPOLYMER VALUE	COPOLYMER VALUE
Specific gravity*	D 792	1.42	1.41
Dielectric strength, 125 mils, short time, V/mil*	D 149	380	380
Dielectric constant*	D 150		
60 Hz		3.7	3.7
10^6 Hz		3.7	3.7
Dissipation factor*	D 150		
60 Hz		na	0.001
10^6 Hz		0.0048	0.006
Volume resistivity, $\Omega\cdot$cm*	D 257	1×10^{15}	1×10^{14}
Arc resistance*	D 495	220	240 (burns)
Water absorption, $^1/8$ in thick specimen, %	D 570		
24 h @ 73°F (23°C)		0.25	0.22
Deflection temperature, 264 psi °F (°C)	D 648	255 (124)	230 (110)
Maximum recommended service temperature °F (°C)		195 (90)	220 (104)
Tensile strength at break, psi*	D 638	10,000	8,800
Izod impact strength	D 256		
Foot pounds per inch notched $^1/8$ in specimen*		1.3	1.4
Flammability ratings			
UL Standard 94, $^1/16$ in specimen			
Standard grade		HB	HB
FR grade		none	none
Oxygen index, %	D 2863		
Standard grade		15	14.9
FR grade		none	none

*At room temperature

Electric/Electronic Uses

Applications of this resin are limited to components where good FR properties and chemical resistance are not required, including cut-out switches by utilities.

Acetals: Copolymer
(Molding Resin)

Acetal copolymer (Celcon™) was introduced in 1961 by Celanese (now Hoechst Celanese) as a competitor to Du Pont's Delrin™, an acetal homopolymer with similar but not identical properties. BASF is now also a producer of the copolymer Ultraform™.

Chemistry

Acetal copolymer is produced from two cyclic ethers: trioxane plus a small proportion of dioxane. Each trioxane molecule yields three oxyethylene linkages, while each dioxane molecule provides two oxyethylene linkages, resulting in a polymer with the structure:

$$\left[\begin{array}{ccc} \underset{H}{\overset{H}{|}}\!\!-\!\!\underset{H}{\overset{H}{|}} & \underset{H}{\overset{H}{|}}\!\!-\!\!\underset{H}{\overset{H}{|}} \end{array} \right]$$

$$-\!\!\left[\begin{matrix}H & H \\ | & | \\ C-C-O- \\ | & | \\ H & H\end{matrix}\begin{matrix}H & H \\ | & | \\ C-C-O \\ | & | \\ H & H\end{matrix}\left(\begin{matrix}H \\ | \\ C-O- \\ | \\ H\end{matrix}\begin{matrix}H \\ | \\ C-O- \\ | \\ H\end{matrix}\begin{matrix}H \\ | \\ C-O \\ | \\ H\end{matrix}\right)_{x}\right]_{n}\!\!-$$

The occasional presence of two adjacent carbons is sufficient to prevent the disastrous "unzipping" of the polymer which might otherwise occur at high temperatures.

Acetal copolymer is classed as an engineering thermoplastic.

Grades

Celcon is available in an internally lubricated injection molding grade, a higher flow grade for parts with thin walls or small gates, a UV-stabilized grade, and a 25% short glass-fibers-reinforced grade for parts requiring high-temperature load-bearing properties. No flame-retardant grades are available. As with the homopolymer, ASTM D 4184 applies.

Processing

All types of injection molding machines may be used in molding acetal copolymer resins. A separate drying step is not required. Shot weight should be in the range of 50–75 percent of rated capacity. Optimum compound temperatures are 370° to 400°F (188°–204°C) with mold temperatures in the 150° to 200°F (66°–93°C) range.

Properties

Compared with acetal homopolymer, the copolymer resin is easier to process and more resistant to attack by chemicals, but mechanical properties are generally lower. Both resins are combustible.

Acetal copolymer resins have good room temperature resistance to:

- Alcohols
- Aliphatic hydrocarbons
- Alkalis
- Carbon tetrachloride
- Detergents
- Ketones
- Mineral oil
- Motor oil

Acetal copolymer resins are attacked by:

- Aromatic hydrocarbons
- Diethyl ether
- Ethyl acetate
- Ethylene dichloride
- Oxidizing agents
- Strong mineral acids

Table 3-18 shows key property values for both acetal copolymer and homopolymer resins.

Electrical/Electronic Uses

Because of its flammability, this resin is not suitable for applications requiring FR properties, which significantly limits electrical/electronic end uses to products such as radio and TV coil forms, control knobs, etc.

Acrylic (PMMA)
(Molding Resin)

Acrylics are noteworthy for superlative optical properties and weatherability. Acrylic molding compounds have been commercially available since their introduction in the late 1930s. The first producers were Rohm and Haas (now AtoHaas) (Plexiglas™) and Du Pont (Lucite™). They are still the largest suppliers, but are now joined by several other companies including ICI Acrylics, Cyro Industries (Perspex™), and Novacor Chemicals (Zylar™).

Chemistry

Acrylics are produced by polymerizing methyl methacrylate alone or combined with other unsaturated monomers. The monomer is the reaction product of acetone and hydrogen cyanide, forming acetone cyanohydrin, which is then heated with methanol in the presence of sulfuric acid. A newer process is based on the oxidation of isobutylene. The homopolymer has the following structure:

$$\left[\begin{array}{cc} \overset{|}{\underset{|}{C}} & \overset{|}{\underset{|}{C}} \\ H & COOCH_3 \end{array}\right]_n$$

(with H, CH₃ on top and H, COOCH₃ on bottom of the C—C unit)

Polymerization methods include bulk, solution, emulsion, and suspension techniques. Molecular weight may be controlled by polymerization method and conditions. Higher molecular weight contributes better mechanical properties, but makes processing more difficult. Improvements in various characteristics are attained by copolymerizing methyl methacrylate with styrene (lower cost), acrylonitrile (higher heat resistance), or ethyl acrylate (easier processing). Improvements are also made by blending methyl methacrylate with vinyls or butadiene-based elastomers (greater impact strength) or polyester resins (easier processing).

Grades

Acrylic resins are supplied as pellets and beads. Grades may differ in molecular weight, with higher molecular weight resins having lower melt flow rates, but better resistance to cracking during ejection. Resins with low molecular weight have high melt flow rates and feature better mold-filling characteristics in complex cavities.

Impact strength up to five times greater than obtainable in general purpose grades is available in grades modified with butadiene or other elastomers. These blends, however, have reduced clarity and weather resistance. Grades with UV stabilizers added have improved UV resistance, but noticeably increased edge color. No flame-retardant molding grades are yet commercially available. Military purchases are made to L-P-380, *Plastic Molding Material, Methacrylate*.

Processing

Acrylic resins may be injection or compression molded using standard techniques. Depending on the formulation and mold design, start-up mold temperatures range from 120° to 200°F (49°–93°C). Recommended cylinder temperatures range from 375° to 430°F (191°–221°C) for single-stage plunger machines and from 310° to 560°F (154°–293°C) for reciprocating screw machines. The newer reciprocating screw machines provide improved feed control, faster cycles, lower pressures, and easier cleaning. Mold shrinkage is in the 2–6 mil per inch range. Optimum processing conditions are unique for each mold, and depend on formulation, machine characteristics, and cavity size and configuration. A drying step is required when not using a vented barrel machine.

Properties

Acrylic resins exhibit outstanding arc resistance, even after immersion in water or long-term outdoor exposure. Impact strength remains unimpaired over a wide portion of the resins' useful temperature range, including subzero temperatures. Clarity is excellent. Surface hardness is much lower than glass. Flammability is relatively high.

Acrylics are resistant to attack by:

- Aliphatic hydrocarbons
- Mineral oils
- Vegetable oils
- Weak acids
- Weak bases

Acrylics are attacked by:

- Alcohols
- Chlorinated hydrocarbons
- Ethers
- Strong inorganic acids

Acrylics are dissolved by:

- Acetone
- Aniline
- Benzene
- Dimethyl formamide
- Ethyl acetate
- Ethylene dichloride
- Toluene

Table 3-19 shows key property values for this resin.

Electrical/Electronic Uses

Major uses for acrylics take advantage of their superior optical properties. They are being used increasingly in fiber optics. High arc resistance makes acrylics suitable for high-voltage applications such as circuit breakers. A decreasing dielectric constant with increasing frequencies makes acrylics attractive candidates for high frequency applications. Acrylics are not suitable where flammability may be a problem.

Amide Polymers: Nylon 6/6
(Molding Resin)

Polyamide polymers were introduced in the U.S. by Du Pont about 1938. They were given the name *nylon*, which is now used generically by all producers, the principal ones being:

PRODUCER	TRADE NAME	TYPE
AlliedSignal	Capron	6, 6/6
Elf Autochem	Rilsan	11, 12
BASF	Ultramid	6, 6/6
Hoechst Celanese	Celanese	6/6
Du Pont	Zytel	6, 6/6, 6/12
Monsanto	Vydyne	6/6

TABLE 3-19 Key Properties of Typical Acrylic Molding Resin

	ASTM TEST METHOD	VALUE
Specific gravity*	D 792	1.19
Dielectric strength, 125 mils, short time, V/mil*	D 149	500
Dielectric constant*	D 150	
60 Hz		3.7
10^6 Hz		2.2
Dissipation factor*	D 150	
60 Hz		0.05
10^6 Hz		0.03
Volume resistivity, $\Omega \cdot$cm*	D 257	1×10^{18}
Arc resistance*	D 495	no tracking
Water absorption, $1/8$ in thick specimen, %	D 570	
24 h @ 73°F (23°C)		0.3
Deflection temperature, 264 psi, °F (°C)	D 648	212 (100)
Maximum recommended service temperature °F (°C)		200 (95)
Tensile strength at break, psi*	D 638	10,000
Izod impact strength	D 256	
Foot pounds per inch notched $1/8$ in specimen*		0.4
Flammability ratings		
UL Standard 94, $1/16$ in specimen		
Standard grade		HB
FR grade		none
Oxygen index, %	D 2863	
Standard grade		na
FR grade		none

*At room temperature.

The type designation, for example 6/6, shows the number of carbon atoms in each part of a nylon molecule formed by a dibasic acid and a diamine, in this case adipic acid and hexamethylene diamine. Type 11 indicates there are 11 carbon atoms in the monomer of a nylon formed by polymerization of the ring compound omega-amino undecanoic acid. Nylon 6/6, the highest volume type, and nylon 11, a specialty, low moisture type, are described in this section and the following section.

Chemistry

There are two basic processes for making nylons: polymerization of ring monomers where the monomer has both acid and amine groups (nylon 6), and condensation of a dibasic acid and a diamine, for example, nylon 6/6, obtained by heating adipic acid (6 carbons) with hexamethylene diamine (6 carbons):

$$H_2N(CH_2)_6NH_2 + HOOC(CH_2)_4COOH \xrightarrow[-H_2O]{heat} -[NH(CH_2)_6NHCO(CH_2)_4CO]_x-$$

Nylon 6/6 is classed as an engineering thermoplastic, although it exhibits an inherent tendency to absorb moisture, resulting in changes in dimensions.

Grades

A large number of grades of nylon 6/6 are available. Nylons are inherently slow burning, but may be made flame retardant with appropriate additives such as halogenated aromatics or cycloaliphatics in conjunction with antimony oxide. Grade modifications to meet specific end uses are obtained through changing molecular weight, copolymerization with other nylon types, and blending with additives and fillers. Modification of nylon 6/6 by copolymerization with nylon 6 provides a resin with improved impact and elongation, but with lower thermal properties. Military purchases are to ASTM D 4066, *Specification for Nylon Injection and Extrusion Materials*.

Processing

For good results, it is essential that nylon resins be dried to less than 0.3 percent moisture content at 140°F (60°C) before molding. Care must be taken to keep resins dry under all conditions of storage and manufacture. Nylons mold readily in conventional equipment that can adequately plasticate and heat the resin to the melt temperatures required of 500° to 580°F (260°–304°C). As noted above, dimensional changes in parts can be expected from changing humidity.

Properties

All nylons are characterized by toughness and abrasion resistance, and are slow burning. Nylon 6/6 has the highest thermal and tensile properties of the common nylon types. It is, however, highly moisture absorbent, as shown in Table 3-20, with the effect on the dielectric constant as shown in Fig. 3-6.

Nylon 6/6 is resistant to attack by:

- Acetone
- Aliphatics
- Ammonia
- Aromatics
- Benzene
- Carbon disulfide
- Carbon tetrachloride
- Chlorobenzene
- Ether
- Ethyl acetate
- Detergents
- Formaldehyde
- Gasoline
- Methyl ethyl ketone

**TABLE 3-20   Key Properties of Typical Nylon 6/6 and
Nylon 11 Molding Resins**

	ASTM TEST METHOD	NYLON 6/6 VALUE**	NYLON 11 VALUE
Specific gravity*	D 792	1.14	1.04
Dielectric strength, 125 mils, short time, V/mil*	D 149	600 (dry)	750 (dry)
Dielectric constant*	D 150		
60 Hz		8.0	3.9
10^6 Hz		4.6	3.1
Dissipation factor*	D 150		
60 Hz		0.2	0.04
10^6 Hz		0.1	0.05
Volume resistivity, $\Omega \cdot$cm*	D 257	1×10^{13}	1×10^{14}
Arc resistance*	D 495	130	123
Water absorption, $^1/_8$ in thick specimen, %	D 570		
24 h @ 73°F (23°C)		1.2	0.3
Deflection temperature, 264 psi, °F (°C)	D 648	194 (90)	131 (55)
Maximum recommended service temperature °F (°C)		265 (130)	149 (65)
Tensile strength at break, psi*	D 638	11,200	8,200
Izod impact strength	D 256		
Foot pounds per inch notched $^1/_8$ in specimen*		2.1	0.75
Flammability ratings			
UL Standard 94, $^1/_{16}$ in specimen			
Standard grade		V-2	V-2
FR grade		V-0	none
Oxygen index, %	D 2863		
Standard grade		31	24.5
FR grade		na	none

*At room temperature.

**At 50% relative humidity unless otherwise indicated.

- Mineral oils
- Silicone fluids
- Soaps
- Tetrachloroethylene
- Vegetable oils
- Vinyl chloride

Nylon 6/6 is attacked by:

- Acids, both organic and inorganic
- Bases (strong)
- Bleaching lye
- Chloroform
- Phenol

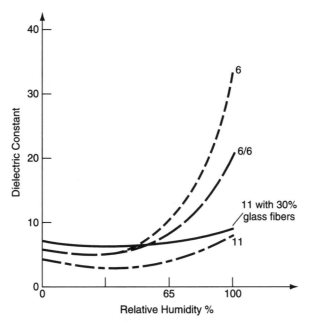

Fig. 3-6   Dielectric constant of selected nylon types as a function of relative humidity at 20°C and under a voltage of 30 V.

Nylon 6/6 dissolves in cresol and cresylic acid.

Table 3-21 shows key property values for both nylon 6/6 and nylon 11 resins.

Electrical/Electronic Uses

The tendency to absorb moisture with harmful effects on dielectric properties limits nylon 6/6 to noncritical electrical/electronic applications. As described in Chapter 7, *Magnet Wire Enamels*, nylon 6/6 is used extensively as a jacket over primary wire insulation. Parts molded from nylon 6/6 include coil forms, insulator blocks, and electrical connectors.

TABLE 3-21   Water Absorption of Selected Nylon Types

NYLON TYPE	% WATER ABSORPTION 24 h (ASTM D 570)
6	1.3–1.9
6/6	1.0–1.3
6/9	0.5
6/12	0.4
11	0.3
12	0.25

Amide Polymers: Nylon 11
(Molding Resin)

Nylon 11, a specialty, low moisture absorption type, was developed in France by ATO and first marketed there in the late 1940s. It is now produced in the U.S. only by Elf Autochem North America, Engineering Polymers, with the trademark "Rilsan."

Chemistry

Nylon 11 is produced by polycondensation from omega–amino undecanoic acid under nitrogen at 419°F (215°C) for several hours. It has the following structure:

$$—[NH(CH_2)_{10}CO]_x—$$

Nylon 11 is classed as an engineering thermoplastic.

Grades

There are several grades of Rilsan:

- Standard grades in a range of flexibility
- Self-lubricated grades (graphite)
- Reinforced grades (glass fibers)
- Temperature-resistant or heat-stabilized grades

Processing

See discussion in previous section on "Amide Polymers: Nylon 6/6."

Properties

Nylon 11 has significantly lower tendency to absorb water than most other nylon types.

The lower susceptibility to moisture results in less change in dielectric properties and greater dimensional stability with increased humidity than for most other nylon types. Thermal and tensile properties of nylon 11 are, however, much lower than for other nylons with fewer carbons in the molecule. Chemical properties are approximately the same as for nylon 6/6.

Table 3-21 shows key properties for both nylon 11 and nylon 6/6.

Electrical/Electronic Uses

Although dielectric properties of nylon 11 qualify it for many electrical/electronic applications, high price and low thermal properties relative to other nylons have limited it to specialty end uses such as jacketing for aircraft control cable and hydraulic tubing and hose.

Imide Polymers: Polyamide-Imide (PAI)
(Molding Resin)

Polyamide-imide, like other imide polymers, is characterized by high heat resistance. It has stable dielectric and mechanical properties up to 500°F (260°C). Amoco Performance Products is the producer of injection-molded grades, trademarked Torlon, first marketed in 1973.

Chemistry

Polyamide-imide is the product of the condensation reaction of trimellitic anhydride and aromatic diamines, with the chemical structure:

$$\left[\!\!-R-\overset{\overset{\displaystyle H}{|}}{N}-\overset{\overset{\displaystyle }{\underset{\underset{\displaystyle O}{\|}}{C}}}-\!\!\left\langle\!\!\bigcirc\!\!\right\rangle\!\!\overset{\overset{\displaystyle O}{\underset{\displaystyle }{\underset{\underset{\displaystyle O}{\|}}{\overset{\|}{C}}}}}{\underset{\underset{\underset{\displaystyle O}{\|}}{C}}{}}\!\!\!N-\!\!\right]_n$$

It is classed as an engineering thermoplastic.

Grades

Injection molding grades include four general purpose grades and reinforced grades with 30% and 40% glass fibers, and up to 30% graphite fibers. Improved processability is obtained by adding a small amount of polytetrafluoroethylene or by copolymerization. Grades approved for military use are listed in MIL-P-46179, included in Chapter 19, *Government Activities*.

Processing

Polyamide-imides are among the most difficult of resins to process, which limits their use to applications where their high temperature properties make them the only choice. Resins are highly shear sensitive and require a low compression screw. Temperature control is difficult because of heat generated by the high viscosity of these thixotropic resins. Molding cycles are very fast. Immediately after molding, parts must be cured in ovens for up to three days to achieve optimum properties. Resins should be thoroughly dried before molding.

Properties

Polyamide-imide has good dielectric properties and radiation resistance, but its outstanding features are superior high-temperature resistance and tensile strength, which is 7,500 psi at 500°F (260°C) and as high as 31,500 psi at –320°F (–160°C). Its UL flammability rating for general purpose grade is V-0, and Oxygen Index is 43 percent. Polyamide-imide does not drip and produces very little smoke on burning. It is, however, adversely affected by high humidity and should not be used in environments where steam could be a problem. It has better thermal and mechanical properties than polyetherimide, but is much more difficult to process.

Polyamide-imide has good resistance to:

- Aliphatic hydrocarbons
- Aromatic hydrocarbons

- Halogenated solvents
- Most dilute acid solutions
- Most dilute base solutions

Polyamide-imide is attacked by:

- High-temperature caustic solutions
- Concentrated sulfuric acid
- *n,n*-dimethyl acetamide
- Ethylene diamine
- Steam

Table 3-22 shows key property values for both polyamide-imide and polyether-imide resins.

TABLE 3-22   Key Properties of Typical Polyamide-Imide and Polyetherimide Molding Resins

	ASTM TEST METHOD	POLYAMIDE-IMIDE VALUE	POLYETHER-IMIDE VALUE
Specific gravity*	D 792	1.40	1.27
Dielectric strength, 125 mils, short time, V/mil*	D 149	580	480
Dielectric constant*	D 150		
60 Hz		4.3	3.2
10^6 Hz		3.9	3.1
Dissipation factor*	D 150		
60 Hz		0.025	0.004
10^6 Hz		0.031	0.006
Volume resistivity, $\Omega \cdot cm$*	D 257	8×10^{16}	6.7×10^{17}
Arc resistance*	D 495	230	128
Water absorption, $^1/_8$ in thick specimen, %	D 570		
24 h @ 73°F (23°C)		0.30	0.25
Deflection temperature, 264 psi °F (°C)	D 648	532 (278)	392 (200)
Maximum recommended service temperature °F (°C)		450 (232)	340 (170)
Tensile strength at break, psi*	D 638	22,000	15,200
Izod impact strength	D 256		
Foot pounds per inch notched $^1/_8$ in specimen*		2.7	1.0
Flammability ratings			
UL Standard 94, $^1/_{16}$ in specimen			
Standard grade		V-0	V-0
FR grade		none needed	none needed
Oxygen index, %	D 2863		
Standard grade		43	47
FR grade		none needed	none needed

*At room temperature.

Electrical/Electronic Uses

Because of high price and processing difficulty, polyamide-imide is considered where no other commercially available resin qualifies for an application. Its flame resistance and low smoke generation make the resin suitable for aerospace applications, including electrical/electronic parts and replacement of metal parts with important weight reduction. Polyamide-imide is used in silicon chip manufacture, where it reduces burn-in time significantly because of its high processing temperature, 482°F (250°C). Other applications include metal replacement in compressor valve plates and in parts for automotive transmissions and engines.

Imide Polymers: Polyetherimide (PEI)
(Molding Resin)

Polyetherimide is one of the most recent engineering plastics developed in General Electric's polymer technology program. G.E. is the only producer of polyetherimide resin "Ultem™," which became commercially available in 1982. Although not as resistant to high temperatures as polyamide-imide, polyetherimide is more readily processible.

Chemistry

Polyetherimide has a basic imide molecule incorporated into an amorphous polymer through ether linkages, resulting in an engineering thermoplastic resin with the following chemical structure:

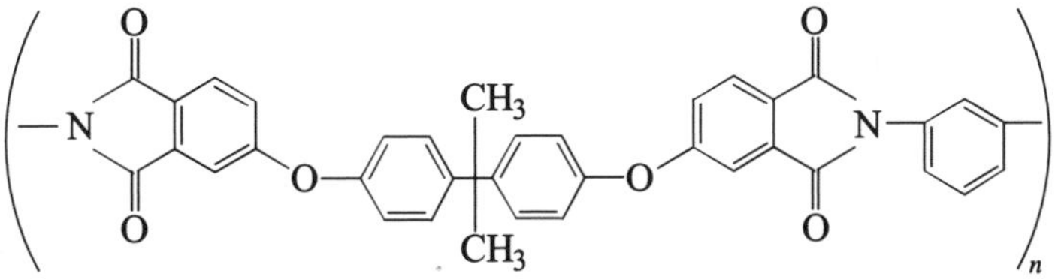

The aromatic imide group provides rigidity, creep resistance, and high deflection temperature. The ether moiety contributes chain flexibility for good flow and processability.

Grades

The grades of polyetherimide resin now available include unreinforced, and 10–40 percent glass-reinforced grades. Military purchases are governed by MIL-P-46184, *Plastic Molding and Extrusion Material, Polyetherimide (PEI)*.

Processing

Polyetherimide resin has outstanding melt flow, melt stability over a wide processing temperature range, and suitability for conversion using all standard injection molding techniques. Prior to processing, moisture should be reduced to 0.05 percent by drying 4 hours at 300°F (149°C), 5 hours at 280°F (138°C), 6 hours at 265°F (129°C), or 7 hours at 250°F (121°C). Mold temperatures should be 150° to 350°F (66°–177°C).

High temperatures improve flow necessary for filling thin-wall and complex parts. Both conventional and laser techniques may be used to machine molded parts. Bonding may be accomplished by adhesives, solvent assembly, and ultrasonic methods.

Properties

Noteworthy properties of polyetherimide include the following:

- High heat resistance
- High dielectric strength
- Stable dielectric constant and dissipation factor over a wide range of temperatures and frequencies
- High mechanical properties
- Outstanding processability
- Inherent flame resistance with low smoke evolution
- UV radiation resistance

Polyetherimide is highly notch-sensitive. Although lower in thermal and mechanical properties, polyetherimide has much better resistance to steam exposure than polyamide-imide and is far easier to process.

Polyetherimide has high gloss/colorability and is resistant to attack by:

- Aircraft and auto fuels, lubricants, and antifreeze
- Transmission and brake fluids
- Alcohols
- Aromatic hydrocarbons
- Detergents
- Ethers
- Freon
- Ketones
- Mineral acids
- Vegetable oils and fats

The resin is soluble in partially halogenated hydrocarbons such as methylene dichloride and trichloroethane.

Table 3-22 shows key property values for both polyetherimide and polyamide-imide resins.

Electrical/Electronic Uses

Heat resistance to wave and vapor-phase soldering, intricate design potential, dimensional stability, and flame resistance make polyetherimide attractive for circuit boards, terminal bases, connectors, fuses, and integrated circuit test devices. The resin is used for automotive and aerospace applications requiring high heat resistance, resistance to attack by fluids, creep resistance, mechanical strength with light weight, dimensional stability under load, and high electrical insulation values. It is well suited for microwave appliances.

Polyarylate (PAR)
(Molding Resin)

First marketed in the U.S. in 1978 by Union Carbide as an import from Japan, electrical grades of polyarylate are now produced in the U.S. by Amoco Performance Products (Ardel™), Du Pont (Arylon™), and Hoechst Celanese (Durel™). These polymers are aromatic polyesters with amophorous molecular structure. The resin lends itself to blending and alloying, and this may be a strong growth area.

Chemistry

There are at least five processes for making the resin, most of which are proprietary, yielding varying degrees of clarity and ranging in color from straw to amber. Common manufacturing ingredients include diphenyl isophthalate and diphenyl terephthalate blends reacted with bisphenyol A.

Polyarylate is classed as an engineering thermoplastic.

Grades

Polyarylate resins are available in unreinforced grades (some meeting requirements for a UL 94 V-0 rating) and in 30 percent glass-reinforced grades.

Processing

Polyarylates require thorough drying before processing. Recommended melt temperatures range from 640° to 735°F (338°–390°C). Mold shrinkage is uniform in all directions, similar to polycarbonate or polysulfone.

Properties

Suitable for high-temperature applications, polyarylate resins, with heat deflection temperature of 310°F (154°C) are positioned between polycarbonate with HDT of 265°F (129°C) and polysulfone with HDT of 345°F (174°C). Polyarylates have inherently good flame retardancy (UL 94V-0) and generate little smoke during burning. They are little affected by humidity under normal service conditions. Good dielectric properties are retained over a broad range of temperature and frequency. Resistance to chemicals and solvents is not noteworthy.

Electrical/Electronic Uses

Flame retardancy and the ability to withstand high operating temperatures while maintaining good dielectric properties make polyarylates suitable for diverse applications including molded three-dimensional wiring boards, wire jacketing, connectors, switchgear, and fuses.

Polybutylene Terephthalate (PBT)
(Molding Resin)

Useful for its good dielectric properties, low moisture absorption, and chemical resistance, this thermoplastic polyester, also known as polytetramethylene terephthalate

(PTMT), became commercially available in 1969. Producers include BASF (Ultradur™), General Electric (Valox™), Hoechst Celanese (Celanex™), and Huls American. Polyethylene terephthalate (PET) resin, used primarily for fiber, filament, and film production, is discussed in Chapter 10, *Dielectric Films*, p. 132.

Chemistry

Polybutylene terephthalate is a product of the polycondensation of 1,4-butanediol and dimethylterephthalate:

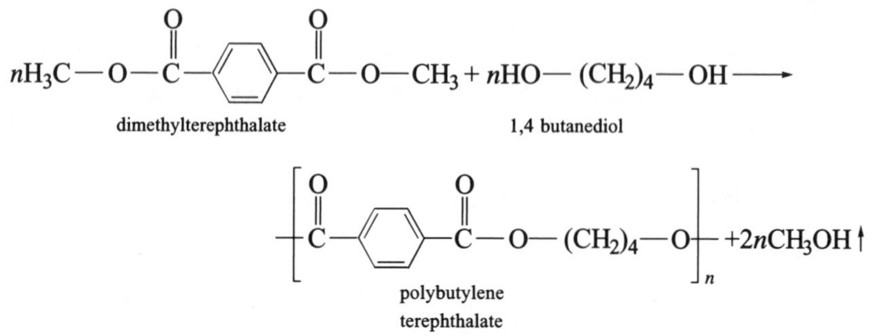

Grades

Commercially available grades include unreinforced general purpose and flame-retardant grades, and reinforced grades for improved stiffness with up to 45 percent glass and/or mineral filler. Special reinforced grades with improved surface appearance over standard reinforced grades are also offered. Alloys with polyarylate to improve dimensional stability are under development.

Processing

Polybutylene terephthalate resins are high crystalline with sharply defined melting points in the 335° to 442°F (168°–228°C) range, depending on grade. Processing temperatures of 450° to 500°F (232°–260°C) facilitate filling of complex molds. Cycle times are relatively short (because of the high degree of crystallinity) compared with other thermoplastics. Conventional screw injection molding machines are preferred because they can operate at higher melt viscosities, hence lower and more homogeneous melt temperatures. Resins should be dried prior to molding for 2–4 hours at 250°F (121°C). A shot of 50–75 percent of rated machine capacity usually produces the best results. Parts assembly may be accomplished by most conventional methods, including solvent and adhesive bonding, and hot plate, ultrasonic, and spot welding.

Properties

Polybutylene terephthalate resin molded parts maintain their dielectric strength, even in high-humidity environments. Dielectric constant remains steady over a wide range of frequencies. Resistance to arcing and tracking is high. With 30 percent glass fiber reinforcement, polybutylene terephthalate resins exhibit outstanding heat deflection temperatures up to 430°F (221°C) at 66 psi compared with 310°F (154°C) for unreinforced general purpose grades. However, with increased load, heat deflection for unreinforced general purpose grades drops rapidly to 130°F (54°C) at 264 psi. The drop at 264 psi with glass-reinforced grades is not nearly as dramatic, amounting to only 15°F (8°C) for some grades. Moisture absorption is low and notched impact resistance high for a thermoplastic material. Polybutylene terephthalate's tendency to warp makes it unsuitable for some end uses.

Polybutylene terephthalate resins have good room temperature resistance to:

- Water
- Aliphatic hydrocarbons
- Aromatic hydrocarbons
- Ethers
- Ketones
- Glycols
- Organic solvents, including degreasing solvents
- Flux cleaners

Polybutylene terephthalate resins are attacked by:

- Aromatics
- Acetone and ketones above ambient temperature
- Strong acids and bases
- Ethyl acetate
- Detergents
- Lubricating oils
- Hot water on prolonged exposure

Solvents include:

- Trifluoroacetic acid
- Phenol/chlorinated aliphatic hydrocarbons
- Hexafluoroisopropanol
- O-chlorophenol

Table 3-23 shows key property values for this resin.

Electrical/Electronic Uses

Unreinforced polybutylene terephthalate grades are used for bobbins, connectors, coilforms, and sensors. Reinforced grades are well suited for distributor caps and rotors, connectors, business machine parts, industrial motor controls, circuit breakers, TV components, terminal blocks, fuse holders, and housings for appliances and hand power tools.

TABLE 3-23 Key Properties of Typical Polybutylene Terephthalate Molding Resin

	ASTM TEST METHOD	VALUE
Specific gravity*	D 792	1.31
Dielectric strength, 125 mils, short time, V/mil*	D 149	420
Dielectric constant*	D 150	
60 Hz		3.3
10^6 Hz		3.1
Dissipation factor*	D 150	
60 Hz		0.002
10^6 Hz		0.02
Volume resistivity, Ω·cm*	D 257	4×10^{16}
Arc resistance*	D 495	184
Water absorption, $1/8$ in thick specimen, %	D 570	
24 h @ 73°F (23°C)		0.08
Deflection temperature, 264 psi, °F (°C)	D 648	130 (54)
Maximum recommended service temperature °F (°C)		300 (150)
Tensile strength at break, psi*	D 638	18,200
Izod impact strength	D 256	
Foot pounds per inch notched $1/8$ in specimen*		1.0
Flammability ratings		
UL Standard 94, $1/16$ in specimen		
Standard grade		HB
FR grade		V-0
Oxygen index, %	D 2863	
Standard grade		22
FR grade		29

*At room temperature

Polycarbonate (PC)
(Molding Resin)

Polycarbonate was discovered in the United States at General Electric and has been commercially available since 1958. It is noted for its exceptionally high impact resistance, good dielectric properties, and corona resistance. Molding resins are produced by Dow Plastics (CALIBRE™), G.E. (Lexan™), and Miles (Makrolon™). Alloys with other polymers are also available from these companies.

Chemistry

Polycarbonate, a polyester of carbonic acid, is obtained from bisphenol A and other polyhydric phenols, reacted with phosgene. It is an engineering thermoplastic with the structure:

$$H\!-\!\left[O\!-\!\langle\bigcirc\rangle\!-\!\underset{\underset{CH_3}{|}}{\overset{\overset{CH_3}{|}}{C}}\!-\!\langle\bigcirc\rangle\!-\!O\!-\!\overset{\overset{O}{\|}}{C}\!-\!OH\right]_x$$

Grades

Resins are available in low, medium, and high melt viscosities for molding and extrusion. There are special grades produced in compliance with Food Additive Regulations for food contact applications (21 CFR 177.1580 and 21 CFR 178.2010). Other special grades are UV-stabilized, and still another grade provides UL Standard 94 rating of V-0. There are glass-fiber-reinforced grades with 20–40 percent glass for improved rigidity, creep resistance, dimensional stability, and heat resistance. Military purchases are made to L-P-393, *Plastic Molding Material, Polycarbonate, Injection and Extrusion.*

Processing

All standard thermoplastic techniques may be employed in injection molding processes. Because of its hydrolytic sensitivity at processing temperatures, the resin should be dried to less than 0.02 percent moisture before processing to avoid deterioration of impact strength and formation of streaks and bubbles in the finished product. Proper design and shop practices minimize the tendency of molded parts to stress crack or craze.

Properties

Polycarbonate has noteworthy attributes: clarity, superior impact strength, dimensional stability over a wide temperature range, and self-extinguishing property.

Polycarbonate has good room temperature resistance to:

- Water
- Dilute organic and inorganic acids
- Oxidizing and reducing agents
- Neutral and acid salts
- Animal and vegetable oils and fats
- Aliphatic and cyclic hydrocarbons

Polycarbonate is attacked by:

- Alkalis
- Amines
- Ketones
- Aromatic hydrocarbons

Solvents for polycarbonate include:

- Methylene and ethylene dichloride
- Cresol
- Dioxane
- Pyridine

Table 3-24 shows key property values for this resin.

TABLE 3-24   Key Properties of Typical Polycarbonate
Molding Resin

	ASTM TEST METHOD	VALUE
Specific gravity*	D 792	1.20
Dielectric strength, 125 mils, short time, V/mil*	D 149	380
Dielectric constant*	D 150	
60 Hz		3.2
10^6 Hz		3.0
Dissipation factor*	D 150	
60 Hz		0.0009
10^6 Hz		0.010
Volume resistivity, $\Omega \cdot cm$*	D 257	8.2×10^{16}
Arc resistance*	D 495	120
Water absorption, $^1/_8$ in thick specimen, %	D 570	
24 h @ 73°F (23°C)		0.15
Deflection temperature, 264 psi, °F (°C)	D 648	265 (129)
Maximum recommended service temperature °F (°C)		240 (115)
Tensile strength at break, psi*	D 638	9,000
Izod impact strength	D 256	
Foot pounds per inch notched $^1/_8$ in specimen*		12
Flammability ratings		
UL Standard 94, $^1/_{16}$ in specimen		
Standard grade		V-2
FR grade		V-0
Oxygen index, %	D 2863	
Standard grade		25
FR grade		35

*At room temperature.

Electrical/Electronic Uses

Because of its colorability and good physical, dielectric, and self-extinguishing properties, polycarbonate housings are used extensively for power tools, appliances, print-
ers, copiers, and other business machines. Polycarbonates are successful in connectors,
circuit boards, wiring blocks, and other component parts. Nonbrominated flame-retardant grades feature processing stability, UL 94V-0 rating, Oxygen Index of 35, low
smoke emission and toxicity, and no corrosive gasses on decomposition.

Polyetheretherketone (PEEK)
(Molding Resin)

Among the resins with the highest useful service temperature, this new aromatic, easily processed resin is produced by ICI Americas under the trademark Victrex.

Chemistry

Polyetheretherketone resin is made by condensation polymerization in an aromatic solvent. It is classed as an engineering thermoplastic.

Grades

Supplied as granules and as a powder in medium and higher viscosity unreinforced grades, and grades reinforced with 20 and 30 percent glass fiber, and 30 percent carbon fiber.

Processing

This resin is readily processed in all commonly used equipment to produce molded parts, film, sheet, wire and cable extrusions, monofilament, and composites.

Properties

The outstanding properties of this resin include:

• Continuous service temperature for electrical/electronic applications of up to 392°F (200°C).

• UL 94V-0 flammability rating with low smoke and toxic gas emission.

• Resistance to attack from up to 60 percent sulfuric acid and 40 percent sodium hydroxide solutions at elevated temperatures. The only common solvent is concentrated sulfuric acid.

• High resistance to hydrolysis—unaffected after long-term immersion in pressurized water at 480°F (250°C).

• No significant reduction in mechanical properties after absorbing 1000 Mrad of irradiation.

Electrical/Electronic Uses

Their easy processing quality make polyetheretherketone resins suitable for a wide range of applications where their outstanding properties justify their relatively high prices.

Polyphenylene Ether, Modified (PPE)
(Molding Resin)

A product of the plastics research and development program at General Electric, modified polyphenylene ether (formerly called polyphenylene oxide) with the trademark Noryl became commercially available in 1966. The resin exhibits good dielectric properties over a wide range of humidity and temperature.

Chemistry

Polyphenylene ether is formed by oxidative coupling in a reaction in which oxygen is passed through 2,6-disubstituted phenol. A copolymer (Prevex™) is formed from 2,6-disubstituted and 2,3,6-trisubstituted phenols. Polyphenylene ether has the basic structure:

Despite its high property values in its unmodified form, polyphenylene ether experienced limited acceptance because of processing difficulties. Modification with styrene polymers, with which it is highly miscible, has greatly improved its processing characteristics and lowered its cost, but at the expense of reduced maximum operating temperature.

Modified polyphenylene ether is classed as an engineering thermoplastic.

Grades

There are at least 16 grades of Noryl, including 12 injection molding grades. In addition to a general purpose grade, other grades emphasize certain properties. There are flame-retardant grades and 20 and 30 percent glass-reinforced grades. A special grade can be molded into parts with superior plating characteristics. Military purchases are made to MIL-P-46115, *Plastic Molding and Extrusion Material, Polyphenylene Oxide; Plastic Molding and Extrusion Material, Polyphenylene Oxide, Modified; Plastic Molding and Extrusion Material, Polyphenylene Oxide, Modified, Glass Fiber Reinforced.*

Processing

Reciprocating screw machines show best results because of reduced pressure through the cylinder, shorter cycle times, and lower molded-in stress. Continuous peripheral venting removes processing gases efficiently and minimizes the possibility of residue formation from the condensation of volatiles. The recommended processing temperature for general purpose grade is 425° to 600°F (218°–316°C). Predrying of the resin is not essential, but produces optimum appearance in molded parts.

Properties

Dielectric constant and dissipation factor of modified resins remain virtually unaffected over wide changes in relative humidity, temperature, and frequency. Water absorption rates are among the lowest for all engineering thermoplastics. Tensile strength and stiffness are retained under load throughout the useful temperature range. Several FR grades are available which satisfy UL 94V-0 standards. Thermally stable phosphorous-based additives improve FR characteristics.

Modified resins have good room temperature resistance to:

- Water
- Dilute inorganic acids and bases
- Saturated solutions of inorganic salts
- Alcohols
- Chlorine bleach
- Mineral oil

Modified resins are attacked by:

- 90 percent sulfuric acid
- Acetone

- Aliphatic hydrocarbons
- Ethers
- Ethyl acetate
- Vegetable oils

Solvents for modified resins include:

- Aromatic hydrocarbons
- Halogenated hydrocarbons

Table 3-25 shows key property values for the modified resin.

Electrical/Electronic Uses

Flame retardancy and good mechanical properties make modified PPE resins attractive for business machine and small appliance housings. Reinforced and other grades are widely used in wall switch plates, electrical connectors, and television, telecommunication, and appliance components.

**TABLE 3-25   Key Properties of Typical Modified
Polyphenylene Ether Molding Resin**

	ASTM TEST METHOD	VALUE
Specific gravity*	D 792	1.06
Dielectric strength, 125 mils, short time, V/mil*	D 149	550
Dielectric constant*	D 150	
60 Hz		2.6
10^6 Hz		2.6
Dissipation factor*	D 150	
60 Hz		0.0004
10^6 Hz		0.0009
Volume resistivity, $\Omega \cdot cm$*	D 257	1×10^{16}
Arc resistance*	D 495	75
Water absorption, $\frac{1}{8}$ in thick specimen, %	D 570	
24 h @ 73°F (23°C)		0.06
Deflection temperature, 264 psi, °F (°C)	D 648	265 (129)
Maximum recommended service temperature °F (°C)		220 (105)
Tensile strength at break, psi*	D 638	9,600
Izod impact strength	D 256	
Foot pounds per inch notched $\frac{1}{8}$ in specimen*		5.0
Flammability ratings		
UL Standard 94, $\frac{1}{16}$ in specimen		
Standard grade		HB
FR grade		V-0
Oxygen index, %	D 2863	
Standard grade		na
FR grade		30

*At room temperature.

Polyphenylene Sulfide (PPS)
(Molding Resin)

Polyphenylene sulfide, developed by Phillips Chemical, became commercially available in 1972 under the trademark "Ryton." Its outstanding feature is resistance to thermal degradation, which occurs completely only at temperatures over 1300°F (704°C). In 1987, Celanese (now Hoechst Celanese) began marketing "Fortran™" and General Electric offered "Supec™" dielectric grades.

Chemistry

Polyphenylene sulfide is produced by reacting dichlorobenzene with sodium sulfide in a polar solvent:

$$Cl-\langle\bigcirc\rangle-Cl + Na_2S \longrightarrow \left[\langle\bigcirc\rangle-S\right]_n + 2\ NaCl$$

Polyphenylene sulfide is classed as an engineering thermoplastic.

Grades

Although unfilled grades of Ryton are available, filled grades are more widely used because of greatly enhanced physical, mechanical, and thermal properties. Fillers include glass fiber and minerals. One grade containing 40 percent glass fiber has outstanding chemical resistance and electrical properties virtually unaffected by moisture. Another grade has high arc resistance and low arc tracking rate. Still another grade is essentially transparent to microwave radiation and is suitable for use at temperatures not to exceed 425°F (218°C). Colors are available. Military purchases are made to ASTM D 4067, *Specification for Reinforced and Filled Polyphenylene Sulfide Injection Molding and Extrusion Materials*.

Processing

Reciprocating screw equipment is recommended for molding polyphenylene sulfide, providing effective temperature control and short cycle capability. Molding compound should be dried for at least 6 hours at 300°F (149°C) prior to molding. Molds should be heated and insulated. Flexural strength is reduced as mold temperature is increased. Recommended melt temperatures are in the 600° to 625°F (316°–329°C) range. Parts may be annealed after molding by heat treatment at 400° to 450°F (204°–232°C) for 2–4 hours. Annealing improves dimensional stability, shear strength, and flexural modulus, but reduces gloss, tensile strength, and impact strength.

Properties

Maximum use temperatures for polyphenylene sulfide molded parts up to 400°F (204°C) are among the highest for all injection-molded thermoplastic resins. Resins are inherently flame retardant. Compounding with fillers significantly improves most mechanical and thermal properties. Hence, unfilled resins are used infrequently.

Mechanical properties for all grades drop sharply with increasing temperatures. Dielectric properties are retained even under high-humidity and high-temperature conditions.

Polyphenylene sulfide is insoluble in any known chemical below 400°F (204°C), and has good resistance to virtually all chemicals with the exception of oxidizing acids such as concentrated hypochlorous, perchloric, nitric, and sulfuric acids.

Table 3-26 shows key property values for this resin.

Electrical/Electronic Uses

Polyphenylene sulfide compounds are used for a wide range of electrical/electronic applications, and are especially suitable for parts exposed to high temperatures. Suggested applications include microwave ovenware, bobbins, coil forms, connectors, sockets, switch components, and brush holders. The high arc resistance and low arc tracking rate of some polyphenylene sulfide compounds make them attractive for high voltage use and for structural components requiring a UL sole support rating. Polyphenylene sulfide is suitable for encapsulating integrated circuits and printed circuit components.

TABLE 3-26   Key Properties of Typical Polyphenylene Sulfide Molding Resin

	ASTM TEST METHOD	VALUE
Specific gravity*	D 792	1.34
Dielectric strength, 125 mils, short time, V/mil*	D 149	380
Dielectric constant*	D 150	
60 Hz		3.1
10^6 Hz		3.2
Dissipation factor*	D 150	
60 Hz		0.0003
10^6 Hz		0.0007
Volume resistivity, $\Omega \cdot cm$*	D 257	1×10^{16}
Arc resistance*	D 495	34
Water absorption, $^1/_8$ in thick specimen, %	D 570	
24 h @ 73°F (23°C)		0.02
Deflection temperature, 264 psi, °F (°C)	D 648	275 (135)
Maximum recommended service temperature °F (°C)		400 (204)
Tensile strength at break, psi*	D 638	9,500
Izod impact strength	D 256	
Foot pounds per inch notched $^1/_8$ in specimen*		0.3
Flammability ratings		
UL Standard 94, $^1/_{16}$ in specimen		
Standard grade		V-0
FR grade		none needed
Oxygen index, %	D 2863	
Standard grade		46
FR grade		none needed

*At room temperature.

Polypropylene (PP)
(Molding Resin)

Polypropylene is a highly versatile resin suitable for processing into molded insulation parts, extruded wire and cable insulation, and dielectric films, all covered separately in this handbook. This section covers polypropylene molding resins, produced by over a dozen large suppliers. Formerly made by Hercules, grades specifically designed for dielectric applications are now produced by Himont U.S.A. with the trademark "Pro-fax."

Chemistry

This subject is covered in "Propylene Polymers" (p. 160) of Chapter 5, *Extrusion Compounds*.

Grades

There are few grades designed specifically for molded dielectric applications. A wide range of properties can be obtained, depending on whether a grade is composed of homopolymer, copolymers, or terpolymers. Only grades with the highest degree of purity are suitable for electrical/electronic uses. Homopolymers generally have the best dielectric properties. Copolymer and terpolymer grades have better thermal and impact properties. Grades with 20–30 percent glass fiber reinforcement show significantly improved tensile strength, impact strength, and deflection temperature. Military purchases are made to L-P-394, *Plastic Molding Material (Propylene Plastics, Injection & Extrusion)* and ASTM D 4101, *Specification for Propylene Plastic Injection and Extrusion Materials*.

Processing

Polypropylene may be processed with all conventional injection molding methods. Melt temperatures generally range from 400° to 450°F (205°–230°C), with mold temperatures between 50° and 105°F (10° and 40°C). Because of its low moisture absorption, it is not necessary to dry polypropylene compounds before molding.

Properties

Polypropylene, the most rigid polyolefin, is one of the lightest plastics, with specific gravity of 0.90, making it more economical to mold than most other resins. Dielectric properties and moisture resistance are among the best for all plastics. Polypropylene is a good gas barrier. It is insoluble in solvents at room temperatures, but will dissolve in aromatic and chlorinated solvents above 175°F (80°C). Polypropylene is attacked by chlorine, fuming nitric acid, and other strong oxidizing agents. However, it is highly resistant to environmental stress cracking, even when it comes into contact with solvents and polar materials. Polypropylene is heat sealable and nontoxic. Without modification, it burns readily and is degraded by heat and light. It has poor impact strength below 16°F (–9°C).

Table 3-27 shows key property values for this resin.

TABLE 3-27   Key Properties of Typical Heat-Resistant Polypropylene Homopolymer Molding Resin

	ASTM TEST METHOD	VALUE
Specific gravity*	D 792	0.90
Dielectric strength, 125 mils, short time, V/mil*	D 149	600
Dielectric constant*	D 150	
60 Hz		2.2
10^6 Hz		2.2
Dissipation factor*	D 150	
60 Hz		<0.0005
10^6 Hz		<0.0005
Volume resistivity, ohm·cm*	D 257	>10^{16}
Arc resistance*	D 495	150
Water absorption, $^1/_8$ in thick specimen, %	D 570	
24 h @ 73°F (23°C)		<0.01
Deflection temperature, 264 psi, °F (°C)	D 648	140 (60)
Maximum recommended service temperature °F (°C)		250 (120)
Tensile strength at break, psi*	D 638	5,400
Izod impact strength	D 256	
Foot pounds per inch notched $^1/_8$ in specimen*		0.5
Flammability ratings		
UL Standard 94, $^1/_{16}$ in specimen		
Standard grade		HB
FR grade		V-0
Oxygen index, %	D 2863	
Standard grade		18
FR grade		27

*At room temperature.

Electrical/Electronic Uses

Polypropylene is used extensively for motor vehicle connectors, fuse blocks, and other electrical components. Automotive battery cases are virtually all made from impact grade polypropylene, which is also used for appliance, tool, and equipment housings.

Styrenics: Polystyrene (PS)
(Molding Resin)

The first general purpose polystyrene resins were marketed in the U.S. in 1939 by Dow Chemical, still the largest producer, using the trademark "Styron." Other producers of homopolymer grades include Amoco Performance Products, ARCO Chemical (Dylite™), Eastman Performance Plastics (Tenite™), Fina Oil & Chemical, Huntsman Chemical, Novacor Chemicals, and Phillips (Marlex™).

Chemistry

General purpose polystyrene is produced by polymerizing styrene monomer, obtained by dehydrogenation of ethyl benzene, into polymers of 100–20,000 monomer units. The reaction is initiated by organic peroxides, such as benzoyl peroxide, or thermally.

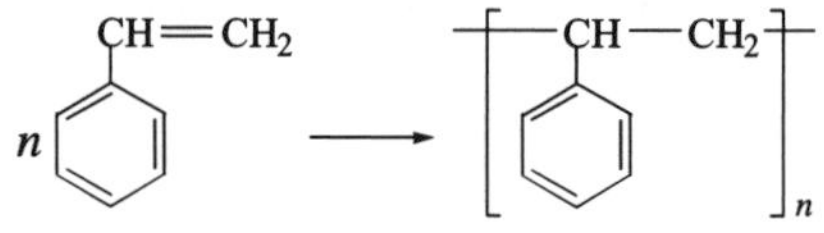

styrene
monomer

polystyrene

Grades

There are many commercially available grades of polystyrene with a wide range of molecular weights and specialized additives. Properties affected include melt flow rate, softening point, deflection temperature, tensile strength, and flammability. Military purchases are to ASTM D 4549, *Specification for Polystyrene Molding and Extrusion Materials*.

Processing

Polystyrene injection molding grades exhibit exceptionally low moisture absorption, and hence usually do not require drying before molding. Conventional injection molding machines are used to process polystyrene. Both molding temperature and molding cycle affect the quality of the molded piece. If the processing temperature is too low, pieces are likely to have a high degree of internal stress. Processing temperatures that are too high will degrade the polymer and its properties. Injection-molding temperatures typically range from 350° to 500°F (177°–260°C), depending on grade, part size, and molding equipment. Oven annealing is advised at 5° to 10°F (2.8°–5.6°C) below the heat distortion temperature.

Properties

General purpose polystyrene resins are noteworthy for their clarity, low moisture absorption, and low cost. Electrical properties are good. Unmodified grades, however, burn readily, have low impact strength, and are attacked by many organic chemicals. Polystyrene has a tendency to stress crack when exposed to certain chemicals while under stress. High temperatures hasten stress cracking. Reagents which cause severe stress cracking include:

- Butter
- Coconut oil
- Dimethyl-dichloro-vinyl phosphate
- Gasoline
- *n*-heptane
- Kerosene
- Turpentine

Polystyrene is resistant to attack by:

- Inorganic acids, except strong oxidizing acids
- Weak organic acids
- Alcohols
- Aliphatic amines
- Bases

Polystyrene is attacked by

- Aldehydes
- Aromatic amines
- Esters
- Chlorinated hydrocarbons
- Insecticides
- Ketones
- Essential oils

Solvents for polystyrene include aliphatic and aromatic hydrocarbons.

Table 3-28 shows key property values for polystyrene, polystyrene-butadiene, styrene acrylonitrile, and acrylonitrile-butadiene-styrene resins.

Electrical/Electronic Uses

The flammability, low impact strength, tendency to craze, and susceptibility to attack by hydrocarbons limit electrical/electronic applications. End uses include computer tape and video-cassette reels, and compact battery cases.

Styrenics: Styrene-Butadiene (Molding Resin)

The impact resistance of polystyrene may be significantly improved by polymerizing it with butadiene, forming a block copolymer. Producers include Dow Plastics and Phillips 66 (K-Resin™).

Chemistry

The major constituent, styrene, is polymerized with 1,3-butadiene in a solution process with an organolithium initiator.

Grades

Commercially available injection molding grades include: general purpose, high heat, easy flow, and flame-retardant modified. Military purchases are to L-P-398, *Plastic Molding Material, Styrene-butadiene.*

Processing

Impact polystyrene may be molded in conventional injection molding equipment without drying, except under high-humidity conditions. Material temperature should not

TABLE 3-28   Key Properties of Typical Polystyrene, Styrene-Butadiene, Styrene-Acrylonitrile, and Acrylonitrile-Butadiene-Styrene Molding Resins

	ASTM TEST METHOD	POLY-STYRENE VALUE	STYRENE-BUTADIENE VALUE	STYRENE-ACRYLO-NITRILE VALUE	ACRYLONITRILE-BUTADIENE-STYRENE VALUE
Specific gravity*	D 792	1.04	1.03	1.07	1.03
Dielectric strength, 125 mils, short time, V/mil*	D 149	500	400	425	425
Dielectric constant*	D 150				
60 Hz		2.5	2.5	2.6	2.6
10^6 Hz		2.4	2.4	2.6	2.6
Dissipation factor*	D 150				
60 Hz		0.0001	0.002	0.007	0.004
10^6 Hz		0.0001	0.002	0.007	0.004
Volume resistivity, $\Omega \cdot$cm*	D 257	$<1 \times 10^{16}$	$<1 \times 10^{16}$	$<1 \times 10^{16}$	2×10^{16}
Arc resistance*	D 495	65	50	115	100
Water absorption, $1/8$ in thick specimen, %	D 570				
24 h @ 73°F (23°C)		0.02	0.06	0.20	0.22
Deflection temperature, 264 psi, °F (°C)	D 648	200 (93)	200 (93)	200 (93)	210** (99**)
Maximum recommended service temperature °F (°C)		160 (70)	150 (65)	180 (80)	200 (95)
Tensile strength at break, psi*	D 638	6,000	4,000	9,500	7,000
Izod impact strength	D 256				
Foot pounds per inch notched $1/8$ in specimen*		0.4	2.5	0.40	6.0
Flammability ratings UL Standard 94, $1/16$ in specimen					
Standard grade		HB	HB	HB	HB
FR grade		V-0	V-0	V-0	V-0/5V
Oxygen index, %	D 2863				
Standard grade		18	18	na	19
FR grade		na	28	na	na

*At room temperature.

**Annealed.

exceed 475°F (246°C), at which point toxic and corrosive gases can be given off, requiring operator protection and other safety precautions. Oven annealing is advised at 5° to 10°F (2.8°–5.6°C) below the heat distortion temperature.

Properties

Significant improvement in impact strength over crystal polystyrene is achieved in impact grades at the expense, however, of lower tensile values. Impact polystyrene will stress crack with certain reagents similar to crystal polystyrene.

Impact polystyrene is resistant to attack by:

- Organic acids
- Aliphatic amines
- Bases
- Salts

Impact polystyrene is attacked by:

- Aldehydes
- Aromatic amines
- Esters
- Aliphatic, aromatic, and chlorinated hydrocarbons

Table 3-28 shows key property values for polystyrene, polystyrene-butadiene, styrene acrylonitrile, and acrylonitrile-butadiene-styrene resins.

Electrical/Electronic Uses

The improved impact resistance of impact polystyrene qualifies it for appliance parts, tape cartridges, cassette covers, and housings. Parts, however, have limited chemical resistance and, except for specialty grades, are flammable.

Styrenics: Styrene-Acrylonitrile (SAN) (Molding Resin)

While retaining excellent clarity, styrene-acrylonitrile has improved toughness, tensile strength, heat distortion, and chemical resistance over crystal polystyrene. Producers include BASF (Luran™), Dow Plastics (Tyril™), and Monsanto (Lustran™).

Chemistry

Styrene-acrylonitrile is produced by emulsion, suspension, or continuous mass copolymerization of styrene and acrylonitrile monomers. With increased molecular weight or acrylonitrile content, physical properties are improved, but processing ease is reduced and polymer color is increased:

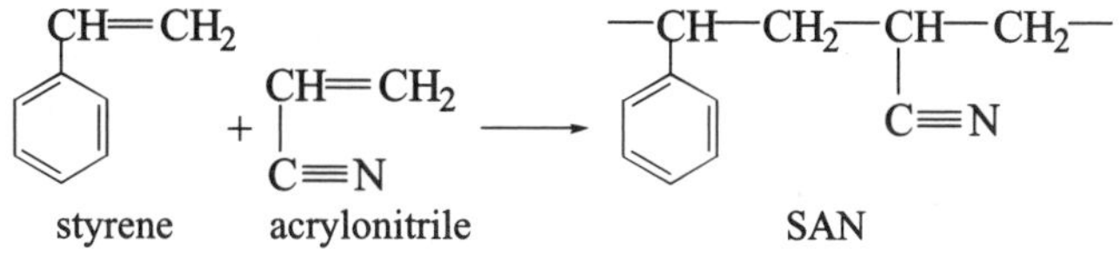

Grades

Commercially available styrene-acrylonitrile grades include general purpose, easy flow, light stabilized, superior chemical resistance, 30 percent glass fiber reinforced, and flame retardant. Military purchases are to L-P-399, *Plastic Molding & Extrusion Material, Styrene Acrylonitrile Copolymers*.

Processing

All conventional injection molding equipment can produce satisfactory styrene-acrylonitrile parts. Resins should be dried prior to molding at 160° to 180°F (71°–82°C) for a period of 2 hours. Oven annealing is advised at 5° to 10°F (2.8°–5.6°C) below the heat distortion temperature.

Properties

Styrene-acrylonitrile resins have excellent transparency and colorability. Parts have hard, scratch-resistant surfaces. Water absorption is higher than crystal styrene or styrene-butadiene. General purpose grades are flammable. Styrene-acrylonitrile resins are compatible with other thermoplastics, and are used to modify and improve flow characteristics of acrylonitrile-butadiene-styrene, polyvinyl chloride, and other resins.

Styrene-acrylonitrile is resistant to attack by:

- Aliphatic hydrocarbons
- Nonoxidizing acids
- Bases
- Formaldehyde
- Gasoline
- Mineral oil
- Vegetable oils
- Most detergents

Styrene-acrylonitrile is attacked by:

- Glacial acetic acid
- Aromatic hydrocarbons
- Chlorinated hydrocarbons
- Esters
- Ketones
- Concentrated sulfuric acid

Solvents for styrene-acrylonitrile include methyl-ethyl-ketone and methylene chloride.

Table 3-28 shows key property values for polystyrene, polystyrene-butadiene, styrene-acrylonitrile, and acrylonitrile-butadiene-styrene resins.

Electrical/Electronic Uses

Applications include transmitter caps, battery cases, instrument covers, and meter lenses.

Styrenics: Acrylonitrile-Butadiene-Styrene (ABS) (Molding Resin)

By varying the proportions of acrylonitrile, butadiene, and styrene, a wide range of properties is obtainable, permitting resins to be tailored to specific end uses with emphasis on toughness. The major U.S. producers of acrylonitrile-butadiene-styrene resins are Dow Chemical (Magnum™), General Electric (Cycolac™), and Monsanto (Lustran™).

Chemistry

Acrylonitrile-butadiene-styrene resins may be made by emulsion, suspension, or mass polymerization methods. The most common process is graft polymerization in which styrene-acrylonitrile copolymer (SAN) is grafted on butadiene polymer. The continuous phase is styrene-acrylonitrile. Monomer components contribute the following properties:

- Acrylonitrile—heat stability, chemical resistance, aging resistance
- Butadiene—low temperature property retention, toughness, impact strength
- Styrene—processing ease, rigidity, gloss, low cost.

Flame retardancy is achieved by incorporating halogen additives or by alloying with polyvinyl chloride.

Acrylonitrile-butadiene-styrene is classed as an engineering thermoplastic.

Grades

Injection-molding grades include easy flow, medium impact, high impact, heat resistant, flame retardant, glass fiber reinforced, and plating. Military purchases are to ASTM D 4673, *Specification for Acrylonitrile Butadiene Styrene (ABS) Molding and Extrusion Materials*.

Processing

Conventional injection molding techniques can be used to process acrylonitrile-butadiene-styrene polymers. Resins should be dried at 180°F (82°C) for 2–4 hours just prior to molding. Melt temperatures should be 450° to 525°F (232°–274°C). Oven annealing is advised at 5° to 10°F (2.8°–5.6°C) below the heat distortion temperature.

Properties

As noted above, a wide range of properties is obtainable by varying the proportions of acrylonitrile, butadiene, and styrene polymers. Physical properties of acrylonitrile-butadiene-styrene resins are high. Resistance to ultraviolet light is poor.

Acrylonitrile-butadiene-styrene is resistant to attack by:

- Inorganic acids except strong oxidizing acids
- Organic acids
- Aliphatic amines
- Bases
- Polyglycols
- Salts

Acrylonitrile-butadiene-styrene is attacked by:

- Aldehydes
- Aromatic amines

- Polyglycol ethers
- Aliphatic, aromatic, and chlorinated hydrocarbons
- Ketones
- Essential oils

Solvents for acrylonitrile-butadiene-styrene include ketones, esters, and some chlorinated hydrocarbons.

Table 3-28 shows key property values for polystyrene, polystyrene-butadiene, styrene-acrylonitrile, and acrylonitrile-butadiene-styrene resins.

Electrical/Electronic Uses

Most applications for acrylonitrile-butadiene-styrene take advantage of its strong mechanical properties. Electrical/electronic uses include business machine housings and components, telephones, conduit, and, when electroplated or vacuum metallized, shielding for electronic equipment.

Sulfone Polymers: Polysulfone
(Molding Resin)

Polysulfone resins were introduced in 1965 by Union Carbide. They are now produced by Amoco Performance Products (Udel™) and by BASF (Ultrason S.™). These resins feature excellent heat- and water-resistant properties.

Chemistry

Polysulfone is made by reacting the sodium salt of bisphenol A with 4,4'-dichlorophenyl sulfone in dimethyl sulfoxide or similar polar solvent. The resulting polymer has the following structure:

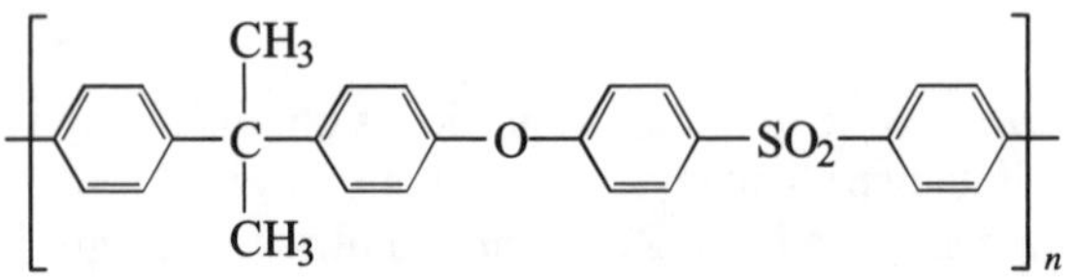

$$\left[\begin{array}{c} \\ \text{—}\bigcirc\text{—}\overset{\overset{\displaystyle CH_3}{|}}{\underset{\underset{\displaystyle CH_3}{|}}{C}}\text{—}\bigcirc\text{—O—}\bigcirc\text{—}SO_2\text{—}\bigcirc\text{—} \\ \end{array}\right]_n$$

Polysulfone is classed as an engineering thermoplastic.

Grades

Udel is available in transparent and opaque grades, and in UL 94V-0 natural or light beige translucent grades. Reinforced and filled grades containing glass fibers and spheres, carbon fibers, and polytetrafluoroethylene are available. Military purchases are to MIL-P-46120, *Plastic Molding and Extrusion Material, Polysulfone.*

Processing

Reciprocating screw injection molding equipment is preferred, although ram machines may also be used with the shot size between 30 and 50 percent of rated capacity. Stock temperatures are usually in the 650° to 750°F (345°–400°C) range with cylinder temperatures 50° to 100°F (10°–40°C) lower in screw machines. Mold temperatures range from 200°F (95°C) for simple designs to 320°F (160°C) for complex parts. Polysulfone resins should be dried below 0.05 percent moisture content before molding. Drying for 3¹/₂ hours at 275°F (135°C) is recommended.

Properties

Polysulfone resins exhibit high continuous use temperatures of up to 300°F (149°C), and are hydrolytically stable with long-term resistance to creep. Dielectric properties are maintained over a wide temperature and frequency range, and resistance to radiation is the highest for all polymers. Compared with polyether-sulfone, polysulfone has superior dielectric values and lower water absorption, but has lower thermal, mechanical, and flammability values.

Polysulfone resins are resistant to:

- Mineral acids (except concentrated strong acids)
- Alkalies in solution
- Salt solutions
- Aliphatic oils

Polysulfone resins are attacked by:

- Esters
- Ketones
- Aromatic and chlorinated hydrocarbons

Polysulfone resins are dissolved by halogenated hydrocarbons. Table 3-29 shows key property values for both polysulfone and polyethersulfone resins.

Electrical/Electronic Uses

Polysulfone resins are preferred for applications requiring high-temperature resistance and hydrolytic stability beyond the capability of most other thermoplastics. Because of stress cracking, they are not suitable for parts exposed to aromatic or halogenated hydrocarbons.

Examples of electrical/electronic applications include printed circuit boards, integrated circuit carriers, coil bobbins, connectors, bushings, terminal blocks, brush holders, alkaline battery cases and cells, TV and stereo components, and business machine parts.

Sulfone Polymers: Polyethersulfone (PES)
(Molding Resin)

Polyethersulfone resins were introduced first in the United Kingdom and then in the United States in 1972–1973. They are produced by BASF with the trademark "Ultrason E" and are also marketed by ICI Americas with the trademark "Vitrex." Continuous use temperature index rating for unfilled resin by Underwriters Laboratories is 355°F (180°C), among the highest for any rated thermoplastic.

TABLE 3-29   Key Properties of Typical Polysulfone and
Polyether-Sulfone Molding Resins

	ASTM TEST METHOD	POLY-SULFONE VALUE	POLYETHER-SULFONE VALUE
Specific gravity*	D 792	1.25	1.37
Dielectric strength, 125 mils, short time, V/mil*	D 149	425	400
Dielectric constant*	D 150		
60 Hz		3.1	3.5
10^6 Hz		3.0	3.5
Dissipation factor*	D 150		
60 Hz		0.0008	0.001
10^6 Hz		0.0034	0.004
Volume resistivity, $\Omega{\cdot}cm$*	D 257	1×10^{16}	$>1 \times 10^{17}$
Arc resistance*	D 495	122	70
Water absorption, $^1/_8$ in thick specimen, %	D 570		
24 h @ 73°F (23°C)		0.3	0.43
Deflection temperature, 264 psi °F (°C)	D 648	345 (174)	397 (203)
Maximum recommended service temperature °F (°C)		300 (150)	355 (180)
Tensile strength at break, psi*	D 638	10,200	12,200
Izod impact strength	D 256		
Foot pounds per inch notched $^1/_8$ in specimen*		1.3	1.2
Flammability ratings			
UL Standard 94, $^1/_{16}$ in specimen			
Standard grade		V-2	V-0
FR grade		V-0	none needed
Oxygen index, %	D 2863		
Standard grade		30	38
FR grade		32	none needed

*At room temperature.

Chemistry

Polyethersulfone is made by Friedel Crafts reaction from diphenyletherchlorosulfone and has the following structure:

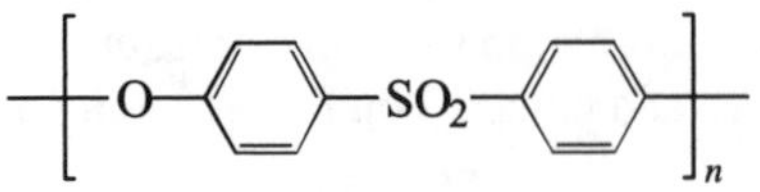

Polyethersulfone is classed as an engineering thermoplastic.

Grades

Injection molding grades of Victrex include the following:

- General purpose amber/transparent grade
- Higher viscosity grade with improved chemical and creep resistance, and slightly lower melt flow properties

- Filled grades with 20–30 percent glass fibers
- Easy-flow grades with 20–30 percent glass fibers

Military purchases are to MIL-P-46185, *Plastic Molding and Extrusion Materials, Polyethersulfone.*

Processing

A reciprocating screw injection molding machine is recommended. Barrel temperatures should be in the range of 644° to 680°F (340°–360°C) for general purpose grade and 680° to 734°F (360°–390°C) for glass-filled grades. Recommended mold temperature range is 248° to –320°F (120°–160°C). Resins should be dried at 300°F (149°C) for 3 hours before molding.

Properties

The outstanding property of polyethersulfone is its high use temperature. Electrical insulating properties are largely retained up to 355°F (180°C). Long-term load-bearing values, toughness, inherent low flammability, and mechanical strength are other noteworthy characteristics. Resistance to radiation is good, but resistance to UV light is poor. Compared with polysulfone, polyethersulfone has superior thermal, mechanical, and flammability values, but has poorer dielectric values and higher water absorption.

Polyethersulfone has good resistance to:

- Mineral acids
- Alkalis in solution
- Alcohols
- Most inorganic aqueous solutions
- Most oils and greases
- Gasolines

Polyethersulfone is attacked by:

- Esters
- Ketones
- Diester and phosphate based oils
- Chloroform
- Polar aromatic chemicals

Polyethersulfone dissolves in:

- N-methylpyrollidone
- Methylene chloride
- 1,1,2-trichloroethane
- Dimethylformamide

Table 3-29 shows key property values for both polyethersulfone and polysulfone resins.

Electrical/Electronic Uses

Polyethersulfone resins are suitable for demanding applications such as terminal blocks for aircraft, printed circuit boards, coil bobbins, switching and control components, and appliance housings subject to possible high temperatures under fault conditions.

4

Thermosetting Molding Compounds

INTRODUCTION

The thermosetting plastics industry got its start in the first decade of the twentieth century as a result of the development work of Dr. Leo Bakeland, the inventor of phenolics, in Germany. As a tribute to him, these early thermosets were referred to as *bakelites*. Until the 1930s, thermoplastics were virtually unknown, so phenolics and succeeding thermosets had no competition until then for molded parts, especially electrical parts. It is unlikely that the electrical industry could have undergone its astonishing early growth without the availability of thermosets to make essential equipment such as switchgear, electrical controls, and wiring devices. The budding automotive industry also became dependent on thermosets for essential components of electrical systems, including magnetos, distributors, and batteries.

This chapter covers the basic thermosets used in forming parts by compression, transfer, and injection molding techniques. Thermosets used in embedding processes are covered in Chapter 16.

As used in this handbook, *thermoset* refers to an organic material capable of undergoing, or that has undergone, a polymerization or polycondensation reaction to form a rigid, infusible substance.

THERMOSETTING PLASTICS CONSUMED IN ELECTRICAL/ELECTRONIC APPLICATIONS

Table 4-1 shows 1993 U.S. consumption of principal thermosetting plastics in dielectric applications.

TABLE 4-1   1993 U.S. Consumption of Principal Thermosetting Plastics in the U.S.

MATERIAL	MILLION POUNDS
Epoxys (electrical laminates)	53
Phenolics	94
Polyesters, reinforced	55
Ureas	54
Total	256

Source: Modern Plastics Magazine.

POLYMERIZATION AND POLYCONDENSATION

Unlike thermoplastics, thermosets, when fully cured, lose their plasticity. Curing rate is controlled by heat and pressure, and may be aided by catalysts, which are unique for each polymer. Except for phenolics and aminos, which cure by polycondensation, other thermosets covered in this chapter cure by addition polymerization and crosslinking, without emission of gaseous byproducts. Where gases are generated, provision for their venting must be incorporated in mold design. See "Polymerization" (p. 41) in Chapter 2, *Thermoplastic Molding Compounds*.

REINFORCEMENTS AND FILLERS, ANTIOXIDANTS, ULTRAVIOLET STABILIZERS

Thermosetting resins, by themselves, are brittle and of little practical value, but are made useful by compounding with reinforcements and fillers, which may constitute 50 percent or more of compound weight. Glass fibers and mineral fillers are most often used in electrical grade recipes. See "Reinforcements and Fillers" (p. 44) in Chapter 3, *Thermoplastic Molding Compounds*. Antioxidants and ultraviolet stabilizers are also covered in that chapter.

COMPRESSION MOLDING

The earliest production method for thermosetting compounds, compression molding, consists of placing a measured charge of material in a mold cavity, closing the mold, and applying heat and pressure to react the resin irreversibly, forming a rigid, infusible part. This process, however, has significant limitations and its use is specialized and mostly for low-volume production. Material charges must be handled individually, often for numerous, small cavities. Curing is slower than in the transfer molding method, especially for parts with thick walls. Pins and inserts in the cavity may be damaged by closing of the press on unheated compound. Compression molding is not well suited for intricate shapes where easy flow is required to fill the mold.

For certain applications, however, compression molding has noteworthy attributes. Tooling and capital equipment costs are lower than for transfer molding, and there is no wastage with sprues and runners. Fillers receive minimal damage in the molding process, and thus mechanical properties of molded parts are optimal. Clamping pressures are lower than for other methods, permitting more cavities per press of given tonnage. Finishing costs are minimal since there are no gate marks to remove.

Compression molding machines are hydraulically operated. Heat is applied by steam, hot oil, high-pressure water, gas flames, or, now most popular, by electrical heating coils, strips, or cartridges which permit higher temperatures and faster cycles.

TRANSFER MOLDING

The process most commonly employed for making parts from thermosetting compounds is transfer molding. A measured charge of suitably preheated material is placed in a chamber, or pot, from which it is forced (transferred) by a plunger through channels (sprues and runners) and past a gate into a closed cavity. Throughout this process, a temperature adequate for rapid curing is maintained. At the end of the curing cycle, the part is ejected attached to the gates, sprues, and runners, which are then trimmed from the part.

Preheating of compound is usually done in dielectric heaters, although heat lamps and ovens are also used. Grades most suitable for transfer molding are of softer plasticity than materials for compression molding.

See the preceding section on "Compression Molding" for a discussion of relative merits of compression and transfer molding methods.

INJECTION MOLDING

The principal difference between injection molding thermoplastics and thermosets is that the former are heated to a fluid state and injected into the mold cavity where the part is shaped and cooled, while thermosets, after being heated to fluid state and injected into the mold cavity, undergo final curing at the specified elected temperature and for the required time period.

For thermosets, the temperature of material in the injection barrel is raised to a level just below the point where precure would occur by hot water or hot oil circulating through a jacket in the barrel and also by the shear friction action of the rotary screw. The last step in injection occurs as the screw rotation is stopped and the mechanism acts as a high-pressure plunger, driving the heated plasticated material through the barrel nozzle, the sprues and runners, and into the cavity. The controlled material flow rate in this stage is very rapid, generating considerable frictional heat which completes curing. The ejection of parts is similar to that of thermoplastic injection machines. See also "Injection Molding" (p. 46) in Chapter 3, *Thermoplastic Molding Compounds*.

STANDARD TESTS

Standard tests for thermosets are the same as for thermoplastics, described in Chapter 3, *Thermoplastic Molding Compounds*. This section covers ranking of thermosets by selected characteristics as shown in the table of key properties for each thermoset. It is important to note that properties listed are for commercially available grades and may vary from producer to producer for the same grade. (See Tables 4-2 through 4-13.)

TABLE 4-2   Specific Gravity of Thermosetting Compounds
(23/23°C)
(ASTM D 792)

COMPOUND	VALUE
Glass-filled allyl MIL-M-14G Type GDI-30	1.70
Glass-filled phenolic MIL-M-14G Type GPI-100	1.74
Glass-filled melamine MIL-M-14G Type MMI-30	1.92
Glass-filled epoxy electrical grade	1.92
Glass-filled alkyd/polyester MIL-M-14G Type MAI-60	2.07

TABLE 4-3   Dielectric Strength of Thermosetting Compounds
(Short Time, 125 mils, volts/mil)
(ASTM D 149)

COMPOUND	VALUE
Glass-filled allyl MIL-M-14G Type GDI-30	400
Glass-filled epoxy electrical grade	390
Glass-filled phenolic MIL-M-14G Type GPI-100	380
Glass-filled alkyd-polyester MIL-M-14G Type MAI-60	375
Glass-filled melamine MIL-M-14G Type MMI-30	340

TABLE 4-4   Dielectric Constant of Thermosetting Compounds
(ASTM D 150)

COMPOUND	60 Hz	10^6 Hz
Glass-filled allyl MIL-M-14G Type GDI-30	4.2	3.5
Glass-filled epoxy electrical grade	5.0	4.6
Glass-filled alkyd/polyester MIL-M-14G Type MAI-60	5.3	4.6
Glass-filled phenolic MIL-M-14G Type GPI-100	6.0	5.0
Glass-filled melamine MIL-M-14G Type MMI-30	8.0	6.2

TABLE 4-5   Dissipation Factor of Thermosetting Compounds
(ASTM D 150)

COMPOUND	10^6 Hz
Glass-filled epoxy electrical grade	0.01
Glass-filled allyl MIL-M-14G Type GDI-30	0.01
Glass-filled phenolic MIL-M-140G Type GPI-100	0.02
Glass-filled alkyd/polyester MIL-M-14G Type MAI-60	0.02
Glass-filled melamine MIL-M-14G Type MMI-30	0.02

TABLE 4-6   Volume Resistivity of Thermosetting Compounds
(ASTM D 257)

COMPOUND	$\Omega \cdot$cm
Glass-filled epoxy electrical grade	1×10^{13}
Glass-filled allyl MIL-M-14G Type GDI-30	1×10^{13}
Glass-filled phenolic MIL-M-14G Type GPI-100	1×10^{13}
Glass-filled alkyd/polyester MIL-M-14G Type MAI-60	1×10^{13}
Glass-filled melamine MIL-M-14G Type MMI-30	2×10^{11}

TABLE 4-7   Arc Resistance of Thermosetting Compounds
(ASTM D 495)

COMPOUND	SECONDS
Glass-filled epoxy electrical grade	187
Glass-filled alkyd/polyester MIL-M-14G Type MAI-60	180+
Glass-filled melamine MIL-M-14G Type MMI-30	180+
Glass-filled phenolic MIL-M-14G Type GPI-100	180
Glass-filled allyl MIL-M-14G Type GDI-30	140

TABLE 4-8   Water Absorption of Thermosetting Compounds
($^1/_8$ in Thick Specimen)
(ASTM D 570)

COMPOUND	% GAIN IN WEIGHT AFTER 24 h @ 73°F (23°C)
Glass-filled alkyd/polyester MIL-M-14G Type MAI-60	0.07
Glass-filled phenolic MIL-M-14G Type GPI-100	0.15
Glass-filled epoxy electrical grade	0.20
Glass-filled allyl MIL-M-14G Type GDI-30	0.20
Glass-filled melamine MIL-M-14G Type MMI-30	0.40

TABLE 4-9   Deflection Temperature of Thermosetting
Compounds
(ASTM D 648)

COMPOUND	FLEXURAL LOAD 264 psi	
	°F	°C
Glass-filled allyl MIL-M-14G Type GDI-30	540	282
Glass-filled phenolic MIL-M-GPI-100	500+	260+
Glass-filled epoxy electrical grade	450	232
Glass-filled melamine MIL-M-14G Type MMI-30	440	227
Glass-filled alkyd/polyester MIL-M-14G Type MAI-60	400+	204+

TABLE 4-10   Maximum Recommended Service
Temperature of Thermosetting Compounds

COMPOUND	°F	°C
Glass-filled allyl MIL-M-14G Type GDI-30	500	260
Glass-filled phenolic MIL-M-14G GPI-100	450	232
Glass-filled alkyd/polyester MIL-M-14G Type MAI-60	400	204
Glass-filled epoxy electrical grade	400	204
Glass-filled melamine MIL-M-14G Type MMI-30	400	204

TABLE 4-11   Tensile Strength of Thermosetting Compounds
(ASTM D 638)

COMPOUND	PSI AT BREAK
Glass-filled epoxy electrical grade	10,500
Glass-filled allyl MIL-M-14G Type GDI-30	10,000
Glass-filled melamine MIL-M-14G Type MMI-30	8,000
Glass-filled phenolic MIL-M-GPI-100	7,500
Glass-filled alkyd/polyester MIL-M-14G Type MAI-60	6,000

TABLE 4-12   Izod Impact Strength of Thermosetting
Compounds
(Foot pounds per inch notched $\frac{1}{8}$ in specimen)
(ASTM D 256)

COMPOUND	VALUE
Glass-filled phenolic MIL-M-14G Type GPI-100	12
Glass-filled alkyd/polyester MIL-M-14G Type MAI-60	9.5
Glass-filled melamine MIL-M-14G Type MMI-30	5.5
Glass-filled allyl MIL-M-14G Type GDI-30	5.0
Glass-filled epoxy electrical grade	0.45

TABLE 4-13   Flammability Ratings of Thermosetting
Compounds
(UL Standard 94, $\frac{1}{16}$ in specimen)
(Standard Grades)

COMPOUND	RATING
Glass-filled epoxy electrical grade	V-0
Glass-filled alkyd/polyester MIL-M-14G Type MAI-60	V-0
Glass-filled phenolic MIL-M-14G Type GPI-100	V-0
Glass-filled melamine MIL-M-14G Type MMI-30	V-0
Glass-filled allyl MIL-M-14G Type GDI-30	HB

RELATIVE COSTS OF THERMOSETTING COMPOUNDS

The difference in specific volumes of two compounds considered equally suitable for an application is an indication of the higher price per pound the compound with lower specific gravity may carry while retaining the same part cost as the lower priced but heavier compound. See discussion of this subject in Chapter 3, *Thermoplastic Molding Compounds*, p. 63.

Table 4-14 shows cubic inches per pound for selected thermosetting compounds, from which are calculated price advantages of lighter compounds. For example, if the price of allyl compound were 21 percent higher than alkyd/polyester, part cost would be equal due to the higher specific volume of allyl (16.3 versus 13.4).

TABLE 4-14   Specific Volume and Cost Factor of Thermosetting Compounds

COMPOUND	CUBIC INCHES/lb*	% PRICE/lb DIFFERENCE FOR EQUAL PART COST
Glass-filled allyl MIL-M-14G Type GDI-30	16.3	21
Glass-filled phenolic MIL-M-14G Type GPI-100	15.9	19
Glass-filled melamine MIL-M-14G Type MMI-30	14.4	7
Glass-filled epoxy electrical grade	14.4	7
Glass-filled alkyd/polyester MIL-M-14G Type MAI-60	13.4	—

*Specific volume.

DEVELOPMENT PROGRAMS

The need for faster production methods for molding thermosetting compounds has resulted in new molding technologies. American Cyanamid has developed *cold sprue molding* especially well suited for molding small parts of alkyd/polyester compounds. Essentially, this method is a version of injection molding with a minimum of material loss. The sprue does not cure with the part and is injected into the cavity in the next cycle. However, there is a small runner which does cure with the part and must be trimmed after ejection. It is claimed that up to 500 cycles/hour can be achieved. Mold design is critical and must be tailored for this method.

Another technology (patented), designed to conserve material, is the *runnerless injection process (RIC)*. In this method, the mold is held open as in compression molding while the material is injected. After filling, the mold is closed, compression occurs, and curing takes place. It is claimed that material savings with this method averages about 30 percent.

MARKET TRENDS

Thermosetting compounds have been used successfully for many years in molding parts for all types of electrical equipment. It is expected that this usage will continue, following the primary trends in the electrical industry, but probably gradually losing market share to high-temperature thermoplastic compounds.

RANKING OF THERMOSETTING COMPOUNDS BY KEY PROPERTIES

Table 4-15 ranks thermosetting compounds according to characteristics listed in the table of key properties for each compound. Rankings range from 1, most favorable from an electrical insulation viewpoint, to 5, least favorable. Note that no compound ranks either highest or lowest in all properties. This table is intended as an aid to design engineers desiring to investigate compounds potentially suitable to meet critical criteria. However, consideration should be given to the fact that properties of each thermosetting type may vary significantly from listed values, depending on characteristics and amount of filler and also on certain additives.

TABLE 4-15   Ranking of Thermosetting Compounds by Key Property Values

	SPECIFIC GRAVITY	DIELECTRIC STRENGTH	DIELECTRIC CONSTANT	DISSIPATION FACTOR	VOLUME RESISTIVITY	ARC RESISTANCE	WATER ABSORPTION	DEFLECTION TEMPERATURE	MAXIMUM SERVICE TEMPERATURE	TENSILE STRENGTH	IZOD IMPACT STRENGTH	UL 94 FLAMMABILITY[a]
Alkyd/polyester	5	4	3	3	1	2	1	5	3	5	2	1
Allyl	1	1	1	1	1	5	3	1	1	2	4	5
Melamine	3	5	5	3	5	3	5	4	3	3	3	1
Epoxy (Novolac)	3	2	2	1	1	1	3	3	3	1	5	1
Phenolic	2	3	4	3	1	4	2	2	2	4	1	1

[a]Number 1 = V-0
        5 = HB

FEDERAL SPECIFICATION MIL-M-14

The title of this specification is *Molding Plastics and Molded Plastic Parts*. This specification is approved for use by all departments and agencies of the Department of Defense, and it covers the basic properties of molding compounds and the methods suited to their satisfactory determination. The appendix covers requirements for parts molded from such compounds, together with procedures for inspection of such parts. The types of compounds for each thermosetting resin are listed in the section for that resin in this chapter.

PRINCIPAL THERMOSETTING COMPOUNDS FOR ELECTRICAL/ELECTRONIC USES

The chemistry, grades, processing, properties, and electrical/electronic uses of principal thermosetting compounds are described in the sections which follow. The values shown in tables are for electrical grades, where possible those defined by MIL-M-14G Specification.

The following resins are covered:

- Alkyds/polyesters
- Allyls
- Aminos
- Epoxies
- Phenolics

Alkyds/Polyesters
(Molding Resins)

Alkyds, also called thermosetting polyesters, are complex resins, many of which are used extensively in electrical/electronic applications. Producers include American Cyanamid, Glastic, Hasite, Premix, Quantum Composites, and Plastics Engineering.

Chemistry

Alkyds are formulated from unsaturated polyesters combined with free-radical-producing peroxide catalysts and unsaturated monomers, for example, diallyl phthalate or vinyl toluene. Polyesters are the condensation reaction products of polyfunctional alcohols and dibasic organic acids. The crosslinking between polyester and monomer is initiated by heat and the catalyst. This is an addition reaction giving off no volatile products. A multitude of complex resins may thus be formulated from numerous suitable and readily available alcohols, acids, and monomers. The terms *alkyd* and *polyester* are often used interchangeably in connection with molding resins. When referring to varnish resins, the term *alkyd* applies to polyesters modified with drying oils.

Grades

MIL-M-14G lists the following grades for alkyd/polyester compounds:

- *Type MAG*. A mineral-filled compound for use where good dielectric properties and arc resistance are required.
- *Type MAI-60*. A glass-fiber-filled compound for use where high-impact strength, *good dielectric properties*, and arc resistance are required.
- *Type MAT-30*. A heat-resistant, track-resistant, flame-resistant, high-impact, mineral-filled glass-reinforced compound. Impact strength is about 3.0 foot pounds per inch notch.
- *Type MAI-30*. A mineral-filled, glass-fiber reinforced compound having excellent handling and molding characteristics. It is an arc-resistant, flame-resistant, heat-resistant, high-impact compound having good mechanical and excellent electrical characteristics.

Processing

Alkyd compounds may be successfully molded by compression, transfer, injection, and runnerless injection equipment. Compounds go through a low-viscosity phase of short duration, enabling the mold to be filled at lower pressures than are required for other thermosetting compounds. The reaction is exothermic and curing is rapid. Molding temperatures depend somewhat on the catalyst employed, and range from 270° to 380°F (132°–193°C). Preheating, while not necessary, improves flow characteristics and curing rate. Preheating may be done in an air circulating oven or in an electronic preheater. Temperature should not exceed 140°F (60°C).

Polymerization reaction of unfilled reactants is highly exothermic, making them unsuitable for molding accurate parts because of excessive shrinkage on cooling. Thus, mineral fillers and glass fiber reinforcements are required to moderate the reaction and control shrinkage.

Properties

Over a wide temperature range, alkyds exhibit excellent electrical properties, mechanical strength, and dimensional stability. Arc resistance is high. As with all thermoset-

TABLE 4-16   Key Properties of Glass-filled Alkyd/Polyester Molding Resin (MIL-M-14G Type MAI-60)

	ASTM TEST METHOD	VALUE
Specific gravity*	D 792	2.07
Dielectric strength, 125 mils, short time, V/mil*	D 149	375
Dielectric constant*	D 150	
60 Hz		5.6
10^6 Hz		4.6
Dissipation factor*	D 150	
60 Hz		0.10
10^6 Hz		0.02
Volume resistivity, $\Omega \cdot$cm*	D 257	1×10^{13}
Arc resistance, seconds*	D 495	180+
Water absorption, $^1/_8$ in thick specimen, %	D 570	
24 h @ 73°F (23°C)		0.07
Deflection temperature, 264 psi, °F (°C)	D 648	400+ (204+)
Maximum recommended service temperature °F (°C)		400 (204)
Tensile strength at break, psi*	D 638	6,000
Izod impact strength	D 256	
Foot pounds per inch notched $^1/_8$ in specimen*		9.5
Flammability ratings		
UL standard 94, $^1/_{16}$ in specimen		
Standard grades		V-0
Oxygen index %	D 2863	
Standard grades		63

*At room temperature.

ting compounds, properties vary significantly with types and amounts of fillers and other ingredients, and also with molding method. For outdoor exposure, recipes should include an ultraviolet light absorber, such as 2-hydrobenzophenone.

The resistance to attack by most chemicals varies from poor to good, depending on compound formulation. All grades are attacked by oxidizing acids, aromatic and halogenated hydrocarbons, esters, and ketones.

Table 4-16 shows key property values for electrical grade alkyd compounds.

Electrical/Electronic Uses

Alkyds are used extensively in electrical/electronic products, including brush holders, distributor caps, circuit breakers, relays, switches, connectors, terminal boards, and housings.

Allyls
(Molding Resins)

Allyl molding compounds are usually sold under military specifications for critical electrical and electronic applications where superior dielectric and thermal properties are required. Principal suppliers include Occidental (Durez™), Rogers, and Quantum Composites.

Chemistry

The two most commonly encountered allyls are diallyl phthalate (DAP) and diallyl isophthalate (DIAP), prepared by esterification of phthalic anhydrides with allyl alcohol in an acid medium:

phthalic   allyl alcohol   diallyl phthalate   water
anhydride

Polymerization is by addition with crosslinking, aided by a peroxide catalyst.

The major difference in properties between diallyl phthalate and diallyl isophthalate is that diallyl isophthalate will withstand significantly higher operating temperatures up to 500°F (260°C) versus 400°F (204°C) for diallyl phthalate. This advantage is obtained at considerable increase in resin price.

Grades

MIL-M-14G lists the following grades for allyl compounds:

• *Type GDI-30*. A glass-fiber-filled diallyl phthalate resin of low loss, high dielectric strength, low shrinkage, excellent moisture resistance, and relatively high impact strength.

- *Type GDI-30F*. A resin for applications where flame retardancy is at a maximum.

- *Type MDG*. A mineral-filled compound for use where good dielectric properties and low shrinkage are required.

- *Type SDG*. A glass-filled compound of low loss, high dielectric strength, low shrinkage, and good moisture resistance. Impact strength is relatively low.

- *Type SDG-F*. A glass-filled compound of low loss, high dielectric strength, low shrinkage, flame resistance, and good moisture resistance. Impact strength is relatively low.

- *Type SDI-5*. An acrylic polymer fiber-filled compound of low loss, high dielectric strength, low shrinkage, excellent moisture resistance, and moderate impact strength.

- *Type SDI-30*. A polyethylene terephthalate fiber-filled compound of low loss, high dielectric strength, low shrinkage, very good moisture resistance, and high impact strength.

A full range of colors is available in most compounds.

Processing

Allyl resins may be processed easily by all techniques suitable for thermosetting compounds. The curing reaction is a chemical crosslinking polymerization (not a polycondensation), without formation of volatiles which occurs with phenolics.

Strength of transfer molded parts is usually somewhat lower and shrinkage is somewhat higher than that of parts molded by compression. Recommended range of mold temperatures is 290° to 320°F (143°–160°C). In molding large parts, intricate parts, or parts with inserts, preheating of compounds is necessary for optimum appearance, minimum cycle time, and low scrap loss. Post-curing is recommended, starting at 275°F (135°C) and rising to at least the maximum temperature encountered in service.

Properties

Allyls have outstanding dielectric properties and selected grades have good arc resistance. They retain these properties at elevated temperatures, and are virtually unaffected by high-humidity conditions. Dimensional stability throughout a wide temperature range is excellent, and post-molding shrinkage is negligible for most parts. Resistance to attack by most chemicals is good. Exceptions are strong inorganic acids and bases, toluene, and some paint removers.

Table 4-17 shows key property values for electrical grade allyl compounds.

Electrical/Electronic Uses

Allyls are used extensively for high reliability connectors, switches, and terminal boards. Other applications include tuner strips, brush holders, motor starter blocks, cases and housings, and parts for aerospace vehicles and missiles.

**TABLE 4-17   Key Properties of Glass-Filled Allyl
Molding Resin
(MIL-M-14G Type GDI-30)**

	ASTM TEST METHOD	VALUE
Specific gravity*	D 792	1.70
Dielectric strength, 125 mils, short time, V/mil*	D 149	400
Dielectric constant*	D 150	
60 Hz		4.2
10^6 Hz		3.5
Dissipation factor*	D 150	
60 Hz		0.004
10^6 Hz		0.01
Volume resistivity, $\Omega \cdot$cm*	D 257	1×10^{13}
Arc resistance, seconds*	D 495	140
Water absorption, $1/8$ in thick specimen, %	D 570	
24 h @ 73°F (23°C)		<0.20
Deflection temperature, 264 psi, °F (°C)	D 648	540 (282)
Maximum recommended service temperature °F (°C)		500 (260)
Tensile strength at break, psi*	D 638	10,000
Izod impact strength	D 256	
Foot pounds per inch notched $1/8$ in specimen*		5.0
Flammability ratings		
UL standard 94, $1/16$ in specimen		
Standard grades		HB
FR grades		V-0
Oxygen index %	D 2863	
Standard grades		26–32
FR grades		36

*At room temperature.

Aminos
(Molding Resins)

The most widely used amino resins are urea formaldehyde (urea) and melamine formaldehyde (melamine). Principal producers of electrical grade compounds include BTL Specialty Resins, ICI Fiberite, Neste Resins, and Plastics Engineering (Plenco™).

Chemistry

One mole of urea combines with two moles of formaldehyde under neutral or mildly alkaline conditions to form dimethylolurea:

$$H_2N-\underset{\underset{urea}{}}{\overset{\overset{O}{\|}}{C}}-NH_2 + 2 \underset{formaldehyde}{HCHO} \longrightarrow HOCH_2NH-\underset{\underset{dimethylolurea}{}}{\overset{\overset{O}{\|}}{C}}-NHCH_2OH$$

Likewise, one mole of melamine combines with up to six moles of formaldehyde to form hexamethylol melamine:

melamine   formaldehyde   hexamethylol
                                                      melamine

These addition products condense in the presence of an acid catalyst to form highly crosslinked urea and melamine resins with water as a byproduct.

Grades

Melamine compounds are generally superior to urea compounds in resistance to acids, alkalis, heat, and boiling water. Parts made of urea shrink on aging and tend to crack, especially around inserts and sharp corners. Melamine exhibits these characteristics to a lesser extent, and is usually preferred for critical electrical applications. MIL-M-14G does not list urea grades, but lists the following grades for melamine compounds:

• *Type CMG.* A cellulose-filled, general purpose compound with good dielectric and mechanical properties, for use where good arc resistance is required.

• *Type CMI-5.* A cellulose-filled, moderate impact compound with good all-around mechanical properties for use where resistance to arcing and moderate impact are required.

• *Type CMI-10.* A cellulose-filled, moderate impact phenol modified compound with good all-around mechanical properties suitable for tableware and similar applications. It is not intended for electrical use.

• *Type MME.* A mineral-filled compound for use where good dielectric properties and arc and flame resistance are required. Of the melamine compounds, this is the most dimensionally stable.

• *Type MMI-5.* A glass-fiber-filled compound of lower impact strength and higher dielectric constant and dissipation factor at 1 megacycle than Type MMI-30. It has superior moldability than Type MMI-30. Impact strength is approximately 0.5 foot-pounds per inch notch.

• *Type MMI-30.* A glass-fiber-filled compound of high impact strength for use where heat resistance, arc resistance, and flame resistance are required.

A full range of colors in most compounds is available.

Processing

Amino compounds are suitable for use with compression and transfer molding equipment. Molding temperatures from 260° to 340°F (127°–171°C) are used for ureas with slightly higher temperatures for melamines. Melamine compounds may be effectively preheated to 200° to 250°F (93°–121°C) by high-frequency heaters. Urea compounds are usually not preheated.

Properties

Alpha cellulose-filled amino resins are among the hardest, most rigid, abrasion-resistant plastics available, and in the case of melamines, these properties are relatively unaffected between −70° and 210°F (−57°–99°C). Ureas are not recommended for use above 170°F (77°C). Melamines, but not ureas, may be used in outdoor environments. Both urea and melamine resins are classed as self-extinguishing. Shrinkage of parts during and after molding, especially for ureas, may present problems with dimensional stability and lead to development of cracks on aging or cycling between dry and wet condition or hot and cold temperatures.

Amino resins are attacked by strong mineral acids and bases, and by sodium hypochlorite solutions. Otherwise, resistance to chemicals is high.

Table 4-18 shows key property values for electrical grade melamine compounds.

TABLE 4-18   Key Properties of Glass-Filled Melamine Molding Resin (MIL-M-14G Type MMI-30)

	ASTM TEST METHOD	VALUE
Specific gravity*	D 792	1.92
Dielectric strength, 125 mils, short time, V/mil*	D 149	300
Dielectric constant*	D 150	
60 Hz		8.0
10^6 Hz		6.2
Dissipation factor*	D 150	
60 Hz		na
10^6 Hz		0.02
Volume resistivity, $\Omega \cdot$cm*	D 257	2×10^{11}
Arc resistance, seconds*	D 495	180+
Water absorption, $1/8$ in thick specimen, %	D 570	
24 h @ 73°F (23°C)		0.40
Deflection temperature, 264 psi, °F (°C)	D 648	440 (227)
Maximum recommended service temperature °F (°C)		400 (204)
Tensile strength at break, psi*	D 638	8,000
Izod impact strength	D 256	
Foot pounds per inch notched $1/8$ in specimen*		5.5
Flammability ratings		
UL standard 94, $1/16$ in specimen		
Standard grades		V-0
Oxygen index %	D 2863	
Standard grades		na

*At room temperature.

na = not available.

Electrical/Electronic Uses

Applications for amino resins include switchgear, wiring devices, appliance housings and knobs, engine ignition parts, terminal strips, and sockets.

Epoxies
(Molding Resins)

Epoxies in many forms are among the most widely used dielectrics. This section covers electrical grade solid epoxy molding compounds. Producers of resins for these compounds include Ciba-Geigy, and Dow Plastics. Electrical grade epoxy compounders include, in addition to the above, ICI, Fiberite, and Rogers. For information on epoxies in embedding compounds, coatings and impregnants, and clad and unclad structures, see appropriate chapters.

Chemistry

The epoxide group, $-CH-CH_2$, serves as a terminal linear polymerization point in all
epoxy resins. Crosslinking of epoxide molecules occurs at the epoxide group and also at hydroxyl groups that may be present in the molecule. The epoxy resins generally used to make solid molding compounds are epoxy novolacs. Novolacs, formed by reaction of phenol or ortho-cresol with formaldehyde under acidic conditions (see discussion in the next section on "Phenolics"), are reacted with epichlorohydrin to form epoxy novolacs:

average value for n, 1.6
epoxy functionality, 3.6

epoxy phenol novolac

For a discussion of hardeners necessary for curing epoxies, see "Epoxies" (p. 183) of Chapter 6, *Embedding Compounds*.

Grades

Both phenol and ortho-cresol epoxy novolacs are available with glass fiber or mineral reinforcements. Military purchases are made to MIL-P-46892, *Plastic Molding Material, Epoxy, Glass Fiber*.

Processing

Conventional preheating, automatic preforming, compression and transfer molding methods are suitable for molding epoxy compounds. Recommended molding temperature range is 290° to 350°F (143°–177°C). Compounds do not require special handling.

Properties

Parts molded from electrical grade epoxy compounds exhibit outstanding dimensional stability and retention of electrical properties with virtually no outgassing at temperatures up to 500°F (260°C).

Epoxies are resistant to attack by most chemicals. Exceptions include phenols, ketones, ethers, and concentrated organic and inorganic acids.

Table 4-19 shows key property values for electrical grade epoxy compounds.

TABLE 4-19   Key Properties of Glass-Filled Epoxy Molding Resin Electrical Grade

	ASTM TEST METHOD	VALUE
Specific gravity*	D 792	1.92
Dielectric strength, 125 mils, short time, V/mil*	D 149	390
Dielectric constant*	D 150	
60 Hz		5.0
10^6 Hz		4.6
Dissipation factor*	D 150	
60 Hz		0.01
10^6 Hz		0.01
Volume resistivity, $\Omega \cdot$cm*	D 257	1×10^{13}
Arc resistance, seconds*	D 495	187
Water absorption, $1/8$ in thick specimen, %	D 570	
24 h @ 73°F (23°C)		<0.20
Deflection temperature, 264 psi, °F (°C)	D 648	450 (232)
Maximum recommended service temperature °F (°C)		400 (204)
Tensile strength at break, psi*	D 638	10,500
Izod impact strength	D 256	
Foot pounds per inch notched $1/8$ in specimen*		0.45
Flammability ratings		
UL standard 94, $1/16$ in specimen		
Standard grades		V-0
Oxygen index %	D 2863	
Standard grades		35

*At room temperature.

Electrical/Electronic Uses

Important applications of epoxy molding compounds include bobbins, connectors, and a variety of electrical and electronic components.

Phenolics/Furfurals (Molding Resins)

Phenolics, developed in Belgium by Leo Baekeland, were the first commercially available plastics and continue to be used in a wide variety of heavy duty electrical applications. Major producers include ICI Fiberite, Occidental Chemical (Durez™), Plastics Engineering (Plenco™), Resinoid Engineering, Rogers, and Valite.

Chemistry and Processing

Pure phenolic resin is the condensation reaction product of phenol and formaldehyde:

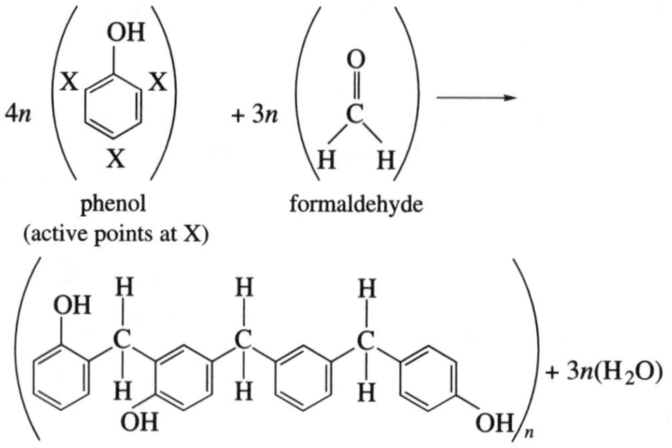

phenol
(active points at X)

formaldehyde

typical (intermediate) condensation product

The molecular ratio of formaldehyde-to-phenol in a typical intermediate is less than one-to-one, for example, three-to-four as in the above illustration. Crosslinking occurs as molecular size increases, tending to make the material hard and infusible.

The problem in making phenolic resins is to control the formaldehyde-to-phenol ratio so that the material is moldable when it goes into the mold, yet is hard, infusible, and serviceable after curing and removal from the mold. The usual method is by a two-step process, which utilizes an acid catalyst. The reaction, once started, is exothermic. The water formed is removed by vacuum distillation. The resulting resin is an amber-colored, brittle, "A-stage" resin, or *novolac* (also *novolak*) which is fusible, thermoplastic, soluble in alcohol and ketones, and can be used as a key ingredient of high-quality synthetic varnishes as well as to make molding resins.

The novolac resin is converted into a molding compound by grinding into a fine powder and then mixing with a filler, a pigment, or a dye; a mold lubricant such as aluminum stearate; and hexamethylenetetramine (hexa). During molding, the hot mold decomposes the hexa into formaldehyde and ammonia. The ammonia acts as a catalyst for crosslink formation, and the formaldehyde begins to react, forming complex three-dimensional networks. The resulting material has the desired characteristics attributed to phenolics. Steam formed in the condensation reaction will cause bubbles, voids, and porosity unless sufficient counterpressure is used. Typical molding conditions are 325°F (163°C) and 1500 pounds per square inch pressure for 2 minutes.

There is also a one-step process in which alkali is used as a catalyst and about 1.2 moles of formaldehyde is introduced per mole of phenol. Condensation reaction proceeds more slowly than with acid catalyst, and is controlled by temperature. To produce a molding resin, the reaction is stopped in the early B-stage by cooling. The resin is then ground and mixed with a filler, a pigment, or a dye, and a lubricant, but not with hexa. The reaction is completed in the mold as in the two-step process. It is claimed by some advocates of this process that it is lower in cost and its resins contain less residual phenol and catalyst, thereby imparting less odor, taste, and tarnish on inserts.

Molding may be carried out in any one of several ways:

• Compression molding with cold powder or radio frequency heated preforms, including use of automatic rotary presses

• Transfer molding with radio frequency heated pills

• Automatic injection molding of parts without inserts, offering advantages in cycle time, material utilization, and other manufacturing costs

• Runnerless injection compression process, a patented method in which the mold is held open while the compound is injected with little or no pressure, after which the mold is closed and compression occurs

Furfural resins are based on furfuraldehyde ($C_4H_4O–CHO$) rather than formaldehyde, and have the unique molding property of having their cure going almost directly from the A-stage to the C-stage without the intermediate B, or gel stage. Thus, they flow and fill a mold readily, then congeal rapidly on reaching cure temperature. This is an important advantage in molding complex objects with metal inserts which must not be deformed or disturbed, as in the molding of large radio cabinets.

Grades

By itself, pure phenolic resin is too brittle, or friable, for practical use. When compounded approximately 50 percent with fibrous fillers, however, utility is enhanced dramatically. The following phenolic compounds are classified by Military Specification MIL-M-14G:

• *Type CFG.* A general purpose, wood-flour-filled compound intended for applications requiring good dielectric properties with mechanical properties better than acceptable minimum. Preheating the molding compound is advisable and improves the electrical properties. Moldability of this material is excellent.

• *Type CFI-15.* A moderate-impact, cotton- or paper-filled compound intended for use where good all-around mechanical properties are required. Impact strength is approximately 0.5 foot-pounds per inch notch.

• *Type CFI-10.* A medium-impact, cotton rag-filled compound providing good finish. Impact strength is approximately 1.0 foot-pounds per inch notch.

• *Type CFI-20.* A high-impact, rag- or cotton-filled compound providing good finish. Impact strength is approximately 2.0 foot-pounds per inch notch.

• *Type CFI-30.* A high-impact, cotton-filled compound, providing good finish. Impact strength is approximately 3.0 foot-pounds per inch notch.

• *Type CFI-40.* The highest impact grade of cotton-filled compound. Impact strength is approximately 4.0 foot-pounds per inch notch.

- *Type MFA-30.* A heat-resistant, arc-resistant, flame-resistant, high-impact, asbestos-filled compound. Impact strength is approximately 3.0 foot-pounds per inch notch.

- *Type MFE.* A low-loss, high-dielectric-strength, low-water absorption, mineral-filled compound intended for applications requiring the best dielectric properties for a phenolic material. To secure optimum dielectric properties, this compound should be preheated immediately before molding. Care should be taken, however, to prevent pre-curing.

- *Type MFG.* A general purpose, asbestos-filled compound intended for applications requiring good mechanical and heat-resistant properties.

- *Type MFH.* A mineral-filled compound intended for applications requiring highest heat resistance. Mechanical properties are relatively low.

- *Type MFI-10.* A heat-resistant, medium-impact, asbestos-filled compound. Impact strength is approximately 1.0 foot-pounds per inch notch.

- *Type MFI-20.* A heat-resistant, high-impact, asbestos-filled compound. Impact strength is approximately 2.0 foot-pounds per inch notch.

TABLE 4-20   Key Properties of Glass-Filled Phenolic Molding Resin (MIL-M-14G Type GPI-100)

	ASTM TEST METHOD	VALUE
Specific gravity*	D 792	1.74
Dielectric strength, 125 mils, short time, V/mil*	D 149	380
Dielectric constant*	D 150	
60 Hz		6.0
10^6 Hz		5.0
Dissipation factor*	D 150	
60 Hz		na
10^6 Hz		0.02
Volume resistivity, $\Omega \cdot cm$*	D 257	1×10^{13}
Arc resistance, seconds*	D 495	180
Water absorption, $1/8$ in thick specimen, %	D 570	
24 h @ 73°F (23°C)		0.15
Deflection temperature, 264 psi, °F (°C)	D 648	500+ (260+)
Maximum recommended service temperature °F (°C)		450 (232)
Tensile strength at break, psi*	D 638	7,500
Izod impact strength	D 256	
Foot pounds per inch notched $1/8$ in specimen*		12
Flammability ratings		
UL Standard 94, $1/16$ in specimen		
Standard grades		V-0
Oxygen index %	D 2863	
Standard grades		45–60

*At room temperature.

• *Type GPI-100.* A glass-fiber-filled compound with high impact strength and good dielectric properties. Impact strength is approximately 10.0 foot-pounds per inch notch.

Military purchases are also made to MIL-P-82650, *Plastic Molding Material, Glass Phenolic*, ASTM D 4617, *Specification for Phenolic Compounds (PF)* includes sections on classification, requirements, and testing.

Properties

As may be noted from the above description of grades, a material specialist has many options in selecting a phenolic resin for a specific end use. In general, phenolics are among the lowest priced resins. Processing costs are also relatively low. Resistance to deformation under load is high, and service temperatures as high as 550°F (288°C) are possible for special grades and certain applications. Dielectric properties are well suited to heavy duty electrical apparatus. Resistance to weathering is excellent. Compounds are inherently flame-retardant. Disadvantages of phenolics include volatiles released during molding and limited color choices in dark shades.

Phenolics are resistant to attack by most chemicals, but have poor resistance to bases and oxidizing acids.

Table 4-20 shows key property values for electrical grade phenolic compounds.

Electrical/Electronic Uses

Long the workhorse for heavy-duty electrical apparatus insulation, such as in switchgear and electrical controls, phenolics are widely used for cases, connectors, wiring devices, panels, bases, and hermetically sealed electrical equipment.

5

Extrusion Compounds

INTRODUCTION

Since World War II, extruded plastics have become the principal dielectrics in the manufacture of wires, cords, and cables. It is convenient to classify these materials into three groups:

- High-temperature insulations, including fluoropolymers and silicone rubber
- Low-temperature insulations, including polyolefins and polyvinyl chloride
- Elastomers, used primarily as jackets but also as insulation on low-voltage cords and cables.

This chapter contains tables comparing key properties of materials within each group. The text also discusses fluoropolymers and elastomers as separate groups. Although there is close likeness in chemical structure of polyethylenes and polypropylene, individual members of this family are discussed separately and at length to highlight differences as well as similarities in characteristics and end uses. Polyvinyl chloride and silicone rubber are also discussed separately with emphasis on the compounding of polyvinyl chloride and because of the unique chemical structure of silicone rubber.

CONDUCTORS

Wires, cords, and cables provide the means for transmitting electrical power from the point of its generation to the point where it energizes electrical and electronic equipment essential to virtually every aspect of modern life.

Copper, electrolytically refined to 99.9 percent purity, has traditionally been the principal conductor metal. In the refining process, traces of copper oxide are removed to produce oxygen-free high-conductivity copper (OFHC). Cast copper bars at temperatures up to 1,700°F (927°C) are rolled into rods, which are then drawn through successively smaller dies to the desired wire gauge. Finally, the wire is annealed to the required ductility. The American Society for Testing and Materials (ASTM) has developed specifications for the wire-making process.

Conductors with greater strength than copper alone are made by fabricating wire with a steel core and copper skin by molten welding of copper to steel, by electroplating copper over steel, or by metallurgically bonding the two metals. Copper-clad wire may be annealed for greater flexibility or hard drawn for greater strength. At high frequencies, conductivity is the same as for copper alone, since high-frequency currents travel along the outer skin of a conductor. At power frequencies, conductivity is 30 to 40 percent of equal gauge copper wire.

At higher cost, copper alloys, with 80 to 85 percent the conductivity of copper, may be used for special applications where weight, strength, and space are critical. Most commonly used are alloys with cadmium, cadmium–chromium, chromium, and zirconium. Cadmium–chromium copper is noted for high conductivity combined with high tensile strength.

Aluminum is also widely used as a conductor metal where weight is the principal consideration. It is 30 percent the weight of an equal volume of copper, but has only 61.8 percent the conductivity of copper. To obtain equivalent current-carrying capacity, the diameters of aluminum conductors must be increased more than 50 percent over copper conductors. The susceptibility to corrosion discourages the use of aluminum in communication, instrumentation, and control cables. Copper conductors are the norm for power plant circuits.

Tensile strength of annealed copper is about 36,000 psi and of annealed aluminum 10,000 psi. Tensile strengths up to 58,000 psi are obtainable from copper-clad steel and copper alloys. Because of the cost of energy to anneal drawn conductors, three-quarters to full hard copper and aluminum are furnished in power cable. Their tensile strengths are correspondingly higher.

Wire sizes are designated by the American Wire Gauge System (AWG), designed by J. R. Brown of Brown & Sharp in 1857. In this system, the higher the AWG number, the smaller the diameter of the wire. Numbers typically range from AWG 44 (0.0019 inch) to AWG 4/0 (0.4554 inch). See also the tables in "Magnet Wire Data" (p. 210) of Chapter 7, *Magnet Wire Enamels*. The convenient part of the system is that successive sizes represent approximately one reduction in die size in the wire drawing operation. Another convenient part is that an increase of two conductor sizes is required when substituting aluminum for copper for comparable conductivities. Note also that an increase (or decrease) of three sizes, halves (or doubles) the cross section in circular mils, and doubles (or halves) the resistance. A circular mil is the area of a circle 1 mil in diameter and is equal to 0.784 mil^2. The area of a solid conductor in circular mils is the square of its diameter in mils.

TEMPERATURE COEFFICIENT OF RESISTANCE

Temperature has a significant effect on the resistance of a conductor, with dc resistance increasing with temperature according to the formula:

$$R_t = R_o[1 + \alpha(t - t_o)]$$

where

R_t = Resistance as measured at temperature t

R_o = Resistance at reference temperature t_o

α = Temperature coefficient of resistance at t_o

t = Temperature at which measurement is made

t_o = Reference temperature

At 20°C (68°F) with copper having 100 percent conductivity, $\alpha = 0.00393/°C$ (0.00218/°F).

AMPACITY

Current-carrying capacity, or ampacity, measures the current an insulated conductor can safely carry without exceeding its insulation and jacket temperature limitations. This property is unique for each wire, cord, or cable construction, and is influenced by conductor material and cross-sectional area, ambient temperature, thermal conductivity and thickness of insulation, ability of construction to dissipate heat after installation, and the number of conductors in the construction. From a practical point of view, this property is a paramount factor in design of insulated conductors and is of greater significance than conductivity values.

INSULATED CONDUCTOR DESIGN

The design of insulated conductors is often complex and involves several elements:

• *Selection of conductor metal.* The choice is usually between copper and aluminum, and is greatly influenced by cost of the completed construction, and not just metal prices. Aluminum, with 61.8 percent the conductivity of copper, requires larger wire sizes for equal current-carrying capacity, with resultant significant increase in overall diameter and usage of other materials. However, aluminum is only 30 percent the weight of copper, requires no plating (copper does at temperatures over 200°C), and may be anodized, adding dielectric properties and mechanical protection.

• *Choice of solid or stranded conductor.* Stranded wires, used for all but small conductor sizes, provide greater flexibility and flex life and are much easier to install. There are several types of stranded conductors. The true concentric and equilay constructions provide the most circular cross section and are preferred for thin wall extruded insulations. These constructions have alternate layers of wires applied helically in opposite directions over a core wire. In the true concentric construction, each successive layer has an increased length of lay, whereas in equilay stranding, the length of lay is the same in all layers. In unidirectional concentric stranding, all the layers of wire have the same direction of lay and an increased length of lay for each successive layer. Unilay stranding employs more than one layer of helically laid wires with the same direction and length of lay for each layer. This provides the smallest diameter, and therefore the lowest weight of all helical types. Rope stranding consists of a stranded central core with one or more layers of helically laid groups of bunched or concentric stranded wires. The lowest cost construction is bunch stranding, which consists of twisting wire strands in the same direction with the same lay. This type is the most difficult to extrusion coat with thin wall insulations. The effect of stranding is to increase the length of the path through which current travels in a conductor. Thus, for a given circular mils cross-sectional area, stranded construction imparts higher dc resistance than solid wire, and because of the spaces between wires in stranding, solid wire would contain more circular mils than stranded wire of the same AWG.

• *Determination of insulation and jacketing materials.* Important considerations, in addition to dielectric requirements, include resistance to moisture, chemicals, and temperatures to be encountered in operation, as well as specific gravity, mechanical properties, flame resistance and combustion products, and cost. The jacket holds all components together as the outer barrier.

• *Selection of shielding.* Shields are designed to protect against dc magnetic fields and ac frequencies and are required for many constructions:

—To filter out stray signals and low-frequency interference from nearby sources, including power lines,

—To contain conductor signals, preventing their radiation to other nearby conductors,

—To serve as a return wire in coaxial cables,

—To act as a second conductor in matched or tuned lines,

—To level out surge impedance in the conductor,

—To physically protect wires and cables from damage by insects, animals, and other natural and man-made hazards.

There are several types of shields and methods of applications. High-permeability metals, such as permalloy and steel, are most effective against dc and low-frequency (60 Hz) fields. Here, the effectiveness of the shield is directly proportional to its thickness, since the objective is to minimize reluctance or resistance to carrying flux. For higher frequencies, copper and aluminum shields are used, since it is desirable to have high conductivity so that voltages, induced by magnetic flux, generate eddy currents which oppose the action of the flux, restricting its penetration through the shield. Bronze tapes are used in underground telephone cables to protect against damage by termites or rodents. Underground power cable shields are usually copper, while aluminum is used for suspended installations because of its weight advantage. Bimetallic shields are toughest of all and provide the best protection from external damage. Shields may be braided or spiral wrapped wire, tapes, or solid metal, including corrugated, providing 100 percent coverage. Semiconductive textile, paper, and plastic shields are used for low-frequency applications where extreme flexibility or light weight are required. Coverage of 75 to 80 percent is usually adequate for low frequencies. At higher frequencies, coverage should be over 85 percent.

FLOODING AND FILLING COMPOUNDS

To prevent the ingress and movement of water in filled telephone cables, including cables with cellular insulation, a waterproofing gel is injected to fill the interstices between insulated conductors. The filling material normally used by independent companies is a polyethylene/petroleum compound usually referred to as PE/PJ. Bell companies use an extruded thermoplastic rubber called FlexGel, a trademark of AT&T. The related ASTM specifications are:

• D 4730, *Specification for Flooding Compounds for Telecommunications Wire and Cable*

• D 4731, *Specification for Hot-Application Filling Compounds for Telecommunications Wire and Cable*

• D 4732, *Specifications for Cool-Application Filling Compounds for Telecommunications Wire and Cable*

Some aircore cable designs are kept dry by pressurizing the core of the cable with dry air or nitrogen.

FIRE HAZARDS

Conductor insulations which burn or decompose in fires may produce lethal volatile products or smoke capable of causing fatalities in certain locations, such as aircraft, telephone exchanges, power stations, and subways. Federal specifications, industry standards, and building codes are under continuing review to prevent or restrict use of all materials which are hazardous in fire environments. An example is FAA Standard 25.1359, which covers electrical system fire and smoke protection in aircraft. Insulation on electrical cable and wire installed in any area of the fuselage must be self-extinguishing when tested at an angle of 60 degrees; the average burn length may not exceed 3 inches for the same size wire or cable used in the aircraft; and the average flame time after removal of the flame source may not exceed 30 seconds. Drippings from the test specimen may not continue to flame for more than an average of 3 seconds after falling.

Improved grades of most plastics are being marketed by most producers and compounders. Even where not required by code or specification, careful consideration should be given to possible exposure to intense heat or fire in the design of insulated conductors. UL 94, *Flammability Test* and ASTM D 2863, *Oxygen Index Test* are described in Chapter 3, *Thermoplastic Molding Compounds*.

TREEING

The complex phenomenon known as treeing occurs in all solid dielectrics under appropriate conditions of electrical stress, and is stimulated by contaminants and voids within the dielectric itself. Treeing is a breakdown of a dielectric along paths which appear to branch out from a central trunk, similar to limbs on a tree. Trees tend to grow in the direction of electrical stress, radiating from minute contaminants or voids toward or away from the cable conductor. Trees may be either nonvented or vented. In nonvented trees, the branches are contained entirely within the insulation. Vented trees, by far the more troublesome, grow from the outer or inner surfaces of the insulation, allowing gases formed by decomposition of the dielectric to escape. This opens paths for penetration of air, moisture, and other undesirable substances which increase conductivity and encourage tree growth. High ac stresses cause the greater damage.

The two stages of tree growth are *initiation* and *propagation*. Factors which stimulate initiation do not necessarily encourage propagation, and vice versa. *Water trees* result from the presence of water and a voltage stress. *Electrical trees* are formed usually by higher voltage stress, but without the presence of water. The subjects of tree initiation and propagation are being studied intensively by, among others, scientists at Union Carbide, Electric Power Research Institute (EPRI), and the Institut de resherche d'Hydro-Quèbec (IREQ).

Although the incidence of electrical failures directly attributable to treeing is low, there is concern that treeing failures may become more common as buried cable installations continue to increase and age. Most of the studies on treeing have involved polyethylene and crosslinked polyethylene, since trees in these materials are visible and they are the most widely used insulations for buried power cables. Treeing may also occur in filled versions of these materials as well as in other insulations. Treeing becomes a more serious problem as voltages increase. It occurs in ac cables over 600 volts.

PRODUCTION PROCESSES FOR POLYMERS

Fluoropolymers are prepared by suspension and emulsion techniques. Polyolefins are produced by gas-phase or liquid-phase processes. Polyvinyl chloride is made by several methods, including suspension, solution, and mass polymerization processes. Elastomers, often copolymers, are made by all of the above processes.

For further discussion of this subject, see individual and group polymer sections of this chapter and the section "Polymerization" (p. 140) in Chapter 3, *Thermoplastic Molding Compounds*.

EXTRUSION PROCESS

Wires, cords, and cables are most frequently insulated and jacketed by the extrusion process, in which a plastic melt is forced under pressure by a screw through a shaping die in equipment similar to that shown in Fig. 5-1.

The screw compresses and imparts shear energy into the plastic, causing it to flow and mix on its way to the die orifice. Screw design differs from one plastic to another.

Extrusion of corrosive compounds, such as fluoroplastics, requires barrels to be lined with a corrosion-resistant liner.

The manufacture of wires, cords, and cables often includes in-line equipment for curing or sintering (in the case of polytetrafluoroethylene) compounds.

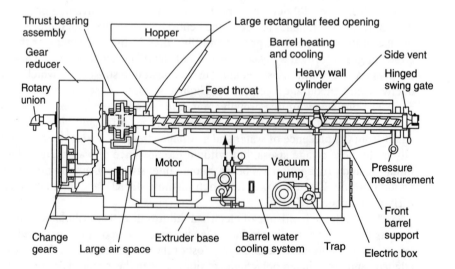

Fig. 5-1 Water-cooled extruder. (*Courtesy HPM Corporation.*)

For further discussion of this subject, see individual and group polymer sections of this chapter.

STANDARD TESTS

Standard tests for extrusion compounds are the same as for thermoplastics, described in Chapter 3, *Thermoplastic Molding Compounds*.

Spark Tester

Extrusion lines for insulated conductors typically include a spark tester. This device applies a voltage to the surface of the insulation to break down and reveal any weak spots, pin holes, or dirt specks. Both ac and dc spark tests are used on 0 to 2,000 volt cables. Spark test voltages vary with insulation wall thickness. For polyethylene insulated cables, the test voltages range from 10,500 volts dc for 45 mils to 42,000 volts ac for 120 mils. For polyvinyl chloride insulated cables, the test voltages range from 7,500 volts dc for 45 mils to 21,000 volts ac for 125 mils. The applicable document is National Electrical Manufacturers (NEMA) Standard WC 5, *Thermoplastic-Insulated Wire and Cable for the Transmission and Distribution of Electrical Energy*.

In most spark testers, the insulated conductor is run between two wet sponges, the conductor being grounded back at the extruder. If a defect exists in the insulation, the voltage on the sponges causes an arc through to the conductor, and a fault registers on the tester. Metal brushes or fine-bead metal chains are often used in place of sponges on testers.

UL STANDARDS FOR WIRE AND CABLE

Underwriters Laboratories has approved the following standards for wire and cable:

- UL 83, *Thermoplastic-Insulated Wires and Cables*
- UL 493, *Thermoplastic-Insulated Underground Feeder and Branch-Circuit Cables*
- UL 543, *Impregnated-Fiber Electrical Conduit*
- UL 854, *Service-Entrance Cables*
- UL 1063, *Machine-Tool Wires and Cables*
- UL 1072, *Medium-Voltage Power Cables*

COMPARATIVE PROPERTIES OF EXTRUDED INSULATIONS

Tables 5-1 through 5-3 show comparative properties for high-temperature insulations, low-temperature insulations, and elastomer insulations.

TABLE 5-1 Comparative Properties of High-Temperature Insulations

PROPERTY	HALON FLUON TEFLON PTFE	TEFLON PFA	TEFLON FEP	HOSTAFLON TEFZEL ETFE	KYNAR PVDF	HALAR ECTFE	KEL-F PCTFE	SILICONE RUBBER[b]
Specific gravity ASTM D 792	2.13–2.20	2.12–2.17	2.14–2.17	1.70	1.76	1.68	2.13	1.15–1.55
Tensile, psi ASTM D 638	4,000	4,300	3,000	7,500	6,000	7,000	5,700	700 (I) 1,800 (J)
Elongation, % ASTM D 638	350	300	300	150	250	200	150	200 (I) 800 (J)
Abrasion resistance[a]	F	P	P	G	G	E	G	F
Cut-thru resistance[a]	F	F	F	E	E	E	na	P
Water resistance[a]	E	E	E	E	G	E	E	G
Max. oper. temp., °C	260	260	205	150	135	165	200	200
Low-temp. embrittlement, °C ASTM D 746	−267	−267	−73	−101	−62	−76	−253	−65
Flame resistance[a]	E	E	E	G	G	E	E	P
Solder iron resistance[a]	E	F	F	G	G	E	na	na
Dielectric constant @ 1 MHz ASTM D 1531	2.1	2.1	2.1	2.6	6.1	2.5	2.5	3.0–3.6 (I)
Dissipation factor @ 1 MHz ASTM D 1531	0.0003	0.0003	0.0007	0.005	0.159	0.009	0.006	na
Volume resistivity, Ω·cm ASTM D 257	$>10^{18}$	$>10^{18}$	$>10^{18}$	$>10^{16}$	2×10^{14}	10^{15}	$>10^{18}$	10^{15} (I)
Dielectric strength, V/mil. 125 mils ASTM D 149	500	550	500	400	260	490	500	na
Relative price per cubic inch[c]	3	5	4	5	2	4	6	1

[a]E—excellent G—Good F—Fair P—Poor

[b](I)—Insulation Grade (J)—Jacketing Grade

[c]1—Lowest 6—Highest

na—not available

TABLE 5-2 Comparative Properties of Low-Temperature Insulations

PROPERTY	POLYVINYL CHLORIDE			POLYETHYLENE				POLYPROPYLENE/ POLYETHYLENE COPOLYMER
	Flexible	Semi-Rigid	Irra-diated	Low Density	High Density	Flame Retard.	Cross-Linked	
Specific gravity ASTM D 792	1.37	1.31	1.34	0.92	0.95	1.30	0.92	0.902
Tensile, psi (min) ASTM D 638	2,200	3,150	4,500	2,200	3,000	1,800	2,200	3,400
Elongation, % (min) ASTM D 638	300	300	150–200	600	400	250	500	400
Abrasion resistance[a]	F-G	G	E	G	G	G	F	G
Cut-thru resistance[a]	G	G	E	G	G	G	F	G
Water resistance[a]	G	G	G	E	E	E	E	E
Max. oper. temp. °C	60–105	80	105	80	90	80	125	90
Brittleness temp. °C (% nonfailure) ASTM D 746	−35 (8/10)	−10 (8/10)	na	−76 (8/10)	−76 (8/10)	−65 (5/10)	−65 (5/10)	−15 (8/10)
Flame resistance[a]	G	G	G	P	P	G	P-G	P
Solder iron resis.[a]	P	P	E	P	P	F	G	P
Dielectric constant @ 1 MHz ASTM D 1531	6.2 max	4.3	2.7	2.28	2.34	2.50	2.30	2.24
Dissipation factor @ 1 MHz ASTM D 1531	na	0.10 max	na	0.0005	0.0007	0.0015	0.0003	0.0003
Volume resistivity Ω·cm ASTM D 257	5×10^{13}	2×10^{14}	2×10^{12}	10^{16}	10^{16}	10^{16}	10^{16}	4×10^{15}
Dielectric strength V/mil ASTM D 149	na	na	na	550	500	na	550	600

[a]E—Excellent G—Good F—Fair P—Poor

na—not available

TABLE 5-3 Comparative Properties of Elastomer Insulations

PROPERTY	NATURAL RUBBER	BUTYL RUBBER	SBR	NEOPRENE	HYPALON	EPDM	TPR
Specific gravity ASTM D 792	0.93	0.90	0.94	1.25	1.18	0.85	0.88
Tensile, psi[a]	1,500 (I) 3,000 (J)	na 2,500 (J)	700 (I) 2,000 (J)	1,250 (I) 3,000 (J)	1,250 (I) 3,000 (J)	700 (I) 2,400 (J)	800 (I) 1,500 (J)
Elongation, %[a] ASTM D 412	350 (I) 800 (J)	na 500 (J)	300 (I) 500 (J)	350 (I) 600 (J)	350 (I) 600 (J)	200 (I) 550 (J)	350 (I) 580 (J)
Abrasion resistance[b]	E	G	G-E	E	E	G	E
Water resistance[b]	F	E	E	F-E	G	G-E	G
Max. oper. temp., °C	82	149	82	116	163	177	125
Min. oper. temp., °C	-51	-46	-51	-40	-40	-51	-70
Flame resistance[b,c]	P	P	P	G	G	P	P
Oil resistance[b]	P	P	P	G	G	P	F
Heat aging[b]	P	F	P	F	G	G+	G+
Resistance to ozone[b]	P	E	P	G+	E	E	E
Dielectric constant @ 60 Hz ASTM D 1531	2.3-3.0	2.1-2.4	3.0-3.5	6-8	6-8	2.5-3.5	2.2
Dissipation factor @ 1 kHz ASTM D 1531	0.0025	0.003	0.003	0.030	0.03-0.07	0.007	na
Volume resistivity Ω·cm ASTM D 257	10^{16}	10^{17}	10^{15}	10^{11}	10^{14}	10^{16}	na
Dielectric strength V/mil 125 mils ASTM D 149	600-800	600	600-800	400-700	400-700	500-800	650
Relative price	Moderate	Moderate	Low	Moderate	Moderate	Low	Low

[a](I)—Insulation Grade (J)—Jacketing Grade

[b]E—Excellent G—Good F—Fair P—Poor

[c]Ratings are for standard electrical grades. Flame-retardant grades are usually available

na—not available

DEVELOPMENT PROGRAMS

Efforts continue to reduce contaminants in extrusion compounds. As an example, Union Carbide is now offering a crosslinkable polyethylene for wire and cable use that is claimed to be at least 75 percent freer from contamination than the firm's existing "clean" grades of polyethylene insulation compounds. The new material, run on modern extrusion lines, is expected to add years of useful life to wire and cable insulated with it.

The compound is made in a facility at Seadrift, TX, which provides enclosed systems and hospital-room cleanliness, a completely integrated process, new equipment and operating procedures, and new contaminant detectors such as the continuous detection Electrical Power Research Institute's pellet inspector which detects contaminants of size less than 0.005 inch.

Intensive development work continues to improve flame-retardant properties of electrical wires and cables. The goal is an insulation with lower rate of fire propagation and with less smoke and toxic gas generation than is currently available.

MARKET TRENDS

Transition is being made from coaxial cable to fiber optic cable for voice and data transmission. The first transoceanic fiber optic cable, TAT-8, has the capacity to carry 40,000 simultaneous voice conversations on two pairs of optic fibers compared with a potential maximum of 9,000 for coaxial cables.

Fiber optic cable thus forestalls the demise of undersea cable predicted with the advent of communications satellites. These satellites, controlled by Communications Satellite Corporation, are a quantum leap forward from coaxial cable. However, because they require an orbit 23,000 miles above the earth, a signal takes a quarter of a second to go from earth to satellite and back to earth, and a response would take as long. Thus, a pause of half a second would occur between "hello" and a reply. This delay becomes significant for data transmission. Fiber optic cable is a much faster method with lower capital cost. Satellites are best suited to transmitting a signal to many widely scattered stations. Fiber optic cable is designed to give the lowest cost and fastest service between two centers with intensive information traffic flow.

Optic fibers are filaments of a high-purity silica glass of 125 μm (about the diameter of a human hair) made by pulling through a die a thin filament drawn from one end of a glass rod heated to its melting point. In one process, the fiber is subsequently coated with polymethyl methacrylate (PMMA) resin and jacketed with a fluoroplastic polymer. In an early process, the primary coating was thermosetting silicone resin and the secondary coating was Nylon 12. Still another, newer, process utilizes in-line technology where a filament is drawn, coated with an ultraviolet (UV) light-curing polymer, the coating cured, and the filament wound on a drum. The coating must have sufficient mechanical strength to withstand high-speed processing, have stable transmission characteristics, and be easily strippable when it is necessary to splice and to test the cable.

In operation, the laser, a high-energy beam of monochromatic light is developed in a cylinder of crystalline material, such as a ruby, or in a tube of gas, for example, carbon dioxide. A high-intensity light source feeds energy into the crystal cylinder or gas tube, where photons are formed. The ends of the device are mirrored to reflect the excited photons, which are finally emitted through a small hole at the "muzzle" end of the device in a single, precise frequency, monochromatic beam. An optical-integrated circuit turns the laser beam on and off thousands of times per second, with pulses formed by a digital code into which a caller's voice or facsimile signal is translated. At the receiving end, a photodetector converts the pulses into electrical signals, which are then separated into two digital codes—one for voice and one for facsimile. Commercial lasers pulse at up to 400 million times a second, enough to transmit 6,000 calls over a single glass fiber. The applicable ASTM guide is D 4967, *Guide for Selecting Materials to be Used for Insulation Jacketing, and Strength Components in Fiber Optic Cables*.

The growing trend toward housing developments, industrial parks, and shopping malls ensures increased underground cable installations at the expense of aerial cables. Underwriters Laboratories has approved the following standards for optical-fiber cables:

• UL 910, *Test Method for Fire and Smoke Characteristics of Electrical and Optical-Fiber Cables Use in Air-Handling Spaces*
• UL 1666, *Flame Propagation Height of Electrical and Optical-Fiber Cables Installed Vertically in Shafts*

PRINCIPAL EXTRUSION POLYMERS FOR ELECTRICAL/ELECTRONIC USES

The chemistry, grades, processing, properties, and electrical/electronic uses of principal extrusion polymers are described in the sections which follow. The values shown in tables are for general purpose grades except as noted.

Polymers are presented in this order:

• Fluoropolymers
• Ethylene Polymers (PE)
• Propylene Polymers (PP)
• Polyurethane Polymers (PUR)
• Polyvinyl Chloride (PVC)
• Silicone Rubber
• Organic Elastomers

Fluoropolymers
(Extrusion Resins)

The first fluorocarbon, polytetrafluoroethylene (PTFE) was discovered at Du Pont in 1938, but because early production was devoted to military products in World War II, it was not commercially available until 1947. Since that time, the family of fluoropolymers has grown to include several materials which are discussed in this section. These materials all have in common the widest operating temperature ranges of all plastics, outstanding resistance to attack by chemicals, excellent dielectric properties, and virtually no tendency to absorb water. These characteristics are attributable to the carbon-fluorine bond, among the strongest known in organic chemistry, and much stronger, for example, than the carbon-chlorine bond.

These plastics are soft, tough materials with room temperature mechanical properties similar to medium density polyethylene. With the exception of polytetrafluoroethylene, fluoroplastics can be processed by conventional thermoplastic molding and extrusion techniques, although processing temperatures are relatively high and equipment must be corrosion resistant. Polytetrafluoroethylene must be processed by sintering at 370°C (700°F) following ram or paste extrusion.

A distinction is sometimes made among the terms *fluoropolymers*, *fluorocarbons*, and *fluoroplastics*. *Fluoropolymers* is the generic term and includes fluorocarbons and fluoroplastics. *Fluorocarbons* consist of polymers with only carbon-fluorine bonds. *Fluoroplastics* may have, in addition to carbon-fluorine bonds, carbon-hydrogen and carbon-chlorine bonds.

In general, fluoroplastics that are not fully fluorinated have a narrower useful temperature range, lower dielectric properties, and less resistance to attack by some chemicals. However, on the plus side, fluoroplastics, within their useful temperature range, are stiffer and have better resistance to cut-through, deformation under load, and creep.

Table 5-4 lists the principal fluoropolymers in common use. Comparative properties are shown in Table 5-1 (p. 142) in this chapter.

ASTM Standards for Fluoropolymer Extrusion Compounds

The American Society for Testing and Materials (ASTM) has developed the following standards:

- D 1430, *Specification for Polychlorotrifluoroethylene (PCTFE) Plastics*
- D 1457, *Specification for PTFE Molding and Extrusion Materials*
- D 2116, *Specification for FEP-Fluorocarbon Molding and Extrusion Materials*
- D 3159, *Specification for Modified ETFE-Fluoropolymer Molding and Extrusion Materials*
- D 3222, *Specification for Unmodified Poly(Vinylidene) Fluoride (PVDF) Molding, Extrusion, and Coating Materials*
- D 3275, *Specification for E-CTFE-Fluoroplastic Molding, Extrusion, and Coating Materials*
- D 3307, *Specification for PFA-Fluorocarbon Molding and Extrusion Compounds*
- D 4745, *Specification for Filled Compounds of Polytetrafluoroethylene (PTFE) Molding and Extrusion Materials*
- D 4894, *Specification for Polytetrafluoroethylene (PTFE) Granular Molding and Ram Extrusion Materials*

TABLE 5-4 Summary of Characteristics of Fluoropolymers

FLUOROPOLYMER	SUPPLIERS	CHEMISTRY	ADVANTAGES	LIMITATIONS
Polytetrafluoro-ethylene (PTFE)	Ausimont Daikan Du Pont Hoechst-Celanese ICI	Tetrafluoroethylene (TFE) made by pyrolysis of chlorodifluoro-methane is polymerized to PTFE in aqueous medium by agitation (for molding uses) or by dispersion (for paste extrusion) using an emulsifying agent. PTFE is 76% fluorine, 24% carbon. $2\ CHClF_2 \xrightarrow{\hspace{1cm}} CF_2{=}CF_2 + 2\ HCL$ $nCF_2{=}CF_2 \xrightarrow{\hspace{1cm}} {-}(CF_2{-}CF_2)_{n}$	Excellent dielectric properties. Widest useful temperature range of all plastics. Unaffected by virtually all chemicals. Low coefficient of friction. Oxygen index over 95%. Excellent solder iron resistance.	Cannot be processed by melt techniques. Must be processed by sintering at 370°C following ram or paste extrusion, or coating on glass fabric or metal. Poor resistance to corona and radiation.
Perfluoroalkoxy (PFA)	Du Pont	The PFA molecular structure is, like PTFE, a straight chain, but with a perfluoroalkane group with an oxygen linkage to the chain. $-\left[(CF_2{-}CF_2)_n\,CF_2{-}CF\right]_m$ $\qquad\qquad\qquad\quad O$ $\qquad\qquad\qquad\quad C_nF_{2n+1}$	Dielectric, mechanical, and chemical properties equal to PTFE. Can be processed by conventional molding and extrusion techniques. Better mechanical properties than FEP above 150°C (302°F).	Higher in price than PTFE.
Fluorinated ethylene-propylene copolymer (FEP)	Du Pont	FEP is a copolymer of hexafluoro-propylene and tetrafluoroethylene, prepared by several methods including aqueous and nonaqueous dispersion polymerizations similar to those used to make PTFE. $-\left[(CF_2{-}CF_2)_n\,CF_2{-}CF\right]_m$ $\qquad\qquad\qquad\qquad\quad CF_3$	Dielectric, mechanical, chemical properties equal to PTFE and PFA except narrower useful temperature range. Can be processed by conventional molding and extrusion techniques. Oxygen index of 95%.	Narrower useful temperature range than PTFE or PFA. Not as tough as PTFE.

Polymer	Manufacturer	Description	Properties	Limitations
Ethylene-tetrafluoroethylene (ETFE)	Ausimont Du Pont	ETFE is a modified copolymer of ethylene and TFE. $\left(\text{CH}_2\text{—CH}_2\text{—CF}_2\text{—CF}_2\right)_n$	Mechanical properties better than other fluoroplastics except PCTFE. Lower maximum operating temperature than PTFE, PFA, FEP, and ECTFE. Can be processed by conventional molding and extrusion techniques.	Dielectric properties not quite as good as PTFE, PFA, and FEP. Oxygen index much lower (28–32%).
Polyvinylidene fluoride (PVDF)	Ausimont Elf Autochem Solvey Polymer	PVDF is made by free radical chain polymerization of vinylidene fluoride using both emulsion and suspension polymerization techniques. $\left(\text{CH}_2\text{—CF}_2\right)_n$	PVDF can be solvated by organic esters and ketones as well as processed by conventional molding and extrusion techniques. High abrasion resistance and resistance to UV and nuclear radiation. Oxygen index 44%. Low price.	Chemical resistance, dielectric properties not as high as other fluoroplastics, with narrowest useful temperature range.
Ethylene-chloro-trifluoro ethylene copolymer (ECTFE)	Ausimont	ECTFE is a copolymer of ethylene and chlorotrifluoroethylene in a 1:1 alternating structure. $\left(\text{CH}_2\text{—CH}_2\text{—CF}_2\text{—CF}\right)_n$ with Cl	Highest tensile strength and abrasion resistance of all fluoroplastics. Good dielectric properties. Can be processed by conventional molding and extrusion techniques. Oxygen index of 60. Lowest specific gravity of all fluoroplastics.	Attacked by hot amines (otherwise good chemical resistance).
Polychlorotri-fluoroethylene (PCTFE)	3M Elf Autochem	PCTFE is produced by suspension or emulsion polymerization of chlorotrifluoroethylene. The monomer is prepared by dechlorination of trichloro-trifluoroethane in liquid phase in the presence of a metallic zinc slurry in methanol, or in a hydrogen gas phase reaction with nickel catalyst at elevated temperature. $\left(\text{CF}_2\text{—CF}\right)_n$ with Cl	Excellent resistance to oxygen, ozone, sunlight, and fuming oxidizing acids. Good abrasion resistance. Excellent gas and liquid barrier. Can be processed by conventional molding and extrusion techniques. Good dielectric properties.	Precise control of temperature required in processing to prevent degradation. Attacked by some halogenated solvents. Price is highest of all fluoroplastics.

Federal Specifications for Fluoropolymer Extrusion Compounds

The following Federal specification has been developed for fluoropolymer extrusion compounds:

 • MIL-P-46122, *Plastic Molding Material and Plastic Extrusion Material, Polyvinylidene Fluoride Polymer and Copolymer*

Federal Specifications for Fluoropolymer Insulated Wire

The following Federal specifications have been developed for fluoropolymer insulated wire:

Polytetrafluoroethylene (PTFE)
 • MIL-W-16878

   /4B Wire, Electrical PTFE Insulated, 200 Deg C, 600 Volts, Extruded Insulation

   /23 Wire, Electrical, PTFE Insulated, 260 Deg C, 250 Volts, Extruded Insulation

   /25 Wire, Electrical, PTFE Insulated, 260 Deg C, 600 Volts, Extruded Insulation

   /27 Wire, Electrical, PTFE Insulated, 260 Deg C, 1000 Volts, Extruded Insulation

 • MIL-W-22759

   /1E Wire, Electric, Fluoropolymer-insulated, TFE and TFE-coated-glass, Silver-coated Copper Conductor, 600 Volts

   /2E Wire, Electric, Fluoropolymer-insulated, TFE and TFE-coated-glass, Nickel-coated Copper Conductor, 600 Volts

   /3D Wire, Electric, Fluoropolymer-insulated, TFE-glass-TFE, Medium Weight, Nickel-coated Copper Conductor, 600 Volts

   /4B Wire, Electric, Fluoropolymer-insulated, TFE-glass-FEP, Silver-coated Copper Conductor, 600 Volts

   /5B Wire, Electric, Fluoropolymer-insulated, Abrasion Resistant, Extruded TFE, Silver-coated Copper Conductor, 600 Volts

   /6B Wire, Electric, Fluoropolymer-insulated, Abrasion Resistant, Extruded TFE, Nickel-coated Copper Conductor, 600 Volts

   /7B Wire, Electric, Fluoropolymer-insulated, Abrasion Resistant, Extruded TFE, Medium Weight, Silver-coated Copper Conductor, 600 Volts

   /8B Wire, Electric, Fluoropolymer-insulated, Abrasion Resistant, Extruded TFE, Medium Weight, Nickel-coated Copper Conductor, 600 Volts

   /9B Wire, Electric, Fluoropolymer-insulated Extruded TFE, Silver-coated Copper Conductor, 1000 Volts

   /10B Wire, Electric, Fluoropolymer-insulated Extruded TFE, Nickel-coated Copper Conductor, 1000 Volts

   /11F Wire, Electric, Fluoropolymer-insulated, Extruded TFE, Silver-coated Copper Conductor, 600 Volts

/12F Wire, Electric, Fluoropolymer-insulated, Extruded TFE, Nickel-coated Copper Conductor, 600 Volts

/20 Wire, Electric, Fluoropolymer-insulated, Extruded TFE, Silver-coated High Strength Copper Alloy Conductor, 1000 Volts

/21 Wire, Electric, Fluoropolymer-insulated Extruded TFE, Nickel-coated High Strength Copper Alloy Conductor, 1000 Volts

/22 Wire, Electric, Fluoropolymer-insulated Extruded TFE, Silver-coated High Strength Copper Alloy Conductor, 600 Volts

/23 Wire, Electric, Fluoropolymer-insulated Extruded TFE, Nickel-coated High Strength Copper Alloy Conductor, 600 Volts

/24 Wire, Electric, Fluoropolymer-insulated, Extruded TFE, Polyimide-coated, Silver-coated Copper Conductor, 600 Volts

/29 Wire, Electric, Fluoropolymer-insulated, Extruded TFE, Polyimide-coated, Nickel-coated Copper Conductor, 600 Volts

/30 Wire, Electric, Fluoropolymer-insulated, Extruded TFE, Polyimide-coated, Silver-coated High Strength Copper Alloy Conductor, 600 Volts

/31 Wire, Electric, Fluoropolymer-insulated, Extruded TFE, Polyimide-coated, Nickel-coated High Strength Copper Alloy Conductor, 600 Volts

Fluorinated Ethylene-Propylene Copolymer (FEP)

• MIL-W-22759

/13D Wire, Electric, Fluoropolymer-insulated, FEP-PVF2, Medium Weight, Tin-coated Copper Conductor, 600 Volts

/14B Wire, Electric, Fluoropolymer-insulated, FEP-PVF2, Lightweight, Tin-coated Copper Conductor, 600 Volts

/15B Wire, Electric, Fluoropolymer-insulated, FEP-PVF2, Lightweight, Silver-Plated High Strength Copper Alloy, 600 Volts

Ethylene-Tetrafluoroethylene (ETFE)

• MIL-W-22759

/17A Wire, Electric, Fluoropolymer-insulated, Extruded ETFE, Medium Weight, Silver-coated High Strength Copper Alloy Conductor, 600 Volts, 150 Deg C

/19A Wire, Electric, Fluoropolymer-insulated, Extruded ETFE, Lightweight, Silver-coated High Strength Copper Alloy Conductor, 600 Volts, 150 Deg C

/32A Wire, Electric, Fluoropolymer-insulated, Crosslinked Modified ETFE, Lightweight, Tin-coated Copper Conductor, 150 Deg C, 600 Volts

/33A Wire, Electric, Fluoropolymer-insulated, Crosslinked Modified ETFE, Lightweight, Silver-coated Copper Conductor, 150 Deg C, 600 Volts

/34B Wire, Electric, Fluoropolymer-insulated, Crosslinked Modified ETFE, Normal Weight, Tin-coated Copper Conductor, 150 Deg C, 600 Volts

/35B Wire, Electric, Fluoropolymer-insulated, Crosslinked Modified ETFE, Normal Weight, Silver-coated High Strength Copper Alloy Conductor, 200 Deg C, 600 Volts

/41A Wire, Electric, Fluoropolymer-insulated, Crosslinked Modified ETFE, Normal Weight, Nickel-coated Copper Conductor, 200 Deg C, 600 Volts

/42 Wire, Electric, Fluoropolymer-insulated, Crosslinked Modified ETFE, Normal Weight, Nickel-coated High Strength Copper Alloy, 200 Deg C, 600 Volts

/43A Wire, Electric, Fluoropolymer-insulated, Crosslinked Modified ETFE, Normal Weight, Silver-coated Copper Conductor, 200 Deg C, 600 Volts

NEMA Standards for Fluoropolymer Insulated Wire

The following NEMA standards have been developed for fluoropolymer insulated wire:

- HP 100, *High Temperature Instrumentation and Control Cables*
- HP 100.1, *High Temperature Instrumentation and Control Cables Insulated with FEP Fluorocarbons*
- HP 100.2, *High Temperature Instrumentation and Control Cables Insulated and Jacketed with ETFE Fluoropolymers*
- HP 100.3, *High Temperature Instrumentation and Control Cables Insulated and Jacketed with Cross-Linked (Thermoset) Polyolefin (XLPO)*
- HP 100.4, *High Temperature Instrumentation and Control Cables Insulated and Jacketed with ECTFE Fluoropolymers*
- HP 3, *Electrical and Electronic PTFE (Polytetrafluoroethylene) Insulated High Temperature Hook-Up Wire; Types (600 Volt), EE (1000 Volt), and ET (250 Volt)*
- HP 4, *Electrical and Electronic FEP Insulated High Temperature Hook-Up Wire, Types K, KK, and KT*

Grades

- Polytetrafluoroethylene is supplied as granular resins, both filled and unfilled, for ram extrusion and compression molding, as aqueous dispersions for coating substrates, and as fine powders for paste extrusion. Suppliers are Ausimont, USA (Halon™, Algoflon™), Daikan (Polyfluon™), Du Pont (Teflon™ TFE), Hoechst Celanese (Hostaflon™), and ICI Americas (Fluon™).

- Perfluoroalkoxy is available as pellets for molding and extrusion, and as powders for coatings. Supplier is Du Pont (Teflon™ FPA).

- Fluorinated ethylene-propylene copolymer is furnished as pellets, as aqueous dispersions, and as powders. Supplier is Du Pont (Teflon™ FEP).

- Ethylene-tetrafluoroethylene is sold as pellets, with and without glass fiber reinforcement, and as powders. Suppliers are Ausimont, USA (Halon ET™), and Du Pont (Tefzel™).

- Polyvinylidene fluoride may be purchased as pellets, both filled and unfilled, as dispersions, and as powders. Suppliers are Autochem (Foraflon™) and Pennwalt (Kynar™).

• Ethylene-chlorotrifluoroethylene copolymer comes as pellets, as powders for rotomolding and coatings, and as a foamable grade. Supplier is Ausimont, USA (Halar™).

• Polychlorotrifluoroethylene is offered as pellets, with and without glass fiber reinforcement. Supplier is 3M Company (Kel-F™).

Processing

Methods used to process fluoropolymers include the following:

• Conventional injection molding and extrusion techniques may be successfully employed with due provision made for higher processing temperatures than are required for other plastics and for the highly corrosive nature of fluoropolymers. Not used for polytetrafluoroethylene.

• Paste extrusion is the process used to convert polytetrafluoroethylene fine powder into thin wire insulation. Particles about 0.2 micron in diameter are blended with naphtha or similar lubricant. In coating wire, the wire is held in a hollow mandrel and fed through the ram of the extruder and not through a crosshead as in the case of melt extrusion. The wire meets the lubricated polymer at the entrance to the die, making it essential that the rate of travel of the wire is the same as the rate of extrusion. The coated wire then passes through an oven in line where the polytetrafluoroethylene is sintered to a homogeneous film with thickness in the range of 0.005 to 0.010 inch. Another method involves extruding a thin film which may be slit into tape and served on wire with subsequent sintering. Sintering temperatures exceed 620°F (326°C).

• Billets, or cylinders, of polytetrafluorethylene weighing up to 1500 pounds may be sintered in ovens at 698°F (370°C) from preforms of granules. These billets may then be precisely skived to films as thin as 0.001 inch or to sheets as thick as 0.1 inch.

• Fluorocarbons, normally not cementable, may be made cementable by special treatment with molten caustic.

Electrical/Electronic Uses

The principal electrical/electronic use for fluoropolymers is for wire and cable insulation where no other materials suffice. Examples include:

• Low-smoke, low-flame-spread communication wire and cable in plenum areas (PTFE), (FEP), (ECTFE)
• Fire alarm cable (PTFE), (FEP)
• Oil well logging cable (FEP)
• Back panel wiring in computers (FEP), (ETFE), (PVDF)
• Hookup wire for aerospace and mass transit applications (ETFE), (ECTFE), (FEP), (PVDF)
• Wiring in nuclear power plants (ETFE), (ECTFE)
• Coaxial cable (ECTFE), (PTFE), (FEP)
• Appliance wire (ECTFE)

- Motor lead wire (ECTFE)
- Down hole cables (FEP)
- Cathodic protection lead wire (ECTFE)

In addition, there are many other demanding applications, if not high volume, for both molded and extruded fluoropolymers.

Ethylene Polymers (PE)
(Extrusion Resins)

From pilot plant production in the early 1930s, polyethylene has become by far the most versatile wire and cable insulation. Polyethylene for this end use is largely the domain of Union Carbide, with Dow Chemical the only supplier of chlorinated polyethylene (Dow also supplies other types). Other suppliers of dielectric grade polyethylene include Du Pont (Alathon™) and Quantum Composites (Petrothene™).

Chemistry

Low-density polyethylene (LDPE), the first polyethylene resin, is produced by polymerizing ethylene gas ($CH_2 = CH_2$) into long polymer chains in an autoclave or tubular reactor. Polymerization is initiated by a small amount of oxygen or other free radical source. The conventional process requires pressures in the 15,000 to 45,000 psi range. A newer process involves gas-phase fluidized-bed reactors or liquid-phase solution process reactors. The material produced by these methods is designated linear low density polyethylene (LLDPE).

High-density polyethylene (HDPE) is a later development, dating from the early 1950s. A catalyzed gas-phase system with pressures under 1,500 psi is one production method. High-density polyethylene may also be produced in solution processes or as dispersions in hydrocarbon diluents. High-density polyethylene resins are characterized by a linear structure and exhibit a high degree of crystallinity.

Medium-density polyethylene (MDPE) may be made by any of the above processes or by blending low-density polyethylene and high-density polyethylene. ASTM has developed definitions for four types of polyethylene (ASTM D 1248, *Specification for Polyethylene Molding and Extrusion Materials*):

TYPE	NOMINAL DENSITY (G/CM³)
I	0.910–0.925
II	0.926–0.940
III	0.941–0.959
IV	0.960 and higher

Polyethylenes are classified not only by density, but also by molecular weight. High-density polyethylene resins with molecular weights averaging in the 300,000–500,000 range are designated high-molecular-weight high-density polyethylene (HMW-HDPE). This version may be made in a gas-phase fluidized-bed reactor or in a gas-phase system using catalysts such as aluminum trialkyl/titanium tetrachloride, or chromium oxide on a silica support. Solution polymerization is not a practical method because when the polymer approaches the high-molecular-weight range, it precipitates. High-molecular-weight high-density polyethylene polymers are predominantly linear. Although the average molecular weight is 300,000–500,000, molecular weight distribution is broad, with a significant proportion of lower molecular weight molecules to facilitate processing.

A high-density polyethylene with molecular weight in the range of 3–6 million is called ultrahigh-molecular-weight polyethylene (UHMWPE). This polymer is produced by a system similar to that used for high-density polyethylene. Ultrahigh-molecular-weight polyethylene, because of its extremely high molecular weight, cannot be processed by conventional techniques.

Thus, a wide range of properties may be obtained by different manufacturing methods, by varying the type and amount of modifier and catalyst, and by controlling process temperature and pressure. A highly crystalline molecular structure, as in high-density polyethylene, results from minimum side chain branching. Branching is controlled by varying process conditions, and also by adding comonomers such as propylene, butene, and hexene. Higher crystallinity increases tensile strength, rigidity, chemical resistance, and opacity, but reduces permeability to liquids and gases.

Polyethylene properties may be modified by additives and fillers, usually mixed with resins during the extrusion process. Antioxidants are important to prevent degradation of polymer properties during manufacture and in service. These include amines, hindered phenols, and phosphites. Compounds for wire and cable insulation may include metal deactivators. Degradation by ultraviolet light is inhibited to a remarkable degree by including in formulations less than 3 percent of carbon black, which in larger amounts (up to 30 percent) can be incorporated to reduce cost. Special carbon blacks provide semiconductive properties for shielding power cables. Inherent flammability of polyethylene is significantly reduced, at the expense of physical properties and processing ease, by additives such as antimony oxide, aluminum trihydrate, and halogenated compounds. Other additives act as slip, antiblock, and antistatic agents.

Polyethylene may be crosslinked (XLPE), or vulcanized, to a thermoset material by addition of $1\frac{1}{2}$ to 2 percent dicumyl peroxide to formulations, or by electron beam radiation, both of which are continuous in-line processes following extrusion of polyethylene on conductors.

Ethylene vinylsilane copolymers are the latest crosslinkable materials designed for wire and cable insulations. The silane is usually a vinylsilane with three functional groups: $H_2C = CHSi(OR)_3$, where the R group can be one of several functional methyl, ethyl, or acetyl groups. The double bond of the vinyl group enables the silane to be attached to or incorporated into the copolymer. The copolymer may be formed either by grafting with a peroxide initiator or by copolymerization directly in a reactor used to produce low density polyethylene. Both graft and copolymerized materials crosslink upon exposure to moisture and a catalyst, such as methyl carboxylate. After extrusion, crosslinking is accomplished in a steam room (requiring a few hours), hot water (16–24 hours at 90°C), or by humidity during storage (which takes up to a few weeks).

High-density polyethylene may be chlorinated in a slurry to a polymer containing 25 to 48 percent chlorine. Chlorinated polyethylene (CPE), in concentrations of 3 to 8 percent, is often blended with polyvinyl chloride to improve impact properties. With over 42 percent chlorine, chlorinated polyethylene is fire retardant.

Ethylene is frequently copolymerized with vinyl acetate. Copolymers are referred to as polyethylene or modified polyethylene when vinyl acetate content is less than 5 percent. With 5 to 50 percent vinyl acetate content, the copolymer is called ethylene-vinyl acetate (EVA). Above fifty percent concentration, the resins are considered vinyl acetate-ethylene compolymers (VAE). Low-density polyethylene jacketing compounds for telephone cable typically contain 2 percent vinyl acetate. As vinyl acetate content increases, density increases and crystallinity decreases. Higher density results in lower stiffness, greater toughness, lower melting point, and higher permeability.

Another technique for changing polyethylene properties is injecting nitrogen into molten polymer or compounding with a blowing agent. The result with either process is lightweight cellular insulation with reduced dielectric constant. This material is used in some filled "exchange" communications cables installed between control offices of the telephone companies. Filled cables are discussed in the section of this chapter entitled "Flooding and Filling Compounds." A potential problem using these compounds with cellular insulation is to prevent the migration of the compound into the insulation where it adversely affects the dielectric and physical properties of the insulation.

ASTM Standards for Extruded Polyethylene Compounds

The American Society for Testing and Materials has developed the following standards for extruded polyethylene and chlorinated polyethylene compounds:

For polyethylene:

- D 1248, *Polyethylene Plastics Molding and Extrusion Materials*
- D 1351, *Polyethylene Insulated Wire and Cable*
- D 2308, *Polyethylene Jacket for Electrical Insulated Wire and Cable*
- D 2647, *Crosslinkable Ethylene Plastics*
- D 2655, *Crosslinked Polyethylene Insulation for Wire and Cable Rated 0 to 2000 V*
- D 2656, *Crosslinked Polyethylene Insulation for Wire and Cable Rated 2001 to 35,000 V*
- D 3485, *Smooth-Wall Coilable Polyethylene (PE) Conduit (Duct) for Pre-assembled Wire and Cable*
- D 3554, *Track-Resistant Black Thermoplastic High-Density Polyethylene Insulation for Wire and Cable*
- D 3555, *Track-Resistant Black Crosslinked Polyethylene Insulation for Wire and Cable*
- D 4020, *Ultra-High-Molecular-Weight Polyethylene Molding and Extrusion Compounds*
- D 4976, *Polyethylene Plastics Molding and Extrusion Materials*

And for chlorinated polyethylene:

- D 4313, *General-Purpose Heavy-Duty and Extra-Heavy-Duty Crosslinked Chlorinated Polyethylene Jackets for Wire and Cable*
- D 4314, *General-Purpose Heavy-Duty and Extra-Heavy-Duty Crosslinked Chlorosulfonated Polyethylene Jackets for Wire and Cable*

Military Specifications for Polyethylene Insulated Wire

The following Military specifications have been developed for polyethylene insulated wire:

- MIL-W-16878

  /10C Wire, Electrical, Polyethylene-Insulated, 75 Deg C, 600 Volts

  /14 Wire, Electrical, Modified Polyethylene (XLPE) Insulated, 125 Deg C, 600 Volts

  /15 Wire, Electrical, Modified Polyethylene (XLPE) Insulated, 125 Deg C, 1000 Volts

  /16 Wire, Electrical, Modified Polyethylene (XLPE) Insulated, 125 Deg C, 3000 Volts

  /33 Wire, Electrical, Polyethylene-Insulated, 75 Deg C, 600 Volts, Polyamide Covering, Abrasion Resistant

- MIL-W-81044

  /6B Wire, Electric, Crosslinked Polyalkene-insulated, Tin-coated Copper Conductor, Normal Weight, 150 Deg C

  /7B Wire, Electrical, Crosslinked Polyalkene-insulated, Silicone-coated, High Strength Copper Alloy Conductor, Normal Weight, 600 Volts, 150 Deg C

  /9B Wire, Electric, Crosslinked Polyalkene-insulated, Tin-coated, Medium Weight, 600 Volts, 150 Deg C

  /10B Wire, Electric, Crosslinked Polyalkene-insulated, Silicone-coated, High Strength Copper Alloy Conductor, Medium Weight, 600 Volts, 150 Deg C

  /12B Wire, Electric, Crosslinked Polyalkene-insulated, Tin-coated Copper Conductor, Light Weight, 600 Volts, 150 Deg C

  /13B Wire, Electric, Crosslinked Polyalkene-insulated, Silicone-coated, Light Weight, 600 Volts, 150 Deg C.

Federal, ICEA/NEMA, AEIC, ANSI, and UL Standards and Listings

Governmental and industry organizations (in addition to ASTM and the Military) have developed the following specifications and listings for polyethylene insulated wire and cable:

- Federal Specification

  L-P-390C: covers PE and copolymers molding and extrusion compounds for use as a dielectric.

- Insulated Cable Engineers Association (ICEA) National Electrical Manufacturers Association (NEMA)

    S-61-402/WC 5 covers thermoplastic-insulated wire and cable products

    S-66-524/WC 7 covers crosslinked-thermosetting-polyethylene-insulated wire and cable products

    S-68/WC 8 covers ethylene-propylene insulated wire and cable products

- Association of Edison Illuminating Companies (AEIC)

    CS5-82: covers thermoplastic and crosslinked PE insulated shielded power cables rated 5,000 to 69,000 volts.

- American National Standards Institute (ANSI)

    C8.35-1975: covers weather-resistant wire and cable—PE type.

- Underwriters Laboratories (UL)

    UL 83: covers thermoplastic-insulated wires and cables.

Grades

As may be inferred from the above discussion, countless grades of polyethylene with its many types, processes, additives, and comonomers are available. A commonly used indicator of average molecular weight in grade descriptions is melt index (MI), which measures the amount in grams of compound that can be forced through an 0.0825 inch orifice when subjected to a 2,160 gram force in 10 minutes at 374°F (190°C), ASTM D 1238. Thus, there is an inverse relationship between MI and molecular weight—a compound with low MI value has high average molecular weight. Density is another indicator included in grade descriptions (see discussion under "Chemistry" (p. 154) in this section).

Grades are available to meet the requirements for wire and cable insulations covered by industry and government specifications listed above.

Halogen-free grades are available which minimize flame propagation, smoke generation, and release of toxic gases. These grades are processed at a relatively low temperature of approximately 347°F (175°C) to avoid degradation of the alumina trihydrate usually blended in at levels of about 65 percent by weight.

Processing

The basic process of extrusion coating polyethylene on a conductor consists of pulling the preheated conductor through an annular die which forms molten compound around the conductor with a coating of 0.005 inch to over 1.0 inch. For good results, it is essential that the coating be of uniform thickness to provide a high degree of concentricity for the conductor. It is also essential that the manufacturing environment be scrupulously clean and free from contaminants.

Linear low-density polyethylene processes more stiffly than does comparable melt index low-density polyethylene. In screw design for linear low-density polyethylene, a longer transition section and shorter metering section are recommended than for low-density polyethylene.

In the extrusion of high-density polyethylene, critical control of melt temperatures is required to produce high-quality insulated conductors. In extruding high-molecular-weight high-density polyethylene, melt viscosities make process controls even more critical. Ultrahigh-molecular-weight polyethylene cannot be extruded or injection molded and must be compression molded.

Following extrusion, the coated conductor passes through a water-cooling trough where the temperature, based on experience, is closely controlled to optimize properties of the finished insulated conductor.

In the manufacture of cable insulated with crosslinkable (vulcanizable) polyethylene, the hot compound must be extruded around the conductor below the rapid reaction temperature in the 230° to 270°F (110°–132°C) range. The conductor coated with the uncured compound and frequently also with an extruded shield is then processed by one of several methods to cure the insulation, including the following:

- The cable is passed into a steam tube with pressure at 200 to 300 psi to bring the insulation to reaction temperature of 350° to 450°F (176°–232°C). All variables must be carefully controlled at this stage for good results. To prevent formation of voids from decomposition of the peroxide crosslinking agent, the cable is finally passed through a chamber where it is cooled under high-pressure water.
- A hot salt bath is used to cure the compound.
- Radiant heat is used in nitrogen atmosphere.
- Electron beam radiation is used for 600 volt cables. High-voltage cables have not been cured by irradiation because of thick walls and treeing inception from trapped electrons.

Properties

Important factors affecting the properties of polyethylene are degree of crystallinity, molecular weight, and molecular weight distribution. In general, polyethylene exhibits excellent dielectric properties which are independent of frequency or temperature. Water vapor permeability is low, but permeability to air and other gases is high. Toughness is retained at low temperatures. However, flammability is high except for special formulations, heat distortion temperatures are quite low, and load-bearing characteristics are poor. Resistance to weathering is low, although this can be improved significantly by addition to recipes of ultraviolet inhibitors and 2.6 percent carbon black.

Chemical resistance is generally good and increases with polymer density. Polyethylene is attacked by organic solvents, halogenated hydrocarbons, and concentrated oxidizing acids.

Crosslinking renders polyethylene infusible and suitable for service temperatures up to 125°C (257°F), compared with 75°C (167°F) for noncrosslinked polyethylene insulation, thus permitting cables to carry higher current densities. Resistance to cold flow and abrasion are superior to conventional polyethylene, while dielectric properties are comparable. Another advantage of crosslinking is that it makes possible higher filler loading without significant loss of physical properties.

Cellular polyethylene may have varying degrees of foaming, with higher gas volumes tending to reduce the dielectric constant toward that of air ($K = 1$). At a gas volume of 50 percent, the dielectric constant becomes about 1.5, significantly lower than even the best solid dielectric, Teflon™ TFE with $K = 2.1$. With cellular polyethylene, care must be taken to prevent cell filling with moisture or cable filler compound.

Comparative properties with other low temperature insulations are shown in Table 5-2 (p. 143) in this chapter.

Electrical/Electronic Uses

By far the principal electrical use for polyethylene is as jacket material and primary insulation for wire and cable. High-molecular-weight low-density polyethylene and increasingly linear low-density polyethylene with carbon black are used extensively for cable jackets. Aging characteristics are excellent. High-molecular-weight polymers and closely controlled molecular weight distribution and chain branching contribute to high stress crack resistance. These compounds meet REA Specification PE-200 for cable jacketing material. High-molecular-weight low-density polyethylene is also used for primary insulation, including crosslinkable and foamable grades.

High-molecular-weight high-density polyethylene is used principally as primary insulation, where it competes with polypropylene. High-molecular-weight high-density polyethylene may be formulated to exhibit excellent oxidative stability, toughness, and abrasion resistance. Resistance to environmental degradation and thermal stress cracking is high, and superior dielectric properties are maintained. Foamable grades are available.

Propylene Polymers (PP)
(Extrusion Resins)

Polypropylene, closely related to polyethylene both chemically and in end uses, was a later development, becoming commercially available in the mid-1950s. The principal supplier of electrical grades of polypropylene, formerly Hercules, is now Himont U.S.A.

Chemistry

Polypropylene is formed by polymerizing propylene ($CH_3CH = CH_2$). The reaction is carried out commercially in several ways. The newest process, introduced by Rexene in mid-1980, is based on high-efficiency catalysts which reduce process energy consumption. This gas-phase process increases yield of isotactic (crystalline) polymer per pound of catalyst with low enough formation of atactic (amorphous) polymer so that its removal is not required. Other steps necessary in other processes are eliminated, including deashing, drying, and solvent recovery.

Diluent and slurry processes are also in general use. In both processes, polymerization, aided by a Ziegler-type catalyst such as aluminum alkyl, is carried out in a reactor vessel and the polymer separates out as an insoluble powder containing about 10 percent of atactic polymer which must be reduced by solvent extraction to less than 7 percent.

Polypropylene homopolymer has a relatively high brittleness temperature. By copolymerizing with 2 to 20 percent ethylene, a resin is formed with significantly more useful properties. For telephone cable, the ratio is 82 to 84 percent propylene, 16 to 18 percent ethylene. This block copolymer has brittleness temperature of –15°C (5°F), over 20°C (36°F) lower than polypropylene homopolymer, and a maximum continuous use temperature of 90°C (195°F). REA accepts this copolymer as an alternative for high-density polyethylene.

Molecular weight is controlled by process variables. High-molecular-weight resins are softer but tougher, and more flexible. Low-molecular-weight resins are harder, more brittle, and stiffer. To protect the polymer from oxidative degradation during fabrication and use, most compounds contain a hindered phenol antioxidant and a hydroperoxide decomposer. Since copper catalyzes oxidative degradation, an inhibitor for this action should be added to recipes for wire coating compounds. Stability to ultraviolet light is achieved by adding UV inhibitors.

As with high-density polyethylene, a cellular insulation with reduced dielectric constant may be produced by injecting nitrogen into molten polymer or compounding with a blowing agent.

Polypropylene Versus High-Density Polyethylene

Since REA and cable manufacturers recognize polypropylene and high-density polyethylene as alternatives for each other, price is a principal factor in determining usage. At equal price, polypropylene has the advantage of lower specific gravity (0.90 versus 0.95) and slightly better dielectric properties, theoretically permitting material savings of about 5 percent.

This is not to say that substitution can be made without problems or adjustments. For example, slightly greater wall thicknesses of high-density polyethylene are needed to compensate for its somewhat higher dielectric constant ($K = 2.32$ versus $K = 2.24$). This requires tighter packing of wires to keep the outer diameter of the high-density polyethylene cable the same as the polypropylene cable. Polypropylene can undergo severe draw downs during line start-up while maintaining a uniform and continuous coating on wire. High-density polyethylene does not act as satisfactorily. High-density polyethylene dielectric properties are affected to a greater extent by changes in other process conditions such as preheat temperature, melt temperature, water trough temperature, and length of water trough. Thus, it is claimed by proponents that polypropylene has significant manufacturing advantages. High-density polyethylene, however, has much lower brittleness temperature (–76°C (–105°F) versus –15°C (5°F)).

Grades

Homopolymer grades are not in general use for electrical end uses because of their high brittleness temperature. Copolymer grades are available for electrical applications and meet the following industry and government specifications:

- Atomic Energy Commission Specification AEC-1C8–10. Specification AEC-NV-RF-1 through -3.
- Western Electric Specification MS 59056 Issue 2 Modified/T.I.M. 434140.
- ASTM Specification D 4101, *Specification for Propylene Plastic Injection and Extrusion Materials*.

• ICEA Standard S-61-402/NEMA Standard WC 5: *Thermoplastic-Insulated Wire and Cable for the Transmission and Distribution of Electrical Energy.*

Processing

Wire preheating at a constant temperature in the 200° to 300°F (93°–149°C) range is necessary to obtain optimum elongation and to control shrinkback. A cooling trough is required. Recommended stock temperature range is 440° to 500°F (227°–260°C) with barrel temperature in the range of 450° to 475°F (232°–240°C).

Properties

For a wide variety of applications, polypropylene possesses an excellent balance of properties, including the lowest specific gravity of all commonly used plastics, high stiffness, good chemical resistance, and superior dielectric properties. Weaknesses of polypropylene homopolymer include high brittleness temperature and low impact resistance below 40°F (4°C). Additives to recipes are required to prevent attack and degradation by free radicals and ultraviolet light. See also preceding discussion of "Polypropylene Versus High Density Polyethylene."

Polypropylene, while having excellent resistance to most chemicals, is attacked by halogenated hydrocarbons and concentrated oxidizing acids.

Comparative properties with other low-temperature insulations are shown in Table 5-2 (p. 143) in this chapter.

Electrical/Electronic Uses

The principal electrical uses of polypropylene/polyethylene copolymer are for extruded primary insulation on wire and cable and for service at under 5,000 volts. It is also used extensively for injection-molded battery cases.

Polyurethane Polymers (PUR)
(Extrusion Resins)

A relatively new thermoplastic polyurethane (Esthane™) suitable for extrusion as halogen-free cable jacketing has been developed by Geon. This compound is reported to feature excellent flame retardancy, low smoke generation, low toxic gas emissions, and to have a V-0 rating in the UL 94 vertical burn flammability test with an oxygen index of 40 percent.

Polyvinyl Chloride (PVC)
(Extrusion Resins)

Polyvinyl chloride was first used as wire and cable insulation and jacketing in the late 1930s. During World War II, its use spread as a replacement for rubber, then in short supply. Polyvinyl chloride is now one of the most widely used materials in extrusion applications, as producers and fabricators have successfully complied with

EPA standards, which in 1974 and 1975 appeared to threaten the survival of the industry.[1] Major producers of insulation grades of polyvinyl chloride include Geon (formerly B. F. Goodrich), Occidental Chemical, and Pantasote Polymers (Kohinor™).

Chemistry

Vinyl chloride, $CH_2 = CHCl$, is a gas at room temperature. There are several commercial polymerization processes, all carried out in pressure vessels.

In the predominating *suspension process*, vinyl chloride droplets are suspended in water by rapid agitation and protective colloids such as polyvinyl alcohol, gelatin, or methyl cellulose. Polymerization is activated by a peroxide initiator, soluble in the monomer, which decomposes with heat to form free radicals. The reaction is exothermic and must be controlled by cooling. At a predetermined point in the process, unreacted monomer is removed by vacuum. Water is extracted by centrifuge from the slurry, which is then dried. Average particle size is in the range of 100 to 160 microns, although process variables may be controlled to produce particles of 1 to 3 microns. A similar process utilizes more emulsifier and less rapid agitation. The resulting particle size is about 0.5 micron, suitable for latexes, plastisols, and organisols.

The *mass polymerization process* is carried out in two stages. Partial polymerization to a viscous liquid occurs in the first stage without the presence of water or other medium. Polymerization is completed to a dry state in a second vessel. Unreacted monomer is stripped by vacuum. Particles range in size from 100 to 160 microns.

In the *solution process*, vinyl chloride monomer is dissolved in a solvent and polymerized with controlled temperature and pressure. With selected solvents, the resin remains in solution or is recovered as a powder by stripping the solvent. With other monomer solvents, the polymer is insoluble and precipitates. Dry resin remains after the solvent is removed.

A wide range of resin properties may be obtained by controlling molecular weight, particle size, and particle configuration. High-molecular-weight resins are more difficult to process, but exhibit superior strength and greater resistance to heat and chemicals. Particle configuration determines the degree of porosity and ability to absorb or not to absorb plasticizers.

PVC typically contains 56.8 percent chlorine, but this ratio may be increased to as much as 67 percent. At this level, there is a significant extension of the upper service temperature limit.

Vinyl chloride may be copolymerized with vinyl acetate to produce resins with improved flow at lower processing temperatures, greater solubility, and better pigment binding. Graft polymers with ethylene/vinyl acetate (EVA) or ethylene/propylene/diene polymer (EPDM) have improved resistance to impact.

Properties of polyvinyl chloride compounds may also be changed by irradiation. This is accomplished by introducing into recipes multifunctional organic chemical additives, or sensitizers, such as tetraethylene glycol dimethacrylate. When this compound is exposed to high-energy electrons, free radicals are created on polyvinyl

---

[1]There are, however, continuing worldwide efforts by Greenpeace, among others, alleging that PVC products are highly toxic and should be phased out of usage.

chloride molecules, which then react with sensitizers to form a network of crosslinked polymer chains. This process imparts higher values for mechanical strength, abrasion and cut-through resistance, and temperature resistance, raising the maximum operating temperature rating of polyvinyl chloride insulated wire to 105°C (221°F).

Plasticizers

Polyvinyl chloride resins alone are inherently hard and brittle at temperatures up to 180°F (82°C). To be made useful, they must be compounded with plasticizers and other additives. A typical electrical grade formulation would contain about 50 percent polyvinyl chloride resin, 25 to 35 percent plasticizer, and the rest other additives. Phthalates have long been the most commonly used plasticizers for all PVC grades, including electrical grades. Trimellitates, with low volatility, are favored for high-temperature wire and cable applications up to 105°C. Adipate, azelate, and sebacate esters improve low-temperature flexibility, a characteristic weakness of polyvinyl chloride compounds. Phosphate plasticizers impart added flame resistance for inside wiring. Epoxidized esters of unsaturated fatty acids give heat and light stability, but tend to impart stiffness at low temperatures.

To achieve the desired balance of properties, two or more plasticizers are usually required. Important considerations in selection of a plasticizer are its compatibility, volatility, tendency to migrate to other materials, toxicity, odor, burning characteristics, dielectric properties, weatherability, and efficiency (concentration required).

Other Additives

There are several other essential ingredients usually added to polyvinyl chloride compounds:

   • *Heat stabilizers*, frequently organometallic compounds, retard thermal degradation, prolonging useful life.

   • *Lubricants*, such as fatty esters, waxes, and metallic soaps, improve processing characteristics and reduce adhesion to metal parts of process equipment.

   • *Impact modifiers* improve impact resistance. Materials include chlorinated polyethylene (CEP), acrylonitrile-butadiene-styrene (ABS), methyl methacrylate-butadiene-styrene, and ethylene/vinyl acetate (EVA) polymers.

   • *Processing aids* include styrene-acrylonitrile (SAN), methyl methacrylate copolymers, and alpha methyl styrene. These materials act to produce uniform melts.

   • *Fillers*, such as calcium carbonate, reduce compound cost and raise heat deflection temperature. Calcined clay improves dielectric properties.

   • *Pigments* impart color and opacity, and improve weatherability.

   • *Flame retardants*, such as antimony oxide, significantly increase the oxygen index (OI) of polyvinyl chloride compounds.

ASTM Standards for Extruded Polyvinyl Chloride Compounds

The American Society for Testing and Materials has developed the following standards for extruded polyvinyl chloride compounds:

- D 1047, *Poly(Vinyl Chloride) Jacket for Wire and Cable*
- D 2219, *Poly(Vinyl Chloride) Insulation for Wire and Cable, 60°C Operation*
- D 2220, *Poly(Vinyl Chloride) Insulation for Wire and Cable, 75°C Operation*

Federal Specifications for Polyvinyl Chloride Insulated Wire and Cable

The following Federal specifications have been developed for polyvinyl chloride insulated wire and cable:

- MIL-W-76, *Wire and Cable, Hookup, Electrical, Insulated*
- MIL-W-5086 (1), *Wire, Electric, PVC Insulated, Copper or Copper Alloy*

  /1B Wire, Electric, PVC Insulated, Nylon Jacket, Tin-coated Copper Conductor, 600 Volts, 105 Deg C

  /2C Wire, Electric, PVC Insulated, PVC-Glass-Nylon, Tin-coated Copper Conductor, 600 Volts, 105 Deg C

  /3B Wire, Electric, PVC Insulated, PVC-Glass-PVC Nylon, Tin-coated Copper Conductor, 600 Volts, 105 Deg C

  /4B Wire, Electric, PVC Insulated, Nylon Jacket, Tin-coated Copper Conductor, 3000 Volts, 105 Deg C

  /6C Wire, Electrical, PVC Insulated, Polyvinylidene Fluoride Jacket, Silver-coated Copper Alloy Conductor, 600 Volts, 110 Deg C

  /7B Wire, Electrical, PVC Insulated, Nylon Jacket, Tin-coated Copper Conductor, Medium Weight, 600 Volts, 105 Deg C

- MIL-W-16878

  /2A Wire, Electrical, PVC Insulated, 105 Deg C, 1000 Volts

  /17 Wire, Electrical, PVC Insulated, 105 Deg C, 600 Volts, Polyamide Jacket

  /18 Wire, Electrical, PVC Insulated, 105 Deg C, 1000 Volts, Polyamide Jacket

  /19 Wire, Electrical, PVC Insulated, 105 Deg C, 3000 Volts, Polyamide Covering

NEMA and UL Standards and Listings

Industry organizations (in addition to ASTM and the Military) have developed the following specifications and listings for polyvinyl chloride insulated wire and cable:

- NEMA Standard WC 5: covers thermoplastic-insulated wire and cable products.
- UL National Electrical Code designations covering building wire:
  T—General Wire
  TW—Wet Locations 60°C Rating

THW—Hot, Wet Locations 75°C Rating
THHN—Hot, Wet Locations, Nylon Jacket 75°C Rating
AWM—80°C, 90°C, 105°C Ratings
NM and UF

Grades

There is a virtually unlimited number of polyvinyl chloride custom and specification grade compounds available, including grades to meet the requirements for wire and cable insulations covered by standards groups and the government.

Processing

In extrusion of polyvinyl chloride compounds on wire, the feed section temperature should be maintained in the 100° to 125°F (38°–52°C) range. Overheating initiates pre-plasticizing, causing the feed throat to plug or bridge. The outer diameter of the insulated wire is controlled by die size and by regulating the speed of the capstan which pulls the wire through the extruder crosshead.

Properties

Polyvinyl chloride may be compounded to produce low-cost wire and cable insulation and jacketing with thermal ratings up to 105°C. Polyvinyl chloride has outstanding tensile and impact strengths and adequate dielectric properties. Low-temperature characteristics are generally poor, and flexibility varies greatly over its useful temperature range. Without careful formulation, plasticizers may volatilize or migrate to other cable materials. Flame resistance is inherently high.

Polyvinyl chloride has good resistance to:

- Aliphatic hydrocarbons
- Oils, fats, and waxes
- Alcohols
- Concentrated inorganic acids and bases
- Dilute organic acids
- Salts
- Most solvents
- Weathering

Polyvinyl chloride is attacked by:

- Aromatic hydrocarbons
- Halogenated hydrocarbons
- Phenols
- Ketones
- Esters
- Ethers
- Concentrated organic acids

Comparative properties with other low temperature insulations are shown in Table 5-2 (p. 143) of this chapter.

Electrical/Electronic Uses

Because of its low cost and low flammability, which can be made lower with flame-retardant additives, polyvinyl chloride is used extensively for inside telephone wire, electronic low-temperature wire, building wire, apparatus wire, and low-tension automotive wire. Polyvinyl chloride is used as jacket material on telephone, power, and apparatus cables.

Silicone Rubber
(Elastomer)

Silicone rubber is a high-temperature insulation applied to specialty types of wire and cable. Dow Corning and General Electric are the principal producers.

Chemistry

The basic polymer for heat-vulcanizable silicone rubber is linear polysiloxane, formed by hydrolysis of dichlorosilane and polycondensation of resulting silanediols. This gum has the following structure:

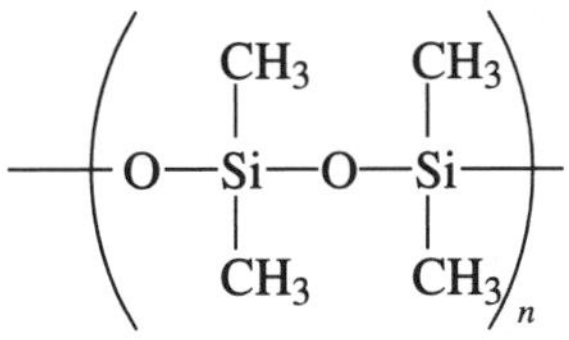

$$\left(\begin{array}{cc} CH_3 & CH_3 \\ | & | \\ -O-Si-O-Si- \\ | & | \\ CH_3 & CH_3 \end{array}\right)_n$$

For wire and cable insulation, some methyl groups are replaced with phenyl and/or vinyl groups to increase tensile strength, impart tear resistance, extend the useful temperature range, and improve resistance to gamma radiation. The extruded gum is crosslinked, or vulcanized, by the action of heat and a vulcanizing agent, such as 2,4 dichlorobenzoyl peroxide.

Physical properties are enhanced by compounding with fillers such as small particle size silica and diatomaceous earths.

Federal Specifications for Silicone Rubber

The following Federal specifications have been developed for silicone rubber:

• MIL-C-915, *Silicone Rubber Insulated Navy Nuclear Plant Cable*
• MIL-W-8777C, *Type MS-25471 (Dacron Jacket) and Type MS-27110 (FEP Jacket) Interconnection Wire, Silicone Rubber Insulated, 600 Volts, –55° To 200°C*
  • MIL-W-16878
  /7A Wire, Electrical, Silicone Rubber Insulated, 200 Deg C, 600 Volts
  /29 Wire, Electrical, Silicone Rubber Insulated, 150 Deg C, 600 Volts
  /30 Wire, Electrical, Silicone Rubber Insulated, 150 Deg C, 1000 Volts

/31 Wire, Electrical, Silicone Rubber Insulated, 150 Deg C, 1000 Volts, Glass
Braid Covering, Abrasion Resistant

/32 Wire, Electrical, Silicone Rubber Insulated, 200 Deg C, 1000 Volts, Glass
Braid Covering, Abrasion Resistant

Grades

As with other elastomers, large wire and cable manufacturers usually compound their
own silicone rubber formulations. Grades are available to meet all Military specifications.

Processing

Extruders should have a screw especially designed for silicone rubber. The gum should
not be allowed to reach a temperature of 130°F (54°C) during extrusion, since higher
temperatures may produce scorching and loss of vulcanizing agent. Extrusions are usually vulcanized in continuous-process equipment, most commonly in a vertical hot air
oven with temperature in the range of 482° to 800°F (250°–427°C), through which the
coated wire is pulled.

Properties

Extruded silicone rubber wire and cable coatings provide long-life insulation at temperatures from −55° to 200°C. When exposed to gamma radiation, silicone rubber will
increase hardiness and decrease in elongation. Dielectric properties do not undergo significant change. Methyl-vinyl silicone polymers are satisfactory for use in the range of
5×10^7 to 1×10^8 rads. Incorporation of phenyl groups increases radiation tolerances,
as shown in Table 5-5.

Heat-curable silicone rubber has good resistance to ozone, corona, and weathering. It is inherently flame retardant, and when burned, forms a nonconductive silica
dioxide ash which maintains circuit integrity. Mechanical properties are relatively low
in comparison with those of organic rubbers at room temperature.

Silicone rubbers have good resistance to:

- Alcohols
- Dilute acids and alkalies

TABLE 5-5   Radiation Resistance—Silicone Rubber

DOSAGE (Rads)	METHYL-VINYL		METHYL-PHENYL-VINYL	
	Elongation %	Tensile Strength (psi)	Elongation %	Tensile Strength (psi)
None	200	1,200	600	1,200
5×10^6	130	1,000	450	1,100
5×10^7	50	900	225	900
1×10^8	20	600	75	850

- Salts
- Oils, fats, and waxes

Silicone rubbers are attacked by:

- Halogenated hydrocarbons
- Aromatic solvents
- Phenol
- Concentrated acids
- Steam

Key property values for heat curable silicone rubber are shown in Table 5-1 (p. 142) in this chapter.

Electrical/Electronic Uses

Because of its relatively high cost, extruded heat-curable silicone rubber insulation is used only when less costly insulation is inadequate. Applications include appliance and fixture wire, shipboard power, control, and hookup cables, aircraft high-tension ignition and control cables, automotive ignition system wiring, nuclear power cable, and flexible multiconductor flat cable.

Organic Elastomers

As used in this handbook, elastomers are polymeric materials which exhibit elastic properties similar to rubber, the original elastomer. When properly processed and within their elastic limits, they tend to return to their original shape after being stretched, compressed, or otherwise deformed. Like natural rubber, some elastomers require vulcanization with sulfur, sulfur compounds, or other polymerizing agents or methods. Other elastomers can be processed directly into finished products. The compounding and processing of all elastomers is an intricate combination of art, technology, and experience, all of which significantly affect the properties of any elastomer.

Table 5-6 lists the principal elastomers used as insulation and jacketing on wires and cables. Comparative properties are shown in Table 5-3 (p. 144) in this chapter.

ASTM Standards for Extruded Organic Elastomer Compounds

The American Society for Testing and Materials has developed the following standards for extruded organic elastomer compounds:

- D 353, *Natural Rubber Insulation for Wire and Cable, 60°C Operation*
- D 469, *Natural Rubber Heat-Resisting Insulation for Wire and Cable, 75°C Operation*
- D 532, *Natural Rubber Jacket for Wire and Cable*
- D 866, *Styrene-Butadiene (SBR) Synthetic Rubber Jacket for Wire and Cable*

TABLE 5-6 Summary of Characteristics of Selected Elastomers

ELASTOMER	SUPPLIERS	CHEMISTRY	ADVANTAGES	LIMITATIONS
Polyisoprene (IR)	Goodyear Chemical Goodrich Chemical Shell Chemical	Synthetic equivalent to natural rubber. Vulcanize cure. $-\!\left(CH_2\!-\!C\!=\!CH\!-\!CH_2\right)_n$ CH_3	Controlled molecular weights. Good dielectric and mechanical properties at room temperature.	Poor aging properties. Low ozone resistance. Attacked by halogenated, aliphatic, and aromatic hydrocarbons, mineral and vegetable oils, oxidizing chemicals. Poor flame resistance.
Butyl Rubber (IIR)	Exxon Chemical	Copolymer of isobutylene with 2 to 3% isoprene. Vulcanize cure. $CH_3 CH_3$ $-\!\left(CH_2\!-\!C\!-\!CH_2\!-\!C\!=\!CH\!-\!CH_2\right)_n$ CH_3	Dielectric properties about the same as IR. Better resistance to aging and oxidation. Resistant to mineral and vegetable oils. Low gas permeability.	Low resilience at room temperature. Attacked by halogenated, aliphatic, and aromatic hydrocarbons, and concentrated acids. Poor flame resistance.
Styrene Butadiene Rubber (SBR) (GR-S)	Goodyear Chemical Goodrich Chemical Phillips Chemical	Copolymer of butadiene (75%) and styrene (25%). Vulcanize cure. $-\!\left(CH_2\!-\!CH\!=\!CH\!-\!CH_2\!-\!CH\right)_n$	Low cost. Dielectric properties about the same as IR. Also mechanical properties.	Lower tensile strength, resilience, and heat resistance than IR. Attacked by halogenated, aliphatic, and aromatic hydrocarbons, mineral and vegetable oils, oxidizing chemicals, and sunlight. Poor flame resistance.
Chloroprene Rubber (CR) (Neoprene™)	Du Pont	Polymerized 2-chlorobutadiene-1,3. Vulcanize cure. H $-\!\left(CH_2\!-\!C\!=\!C\!-\!CH_2\right)_n$ Cl	Better resilience than IR to high temperatures, weathering, oxidation, aliphatic hydrocarbons, bases, oils. Good mechanical properties and flame resistance.	Lower dielectric properties than IR. Attacked by halogenated and aromatic hydrocarbons, esters, and concentrated oxidizing acids and agents.

Material	Manufacturer	Description	Properties	Limitations
Chlorosulfonated Polyethylene (CSM) (Hypalon™)	Du Pont	Made by reacting PE with sulfur dioxide and chlorine. Vulcanize cure. $\left[\left(CH_2-CH_2-CH_2-\underset{\underset{Cl}{\mid}}{CH}-CH_2-CH_2\right)_{12}-\underset{\underset{Cl}{\mid}}{C}-SO_2\right]_n$	Excellent resistance to abrasion, moisture, high humidity, ozone, oils, weathering, underground environments. Superior flexibility. Good flame resistance.	Lower dielectric properties than IR. Attacked by halogenated, aliphatic, and aromatic hydrocarbons, gasoline, aldehydes, and ketones.
Ethylene Propylene Rubber (EPR), (EPM) and Ethylene Propylene Diene Rubber (EPDM)	Du Pont Uniroyal Exxon Chemical Goodrich Chemical	Copolymer of ethylene and propylene. Terpolymer of ethylene, propylene, and 1,4 hexadiene or ethylidene norbornene. Vulcanize cure.	Low specific gravity. Superior resistance to ozone, corona cutting, weathering, heat, hot water, steam, acids, bases, alcohols, esters, ketones.	Poor flame resistance. Attacked by halogenated, aliphatic, and aromatic hydrocarbons, turpentine, and oils.
Chlorinated Polyethylene (CPE)	Dow Chemical	Made by randomly chlorinating HDPE to resin containing 25–48% chlorine. Vulcanize cure.	Excellent resistance to ozone, weathering, heat aging, oils, bases, most acids, alcohols. Low cost. Blends well with other polymers.	Lower dielectric and mechanical properties than most other elastomers. Forms toxic fumes on burning. Attacked by halogenated and aromatic hydrocarbons.
Thermoplastic Rubber (TPR)	Reichhold Advanced Elastomer Systems	Physical blend of EPDM (5 parts) and polyolefins (3 parts). Reichhold's TPR is partially prevulcanized. Santoprene™ (Advanced Elastomer Systems) is fully prevulcanized. Both are thermoplastic.	Low specific gravity. Excellent dielectric properties. Good resistance to heat aging, ozone, UV radiation, organic acids and bases. Low temperature flexibility. Thermoplastic processing.	Relatively low physical properties. Attacked by halogenated and aromatic hydrocarbons.

- D 1352, *Ozone-Resisting Butyl Rubber Insulation for Wire and Cable*
- D 1523, *Synthetic Rubber Insulation for Wire and Cable, 90°C Operation*
- D 1679, *Synthetic Rubber Heat- and Moisture-Resisting Insulation for Wire and Cable, 75°C Operation*
- D 2768, *General-Purpose Ethylene-Propylene Rubber Jacket for Wire and Cable*
- D 2770, *Ozone-Resisting Ethylene-Propylene Rubber Integral Insulation and Jacket for Wire and Cable*
- D 2802, *Ozone-Resistant Ethylene-Propylene Rubber Insulation for Wire and Cable*
- D 4203, *Styrene-Acrylonitrile (SAN) Injection and Extrusion Materials*
- D 4244, *General-Purpose, Heavy-Duty, and Extra-Heavy-Duty Acrylonitrile-Butadiene/Polyvinyl Chloride (NBR/PVC) Jackets for Wire and Cable*
- D 4245, *Ozone-Resistant Thermoplastic Elastomer Insulation for Wire and Cable, 90°C Dry/75°C Wet Operation*
- D 4246, *Ozone-Resistant Thermoplastic Elastomer Insulation for Wire and Cable, 90°C Operation*
- D 4247, *General-Purpose Black Heavy-Duty and Black Extra-Heavy-Duty Polychloroprene Jackets for Wire and Cable*

ICEA/NEMA Standard for Rubber Insulated Wire and Cable

The Insulated Cable Engineers Association (ICEA) and the National Electrical Manufacturers Association (NEMA) have jointly approved the following standard for Rubber-Insulated Wire and Cable for the Transmission and Distribution of Electrical Energy:

- NEMA WC 3/ICEA S-19

UL Standard for Insulated Wires and Cables

Underwriters Laboratories (UL) has approved the following standard:

- UL 44, *Rubber-Insulated Wires and Cables*

Grades

Electrical grades are available from several, but by no means all, suppliers of elastomers and elastomer compounds. Compounding is generally done at the fabricator's plant with due consideration given to recommendations by suppliers, based on intended application and type of process equipment to be used.

Processing

The compounding of elastomers and blends of elastomers is very complex, often involving recipes with 10–15 or more ingredients. Depending on the elastomer(s) and curing system, ingredients may include:

- The elastomer(s), up to 50 percent by weight
- Sulfur or sulfur compounds as crosslinking agents, 2 percent or less.
- A peroxide curing agent, such as dicumyl peroxide, 3 percent or less.
- A rubber accelerator, such as dipentamethylene thiuram hexasulfide (Du Pont's Tetrone A™ or 2-mercaptobenzothiazole), 2 percent or less.
- A curing promoter to improve scorch and insulation resistance, such as N,N'-*m*-phenylenedimaleide (Du Pont's HAV-2™), 1 percent or less.
- A metal oxide acid acceptor and vulcanizing agent, such as sublimed letharge (PbO) or dibasic lead phthalate, up to 5 percent.
- A hydrogenated wood rosin to activate compounds containing litharge, such as Stabelite Resin™ (Hercules), 1 percent or less.
- A stabilizer to improve heat resistance, such as nickel dibutyldithiocarbamate (Du Pont's NBC™), 2 percent or less.
- Carbon black to improve weathering properties, up to 25 percent.
- A lubricant, such as paraffin or petrolatum, up to 12 percent.
- A hindered phenol antioxidant, 3 percent or less.
- A filler, such as calcined clay or kaolin, up to 50 percent.
- A plasticizer, such as phthalate ester, chlorinated paraffin, or alkyl substituted trimellitate.

There are three basic curing systems in general use for elastomers:

- *Conventional curing* using vulcanizing agents such as:
  —Sulfur or sulfur-containing compounds, often with metal/organic activators.
  —Organic peroxides, which usually produce better dielectric properties and water resistance with no tendency to form metal sulfides.
- *Lead sheath curing*, which prevents sponging as compound is vulcanized.
- *Radiation curing*, with less tendency to scorch, faster extrusion rates, and reduced heat build-up during cure.

The first of these uses continuous vulcanization processing in equipment 100–350 feet in length (30–108 meters), incorporating pay-off reel, extruder, pressurized steam tube, pressurized water cooling chamber, capstan, cooling trough, and take-up reel. The system may be arranged so the cable travels horizontally, vertically, or with the pay-off reel and extruder on a level above the water-cooling chamber, permitting the cable to hang loose in the slanted steam tube (the catenary system). Production rate depends on cable size, thickness of extrudate, steam pressure, length of curing tube, and cure rate of compound.

Lead sheath curing involves formation by lead presses of a tight lead covering over the insulated or jacketed cable. Processing may be continuous or batch operation, although storing cable at least 24 hours before applying the lead sheath improves quality. After leading and reeling, cable ends are sealed and the cable is cured in steam vulcanizer at from $1^1/_2$ to $4^1/_2$ hours at 235°–265°F (113°–130°C). After curing, the cable is rapidly cooled with water to reduce internal pressure and the possibility of rupture of the lead sheath. Finally, the lead is stripped from the cable, cut into chips, and reused.

Radiation, or electron beam curing, covered by U.S. Patent 3.990.479 assigned to Samuel Moore Company, utilizes high-energy radiation to crosslink polymers through formation of free radicals. Radiation dosage controls the density of crosslinking. However, required dosage levels are unique with each polymer. Choice of compound is very selective with this method. Highly unsaturated or highly aromatic plasticizers should be avoided.

While electrical applications for elastomers are mostly for cable jackets, there is also some usage as primary insulation on low-voltage conductors. Here, and also in the lead-press curing method where compound comes in direct contact with metal, vulcanizing with sulfur and sulfur-containing compounds may result in the formation of undesirable metal sulfides.

Electrical/Electronic Uses

Elastomers are used extensively as jackets on wires and cables, and to a lesser extent as primary insulation in certain applications:

- Welding cable jackets (EPDM), (CR), (TPR)
- Motor lead wire insulations in corrosive environments (CSM)
- X-ray cables (CSM)
- 600 volt power cable insulation/jacket
- Semiconducting shield (CPE)
- Jackets for mine excavator trailing cables (CR)
- Building wire and cable (CR), (EPDM), (SBR)
- Control cable jackets (CR), (CSM), (EPDM), (SBR)
- Trailer cable (TPR)
- Apparatus and portable cords (EPDM), (SBR), (CR), (CSM)
- Appliance cord (CSM), (TPR), (EPDM)
- Roadway heating cables (CSM)
- Primary wiring for trucks (CSM), (TPR), (EPDM)

In addition, there are electrical applications for molded elastomers:

- Connectors (CR), (TPR)
- Direct-burial telephone cable termination enclosures (CR)
- Encasements for underground transformers (CR)

6

Embedding Compounds

INTRODUCTION

As used in this handbook, the term *embedding* includes encapsulating, potting, casting, and dipping processes employing a compound that may be cured either at room or elevated temperature. Coatings and impregnants are covered in Chapter 8.

All embedding processes involve encasing electrical/electronic parts in plastic with or without the use of molds. The parts are thus protected dielectrically and physically, ensuring greater reliability and longer life than otherwise would be the case.

This chapter covers the major embedding resins, namely, epoxies, polyesters, polyurethanes, and silicones. Other resins are also used to embed electronic parts employing injection molding and low-pressure transfer molding processes. For information on these resins, see Chapters 3 and 4, *Thermoplastic Molding Compounds* and *Thermosetting Molding Compounds*.

CRITICAL PROPERTIES

To perform the processing, protective, and dielectric functions of a good encapsulant, critical properties include:

- A high degree of purity ensuring the absence of ionic and other contaminants
- Long pot life
- Low toxicity
- Ease of application and clean up
- Initial low viscosity to facilitate flow and penetration into the interstices of the part being encapsulated
- Ability to cure with low exotherm and without formation of water, corrosive, or volatile reaction products
- Low shrinkage on curing
- Good adhesion to materials to be encapsulated
- Toughness and resiliency to accommodate differences in thermal expansion within a system
- Adequate dielectric properties
- Good thermal properties to dissipate heat quickly
- Satisfactory resistance to thermal shock and thermal cycling
- High-moisture, chemical, and solvent resistance
- Low degradation by ultraviolet radiation
- Wide useful temperature range
- Long service life without loss of properties
- Low cost

Epoxies generally satisfy these criteria best and are most widely specified. Urethanes, once widely accepted by the Military, fell from favor for critical applications because of poor hydrolytic stability. Subsequently, however, their hydrolytic stability was improved and their use has increased significantly. Other encapsulants in common use are polyesters and silicone.

FILLERS

While unfilled systems have the least water absorption—a desirable dielectric feature—there are important advantages to be realized by judicious use of fillers in embedding applications for electrical/electronic parts. With most systems, the tendency of a casting to crack can be a major problem. This is caused by the difference in thermal expansion of the cast part and the embedding material. This disparity in expansions can be significantly reduced or eliminated by adding the proper filler in sufficient quantities to formulations.

Certain fillers increase thermal conductivity in resin systems, lessening the likelihood of overheating, weight loss, and shrinkage. Pot life on the system can also be extended by fillers. Dielectric properties, however, are usually not improved by fillers and may be adversely affected in humid environments by fillers which readily absorb water.

Fillers may be classified as follows:

- Bulk fillers
  - Sand (lowest cost, good thermal conductivity and compressive strength, and low shrinkage)
  - Silica (same features as sand, good dielectric properties)
  - Talc (excellent machinability, low shrinkage, low cost)
  - Clay (excellent machinability, good thermal shock resistance and compressive strength, low shrinkage, low cost)
  - Calcium carbonate (excellent machinability, low shrinkage, low cost)
  - Anhydrous calcium sulfate (excellent arc resistance, good thermal conductivity and compressive strength, low shrinkage, low cost)
- Reinforcing fillers
  - Mica (excellent impact resistance, good thermal shock resistance, low cracking, low cost, good dielectric properties)
  - Asbestos (same features as mica, but a known health hazard)
  - Wollastonite (excellent impact resistance, good thermal conductivity and thermal shock resistance, low cost)
  - Wood flour (low density, good thermal shock resistance, low cost)
  - Sawdust (same features as wood flour)
- Specialty fillers
  - Quartz (low coefficient of expansion, good thermal conductivity and mechanical properties, low shrinkage, low cracking, good dielectric properties)
  - Alumina (low coefficient of expansion, good thermal properties and compressive strength, low shrinkage)
  - Alumina trihydrate (excellent arc resistance, flame retardant)

—Lithium alumina silicate (low coefficient of expansion and shrinkage, high conductivity and compressive strength)
—Beryl (same features as lithium alumina silicate)
—Silica aerogel (excellent nonsettling)
—Bentonite (excellent nonsettling)
—Graphite (high electrical conductivity, good machinability)
—Ceramic spheres (low density)

Fillers should mix well in compounds without settling or excessively increasing viscosity. Most fillers tend to reduce compound cost, although not in direct ratio to the amount of filler incorporated because compound density is increased more than compound volume.

FEDERAL SPECIFICATION MIL-I-16923

The title of this specification is *Insulating Compound, Electrical, Embedding*. This specification is approved for use by all departments and agencies of the Department of Defense. It covers a general purpose casting and potting compound which has excellent electrical and hydrolytic stability characteristics, and which can preserve the electrical properties of the equipment to which it is applied by sealing against such environmental conditions as moisture, dirt, fumes, fungus, or other deleterious substances which may be encountered in military service. This is a performance specification, and all products that qualify are included in a qualified products list.

Key qualitative requirements include the following:

• The material shall cure throughout to a permanently infusible state without volatile loss. It shall be of sufficiently low viscosity, and shall have such wetting properties at the pouring temperature that it will flow between elements of the part

**TABLE 6-1   Key Properties Required of
Embedding Compounds
(MIL-I-16923)**

PROPERTY	VALUE
Dielectric strength, volts/mil, minimum, $^1/_8$ in	325
Dielectric constant	
60 Hz, maximum	5.0
1 MHz, maximum	4.5
Dissipation factor	
60 Hz, maximum	0.04
1 MHz, maximum	0.05
Volume resistivity, $\Omega \cdot cm$, minimum	
As received	1×10^{13}
After conditioning	1×10^{12}

Note: The above are room temperature values and were obtained in accordance with Federal Specification FED-STD-406

or parts of the assembly to be embedded, displacing all air and leaving no voids, and both the pouring and curing temperature shall not exceed 266°F (130°C). The material shall be capable of mechanically supporting the part or assembly embedded and shall not require a case or other external support after curing has been completed, and shall not cause deterioration of materials in electronic assemblies and parts.

• Cast specimens shall be self-extinguishing when tested according to the specified procedure.

• Cast specimens shall not exhibit softening, blistering, or warping after moisture conditioning at 158°F (70°C), 95 percent relative humidity, for ten days.

Table 6-1 shows values for key properties required by this specification.

ASTM STANDARD FOR EMBEDDING COMPOUNDS

The American Society for Testing and Materials (ASTM) formerly established the following standard, now discontinued, but which contains a listing of current applicable property tests:

• D 1674, *Polymerizable Embedding Compounds Used for Electrical Insulation.* These compounds must cure without pressure and may or may not require heat to accomplish the reaction. When cured, they are required to electrically insulate and mechanically protect and support the part or assembly without a case, pot, or other external support. Table 6-2 lists test methods.

TABLE 6-2   Property Tests for Embedding Compounds
(ASTM D 1674)

PROPERTY MEASURED	ASTM TEST METHOD
Sampling	D 1674
Specimen preparation and conditioning	D 618
Hardness	D 2240
Coefficient of linear thermal expansion	D 696
Specific gravity	D 792
Coefficient of thermal conductivity	D 1674
Thermal shock resistance	D 1674
Dielectric constant and dissipation factor	D 1674
Dielectric strength	D 1674
Arc resistance	D 495
Effect of high humidity	D 1674
Flame resistance (ignition time and burning time)	D 229
Dielectric strength of embedded electrodes	D 149

EMBEDDING PROCESSES

There are several embedding processes in general use, with curing accomplished either at room or elevated temperature:

• *Bench casting* involves pouring the casting compound around the part to be protected placed in a mold. The cast part is removed from the mold after the material is cured. Precautions must be taken to eliminate air bubbles, including careful pouring, slowly warming the liquid to permit bubbles to escape before curing occurs, vibration, and vacuum degassing prior to pouring.

• *Vacuum casting* is best accomplished by pouring under vacuum, thus permitting escape of entrapped air and moisture from the part as well as air bubbles from the liquid. When pouring is done at room temperature, the liquid is likely to foam and froth, requiring precise control of the evacuation process.

• *Pressure casting*, often used after vacuum casting, involves pressurizing the chamber. Minute bubbles may form from somewhat volatile curing agents, making it necessary to increase the pressure by means of centrifugal action to 1,000 psi.

• *Dip casting* is simply dipping the part to be encased in the liquid, which is formulated to control runoff as the part is removed.

• *Spraying* is a suitable method where the configuration of a part permits a uniform coat to be applied.

• *Dropping* a precisely measured amount of material using a microsyringe, or "eyedropper," is a commonly employed technique for embedding small electronic parts. This process may be automated by moving the components on conveyors past automatic dispensers.

• *Potting* is similar to bench casting, but requires a permanent case.

• *Pot-on-sand* technique is employed for high-volume automotive parts. A two-component epoxy is used with a recommended sand in the ratio of approximately four parts of sand to one part mixed epoxy. The component case is vibrated to pack the sand to maximum density. Then, the case is heated to the curing temperature of the epoxy, after which the resin is poured on the sand in the desired amount. Curing is completed in a circulating air oven.

• *Fluidized-bed, or powder coating* is described under "Development Programs" (p. 206) in Chapter 7, *Magnet Wire Enamels*.

• *Compression molding* is described in Chapter 4, *Thermosetting Molding Compounds*.

• *Transfer molding* is described in Chapter 4, *Thermosetting Molding Compounds*.

• *Injection molding* is described in Chapter 4, *Thermosetting Molding Compounds*.

Barrier or stress-relief coatings are often applied first to attach electronic components prior to embedment. They serve as protection from mechanical and thermal shock and to relieve stresses from differences in coefficients of expansion between the encapsulant and substrate. These coatings must be flexible and possess a low modulus of elasticity as well as providing protection from moisture, gases, and corrosion. Silicones are often used as barrier coatings for parts subsequently embedded in epoxies.

CURING SYSTEMS

In addition to traditional curing at room temperature or with applied heat, low-intensity ultraviolet (UV) light curing promoted by Polychem Corporation is suitable for specially formulated compounds, including epoxies, acrylics, polyurethanes, and blends. Ultraviolet light can be artificially produced by an arc lamp equipped with special glass transparent to UV rays. UV curing compounds for this system are single-component products requiring no mixing. Photoinitiators in the compound react to a specific wavelength of low-intensity UV light. The speed of the reaction is governed by the intensity of the light source, although the relationship between speed of curing and UV light intensity is not linear (doubling the intensity does not necessarily cut the curing time in half). This system is reported to be effective for deep curing sections of over 1 inch and is less than 1 minute with certain formulations.

DEVELOPMENT PROGRAMS

Development of improved embedding compounds focuses on materials with low toxicity that are easy to apply, have minimal shrinkage, and cure rapidly free of voids and volatile components.

MARKET TRENDS

There is increasing usage of polyurethanes, both for military and nonmilitary applications, now that hydrolytic stability of these resins has been improved.

While there may be pauses in the enactment and enforcement of environmental and worker safety regulations, the trend toward use of lower toxicity embedding materials will continue.

TABLE 6-3   Comparative Characteristics of Encapsulants

	CHARACTERISTIC					
Encapsulant	Dielectric	Adhesion	Shrinkage	Maximum Service Temp. °C	Chemical Resist.	Cost
Epoxy	Excellent*	Excellent	Low	155	Good	Moderate
Polyester	Good	Fair	High	180	Poor	Low
Polyurethane	Good*	Good	Moderate	130–155	Fair	Moderate
Silicone	Excellent	Low	Low	200+	Good	High

*PURs retain their dielectric properties better than epoxies at the upper end of their useful temperature range.

COMPARATIVE CHARACTERISTICS OF ENCAPSULANTS

Table 6-3 compares key properties of principal encapsulants. In terms of usage, epoxies predominate.

PRINCIPAL EMBEDDING POLYMERS FOR ELECTRICAL/ELECTRONIC USES

The chemistry, grades, properties, and electrical/electronic uses of principal embedding polymers are described in the sections which follow in this order:

- Epoxies
- Polyesters
- Polyurethanes
- Silicones

Epoxies
(Embedding Resins)

Epoxies have been in general use since 1947, and have since gained a predominant position as embedding resins. This section covers electrical grade epoxy embedding resins. Producers of epoxy polymers include Ciba Geigy (Araldite™), J. C. Dolph (Dolphon™), Dow Chemical (D.E.R.™, D.E.N.™, Quatrex™), P. D. George (Sterling—Thermopoxy™), Hoechst Celanese (Epi-Rez™), Shell (Epon™), and Union Carbide (Bakelite™). Principal formulators include Conap (Conapoxy™), Emerson & Cuming (Stycast™), Epoxylite, Epic Resins Division General Fiberglass Supply, Fiberite, Hysol Division The Dexter Corp., Rexnord (Norbak™), and Thermoset Plastics. Major users often formulate their own epoxies. For information on epoxies in coatings and impregnants, thermosetting molding compounds, and clad and unclad structures, see appropriate chapters.

Chemistry

Although there are many variations, the basic epoxy resin is the polycondensation product of bisphenol-A and epichlorohydrin:

basic epoxy resin

Cycloaliphatic epoxies are produced by epoxidation of cyclic olefinic compounds with peracetic acid:

alicyclic diepoxy carboxylate

Epoxy novolacs are covered in "Epoxies" (p. 126) in Chapter 4, *Thermosetting Molding Compounds*.

To be made into useful end products, epoxies must be cured by the action of hardeners. Some hardeners cause curing by reacting with the epoxide group and becoming part of the polymer. Other hardeners promote the resin's self-polymerization by catalytic action. There are several types of hardeners:

• *Primary, secondary, and tertiary aliphatic amines* are the most commonly used hardeners in concentrations of 4 to 20 parts per hundred parts of resin (phr). Curing proceeds at room temperature. Properties may be improved by postcuring for 2 hours at 200°F (93°C). Systems may be toxic.

• *Primary, secondary, and tertiary aromatic amines* require elevated temperatures for curing. They impart longer pot life and improved thermal, mechanical strength, and chemical resistance properties. They produce harder cured systems than aliphatic amines. Systems may be toxic.

• *Polyamides* affect curing at room temperature, producing tough, resilient systems with good moisture resistance and excellent adhesive properties in low-heat applications. Polyamides are the condensation polymers of dimerized or trimerized vegetable oils, or of unsaturated fatty acids, with polyamines. Ratio of epoxy to polyamide may range from 80/20 to 40/60 by weight.

• *Anhydrides*, derived from dicarboxylic acids, require a temperature of 250°F (121°C) for rapid curing. These hardeners in concentrations of 30 to 140 phr produce cured systems with good thermal, mechanical, and dielectric properties. Systems have long pot life and are generally not toxic.

• *Tertiary amines* may be used as accelerators for other hardeners, but tend to increase toxicity.

• *Urea-formaldehyde or phenol-formaldehyde resins* may be crosslinked with epoxies, usually by baking, to produce high-molecular-weight polymers.

Flame resistance of epoxy systems may be improved by incorporating into recipes phenols and anhydrides containing bromine, chlorine, and phosphorous, as well as antimony oxide (antimony trioxide).

The viscosity of liquid epoxies may be reduced by reactive diluents such as allyl, butyl, and phenyl glycidyl ether, or styrene oxide. Unreactive diluents such as dibutyl phthalate are also used for this purpose, acting also to plasticize and add flexibility to cured systems.

Grades

For large-scale production, compounds are usually prepared in house. Formulations are tailored for each end use with little or no duplication because of the virtually unlimited possible combinations of resins, hardeners, accelerators, fillers, and other additives.

Lower volume production requirements are served by scores of merchant compounds offering a wide range of formulations, including one- and two-part, room temperature and oven curing. Producers of basic resins provide technical service and make referrals to assist manufacturers with embedding applications suitable for epoxies.

Processing

Two-part epoxy systems, generally preferred for dielectric applications, may be either room temperature curing or oven curing. The pot life of room temperature curing systems is typically in the 30 minute to 2 hour range, although special formulations may extend pot life to one working day. The most commonly employed ratio of epoxy resin to hardener is one to one. However, practical ratios may be as high as 100 parts epoxy resin to one part hardener. To facilitate uniform mixing, resins and hardeners are usually of different colors. Mixing continues to the point where a uniform color is obtained.

For high-volume production, a heat curing system consisting of resin, hardener, and accelerator, with pot life of 2 to 4 days, is most satisfactory. Oven curing is accomplished at 250°F (121°C) for 2 to 4 hours.

Proper precautions are required to handle epoxy systems safely. Included in these systems are solvents used to clean equipment. Hazards include the following:

- Skin irritation—some workers are highly susceptible
- Ingestion—materials are toxic
- Inhalation—vapors and dust from materials are toxic
- Precutaneous absorption—materials are toxic when absorbed through the skin, as by direct contact with contaminated clothing
- Eyes—materials may cause severe eye burns
- Flammability—solvents are likely to be highly flammable

Some room temperature curing epoxy formulations, especially those with polyamines, are characterized by high exothermic reactions on curing, and measures should be taken to ensure that the temperature rise is adequately controlled. Thus, there may be an epoxy mass limit for some systems over which exotherm control becomes impractical.

See also "Embedding Processes" (p. 181) in this chapter.

Properties

Epoxies are the most widely used embedding resins because of their many outstanding characteristics:

- Excellent dielectric properties
- Superior adhesion to most surfaces
- Low shrinkage during cure

- Good thermal properties
- Good chemical resistance
- Good moisture resistance, depending on fillers
- Properties may be varied widely to suit specific end uses

Cycloaliphatic epoxies are characterized by superior arc-track resistance, high heat deflection temperatures, and good weathering properties.

The dielectric properties of epoxies tend to decrease more rapidly than for other polymers at the upper end of their useful temperature range.

Table 6-4 shows key property values for a typical electrical grade embedding resin.

Electrical/Electronic Uses

Epoxies are widely used for embedment of components for transformers, motors, generators, switchgear, coils, capacitors, resistors, and electronic modules. Cycloaliphatic epoxies are used for large, high-voltage bushings.

TABLE 6-4   Key Properties of Typical Two-Part Oven Curing Electrical Grade Epoxy Embedding Resin

	ASTM TEST METHOD	VALUE
Specific gravity*	D 792	1.14
Dielectric strength, 125 mils, short time, V/mil*	D 149	500
Dielectric constant*	D 150	
60 Hz		3.9
10^6 Hz		3.2
Dissipation factor*	D 150	
60 Hz		0.04
10^6 Hz		0.03
Volume resistivity, $\Omega \cdot cm$*	D 257	3.9×10^{14}
Arc resistance*	D 495	45
Water absorption, $^1/_8$ in thick specimen, %	D 570	
24 h @ 73°F (23°C)		0.16
Deflection temperature, 264 psi, °F (°C)	D 648	
Maximum recommended service temperature °F (°C)		311 (155)
Tensile strength at break, psi*	D 638	2,400
Izod impact strength	D 256	
Foot pounds per inch notched $^1/_8$ in specimen*		0.40
Flammability ratings		
UL Standard 94, $^1/_{16}$ in specimen		
Standard grade		na
Oxygen index, %	D 2863	
Standard grade		na

*At room temperature.

na = not available.

Polyesters
(Embedding Resins)

The term *polyester* encompasses a variety of materials used in many forms as dielectrics, including thermoplastic and thermosetting molding compounds, embedding compounds, wire enamels, coatings and impregnants, and films, all covered in appropriate chapters in this handbook. Use as an encapsulant, while not uncommon, is overshadowed by epoxies for most dielectric applications. The principal supplier of polyester for embedding electrical and electronic parts is P. D. George (Sterling—Thermopoxy™).

Chemistry

For a discussion of the chemistry of polyesters used to make embedding resins, see "Alkyds/Polyesters" (p. 119) in Chapter 4, *Thermosetting Molding Compounds* and "Unsaturated Polyesters" (p. 248) in Chapter 8, *Insulating Coatings and Impregnants*.

Processing

High shrinkage, an inherent problem with polyesters, may be somewhat controlled by judicious use of selected fillers. Relatively high hydrolytic instability of polyesters compared with epoxies is another concern when high-humidity environments are encountered. When peroxide catalysts are used and the embedding compound comes into direct contact with exposed copper, corrosion may occur. See also "Embedding Processes" (p. 181) in this chapter.

Electrical/Electronic Uses

Polyesters are low cost, and are satisfactory for use in many applications not requiring the generally superior properties of epoxies and polyurethanes.

Polyurethanes (PURs)
(Embedding Resins)

The basic work on polyurethanes was done in Germany in the late 1930s, but commercial products were not available in the U.S. until after World War II. The principal ingredients of polyurethane resins and their suppliers include the following:

• Diphenylmethane diisocyanate (MDI): Mobay, Rubicon, Upjohn

• Toluene diisocyanate (TDI): Allied, BASF, Dow Chemical, Du Pont, Mobay, Olin Chemicals, Rubicon, Spencer Kellogg, Union Carbide

• Aliphatic isocyanates: Mobay

• Polyols: about 20 producers, the majors being Du Pont, Dow, Olin Chemicals, Union Carbide, Upjohn, Witco, and Wyandotte

• Castor oil: Baker, Spencer Kellogg

Major suppliers of polyurethane embedding compounds are Conap (Conathane™), Dexter (Hysol™), and P. D. George (Sterling).

Chemistry

The basic building blocks of polyurethane resins are di- or poly-isocyanates, commonly a mixture of 80 percent toluene-2,4-diisocyanate and 20 percent toluene-2,6-diisocyanate:

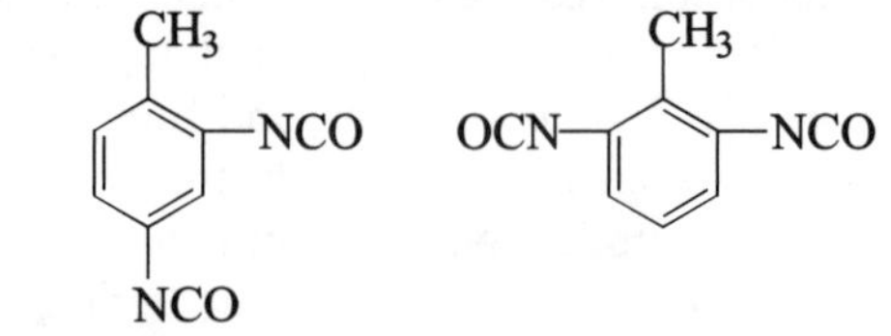

toluene-2,4
diisocyanate

toluene-2,6
diisocyanate

or

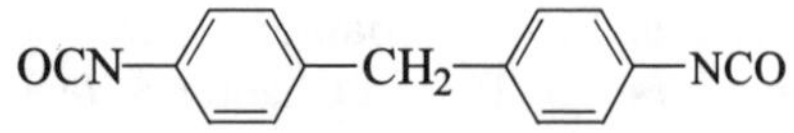

p,p'-diphenylmethane diisocyanate

Because of the high toxicity of toluene diisocyanate, it is usually formed into a prepolymer by reaction with a polyol, such as triethylolpropane, which combines with three of the six available isocyanate groups to form a low-vapor-pressure adduct safe to work with:

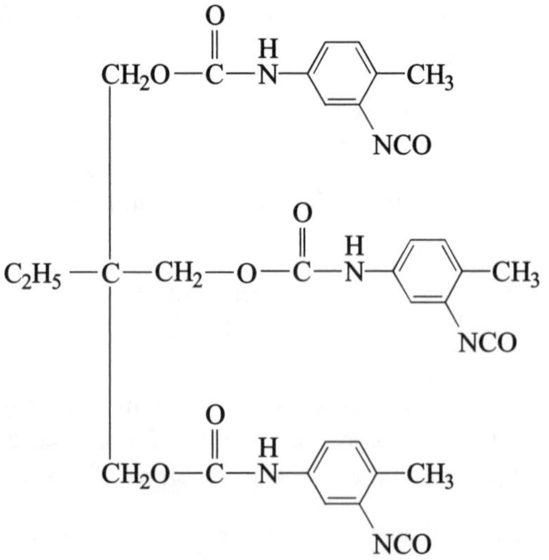

urethane adduct

The urethane linkage occurs when a hydroxyl group reacts with an isocyanate group:

$$R—N{=}C{=}O + HO—R' \longrightarrow R—N—C—O—R'$$

isocyanate   hydroxyl   urethane resin
                compound

where R is an aliphatic or aromatic hydrocarbon group.

When polyisocyanates combine with polyols, the resulting products are polyurethanes. The polyurethanes used as embedding resins are usually two-component polyol-cured resins ASTM Type 5. With increasing ratio of triols to diols, crosslinking is increased. Commonly used materials include hydroxyl-terminated polyesters, polyethers, and castor oil, as well as epoxies with free hydroxyl groups in their structure.

Polyurethane Classification

The principal types of polyurethanes are classified by the American Society for Testing and Materials (ASTM) as follows:

Type 1: One-component urethane alkyds, or oil modified urethanes, are formed by the reaction of a diisocyanate with vegetable oils, or their fatty acids, and polyhydric alcohols. They cure rapidly by oxidation in as few as 10 minutes. Uses include insulation and conformal coatings, wood, metal, and marine finishes.

Type 2: One-component moisture-cured urethanes are isocyanate prepolymers made by reacting polyols with excess isocyanate. When exposed to humid air, the amino groups initially formed react with more isocyanate generating urea linkages and releasing carbon dioxide. Films, which must be thin to prevent entrapment of gas, are extremely tough and find use in seamless floors, bowling alleys, gym floors, industrial floors, and insulating coatings.

Type 3: One-component heat-cured urethanes contain phenol-blocked isocyanates. When heated to 160°C (320°F), the phenol is expelled and the liberated isocyanate then reacts with a polyol. This type is widely used in magnet wire enamels.

Type 4: Two-component catalyst cured urethanes are similar to Type 2 urethanes, but are cured with catalysts, such as tertiary amines, forming urea linkages. This type is used for textile finishes and also for floor coatings.

Type 5: Two-component polyol-cured urethanes are made by reacting isocyanate prepolymers with hydroxyl-terminated polyesters, or polyols such as castor oil. This is the principal type used in solvent-base and solventless insulating coatings.

Type 6: Thermoplastic urethanes are the reaction product of isocyanates with polyester or polyether diols. They have high molecular weight and are supplied as lacquers in ketone or ester solvents. They are used as textile coatings to give the "wet look."

Federal Specification MIL-M-24041

The title of this specification is *Molding and Potting Compound, Chemically Cured, Polyurethane*, and it covers the requirements of two-component, chemically cured, polyurethane compounds for the molding, encapsulating, and potting of watertight electrical connectors, cables, cable end seals, circuit boards, and other electrical and electronic components.

Compounds covered are of the following categories and types:

Category A—(4,4′ Methylene BIS (2-chloroaniline) cured)
Category B—(Non-4,4′ Methylene BIS (2-chloroaniline) cured)
Type I—Two-component units
Type II—Premixed, degassed, and frozen

Referenced standards include the following standards of the American Society for Testing and Materials (ASTM):

• ASTM D 149, *Dielectric Breakdown Voltage and Dielectric Strength of Electrical Insulating Materials at Commercial Power Frequencies*
• ASTM D 150, *A-C Loss Characteristics and Permittivity (Dielectric Constant) of Solid Electrical Insulating Materials*
• ASTM D 257, *D-C Resistance of Conductance of Insulating Materials*
• ASTM D 395, *Rubber Property—Compression Set*
• ASTM D 412, *Rubber Properties in Tension*
• ASTM D 495, *High-Voltage, Low-Current, Dry Arc Resistance of Solid Electrical Insulation*
• ASTM D 518, *Rubber Deterioration-Surface Cracking*
• ASTM D 624, *Rubber Property—Tear Resistance*
• ASTM D 792, *Specific Gravity and Density of Plastics by Displacement*
• ASTM D 1149, *Rubber Deterioration-Surface Ozone Cracking in a Chamber (Flat Specimens)*
• ASTM D 2240, *Rubber Property—Durometer Hardness*

Table 6-5 shows values for dielectric properties required by this specification.

TABLE 6-5   Dielectric Properties Required of Polyurethane Embedding Compounds (MIL-M-24041)

PROPERTY	VALUE
Dielectric strength, volts/mil, minimum, $^1/_8$ in	250
Volume resistivity, $\Omega \cdot$ cm, minimum	4×10^{11}
@ 250°F (121°C)	2.5×10^9
Surface resistivity, ohms, minimum	2.4×10^{12}
@ 250°C (121°)	2.4×10^9
Dielectric constant	
1 kHz, maximum	10
1 MHz, maximum	8
Power factor (approximate dissipation factor)	
1 kHz, maximum	0.08
1 MHz, maximum	0.10
Arc resistance, seconds, minimum	50

Note: The above are room temperature values unless otherwise stated.

Processing

A major concern in handling the ingredients for polyurethanes is their hygroscopic nature. When isocyanate and water react, an amine is formed, thereby decreasing ure-thane linkages. It is, therefore, essential to ensure that all ingredients and the storage environment are free of moisture. Isocyanates are also highly toxic, and precautions must be taken to control the environment where they are present and to provide adequate protection for workers handling them. See also "Embedding Processes" (p. 181) in this chapter.

Grades

Commercially available embedding products include Types 1, 2, and 5. Type 3 is used in wire enamel. Polyurethane modified polybutadiene grades are used where hydrolytic stability is a concern. Mixing ratios vary widely for electrical grades (13/100–100/39 by weight).

Properties

The outstanding feature of polyurethanes is their toughness and resistance to abrasion. Chemical resistance is fair to good, except for chlorinated solvents. Methylene chloride, for example, will strip polyurethanes off parts. Dielectric properties are retained better than those of epoxies at the upper end of their useful temperature range.

The formerly poor hydrolytic stability of polyurethanes has been overcome with the development and commercial availability of polyurethane-butadiene polymers.

TABLE 6-6   Key Properties of Selected Electrical Grade Polyurethane Embedding Resins

	ASTM TEST METHOD	TYPE 5 2-PART SOLVENTLESS FLEXIBLE VALUE	2-PART SOLVENTLESS POLYBUTYLENE-BASED VALUE
Specific gravity*	D 792	1.04	0.98
Dielectric strength 62.5 mils			
Short time, volts/mil*	D 149	645	610
Dielectric constant*	D 150		
100 Hz		5.71	3.14
10^6 Hz		3.42	2.87
Dissipation factor*	D 150		
100 Hz		0.123	0.030
10^6 Hz		0.038	0.011
Volume resistivity, $\Omega \cdot cm$*	D 257	3.1×10^{13}	2.1×10^{15}
Arc resistance, seconds*	D 495	180	>120
Water absorption, 24 h, %	D 570	0.40	0.40
Thermal class, °C		130	130
Tensile strength at break, psi*	D 638	800	1950
Cure time, days*		3–5	7

*At room temperature.

These materials pass government tests for 70 days at 100°C (212°F), 95 to 100 percent relative humidity, and for 120 days at 85°C (185°F), 95 to 98 percent relative humidity. Polyester polyurethanes are poorest in hydrolytic stability.

Table 6-6 shows key property values for typical electrical grade embedding resins.

Electrical/Electronic Uses

Polyurethanes are used for embedment of transformers, coils, switches, inductors, solid-state ignition systems, voltage regulators, ballasts, microcircuits, rectifiers, and printed circuit assemblies.

Silicones
(Embedding Resins)

Silicones are used where parts are subject to higher or lower temperatures outside the operating limits of lower cost encapsulants. Electrical grade silicone encapsulating resins and room temperature vulcanizing (RTV) elastomers are made by Dow Corning, General Electric, Stauffer Chemical, and Thermoset Plastics.

Chemistry

Silicones for encapsulation of electrical/electronic parts are usually prepared by a process in which no water or other contaminant is produced during cure. Siloxane polymers containing methyl, vinyl, or allyl groups are polymerized by free-radical mechanism initiated by catalysts such as benzoyl, dichlorobenzoyl, or di-tertiary butyl peroxides. For example, it is hypothesized that when methyl vinyl silicone resin is heated with dichlorobenzoyl peroxide, two dichlorophenyl free radicals are formed as carbon dioxide is driven off. These free radicals add to the vinyl group of the silicone resin, thereby forming other free radicals which, in a complex exchange, develop into a methyl free radical on one polymer chain and an ethyl free radical on another. These chains now join, forming a propyl linkage with regeneration of dichlorophenyl free radicals:

$$\left(\begin{array}{c} CH_3 \\ | \\ -Si-CH_2-CH_2-CH_2-Si- \\ | \\ O \\ | \end{array}\begin{array}{c} CH_3 \\ | \\ \\ | \\ O \\ | \end{array}\right)_n$$

Silicones which cure to room temperature vulcanizing (RTV) elastomers are prepared from mono-, di-, and trichlorosilanes, or mixtures thereof. These chemicals react with water to form silanols which may be polymerized by condensation with other hydroxy silicones or alkoxy silanes. Water or alcohol is a byproduct. Catalysts used to facilitate this reaction include triethanolamine, dibutyltin dilaurate, and metal salts of organic acids. Improved properties and shorter cures may be achieved by elevating curing temperatures. The structure of these cured silicones may be represented as follows:

Grades

Grades of the first type are available as clear or black, room temperature or heat curing. High-tear-strength and flame-retardant grades are also available. Grades are two-part, except for a one-part material, and are solventless.

RTV grades include pigmentable bases which are reversion-resistant. These grades are two-part.

Transparent dielectric gels are available for stress-free and vibration-dampening applications.

Federal Specification MIL-S-23586

The title of this specification is *Sealing Compound, Electrical, Silicone Rubber, Accelerator Required*. This specification is approved for use by all departments and agencies of the Department of Defense, and it covers the requirements for room temperature curing nonfuel-resistant silicone rubber compounds intended for use at temperatures ranging from –80° to 400°F (–62°–204°C) for various electrical applications on Military Weapons Systems.

There are three viscosities, low, intermediate, and high, and three curing rates, fast, medium, and slow. Three kinds of cure are specified:

Grade A—Condensation cure

Grade B1—Reversion resistant, condensation cure

Grade B2—Reversion resistant, addition cure

Referenced standards include the following standards of the American Society for Testing and Materials (ASTM)

- ASTM D 149, *Dielectric Breakdown Voltage and Dielectric Strength of Electrical Insulating Materials at Commercial Power Frequencies*
- ASTM D 150, *A-C Loss Characteristics and Permittivity (Dielectric Constant) of Solid Electrical Insulating Materials*
- ASTM D 257, *D-C Resistance or Conductance of Insulating Materials*
- ASTM D 412, *Rubber Properties in Tension*
- ASTM D 471, *Rubber Property—Effect of Liquids*
- ASTM D 495, *High-Voltage, Low-Current, Dry Arc Resistance of Solid Electrical Insulation*

TABLE 6-7   Dielectric Properties Required of Silicone Rubber Embedding Compounds (MIL-S-23586)

PROPERTY	LOW VISCOSITY GRADES		INTERMEDIATE VISCOSITY GRADES		HIGH VISCOSITY GRADES	
	A	B1, B2	A	B1, B2	A	B1, B2
Dielectric strength, volts/mil minimum*	400	400	400	400	400	400
Volume resistivity, $\Omega \cdot$cm, minimum	2×10^{14}	5×10^{13}	1×10^{14}	3×10^{13}	1×10^{14}	1×10^{14}
@ 300°F (140°C)	2×10^{12}	5×10^{11}	1×10^{13}	5×10^{11}	2×10^{12}	2×10^{12}
Surface resistivity, ohms, minimum	1×10^{15}	3×10^{14}	1×10^{15}	1×10^{15}	1×10^{15}	1×10^{15}
@ 300°F (149°C)	1×10^{14}	3×10^{13}	1×10^{14}	1×10^{14}	1×10^{14}	1×10^{14}
Dielectric constant						
1 kHz, maximum	4.5	4.5	4.5	4.5	4.5	4.5
1 MHz, maximum	4.5	4.5	4.5	4.5	4.5	4.5
Dissipation factor						
1 kHz, maximum	0.020	0.020	0.020	0.020	0.020	0.020
1 MHz, maximum	0.010	0.010	0.010	0.010	0.010	0.010
Arc resistance, seconds, minimum	100	100	100	100	100	100

*0.075 in thick sample.

Note: The above are room temperature values unless otherwise stated.

- ASTM D 746, *Brittleness Temperature of Plastics and Elastomers by Impact*
- ASTM D 2240, *Rubber Property—Durometer Hardness*

Table 6-7 shows values for dielectric properties required by this specification.

Processing

Working times depend primarily on resin composition, and catalyst type and concentration, and may vary from 1 minute to several hours. Materials are adaptable to automatic dispensing equipment. Grades which cure at room temperature require about 24 hours. Oven curing grades require 4 hours at 149°F (65°C). See also "Embedding Processes" (p. 181) in this chapter.

Properties

The two outstanding properties of silicone encapsulants are (1) their wide useful temperature range, −86° to 482°F (−65°–265°C), over which both dielectric and mechanical properties are quite stable; and (2) their extended life expectancy, which may be 100 times that of other suitable encapsulants. Class 180 insulation applies specifically

TABLE 6-8   Key Properties of Basic Clear Two-Part Heat Curing Electrical Grade Silicone Embedding Resin

	ASTM TEST METHOD	VALUE
Specific gravity*	D 792	1.05
Dielectric strength, 125 mils, short time, V/mil*	D 149	550
Dielectric constant*	D 150	
60 Hz		2.7
10^6 Hz		2.7
Dissipation factor*	D 150	
60 Hz		0.001
10^6 Hz		0.001
Volume resistivity, $\Omega \cdot$cm*	D 257	2×10^{14}
Arc resistance*	D 495	na
Water absorption, $^1/_8$ in thick specimen, %	D 570	
24 h @ 73°F (23°C)		<0.10
Deflection temperature, 264 psi, °F (°C)	D 648	na
Maximum recommended service temperature °F (°C)		392 (200)
Tensile strength at break, psi*	D 638	900
Izod impact strength	D 256	
Foot pounds per inch notched $^1/_8$ in specimen*		na
Flammability ratings		
UL Standard 94, $^1/_{16}$ in specimen		
Standard grade		V-1
Oxygen index, %	D 2863	
Standard grade		na

*At room temperature.

na = not available.

to silicones, indicating they may be used for long periods at 180°C (356°F). Dielectric properties are excellent.

The mechanical properties of silicones are lower than for other encapsulants and adhesion to most materials is inherently poor. The initial cost of silicones is significantly higher than for other encapsulants.

Table 6-8 shows key property values for a basic electrical grade embedding resin.

Electrical/Electronic Uses

The relatively high cost of silicones limits their usage to applications too demanding for other encapsulants. Uses include power supplies, relays, capacitors, amplifiers, coils, cores, connectors, circuit boards, subassemblies, and military components.

Magnet Wire Enamels

INTRODUCTION

Magnet wire enamels are coatings which electrically insulate magnet coils used to energize electrical and electronic equipment, including motors, generators, transformers, solenoids, relays, and encapsulated units.

Because of their internal requirements, General Electric and Westinghouse were the pioneers in the early development of wire enamels. As the electrical industry rapidly expanded, Schenectady Varnish (now Schenectady International) and P. D. George became the principal enamel suppliers to magnet wire manufacturers without captive facilities. Today, most magnet wire companies produce at least some of their own enamel requirements.

COIL MANUFACTURE

To make field coils for motors and generators, magnet wire is wound to the desired shape on a mandrel or form. The coils are then bound by cord or tape. Coils for transformers are wound on tubes of insulating material. Layers of wire are separated by insulating grade paper or other dielectric material. Magnet wire is wound on bobbins specifically designed for each end use to make coils for electronic equipment. Most large coils are subsequently treated with insulating varnishes, which, on curing, provide additional insulation and give structural integrity to the coils. Small coils are often encapsulated with epoxy or polyester compounds.

BONDABLE MAGNET WIRE

Varnish impregnation of coils may be eliminated through the use of bondable top-coats such as polyvinyl butyral, polyamides, epoxies, and polyesters. The adhesive type is selected to retain bond strength at the rated thermal class of the magnet wire. Bondable magnet wire is formed into coils using the same equipment and techniques employed for standard magnet wire with the additional provision for activating the adhesive.

Adhesives are activated either by solvents or heat. In solvent bonding, the adhesive is activated either by passing the wire through a solvent-saturated wick or by spraying or brushing the solvent on the coil. The adhesive hardens as the solvent evaporates. In large, multilayer coils where evaporation paths are impeded, this process takes place slowly. This bonding method is for use in Classes 105 and 130 magnet wire. It is not recommended for higher classes. For Class 105, solvents include ethyl or methyl alcohol. Ethylene dichloride is suitable for Class 130 magnet wires.

Adhesives may also be activated by thermally curing coils formed with adhesive coated magnet wire. Baking temperatures are determined by the adhesive type and the thermal rating of the magnet wire. Baking time depends on coil configuration, and must be sufficient for heat to penetrate to inside windings. This can usually be accomplished within an hour.

Another method for heat bonding is to pass a high electric current, usually direct current, through the coil long enough to allow the temperature of the coil to rise to the flow point of the bonding agent. Care should be taken to avoid hot spots within the coil. Where solderable insulation is used, temperatures throughout the coil should not exceed 180°C.

All these bonding methods produce fumes which must be effectively removed from the workplace with the minimum exposure to workers.

MAGNET WIRE MANUFACTURE

Magnet wire is composed of a copper or aluminum conductor covered with an insulating material, predominantly wire enamel. Other insulations include servings of paper, textiles, nylon, Dacron™, and glass with or without impregnants or coatings. Besides their dielectric function, these coverings provide protection against abrasion and winding stresses. Desirable properties in wire coverings include flexibility, windability, solderability without stripping, abrasion resistance, compatibility with insulating varnishes and compounds, and long service life.

In addition to round wire, conductors may have square or rectangular cross sections. When the thickness of rectangular conductors is less than 0.008 inch, they are termed *foils*. Above this thickness and with width-to-thickness ratios greater than 50 to 1, conductors are called *strip*. Insulated aluminum strip conductors are used in distribution and power transformers, field coils in rotating electrical equipment, lift magnet coils, and in many other electromagnetic devices where they provide better heat transfer and stronger coil structure than round wire.

Although traditionally comprising little more than 10 percent of the weight of all magnet wire conductors, aluminum has some advantages over copper, including:

* Aluminum may be anodized, adding insulation properties and mechanical protection to conductors.

* Aluminum is only 30 percent the weight of an equal volume of copper, a possible advantage in some applications. However, aluminum, with 61.8 percent of the conductivity of copper, requires larger wire sizes for the same current-carrying capacity, with resultant greater coil bulk, possibly requiring a larger equipment housing.

* Copper wire, which oxidizes at temperatures over 200°C, needs silver or nickel coatings for use at these temperatures. Nickel plating permits use up to 260°C, and nickel cladding is needed for use up to 400°C.

* The same insulation usually has a thermal rating of one class higher on aluminum.

Copper magnet wire is manufactured from electrolytic tough pitch copper for which these specifications are applicable:

* ASTM B 49, *Hot-Rolled Copper Rods for Electrical Purposes*
* ASTM B 115, *Electrolytic Cathode Copper*

Aluminum magnet wire is derived from high-purity electrical grade aluminum covered by ASTM B 230, *Aluminum 1350-H19 Wire, for Electrical Purposes*.

Conductor sizes for magnet wire typically range from 0.002 inches nominal diameter for American Wire Gauge size 44 to 0.0641 inches nominal diameter for size AWG

14, with other sizes available. It is desirable to keep insulation thickness to the necessary minimum. A measure of design efficiency is the *space factor*, or ratio of the space occupied by the bare wire of an insulated conductor to the space occupied by the insulated wire. A higher space factor means more conductor for a given volume, imparting higher electrical conductance and thermal conductivity to the coil and permitting more current to flow with less temperature rise.

NEMA STANDARD AND FEDERAL SPECIFICATION FOR MAGNET WIRES

The National Electrical Manufacturers Association (NEMA) has developed standard MW 1000 for magnet wires. Coating thicknesses are designated as single film, heavy film, and triple film. The related Federal Specification is J-W-1177, *Wire, Magnet.*

ASTM STANDARDS FOR FILM-INSULATED MAGNET WIRE

The American Society for Testing and Materials has developed the following test methods for enameled magnet wire:

- D 1676, *Film-Insulated Magnet Wire*

Table 7-1 lists test methods described in this standard.

TABLE 7-1   Property Tests for Film-Insulated Magnet Wire (ASTM D 1676)

PROPERTY MEASURED
Film flexibility and adherence
Stiffness
Oiliness
Continuity
Heat shock
Unidirectional scrape resistance
Elongation
Dielectric breakdown voltage
Resistance to softening in liquids
Solderability
Electrical resistance of conductors
Measure of diameter and film addition of round magnet wire
Cut-through temperature of film-coated round wire
Dielectric breakdown voltage at elevated temperature
Percent extractables of magnet wire insulation by refrigerants
Dielectric breakdown voltage after conditioning in refrigerant atmosphere
High voltage dc continuity bench test
Bond strength of round magnet wire self-bonding coating by the helical coil text

- D 4880, *Salt Water Proofness of Insulating Varnishes Over Enamelled Magnet Wire*
- D 4881, *Thermal Endurance of Varnished Fibrous or Film Wrapped Magnet Wire*
- D 4882, *Bond Strength of Electrical Insulating Varnishes by the Twisted Coil Test*

PROCESSING WIRE ENAMEL

In the production of magnet wire, it is essential that the conductor present a smooth, clean surface, as free as possible of oxides. Otherwise, coatings will not have good adhesion and will tend to have pinholes and other imperfections that degrade dielectric and mechanical properties.

Traditionally, magnet wire is made by passing copper to aluminum wire through tanks containing a solution of insulating enamel. Enameling dies regulate film thickness after which the coated wire passes through a vertical oven with controlled temperature where solvent removal and curing take place. Multiple passes permit complete curing and solvent removal, and reduce likelihood of pinholes in the insulation. Solvent vapors must be burned or recycled so that emissions are in compliance with EPA and OSHA standards and regulations. See "Federal Air Pollution Program," p. 517.

Bondable, nylon, polyamide-imide, or modified polyester topcoats are applied over basecoats in much the same way in a tandem or separate tower.

MAGNET WIRE THERMAL CLASSIFICATION

The thermal classification of a magnet wire insulation is determined by the test method described by American Society for Testing and Materials Publication 2307, *Thermal Endurance of Film-Insulated Round Magnet Wire*. Twisted pairs of enameled wire are subjected to increasing temperature levels for a specified number of days at each temperature level. Factors influencing test results include wire size and film thickness and oven operating properties. To be qualified for a given temperature class, an enamel should remain effective for a period of 5,000 hours when exposed to a temperature 20°C above the class rating, and its extrapolated life must be at least 20,000 hours at the claimed thermal rating. This test method notes: "The temperature index determined by this test method is a nominal or relative value expressed in degrees Celsius at 20,000 hours. It is to be used for comparison purposes only and is not intended to represent the temperature at which the film insulated wire could be operated."

Since copper oxidizes rapidly at temperatures above 200°C, magnet wire enamels coated on copper wire tend to lose their adhesion at these temperatures, thereby lowering useful operating life. Therefore, testing performed on enameled copper wires below 200°C produces more meaningful results than testing at higher temperatures.

Following are the generally recognized thermal classes:

CLASS	FOR PROLONGED LIFE AT °C
A or 105	105
B or 130	130
F or 155	155
H or 180	180
200	200
220	220
Over 220	Over 220

Since temperatures vary at different points within operating electrical apparatus, the thermal class of every insulation in the system does not necessarily have to meet hot spot requirements, and in practice often does not. Where there is no possibility that an insulation will be exposed to high temperatures existing at some point in the apparatus, a lower thermal class insulation may be sufficient. Exceptions to this practice are found in apparatus manufactured to government specifications or approved by Underwriters Laboratories.

See also "Temperature Index of Dielectric Materials," p. 21.

MAGNET WIRE ENAMEL TESTING

Standards and testing procedures for wire enamels are described in ASTM Publication D 3288, *Magnet Wire Enamels*.

Federal Specification J-W-1177 describes tests for the following properties:

Adhesion and flexibility	Low-voltage continuity
Heat shock	High-voltage continuity
Elongation	Solubility
Scrape resistance	Completeness of cure
Springback	Solderability
Thermoplastic flow	Heat and solvent bonding
Dielectric strength (at room and rated temperature)	

NEMA testing procedures for essentially the same properties are contained in Standard MW 1000. The significance of selected lists and brief descriptions follow.

Adhesion, necessary to prevent cracks in the film as the magnet wire is stretched, is determined by elongating a 10 inch specimen, sizes 13 AWG and heavier, 25–30 percent at a rate of 12 inches per minute. Normal visual inspection should reveal no cracks. For sizes 14 AWG and finer, the wire is given a sudden jerk with minimum elongation of 15–20 percent (depending on the wire size), or to the breaking point, whichever is less. Magnification of 6× to 15× (depending on wire size) is used to determine cracks in the film for 31 AWG and finer wires.

Flexibility is the ability of magnet wire after preelongation of 15–30 percent (depending on wire size) to be wrapped not more than ten turns around a mandrel one–five times (depending on wire size) the AWG size of the wire without film failure. The smaller the mandrel diameter without film failure, the higher the flexibility rating for a given size wire. Normal visual inspection is used to determine film failure for 30 AWG and heavier wires. Magnification of 6× to 15× (depending on wire size) is used for 31 AWG and finer wires.

Heat shock resistance indicates how well an insulation holds up when elongated magnet wire is wrapped not more than ten times around a mandrel typically one–three times the AWG size of the wire. The specimen is then removed from the mandrel and placed in a circulating air oven for $^1/_2$ hour at 20°C above the specified thermal rating of the magnet wire. There should be no cracks on subsequent normal visual inspection for sizes 30 AWG and heavier wire or under magnification of 6× to 10× for 31 AWG and finer wires. Polymers with low heat shock resistance require gradual preheating before varnishing or encapsulation.

Scrape or abrasion resistance indicates the mechanical abuse a magnet wire will withstand. This property is determined by a procedure in which a special needle is rubbed on the coated wire at a specified rate with a constantly increasing load until dielectric failure occurs at a potential of 7.5 volts between the needle and the wire. The load weight of failure measures the scrape resistance. An alternative method measures the number of strokes to failure with a constant load.

Springback measures the tendency of a coil to spring open when removed from the form. A low springback indicates high formability. A specimen of film-coated wire is wound three turns around a mandrel having a specified diameter ranging from $^3/_4$ inch for size 30 AWG wire to $3^1/_4$ inch for size 14 AWG wire, under tension ranging from 2 ounces for size AWG 30 wire to 16 ounces for size AWG 14 wire. A special device measures the degree of springback when tension is removed.

Thermoplastic flow, or *cut-through resistance*, indicates the temperature at which the insulation film fails. Two lengths of the specimen positioned at right angles to each other are loaded with 2,000 grams for size 18 AWG or 100 grams for size 36 AWG. The test equipment applies a voltage of 115 volts ac through the crossover of the wires as the temperature is increased at a rate of less than 5°C per minute to the point of dielectric failure.

Dielectric strength is an indication of the film thickness needed to withstand operating and surge voltages. The specimen may be either two layers of magnet wire or two twisted wires with specified number of twists. The ac voltage applied between the wire layers or twisted wires is increased gradually so that breakdown does not occur in less than 5 seconds. The voltage at breakdown is the basis for determining the dielectric strength of the specimen.

Low-voltage continuity measures the film defects of magnet wire sizes 31–56 AWG through dielectric failure at low voltage. Formerly, a 100-foot specimen of magnet wire was passed through a 1 inch long mercury bath at a speed of 100 feet per minute. A direct-current potential of 20–75 volts (depending on wire size) was applied between the bath and the wire. A discontinuity indicating device operated when the resistance between the bath and the wire was less than 5,000 ohms but did not operate when the resistance was 10,000 ohms or more. The number of faults per minute indicated the low-voltage continuity. This test is being revised to eliminate mercury with its potential health hazard.

High-voltage continuity measures the film defects of magnet wire sizes 14–30 AWG through dielectric failure at high direct-current voltage. A 100-foot specimen of magnet wire is passed with contact over an energized sheave at a speed of 60 feet per minute. Before and after the contact sheave is a similar grounded guide sheave. The sheaves are arranged so that the length of wire contacting the energized center sheave is one inch. The specimen wire is grounded. The specified open-circuit voltage applied to the contact sheave ranges from 500 volts for sizes 25–30 AWG single film to 2,000 volts for sizes 14–24 AWG triple film. The sensitivity of the fault detection device is such that the circuit is capable of detecting any fault having a resistance of less than 30 megohms, but will not operate when the insulation resistance of the test wire film exceeds 180 megohms. The number of faults per minute indicates the high-voltage continuity.

Solubility in certain liquids is tested by immersing stress-relieved specimens of magnet wire 12 inches long at least 6 inches in the liquids specified. After removal from the liquids, specimens are promptly drawn once between the folds of cheesecloth held firmly between the forefinger and the ball of the thumb. Removal of the coating indicates failure. Another test for solubility is performed on samples removed from liquids and placed in a device in which a needle scrapes the surface of the film at right angles to the lengths of wire. The needle is loaded with 580 grams when testing copper magnet wire and 340 grams when testing aluminum magnet wire. Film failure is indicated by shorting in a circuit having a potential of 7.5 volts between the needle and the conductors. Perhaps the simplest method is to immerse the magnet wire specimens in specified liquids for 24 hours at 25°C. Following removal of specimens, the coating is rated *pass* (no effect), *fair* (some surface softening or crazing), or *fail* (severe surface softening or crazing).

Solderability without the necessity of stripping insulation is a highly desirable property which facilitates assembly operations. Soldering tests are made for wire sizes 14–23 AWG by forming a loop of a 12 inch length of magnet wire and twisting the ends together for a distance of $^{3}/_{4}$–1 inch with five–ten turns. The loop is first immersed in a rosin-alcohol flux, and then immersed in a soldering pot of 50/50 tin–lead solder for 8–10 seconds at 430°C. For sizes 24–46 AWG, a magnet wire specimen is wound with five–ten turns for a distance of $^{1}/_{2}$–$^{3}/_{4}$ inch around the end of a 6 inch length of 20 AWG tinned, hot-dipped copper wire. The sample is then immersed in a rosin-alcohol flux, and then immersed in a soldering pot of 50/50 tin–lead solder at 360°C for 4–6 seconds. Magnification of 6× to 10× is used to inspect sizes 37–46 AWG wire. Normal vision is used to examine larger sizes.

Heat and solvent bonding are methods used to bond coils of magnet wire coated with bonding agents, thus eliminating the necessity of subsequent varnish treatment or encapsulation. Specimens are prepared from sizes 18, 26, and 36 AWG wires. Size 18 and 26 AWG wires are formed into a 3-inch single-layer coil around mandrels with diameters of 0.250 and 0.157 inch, respectively. Size 36 AWG wire is formed into a coil of 50 turns around a mandrel with diameter of 0.0394 inch. *Heat bonding* is accomplished in a forced air oven for 1 hour at 150°C, after which specimens are cooled to room temperature. *Solvent bonding* is performed by dipping specimens in specified solvent for 5 seconds with subsequent drying for 1 hour at room temperature.

Bond strength is measured in a special device which applies a specified load or, alternatively, an increasing load to the center of the coil suspended at its ends.

DEVELOPMENT PROGRAMS

One of the prime objectives of magnet wire development programs is the reduction or elimination of organic solvents, with resultant material cost savings, significantly fewer pollution control problems, energy and feedstock conservation, and lower fire or explosion hazard. It is well to remember, however, the key role solvents have played in the improvement in magnet wire technology over the years. Solvent systems not only provide a practical means for applying resins to conductors, but function to wet the conductor to ensure strong film adhesion and provide for timely solvent release during film formation, thereby minimizing pinholes, blistering, and other film defects. Eliminating solvents thus requires changes in resin composition as well as processing methods.

The following are programs under development for several years, but which so far have not produced results to encourage widespread commercialization.

High solids coatings, with fewer than 25 percent solvents by volume, have long been a goal in the magnet wire industry. The problem is to maintain practical viscosity and flow characteristics within the limitations of present processing equipment. Solvent content in the 55–70 percent range is about the present limit in the U.S. It is reported that 50–60 percent solids enamels are in use in Germany. Preheating would lower the viscosity, but would introduce problems of pot life and solvent loss.

Water-borne systems development has been suspended or abandoned. A major problem with these systems is the high surface tension of water which impedes proper wetting of the conductor by the enamel, causing borderline mechanical and dielectric properties.

Hot melt enamels have also not met expectations. The manufacturing process consists essentially of extruding a heated thermoplastic resin on a preheated conductor. Although available equipment can be used with these enamels, extensive modification would be required to achieve high productivity with resultant profit improvement. Since hot melts have much higher viscosities than solvent enamels, they produce thicker films per pass. Two passes using hot melts build film thickness equal to six passes with solvent enamels. However, fewer passes increase the likelihood of film defects. Polyester hot melts have a tendency to craze and crack with shelf aging. Nylon hot melts are better in this respect, but have high water absorption with significant reduction in dielectric properties. Thermal properties of hot melt films could be improved by crosslinking, which should also enhance physical and electrical characteristics, but this could present production problems of adequate pot life and maintaining constant viscosity.

Powder coating is another intriguing technology which would eliminate solvents and permit higher operating speeds with a single coating process. Hysol, 3M, and Westinghouse have developed powder coatings and techniques for applying them. Epoxies and polyesters are the base resins employed. In the electrostatic fluidized bed application method, a fluidized bed is formed by charged dry air passing through powder on a porous membrane. This air has been ionized by passing through a high-voltage charging medium. The charged powder forms a cloud and is attracted to the grounded wire moving through the chamber. The coating thickness is controlled by the amount of charged air, the charging voltage, and the speed of the wire. Drawbacks to powder coatings include difficulty in producing uniform thin films, poor corona resistance, relatively high cost for finer powders, and a potential explosion hazard when using fine powders. The main end use so far for powder coated magnet wire has been in liquid filled transformers.

Radiation curing employing ultraviolet (UV) or electron beam (EB) technology can be used to produce crosslinked, tough, flexible films. This approach would employ relatively compact production equipment in comparison with conventional ovens, and higher throughput should improve productivity. Development using this technology still has a long way to go to produce magnet wire with properties equal to those resulting from conventional manufacturing methods. The achievement of a concentric coating and film homogeneity are major problems with radiation curing.

MARKET TRENDS

The earliest use of magnet wire insulated with silk, cotton, or paper dates back to the late 1830s. Since then, magnet wire development has kept pace with the requirements of the electrical, and more recently, electronic industries. For many decades, the principal usage for magnet wire has been in rotating electrical equipment and transformers. Changes are now underway in motor design to achieve greater energy efficiency with lower operating temperature. This may permit use of lower thermal class insulation in some cases. The desire to achieve high space factor will probably also influence selection of insulating film to some degree.

The trend toward smaller cars implies use of smaller electrical and electronic components requiring less magnet wire. Imported cars use little, if any, U.S. made magnet wire. Thus, the near-term outlook is for decreasing usage of domestic magnet wire in motor vehicles. The advent of battery powered electric vehicles incorporating dc motor drives would, of course, require more magnet wire than present vehicles, but production of electric vehicles in volume is not likely to occur, if it occurs at all, before the late 1990s.

In other industries, there is also a trend toward smaller units, but magnet wire usage may remain stable due to increasing volume of units and new applications as, for example, in small motors for computer peripheral equipment.

Bondable magnet wire with thermosetting coating is reported to be used increasingly in Germany, and may gain greater favor in the U.S. than is now the case. Bondable wire avoids the necessity of varnish impregnation.

COMPARATIVE CHARACTERISTICS OF MAGNET WIRES

Table 7-2 shows advantages and limitations of commercially available enameled magnet wires.

TABLE 7-2 Summary of Enameled Magnet Wire Characteristics

INSULATION TEMP. CLASS, °C	ADVANTAGES	LIMITATIONS
Plain enamel (oleoresinous) 105	Low cost Good film continuity Good cut-thru resistance Good overload resistance Ease of mechanical and chemical stripping Tight dimensional tolerances	Low abrasion resistance Not for heavy duty winding Limited compatibility with other varnishes
Modified polyvinyl formal resin 105	Excellent windability High adhesion rating Compatible with most varnishes, waxes, and impregnating compounds Suitable for use with transformer oil Good overall chemical and physical properties	Must be stripped before soldering Crazes when exposed to varnish solvents unless stress relieved first
Polyurethane 105, 130, 155	Solderable at 360°C–425°C (film vaporizes) Self fluxing Superior Q characteristics at high frequencies and humidities Good film adhesion and flexibility Will operate at 120°C continuously—highest in Class 105—with newer versions qualifying for Classes 130 and 155	Not for use where severe overloads may occur Lower abrasion resistance than Formvar or nylon jacketed wires Susceptible to softening through prolonged exposure to solvents Poor radiation stability
Polyurethane with nylon overcoat 130	Solderable with better windability than straight polyurethane Excellent abrasion resistance High heat shock resistance Compatible with most varnishes and encapsulants Excellent solvent resistance	Not for use where severe overloads may occur Dielectric strength lower than straight PUR Not for use in hot transformer oil or fluorocarbon gases Poor radiation stability
Epoxy 130	Outstanding oil and Askarel resistance (wet transformer usage) Good adhesion and flexibility Good resistance to chemicals and moisture Good corona resistance	Must be stripped before soldering
Polyester 155	Good overload characteristics Excellent dielectric properties	Must be stripped before soldering Not for use in oil-filled transformers May hydrolyze in sealed systems
Solderable polyester 155	Solderable at 482°C Long thermal endurance at 175°C Good heat shock resistance Good overload characteristics for solderable wire Good radiation stability	Not for use in oil-filled transformers Not for use with systems using amine type catalysts
Solderable nylon polyester 155	Solderable at 482°C Better windability than unjacketed grade Good resistance to heat and solvent shock	Not for use in oil-filled transformers Not for use with systems using amine type catalysts Not as resistant to moisture as unjacketed wire
Isonel* polyester with polyamide-imide overcoat 200	Physical and electrical properties almost equal to Formvar Excellent cut-thru resistance Excellent resistance to Freon	Must be stripped before soldering Not for use in oil-filled transformers Not for use in systems containing chlorine compounds

TABLE 7-2   (*Continued*)

INSULATION TEMP. CLASS, °C	ADVANTAGES	LIMITATIONS
	Good solvent resistance Compatible with most varnishes and impregnating compounds	Not for use with systems using amine catalysts
Polyester with nylon overcoat 155–180	Excellent windability Resistant to heat and solvent shocks Excellent in flexibility, scrape abrasion and film adhesion Good overload characteristics	Must be stripped before soldering Not for use in enclosed equipment Poor radiation stability
Polyester/ polyamide-imide 200, 220	Polyester base coat with amide-imide topcoat Good windability More resistant to heat and solvent shocks than conventional polyesters Excels regular polyesters in resistance to cut-thru and abrasion resistance High resistant to Freon and solvents Excellent dielectric properties Suitable for use in hermetic motors	Must be stripped before soldering Not for use in enclosed equipment where moisture or chlorine compounds are present
Polyesteramide-imide 200	Single film coating Excellent overload characteristics Compatible with epoxy casting and encapsulating compounds Compatible with most varnishes and impregnating compounds Long thermal endurance at 180–210°C For use in oil-filled transformers Withstands high speed winding applications	Must be stripped before soldering High price
Isomid* polyester-polyimide 180, 200	Single film coating Excellent cut-thru, flexibility and adhesion properties Excellent wet and dry dielectric properties Highly resistant to Freon and solvents Improved resistance to heat shock Compatible with most varnishes and impregnating compounds	Must be stripped before soldering High price
Pyre-ML** polyimide 220 (240 pe MW 1000)	Single film coatings Retains high dielectric properties at operating temperatures of 220°C Highest heat shock resistance and thermal stability of all insulations Highest resistance to radiation High dielectric strength and low loss characteristics Compatible with most varnishes and impregnating compounds May be used in hermetic motors and oil-filled transformers Highly resistant to attack by solvents and chemicals Highest overload rating	Very difficult to strip before soldering Subject to hydrolysis or sealed systems containing moisture Will solvent craze unless stress relieved Highest price film coated wire

*Schenectady International.

**DuPont™.

MAGNET WIRE DATA

Tables 7-3–7-10 show useful data for magnet wires:

TABLE	TITLE	UNITS
7-3	Round Copper Film Coated Wire Data	English
7-4	Round Copper Film Coated Wire Data	Metric
7-5	Round Aluminum Film Coated Wire Data	English
7-6	Round Aluminum Film Coated Wire Data	Metric
7-7	Square Copper Film Coated Wire Data	English
7-8	Square Copper Film Coated Wire Data	Metric
7-9	Square Aluminum Film Coated Wire Data	English
7-10	Square Aluminum Film Coated Wire Data	Metric

TABLE 7-3   Round Copper Film Coated Wire Data
(English Units)

SINGLE FILM

WIRE SIZE AWG*	DIAMETER OVER INSULATION MINIMUM inch	DIAMETER OVER INSULATION MAXIMUM inch	MINIMUM FILM ADDITION inch	WEIGHT lb/M ft	WEIGHT ft/lb	dc RESISTANCE AT 20°C (68°F) Ω/lb
14	0.0651	0.0666	0.0016	12.5	80.0	0.202
15	0.0580	0.0594	0.0015	9.95	101	0.321
16	0.0517	0.0531	0.0014	7.78	127	0.511
17	0.0462	0.0475	0.0014	6.27	159	0.803
18	0.0412	0.0424	0.0013	4.97	201	1.28
19	0.0367	0.0379	0.0012	3.94	254	2.04
20	0.0329	0.0339	0.0012	3.14	318	3.21
21	0.0293	0.0303	0.0011	2.49	402	5.15
22	0.0261	0.0270	0.0011	1.97	508	8.23
23	0.0234	0.0243	0.0010	1.58	633	12.8
24	0.0209	0.0217	0.0010	1.24	806	20.7
25	0.0186	0.0194	0.0009	0.988	1,010	32.8
26	0.0166	0.0173	0.0009	0.780	1,280	52.6
27	0.0149	0.0156	0.0008	0.623	1,610	82.5
28	0.0133	0.0140	0.0008	0.492	2,030	133
29	0.0119	0.0126	0.0007	0.396	2,530	205
30	0.0106	0.0112	0.0007	0.311	3,220	334
31	0.0094	0.0100	0.0006	0.247	4,050	530
32	0.0085	0.0091	0.0006	0.200	5,000	810
33	0.0075	0.0081	0.0005	0.158	6,330	1,300
34	0.0067	0.0072	0.0005	0.124	8,060	2,100
35	0.0059	0.0064	0.0004	0.0980	10,200	3,380
36	0.0053	0.0058	0.0004	0.0782	12,800	5,310
37	0.0047	0.0052	0.0003	0.0634	15,800	8,080
38	0.0042	0.0047	0.0003	0.0501	20,000	12,900
39	0.0036	0.0041	0.0002	0.0385	26,000	22,000
40	0.0032	0.0037	0.0002	0.0302	33,100	35,700
41	0.0029	0.0033	0.0002	0.0246	40,700	53,700
42	0.0026	0.0030	0.0002	0.0196	51,000	84,700
43	0.0023	0.0026	0.0002	0.0153	65,400	140,000
44	0.0020	0.0024	0.0001	0.0126	79,400	206,000

TABLE 7-3   (Continued)

HEAVY FILM

| WIRE SIZE AWG* | DIAMETER OVER INSULATION | | MINIMUM FILM ADDITION | WEIGHT | | dc RESISTANCE AT 20°C (68°F) |
	MINIMUM inch	MAXIMUM inch	inch	lb/M ft	ft/lb	Ω/lb
4	0.2060	0.2098	0.0037	127.5	7.843	0.001949
5	0.1837	0.1872	0.0036	101.0	9.901	0.003103
6	0.1639	0.1671	0.0035	80.09	12.49	0.004934
7	0.1463	0.1491	0.0034	63.55	15.74	0.007838
8	0.1305	0.1332	0.0033	50.40	19.84	0.01246
9	0.1165	0.1189	0.0032	39.97	25.02	0.01983
10	0.0140	0.1061	0.0031	31.73	31.52	0.03148
11	0.0928	0.0948	0.0030	25.2	39.7	0.05004
12	0.0829	0.0847	0.0029	20.0	50.0	0.07950
13	0.0741	0.0757	0.0028	15.9	62.9	0.126
14	0.0667	0.0682	0.0032	12.6	79.4	0.209
15	0.0595	0.0609	0.0030	10.0	100	0.318
16	0.0532	0.0545	0.0029	7.94	126	0.510
17	0.0476	0.0488	0.0028	6.32	158	0.798
18	0.0425	0.0437	0.0026	5.02	199	1.27
19	0.0380	0.0391	0.0025	3.99	251	2.02
20	0.0340	0.0351	0.0023	3.17	315	3.19
21	0.0304	0.0314	0.0022	2.52	397	5.08
22	0.0271	0.0281	0.0021	2.00	500	8.10
23	0.0244	0.0253	0.0020	1.60	625	12.7
24	0.0218	0.0227	0.0019	1.26	794	20.4
25	0.0195	0.0203	0.0018	1.00	1,000	32.4
26	0.0174	0.0182	0.0017	0.794	1,260	51.6
27	0.0157	0.0164	0.0016	0.634	1,580	81.1
28	0.0141	0.0147	0.0016	0.501	2,000	130
29	0.0127	0.0133	0.0015	0.404	2,480	201
30	0.0113	0.0119	0.0014	0.317	3,150	328
31	0.0101	0.0108	0.0013	0.252	3,970	520
32	0.0091	0.0098	0.0012	0.204	4,900	794
33	0.0081	0.0088	0.0011	0.161	6,210	1,280
34	0.0072	0.0078	0.0010	0.127	7,870	2,050
35	0.0064	0.0070	0.0009	0.101	9,900	3,280
36	0.0057	0.0063	0.0008	0.0805	12,400	5,160
37	0.0052	0.0057	0.0008	0.0652	15,300	7,830
38	0.0046	0.0051	0.0007	0.0516	19,400	12,600
39	0.0040	0.0045	0.0006	0.0397	25,200	21,300
40	0.0036	0.0040	0.0006	0.0311	32,200	34,700
41	0.0032	0.0036	0.0005	0.0254	39,400	52,000
42	0.0028	0.0032	0.0004	0.0203	49,300	81,800
43	0.0025	0.0029	0.0004	0.0159	62,900	135,000
44	0.0023	0.0027	0.0004	0.0131	76,300	198,000

*Half sizes are also available.

TABLE 7-4   Round Copper Film Coated Wire Data
(Metric Units)

SINGLE FILM

WIRE SIZE AWG	DIAMETER OVER INSULATION		MINIMUM FILM ADDITION	WEIGHT		dc RESISTANCE AT 20°C
	MINIMUM mm	MAXIMUM mm	mm	kg/m	m/g	Ω/kg
14	1.654	1.692	0.041	18.6	0.0538	0.445
15	1.473	1.509	0.038	14.8	0.0679	0.708
16	1.313	1.349	0.036	11.7	0.0853	1.13
17	1.173	1.206	0.036	9.33	0.107	1.77
18	1.046	1.077	0.033	7.40	0.135	2.82
19	0.932	0.963	0.030	5.86	0.171	4.50
20	0.836	0.861	0.030	4.67	0.214	7.08
21	0.744	0.770	0.028	3.71	0.270	11.3
22	0.633	0.686	0.028	2.93	0.341	18.1
23	0.594	0.617	0.025	2.35	0.425	28.2
24	0.531	0.551	0.025	1.85	0.542	45.6
25	0.472	0.493	0.023	1.47	0.679	72.3
26	0.422	0.439	0.023	1.16	0.860	11.6
27	0.378	0.396	0.020	0.927	1.08	182
28	0.338	0.356	0.020	0.732	1.36	293
29	0.302	0.320	0.018	0.589	1.70	455
30	0.269	0.284	0.018	0.463	2.16	736
31	0.239	0.254	0.015	0.368	2.72	1,170
32	0.216	0.231	0.015	0.298	3.36	1,790
33	0.190	0.206	0.013	0.235	4.25	2,870
34	0.170	0.183	0.013	0.185	5.42	4,630
35	0.150	0.163	0.010	0.1458	6.85	7,450
36	0.135	0.147	0.010	0.1164	8.60	11,700
37	0.119	0.132	0.008	0.0943	10.6	17,800
38	0.107	0.119	0.008	0.0746	13.4	28,400
39	0.091	0.104	0.005	0.0573	17.5	48,500
40	0.081	0.094	0.005	0.0449	22.2	78,700
41	0.074	0.084	0.005	0.0366	27.4	118,000
42	0.066	0.076	0.005	0.0292	34.3	187,000
43	0.058	0.066	0.005	0.0227	44.0	309,000
44	0.051	0.061	0.0025	0.0188	53.4	454,000

TABLE 7-4 (*Continued*)
(Metric Units)

HEAVY FILM

WIRE SIZE AWG	DIAMETER OVER INSULATION		MINIMUM FILM ADDITION	WEIGHT		dc RESISTANCE AT 20°C
	Minimum mm	Maximum mm	mm	kg/m	m/g	Ω/kg
4	5.232	5.329	0.094	189.7	0.00527	0.004297
5	4.666	4.755	0.091	150.3	0.00665	0.006841
6	4.163	4.244	0.089	119.2	0.00839	0.01088
7	3.716	3.787	0.086	94.57	0.01057	0.01728
8	3.315	3.383	0.084	75.00	0.01333	0.02747
9	2.959	3.020	0.081	59.48	0.01681	0.04372
10	2.642	2.695	0.079	47.22	0.02118	0.06940
11	2.357	2.408	0.076	37.5	0.0267	0.1103
12	2.106	2.151	0.074	29.8	0.0336	0.1753
13	1.882	1.923	0.071	23.7	0.0423	0.278
14	1.694	1.732	0.081	18.8	0.0534	0.461
15	1.511	1.547	0.076	14.9	0.0672	0.701
16	1.351	1.384	0.074	11.8	0.0847	1.12
17	1.209	1.240	0.071	9.41	0.106	1.76
18	1.080	1.110	0.066	7.47	0.134	2.80
19	0.965	0.993	0.064	5.93	0.169	4.45
20	0.864	0.892	0.058	4.71	0.212	7.03
21	0.772	0.798	0.056	3.75	0.267	11.2
22	0.688	0.714	0.053	2.97	0.336	17.9
23	0.620	0.643	0.051	2.38	0.420	28.0
24	0.554	0.577	0.048	1.88	0.534	45.0
25	0.495	0.516	0.046	1.49	0.672	71.4
26	0.442	0.462	0.043	1.18	0.847	113
27	0.399	0.417	0.041	0.943	1.06	179
28	0.358	0.373	0.041	0.746	1.34	287
29	0.322	0.338	0.038	0.601	1.67	443
30	0.287	0.302	0.036	0.472	2.12	723
31	0.257	0.274	0.033	0.375	2.67	1,150
32	0.231	0.249	0.030	0.303	3.29	1,750
33	0.205	0.224	0.028	0.204	4.17	2,820
34	0.183	0.198	0.025	0.189	5.29	4,520
35	0.162	0.178	0.023	0.150	6.65	7,230
36	0.145	0.160	0.020	0.120	8.33	11,400
37	0.132	0.145	0.020	0.0970	10.3	17,300
38	0.117	0.130	0.018	0.0768	13.0	27,800
39	0.102	0.114	0.015	0.0591	16.9	47,000
40	0.091	0.102	0.015	0.0463	21.6	76,500
41	0.081	0.091	0.013	0.0378	26.5	115,000
42	0.071	0.081	0.010	0.0302	33.1	180,000
43	0.064	0.074	0.010	0.0237	42.3	298,000
44	0.058	0.069	0.010	0.0195	51.3	437,000

TABLE 7-5   Round Aluminum Film Coated Wire Data
(English Units)

SINGLE FILM

WIRE SIZE AWG	DIAMETER OVER INSULATION		MINIMUM FILM ADDITION	WEIGHT		dc RESISTANCE AT 20°C (68°F) 61.8% CONDUCTIVITY
	MINIMUM inch	MAXIMUM inch	inch	lb/M ft	ft/lb	Ω/lb
14	0.0651	0.0666	0.0016	3.87	258	1.33
15	0.0580	0.0594	0.0015	3.08	325	1.67
16	0.0517	0.0531	0.0014	2.44	410	2.66
17	0.0462	0.0475	0.0014	1.95	512	4.19
18	0.0412	0.0424	0.0013	1.55	645	6.6
19	0.0367	0.0379	0.0012	1.23	813	10.6
20	0.0329	0.0339	0.0012	0.983	1,017	16.7
21	0.0293	0.0303	0.0011	0.778	1,285	26.6
22	0.0261	0.0270	0.0011	0.620	1,612	42.2
23	0.0234	0.0243	0.0010	0.501	1,996	65.7
24	0.0209	0.0217	0.0010	0.391	2,556	106
25	0.0186	0.0194	0.0009	0.313	3,195	167

HEAVY FILM

WIRE SIZE AWG	DIAMETER OVER INSULATION		MINIMUM FILM ADDITION	WEIGHT		dc RESISTANCE AT 20°C (68°F) 61.8% CONDUCTIVITY
	MINIMUM inch	MAXIMUM inch	inch	lb/M ft	ft/lb	Ω/lb
4	0.2060	0.2098	0.0037	39.60	25.25	0.0102
5	0.1837	0.1872	0.0036	31.27	31.98	0.0162
6	0.1639	0.1671	0.0035	24.80	40.32	0.0258
7	0.1463	0.1491	0.0034	19.68	50.81	0.0409
8	0.1305	0.1332	0.0033	15.62	64.02	0.0650
9	0.1165	0.1189	0.0032	12.40	80.64	0.103
10	0.1040	0.1061	0.0031	9.86	101	0.163
11	0.0928	0.0948	0.0030	7.87	127	0.259
12	0.0829	0.0847	0.0029	6.22	161	0.413
13	0.0741	0.0757	0.0028	4.97	201	0.651
14	0.0667	0.0682	0.0032	3.97	252	1.03
15	0.0595	0.0609	0.0030	3.13	319	1.64
16	0.0532	0.0545	0.0029	2.50	400	2.60
17	0.0476	0.0488	0.0028	2.00	500	4.09
18	0.0425	0.0437	0.0026	1.60	625	6.50
19	0.0380	0.0391	0.0025	1.28	781	10.1
20	0.0340	0.0351	0.0023	1.01	990	16.2
21	0.0304	0.0314	0.0022	0.808	1,238	25.6
22	0.0271	0.0281	0.0021	0.650	1,538	40.3
23	0.0244	0.0253	0.0020	0.521	1,919	63.2
24	0.0218	0.0227	0.0019	0.411	2,433	100
25	0.0195	0.0203	0.0018	0.325	3,077	161

TABLE 7-6   Round Aluminum Film Coated Wire Data
(Metric Units)

SINGLE FILM

WIRE SIZE AWG	DIAMETER OVER INSULATION		MINIMUM FILM ADDITION	WEIGHT		dc RESISTANCE AT 20°C (68°F) 61.8% CONDUCTIVITY
	MINIMUM mm	MAXIMUM mm	mm	kg/km	m/g	Ω/kg
14	1.654	1.692	0.041	5.76	0.173	2.93
15	1.473	1.509	0.038	4.58	0.218	3.69
16	1.313	1.349	0.035	3.63	0.276	5.88
17	1.173	1.206	0.035	2.90	0.344	9.23
18	1.046	1.077	0.033	2.31	0.433	14.6
19	0.932	0.963	0.030	1.83	0.546	23.3
20	0.836	0.861	0.030	1.46	0.683	36.7
21	0.744	0.770	0.028	1.16	0.863	58.6
22	0.633	0.686	0.028	0.923	1.083	93.1
23	0.594	0.617	0.025	0.746	1.341	144
24	0.531	0.551	0.025	0.581	1.718	234
25	0.472	0.493	0.023	0.453	2.147	369

HEAVY FILM

WIRE SIZE AWG	DIAMETER OVER INSULATION		MINIMUM FILM ADDITION	WEIGHT		dc RESISTANCE AT 20°C (68°F) 61.8% CONDUCTIVITY
	MINIMUM mm	MAXIMUM mm	mm	kg/km	m/g	Ω/kg
4	5.232	5.329	0.094	58.93	0.0170	0.0225
5	4.666	4.755	0.091	46.53	0.0215	0.0357
6	4.163	4.244	0.089	36.91	0.0271	0.0569
7	3.716	3.787	0.086	29.29	0.0341	0.0902
8	3.315	3.383	0.084	23.24	0.0430	0.143
9	2.959	3.020	0.081	18.45	0.0542	0.228
10	2.642	2.695	0.079	14.67	0.0679	0.359
11	2.357	2.408	0.076	11.71	0.0853	0.571
12	2.106	2.151	0.074	9.26	0.108	0.910
13	1.882	1.923	0.071	7.40	0.135	1.43
14	1.694	1.732	0.081	5.91	0.169	2.27
15	1.511	1.547	0.076	4.66	0.214	3.62
16	1.351	1.384	0.074	3.72	0.269	5.73
17	1.209	1.240	0.071	2.98	0.336	9.02
18	1.080	1.110	0.066	2.38	0.420	14.3
19	0.965	0.993	0.064	1.90	0.525	22.4
20	0.864	0.892	0.058	1.50	0.665	35.8
21	0.772	0.798	0.056	1.20	0.832	56.5
22	0.688	0.714	0.053	0.967	1.033	88.8
23	0.620	0.643	0.051	0.775	1.290	139
24	0.554	0.577	0.048	0.612	1.635	222
25	0.495	0.516	0.046	0.484	2.068	355

TABLE 7-7 Square Copper Film Coated Wire Data
(English Units)

HEAVY FILM

WIRE SIZE AWG	DIAMETER OVER INSULATION MINIMUM inch	MAXIMUM inch	MINIMUM FILM ADDITION inch	WEIGHT lb/M ft	ft/lb	dc RESISTANCE AT 20°C (68°F) Ω/lb	WIRE SIZE AWG
1	0.2894	0.2972	0.0030	318.4	3.141	0.0003106	1
2	0.2580	0.2652	0.0030	251.5	3.976	0.0004988	2
3	0.2301	0.2367	0.0030	198.4	5.040	0.0008009	3
4	0.2053	0.2113	0.0030	156.3	6.398	0.001291	4
5	0.1831	0.1887	0.0030	122.8	8.143	0.002089	5
6	0.1634	0.1686	0.0030	98.25	10.18	0.003268	6
7	0.1469	0.1507	0.0030	77.34	12.93	0.005280	7
8	0.1302	0.1348	0.0030	60.74	16.46	0.008575	8
9	0.1163	0.1205	0.0030	48.67	20.55	0.01338	9
10	0.1039	0.1079	0.0030	38.17	26.20	0.02177	10
11	0.0927	0.0967	0.0030	30.7	32.5	0.0336	11
12	0.0828	0.0868	0.0030	24.0	41.7	0.0549	12
13	0.0740	0.0780	0.0030	19.3	51.8	0.0850	13
14	0.0661	0.0701	0.0030	15.2	65.8	0.138	14

TABLE 7-8 Square Copper Film Coated Wire Data
(Metric Units)

HEAVY FILM

WIRE SIZE AWG	DIAMETER OVER INSULATION MINIMUM mm	MAXIMUM mm	MINIMUM FILM ADDITION mm	WEIGHT kg/km	m/g	dc RESISTANCE AT 20°C Ω/kg	WIRE SIZE AWG
1	7.351	7.549	0.076	473.8	0.002111	0.0006847	1
2	6.553	6.736	0.076	374.3	0.002672	0.001100	2
3	5.845	6.012	0.076	295.2	0.003387	0.001766	3
4	5.215	5.367	0.076	232.6	0.004300	0.002846	4
5	4.651	4.793	0.076	182.7	0.005472	0.004605	5
6	4.150	4.282	0.076	146.2	0.006841	0.007205	6
7	3.731	3.828	0.076	115.1	0.008689	0.01164	7
8	3.307	3.424	0.076	90.4	0.01106	0.01890	8
9	2.954	3.061	0.076	72.4	0.01381	0.02950	9
10	2.639	2.741	0.076	56.8	0.01761	0.04667	10
11	2.355	2.456	0.076	45.7	0.0218	0.0741	11
12	2.103	2.205	0.076	35.7	0.0280	0.121	12
13	1.880	1.981	0.076	28.7	0.0348	0.187	13
14	1.679	1.781	0.076	22.6	0.0442	0.304	14

TABLE 7-9   Square Aluminum Film Coated Wire Data
(English Units)

HEAVY FILM

WIRE SIZE AWG	DIAMETER OVER INSULATION		MINIMUM FILM ADDITION	WEIGHT		dc RESISTANCE AT 20°C 61.8% CONDUCTIVITY	WIRE SIZE AWG
	MINIMUM inch	MAXIMUM inch	inch	lb/M ft	ft/lb	Ω/lb	
1	0.2894	0.2972	0.0030	97.26	10.28	0.001646	1
2	0.2580	0.2652	0.0030	76.92	13.00	0.002637	2
3	0.2301	0.2367	0.0030	60.76	16.46	0.004233	3
4	0.2053	0.2113	0.0030	48.00	20.83	0.006802	4
5	0.1831	0.1887	0.0030	37.76	26.48	0.01101	5
6	0.1634	0.1686	0.0030	30.32	33.10	0.01720	6
7	0.1469	0.1507	0.0030	23.85	41.93	0.02771	7
8	0.1302	0.1348	0.0030	18.80	53.19	0.04485	8
9	0.1163	0.1205	0.0030	15.13	66.09	0.06966	9
10	0.1039	0.1079	0.0030	11.88	84.18	0.1132	10
11	0.0927	0.0967	0.0030	9.51	105.2	0.176	11
12	0.0828	0.0868	0.0030	7.41	135.0	0.287	12
13	0.0740	0.0780	0.0030	5.99	166.9	0.444	13
14	0.0661	0.0701	0.0030	4.75	210.5	0.714	14

TABLE 7-10   Square Aluminum Film Coated Wire Data
(Metric Units)

HEAVY FILM

WIRE SIZE AWG	DIAMETER OVER INSULATION		MINIMUM FILM ADDITION	WEIGHT		dc RESISTANCE AT 20°C 61.8% CONDUCTIVITY	WIRE SIZE AWG
	MINIMUM mm	MAXIMUM mm	mm	kg/km	m/g	Ω/kg	
1	7.351	7.549	0.076	144.7	0.006908	0.003629	1
2	6.553	6.736	0.076	114.5	0.008736	0.005816	2
3	5.844	6.012	0.076	90.42	0.011061	0.009332	3
4	5.215	5.367	0.076	71.43	0.013997	0.014996	4
5	4.651	4.793	0.076	56.19	0.017794	0.02427	5
6	4.150	4.282	0.076	44.96	0.02224	0.03792	6
7	3.731	3.828	0.076	35.49	0.02818	0.06109	7
8	3.307	3.424	0.076	27.98	0.03574	0.09888	8
9	2.954	3.061	0.076	22.52	0.04441	0.1536	9
10	2.639	2.741	0.076	17.68	0.05657	0.250	10
11	2.355	2.456	0.076	14.15	0.0707	0.388	11
12	2.103	2.205	0.076	11.03	0.0907	0.633	12
13	1.880	1.981	0.076	8.91	0.112	0.979	13
14	1.679	1.781	0.076	7.07	0.141	1.57	14

MAGNET WIRE MANUFACTURERS

Major broad line magnet wire manufacturers include: Elekrisola, Essex, Phelps Dodge, Rea, and Westinghouse.

All of these companies formulate or modify some of the enamels used.

BASIC WIRE ENAMEL TYPES

The chemistry, grades, processing, properties, and electrical/electronic uses of basic wire enamel types are described in the sections which follow and are covered in this order:

- Plain enamel
- Polyvinyl formal
- Polyurethane (PUR)
- Nylon
- Epoxy
- Polyester
- Polyamide-imide (PAI)
- Polyester-amide-imide
- Polyester-imide (PEI)
- Polyimide (PI)

Plain Enamel
(for Wire Enamel Class 105)

This original wire enamel has been largely superseded by enamels with significantly superior performance.

Chemistry

The first enamels were oleoresinous types made from drying oils and natural resins such as rosin and copal. These were superseded by proprietary improvements based on isophthalic polyesters with alkyd modifiers.

Grades

A typical enamel is supplied in 50 percent solution in xylol or other aromatic solvent.

Processing

Plain enamel coated magnet wire is covered by Federal Specification J-W-1177/1, Type E. The enameling process is described in the section "Processing Wire Enamel" in this chapter.

Properties

This low-cost enamel has adequate dielectric, physical, mechanical, and chemical resistant properties for many applications. The film is easy to strip mechanically or chemically for soldering. Dimensional tolerances on fine wires insulated with plain enamel are easier to control than for most other enameled wires. Care should be taken to ensure that the plain enamel film is compatible with varnishes and encapsulating compounds used in subsequent operations.

Electrical/Electronic Uses

Principal uses include small transformers, and telephone, automatic ignition, and solenoid coils.

Polyvinyl Formal
(for Wire Enamel Class 105)

The first successor to plain enamel for magnet wire insulation was polyvinyl formal enamel, in use since 1938 and still used extensively where solderability is not required in Class 105 applications. Monsanto is the developer of polyvinyl formal resin, Formvar™, marketed in solution by Schenectady International and P. D. George.

Chemistry

Polyvinyl formal is a condensation product of formaldehyde and hydrolyzed polyvinyl acetate. The molecular structure contains vinyl formal, vinyl alcohol, and vinyl acetate units:

vinyl formal   vinyl alcohol   vinyl acetate

The inherent poor solvent resistance and soft film characteristics of polyvinyl formal are significantly improved by incorporation into the enamel of phenolic resin, polyisocyanate, or butylated melamine-formaldehyde resin.

Grades

Solutions of polyvinyl formal resin for wire enamel are supplied at 15–17 percent solids. Weight is 8.3 pounds per gallon.

Processing

The applicable NEMA MW 1000 and Federal J-W-1177 designations for various wire constructions are:

WIRE CONSTRUCTION	MW 1000	J-W-1177
Round	15	/4, T, T2, T3
Square and rectangular	18	/6, T, T2, T3
Nylon overcoat	17	/5, TN, TN2
Bondable	19	/6, TB, T2B, T3B

The enameling process is described in the section "Processing Wire Enamel" in this chapter.

Properties

Modified polyvinyl formal film, with thermal rating of 105°C, has high adhesion to base wire and provides excellent windability, abrasion resistance, and compatibility with most varnishes and encapsulating compounds. It exhibits good overall dielectric and physical properties, and may be used with automatic winding equipment without loss of dielectric properties. It resists attack by most solvents, but is soluble in cresylic acid and other coal tar solvents and reactants. It withstands thermal shock very well. A nylon jacket improves physical properties and solvent resistance.

Polyvinyl formal based coatings tend to craze under stress and tension in winding operations when exposed to toluol or xylol. This may be prevented by annealing at 125°C for 2–4 hours before varnishing. The film must be stripped before soldering.

Electrical/Electronic Uses

Polyvinyl formal magnet wire is still used extensively in random wound stators and armatures, motor and generator windings, and dry and oil-filled transformers.

Polyurethane (PUR)
(for Wire Enamel Classes 105, 130, 155)

Polyurethane coatings, primarily because they permit soldering without stripping, are the most widely used coatings for magnet wire suitable for temperatures up to 130°C. They have taken a significant market share from Formvar™, which requires stripping before soldering and is limited to an operating temperature of 105°C. The leading supplier of polyols and toluene diisocyanate, the basic ingredients for polyurethane resins, is Mobay Chemical. Enamels are made from these and other components by Schenectady International and P. D. George.

Chemistry

Polyurethane resins used in wire enamel are one-component ASTM Type 3 derived from polyester or polyesterimide polyols and toluene diisocyanate, which has been blocked with phenol to prevent premature reaction. When heated to 320°F (160°C), the phenol adduct unblocks, freeing the isocyanate group to polymerize with a hardener such as Mobay's Multron R-2. Recipes often contain polyvinyl formal, which reacts with isocyanate, improving film physical properties. See also "Polyurethanes (Embedding Resins)" in Chapter 6, *Embedding Compounds*.

Grades

A typical polyurethane enamel is supplied 25 percent solids in solvent which is two parts cellulose acetate, one part xylene. Weight is 8.5 pounds per gallon.

Processing

The applicable NEMA MW 1000 and Federal J-W-1177 designations for various wire constructions are:

WIRE CONSTRUCTION	THERMAL CLASS	MW 1000	J-W-1177
Round	105	2	SU/2
Round	155	79	SPU/41
Round (improved solderability)	130	75	SP/37
Bondable (butyral overcoat)	105	3	SUB/44
Bondable (polyurethane/nylon/butyral)	130	29	SUN/30
Nylon overcoat			
(14–24 AWG)	130	28	SUN/9
(25–46 AWG)	155	80	SPUN/42

The enameling process is described in the section "Processing Wire Enamel," p. 202 in this chapter.

Properties

The property which has led to the popularity of polyurethane enameled magnet wire, with or without nylon overcoat, is solderability at 680° to 797°F (360°–425°C) without the necessity of fluxing or stripping the film, which vaporizes in this temperature range. Soldering at higher temperatures produces unsatisfactory results. Polyurethane-coated magnet wire without overcoat or with polyvinyl overcoat has a thermal rating of 105°C, although it will operate continuously at 120°C, highest in its class. With nylon overcoat, polyurethane-enameled magnet wire has a thermal rating of 130°C. This topcoat improves windability and resistance to abrasion, heat shock, and solvents.

Another feature of polyurethane insulated wire is its superior "Q" characteristics at high frequencies and humidities, making it attractive for many electronic applications. The "Q" value is the reciprocal of the dissipation factor. The film also has good adhesion and flexibility.

Without topcoat, polyurethane film has lower abrasion resistance than polyvinyl formal film or nylon-jacketed wires, and is susceptible to softening on prolonged exposure to alcohol, acetone, and methylethyl ketone. Certain catalysts and hardeners used in encapsulated systems may also cause softening. Radiation stability is poor.

An overcoat of polyvinyl butyral provides a surface which is bondable with heat or solvents.

With or without overcoat, polyurethane-enameled magnet wire should not be used where there is the possibility of severe overload resulting in high temperature causing the film to vaporize.

Electrical/Electronic Uses

Polyurethane insulated magnet wire with or without nylon jacket is the logical candidate where soldering is required, with consideration given to its temperature and other limitations discussed above. Its superior 'Q' characteristics make it the leading choice for electronic applications where this feature is required.

Nylon
(for Wire Enamel Classes 105, 130, 155)

Nylon has limited use as primary insulation for magnet wire, but is used extensively as an overcoat to improve physical, mechanical, and solvent resistant properties of polyvinyl formal, acrylic, polyurethane, and polyester-enameled wires. Du Pont is the only U.S. supplier of Nylon 6/6 used in wire enamel. Solutions of this nylon are marketed by Du Pont, Schenectady International, and P. D. George.

Chemistry

Nylon 6/6 is obtained by heating hexamethylene diamine with adipic acid:

$$n\text{H}_2\text{N}(\text{CH}_2)_6\text{NH}_2 + n\text{HOOC}(\text{CH}_2)_4\text{COOH} \xrightarrow[-\text{H}_2\text{O}]{\text{heat}}$$

$$-[\text{NH}(\text{CH}_2)_6\text{NHCO}(\text{CH}_2)_4\text{CO}]-_n$$

Grades

Solutions of Nylon 6/6 are 15–20 percent solids, 8.7 pounds per gallon.

Processing

Magnet wire with a base coat of nylon is covered by NEMA Designation MW 6 and Federal Specification J-W-1177/3, Types N, N2. This magnet wire has a thermal rating of 105°C. When used as a jacket, where it amounts to 10–15 percent of total coating weight, the following standards, specifications, and thermal classes apply:

BASE COAT	THERMAL CLASS	NEMA DESIGNATION	FEDERAL SPECIFICATION
Polyvinyl formal	105	MW 17	J-W-1177/5, Types TN, TN2
Solderable acrylic	105	MW 39	J-W-1177/8 Types SAN, SAN2
Polyurethane	130	MW 28	J-W-1177/9 Types UN, UN2
Polyester	155	MW 24	J-W-1177/11 Types LN, LN2, LN3

The enameling process is described in the section "Processing Wire Enamel" in this chapter.

Properties

The outstanding features of nylon film are excellent windability and high abrasion and solvent resistance. The coating is compatible with most varnishes and encapsulating compounds. The nylon coating is solderable without mechanical or chemical stripping.

Nylon should not be used in humid environments as it has relatively high moisture absorption.

Electrical/Electronic Uses

Nylon-jacketed magnet wire is suitable for use where physical and mechanical requirements are demanding, such as in rotating equipment and transformer coils.

Epoxy
(for Wire Enamel Class 130)

Epoxy-enameled magnet wire, once widely used in wet transformers because of its outstanding resistance to Askarel liquids, has been phased out of general use along with the withdrawal of Askarel from the market for environmental reasons.

Polyester
(for Wire Enamel Classes 155, 180, 200)

Polyesters of several types are the most widely used resins in wire enamels. The most recent improvements are covered by patents held by Schenectady International, and are supplied under the trademark Isonel. P. D. George also makes proprietary polyesters.

Chemistry

Polyesters are the reaction products of polyhydric alcohols with polybasic acids. The ingredients for making Isonel are tris (2 hydroxyethyl isocyanurate (Theic™, Allied Corporation, now AlliedSignal Corporation), terephthalic acid, and glycol). Theic confers superior thermal properties over earlier polyesters made with glycerin. The polyhydric structure of Theic with its cyclic urethane linkages is shown here. Thus, although Isonel is classed as a polyester, it may also be considered a polyurethane.

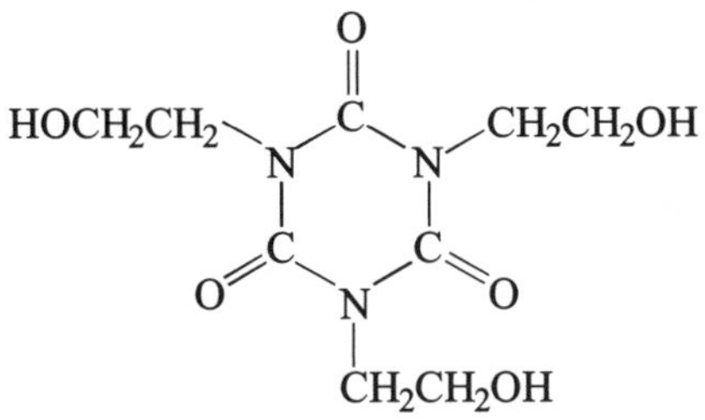

Theic

Grades

Isonel enamel is supplied in 31 percent solids solution, 9.2 pounds per gallon.

Processing

The applicable NEMA MW 1000 and Federal J-W-1177 designations for various wire constructions are:

WIRE CONSTRUCTION	THERMAL CLASS	MW 1000	J-W-1177
Polyester-imide solderable	155	26	SPE/28
	180	77	SPE/39
Above with nylon overcoat	155	27	SPEN/29
	180	78	SPEIN/40
Polyester, polyester-imide	180	30	H/12
	200	74	PEAI/43
Polyester with nylon overcoat	155	24	LN/11
	180	76	PEIN/38
Polyester with polyamide-imide overcoat	200	35, 73	K14
Above with bondable overcoat	180/200	—	—

The enameling process is described in the section "Processing Wire Enamel," p. 202 in this chapter.

Properties

Tris-based polyester has a thermal rating of up to 200°C, depending on its formulation and topcoat, if any. Physical and dielectric properties approach Formvar. Resistance to cut-through is excellent, as is resistance to Freon™. Solvent resistance is good, and the enamel is compatible with most varnishes and encapsulating compounds, with the exception of some epoxies and epoxy hardeners. Radiation stability is good, but not as high as polyimide enamel.

This magnet wire enamel will hydrolyze, and is not for use in oil-filled transformers or in systems containing chlorine compounds or amine catalysts. The film has a tendency to heat shock, and parts wound with this magnet wire should be gradually preheated before varnishing or encapsulating. The film must be stripped before soldering.

Electrical/Electronic Uses

Tris-based polyester-enameled magnet wire is suitable for a variety of applications where operating temperatures are as high as 200°C. Good radiation resistance qualifies it for use where most other enamels (except polyimide) would fail.

Polyamide-Imide (PAI)
(for Wire Enamel Classes 200, 220)

Polyamide-imide enamels are used primarily as topcoats to improve the temperature rating of magnet wires with polyester basecoats. Proprietary polyamide-imide resins have been developed by magnet wire manufacturers and by Schenectady International.

Chemistry

Polyamide-imide resin is the reaction product of trimellitic anhydride and methylenedianiline:

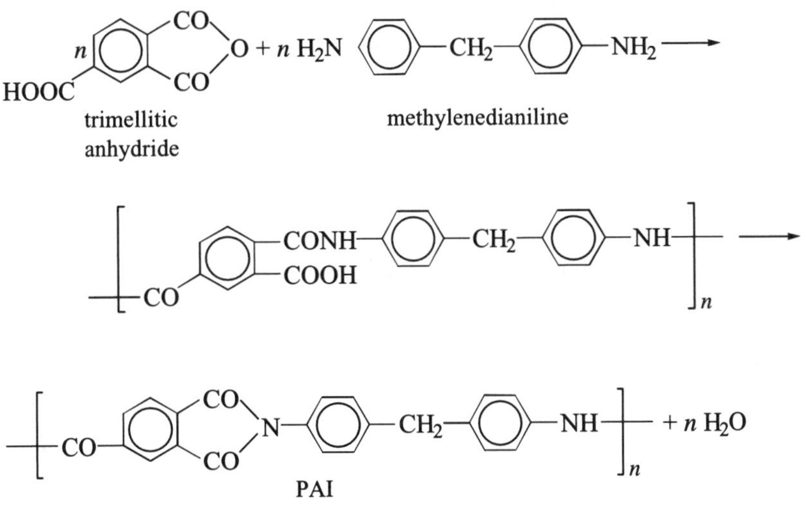

trimellitic
anhydride

methylenedianiline

PAI

Grades

Grades are supplied as 26 percent solids solution in an N-methylpyrrolidone/xylene blend.

Processing

Polyamide-imide overcoated magnet wire with a modified polyester basecoat is covered by NEMA Designation MW 35 and by Federal Specification J-W-1177/14 Types K, K2. The enameling process is described in the section "Processing Wire Enamel" in this chapter.

Properties

Polyamide-imide overcoated magnet wire with a basecoat of modified polyester enamel has a thermal rating of 200°C on copper conductor and 220°C on aluminum conductor. The polyamide-imide contributes toughness and good windability which withstands automatic winding operations. It is more resistant to heat and solvent shock than conventional polyesters, and has less tendency to solvent craze. It exceeds regular polyesters in cut-through and abrasion resistance, and may be used continuously up to its thermal rating. It is highly resistant to attack by Freon™ and most solvents. Wet and dry dielectric properties are excellent.

Polyamide-imide overcoated magnet wire should not be used in closed systems with materials containing moisture or chlorine. Best results are obtained by gradual preheating wound components before varnish operations. Polyamide-imide enamel requires mechanical or chemical stripping before soldering.

Electrical/Electronic Uses

Polyamide-imide overcoated magnet wire is suitable for use in hermetic motors and where resistance to attack by Freon or most solvents is required. Its superior temperature resistance and mechanical properties give it advantages over conventional polyesters for many applications.

Polyester-Amide-Imide
(for Wire Enamel Class 200)

This high-temperature resin, first developed by Westinghouse, is now also produced by P. D. George and Schenectady International.

Chemistry

Polyester-amide-imide is prepared by reacting fewer than 2 mols of trimellitic anhydride with 1 mol of a diamine, followed by esterification of the carboxy groups.

Processing

Polyester-amide-imide insulated wire is covered by NEMA Designation MW 35. The enameling process is described in the section "Processing Wire Enamel" in this chapter.

Properties

Magnet wire insulated with polyester-amide-imide has a thermal rating of 200°C. It is reported to be equal to two-coat polyester systems in solvent shock, heat shock, and thermal stability properties. It is compatible with most insulating varnishes and encapsulating compounds. This coating withstands high-speed winding and is resistant to attack by R22 refrigerant. The film requires stripping before soldering.

Electrical/Electronic Uses

This magnet wire is especially suited for use with high-speed winding devices. High physical properties encourage its use in demanding applications for dry and oil-filled transformers, and in rotating equipment.

Polyester-Imide (PEI)
(for Wire Enamel Classes 180, 200)

Polyester-imide for high-temperature wire enamel is produced in the United States by Schenectady International (Isomid™). P. D. George is licensed by Schenectady International to make polyester-imide using the trademark Teramid. In Europe, there are four producers, all in Germany: Kurt Herberts, Schramm, Dr. Beck, and Cella Lackfabrik.

Chemistry

The process for making Isomid is covered by British Patent #1,082,818 (1967) and French Patent #1,478,938. Isomid is the complex reaction product of Theic, tris (2-hydroxyethyl) isocyanurate, with dimethylterephthalate, ethylene glycol, trimellitic anhydride, and aromatic diamine. It has the following structure:

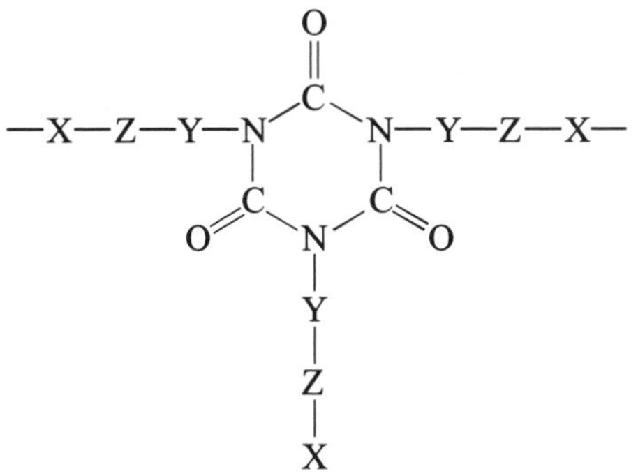

where *X* is:

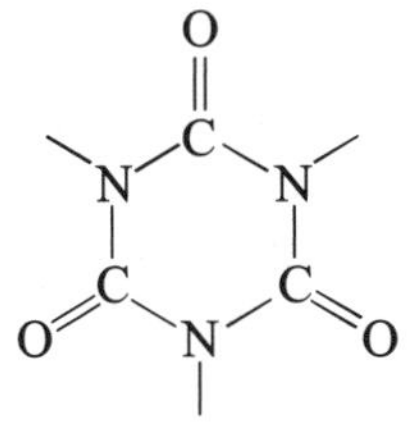

Y is:

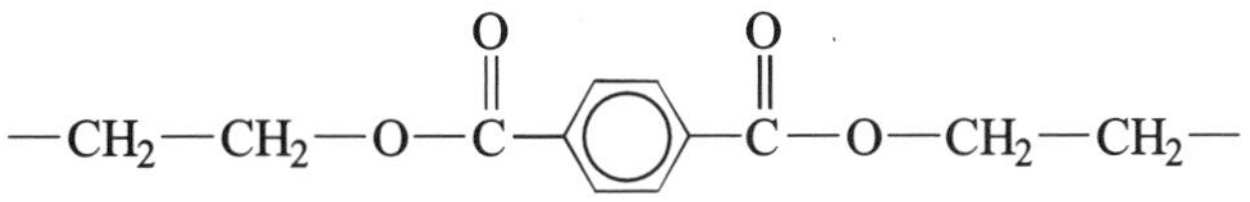

and *Z* is:

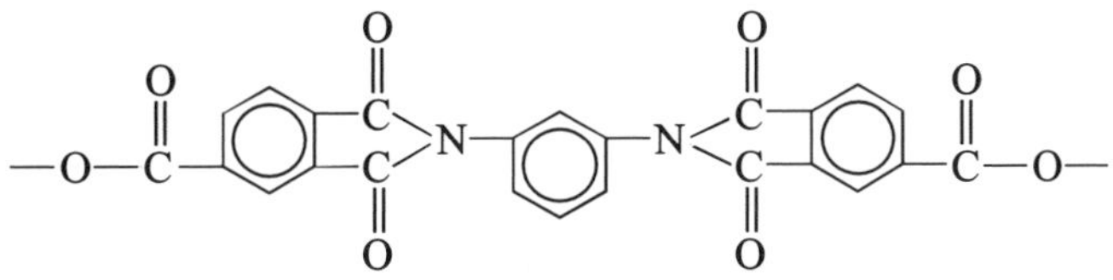

Grades

Polyester-imide is supplied in 31 percent solids solution in cresylic acid, 9.2 pounds per gallon.

Processing

Polyester-imide enameled wire is covered by NEMA Designation MW 30 for thermal class 180 and MW 74 for thermal class 200. The enameling process is described in the section "Processing Wire Enamel" in this chapter.

Properties

Polyester-imide insulated wire has thermal ratings of 180°C/200°C. Because of its excellent cut-through properties and toughness, polyester-imide enameled wire does not require a jacket. It has high resistance to Freon, solvents, and heat shock. The coated wire is highly flexible with good adhesion of enamel to wire. Dielectric properties, both wet and dry, are excellent. The coating is compatible with most varnishes and encapsulating compounds, but must be stripped before soldering.

Electrical/Electronic Uses

Polyester-imide provides a coating with outstanding features, and is suitable for a wide range of electrical/electronic high-temperature applications, although not as high temperature as polyimide (220°C). Polyester-imide is, however, much lower in price.

Polyimide (PI)
(for Wire Enamel Class 220)

Of the many varieties of polyimides, the unmodified polyimide widely used in high-temperature wire enamels is produced by Du Pont, trademarked Pyre-ML. Major magnet wire manufacturers supply Pyre-ML coated wire as their top-of-the-line product. With the expiration in 1982 of Du Pont's basic patent, Schenectady International and P. D. George have introduced competitive products to Pyre ML.

Chemistry

Pyre-ML resin is the condensation product of polyamic acid when exposed to heat in enameling ovens:

polyamic acid

polyimide

Grades

Polyamic acid is offered to magnet wire manufacturers in 15–17 percent solids solution in N-methylpyrrolidone and dimethyl-acetamide blended in a 2:1 ratio. Weight per gallon is 8.7 pounds.

Processing

Polyimide-coated wire is covered by NEMA Designation MW 16 for round wires and MW 20 for square and rectangular wires, and by Federal Specification J-W-1177/15, Types M, M2, M3, and J-W-1177/18, Types M2, M4. The enameling process is described in the section "Processing Wire Enamel" in this chapter.

Properties

Polyimide-enameled wire has the highest thermal stability, and is the most resistant to cut-through, chemicals, and radiation of all enameled wires. It has a thermal rating of 220°C and retains high dielectric values at that temperature. Despite its high price, its high space factor may permit the total cost of finished products to be competitive with equivalent products made with bulkier high-temperature insulations. The coated wire is chemically compatible with a wide range of varnishes and encapsulating compounds.

On the negative side, polyimide insulated wire is not as abrasion resistant as Formvar, or nylon overcoated wire, which, however, have a much lower temperature rating. Polyimide film will hydrolyze in sealed systems containing moisture, and will solvent craze unless winding stresses are relieved by preheating $1/2$ hour at 150°–170°C. Because of its high resistance to attack by virtually all chemicals, polyimide-coated wire is very difficult to strip, requiring highly corrosive strippers.

Electrical/Electronic Uses

Polyimide-insulated wire has the highest temperature rating of all enameled wires. This property, plus highest cut-through, chemical, and radiation resistance, permit it to be used in applications and environments where other enameled wires would be unsuitable. Its high price limits its use where these outstanding properties are not required. It is recommended for use in hermetic units with no entrapped moisture, high-temperature rotating equipment, and aerospace applications.

Insulating Coatings and Impregnants

INTRODUCTION

Insulating coatings and impregnants are of three principal types:

- Varnishes composed of resins or polymers in solution, dispersion, or emulsion in nonreactive liquids, with or without thixotropes or other additives. (A *thixotrope* is an additive that permits varnish viscosity to decrease under shear and recover its original viscosity when the shear force is removed.)
- Varnishes composed of resins or polymers liquefied by a reactive solvent, with or without additives, and without volatile, nonreactive liquids.
- Powder coatings.

Insulating coatings and impregnants covered in this chapter have these principal uses:

- To insulate and bond coils in rotating electrical equipment and transformers
- To insulate circuit materials and electronic inductive components (transformers, chokes, coils, yokes, etc.) from harmful environments.

Magnet wire enamels are covered in Chapter 7.

An important consideration in selecting a coating or impregnant is its compatibility with other system components, especially magnet wire. Since no two systems are exactly alike even for equivalent equipment made by different manufacturers, successful systems are the result of evolution over a long period of use.

While manufacturers are reluctant to adopt new materials and techniques for small economic or quality benefits, changes are sometimes necessary to comply with deadlines for implementation of standards and regulations of the Environmental Protection Agency (EPA), the Occupational Safety and Health Administration (OSHA), and state and local governments.

SUPPLIERS OF COATINGS AND IMPREGNANTS

There are four major broad line producers of insulating coatings and impregnants:

- *John C. Dolph* is a major supplier to apparatus service shops, as well as to original equipment manufacturers. Products include polyesters, epoxies, and polyurethanes.
- *P. D. George (Sterling)* is a principal producer of magnet wire enamels, as well as polyester and epoxy varnishes and enamels.
- *Schenectady International* features polyesters in its magnet wire enamel and varnish product lines. Some epoxies for hermetic motors are also produced.
- *Westinghouse* produces varnishes and wire enamels for its internal consumption, and also supplies these to other equipment manufacturers. The product line includes polyesters, epoxies, and polyurethanes.

Companies which do not produce a broad line of products, but which make significant quantities of special types of coatings and impregnants include:

- *Conap*, produces conformal coatings;
- *Dow Chemical*, produces epoxies;
- *Dow Corning*, the largest producer of silicones;
- *Emerson and Cuming*, produces epoxies, including powder resins;
- *Epolylite Corporation*, produces epoxies;
- *General Electric*, produces silicones;
- *Hysol Division, The Dexter Corporation*, produces epoxies, including powder resins;
- *3M*, produces epoxies, including powder resins;
- *Polymer Corporation*, produces polyester and epoxy powders;
- *Union Carbide*, produces silicone and parylene resins;
- *Viking Products*, produces phenolic and modified epoxy coatings.

DIELECTRIC STRESSES

There are several factors which tend to shorten the useful life of coatings and impregnants:

- Aging in service of organic insulations causes loss of plasticizers and flexible ingredients through volatilization or migration, leading to embrittlement and ultimately to mechanical breakdown.
- Where polymers contain unsaturated linkages, oxidation and subsequent chain cleavage cause embrittlement and eventual disintegration.
- Prolonged exposure to voltage stress through a solid insulation or along its surface will cause dielectric failure. Corona discharge occurring at high ac voltages, with its formation of ozone, will greatly accelerate dielectric breakdown, focusing on voids and contaminants in the insulation.
- Harsh environments containing chemicals, water or water vapor, radiation, and abrasive materials will encourage dielectric failure.
- Strong vibration and mechanical shock will reduce the useful life of a dielectric.
- Thermal cycling and exposure to temperatures above its rated thermal class will cause rapid deterioration of an insulation.

CRITICAL PROPERTIES

Designers of electrical apparatus and electronic components are performance oriented and are continually searching for dielectrics, whatever their composition, which are effective, easy to apply, durable, energy efficient, free from environmental and worker health and safety problems, and low cost. To meet these requirements, formulators of insulating coatings and impregnants are challenged to develop products with these properties:

- Better temperature resistance to prolong operating life
- Improved moisture resistance both before and after curing
- Greater thermal conductivity for lowering equipment operating temperature
- Higher bonding strength to permit rotating equipment to operate faster and to be more resistant to shock
- Greater dielectric strength to permit thinner insulation films for equivalent protection
- Improved resistance to corona damage
- Lower dielectric losses to reduce internal heating of insulations
- Higher impulse strengths to withstand current surges without damage
- Improved flame resistance to reduce fire hazard
- Greater resistance to environmental attack
- Decomposition products that are nonconducting and are not harmful to health, to other insulations, or to the environment
- Reduction or elimination of solvents to reduce voids in insulations, and to conform to environmental and health legislation
- Reduction or elimination of objectionable or harmful odors during storage or processing
- Shorter curing cycles to reduce manufacturing time
- Greater stability in storage and in processing equipment.

THERMAL CLASSIFICATION OF COATINGS AND IMPREGNANTS

See "Magnet Wire Thermal Classification" in Chapter 7, p. 202 *Magnet Wire Enamels*.

FEDERAL SPECIFICATIONS AND PRACTICES

There are several Federal specifications for insulating varnishes, compounds, and enamels, among which the following are significant:

- MIL-1-24092, *Insulating Varnish, Electrical, Impregnating, Solvent Containing*. This performance specification covers clear, air drying, and baking varnishes, thermal classes 130–220. The intended use is on all types of electrical coils and windings. There are three grades:

        Grade CA—clear, air drying varnish, classes 130 and 155
        Grade CB—clear, flexible, baking varnish, classes 155, 180, 200, and 220
        Grade CBH—clear, semi-rigid, baking varnish, classes 155, 180, and 200

In practice, the preponderance of usage is for modified polyester baking varnish, class 155, with minimum nonvolatile content of 40 percent. Silicone varnishes may not be substituted where other varnishes are required.

• MIL-I-46058, *Insulating Compound, Electrical (for Coating Printed Circuit Assemblies)*. This specification is approved for use by all departments and agencies of the Department of Defense. Covered are conformal coatings suitable for application to printed circuit assemblies by dipping, brushing, spraying, or vacuum deposition. The following types of conformal coatings are included:

TYPE	MAXIMUM CURING TIME (hours) AT 125°C
AR—Acrylic Resin	4
ER—Epoxy Resin	8
SR—Silicone Resin	24
UR—Polyurethane Resin	24
XY—Para-xylylene	vacuum deposition

The specification contains specific requirements for:
- Compatibility
- Appearance
- Coating thickness on test specimens
- Fungus resistance
- Shelf life
- Insulation resistance
- Dielectric withstanding voltage
- Q (resonance)
- Thermal shock
- Moisture resistance
- Flexibility
- Hydrolytic stability
- Flame resistance

• MIL-E-22118, *Enamel, Electrical, Insulating*. This specification is approved for use by all departments and agencies of the Department of Defense. Requirements for the coating include the following:
- The color shall be characteristic of red iron oxide pigment
- The vehicle shall be phthalic alkyn resin, together with necessary addition of volatile solvents and driers
- The volatile content shall have:
  - a. 5 percent maximum solvents having an olefinic or cyclo-olefinic type of unsaturation
  - b. 8 percent maximum combination of aromatic compounds with eight or more carbon atoms to the molecule except ethyl benzene

    c. 20 percent maximum combination of ethyl benzene, ketones having branched hydrocarbon structures, or toluene

    d. Total of 20 percent maximum a + b + c.

Quantitative requirements are shown in Table 8-1.

TABLE 8-1   Quantitative Requirements for Red Insulating Enamel (MIL-E-22118)

CHARACTERISTICS	MINIMUM	MAXIMUM
Total solids, percent by weight of enamel	58	—
Pigment, percent by weight of enamel	25	—
Vehicle solids, percent by weight of enamel	32	—
Phthalic anhydride, percent by weight of nonvolatile vehicle	30	—
Iron oxide (Fe_2O_3), percent by weight of pigment	28	—
Calcium oxide, percent by weight of pigment	—	1.5
Coarse particles and skins, percent by weight of pigment	—	0.5
Extenders, percent by weight of pigment	—	70
Water, percent by weight of enamel	—	0.5
Viscosity, Krebs Stormer	61	69
Weight per gallon, pounds	9	—
Drying time, air drying		
Tack free, hours	—	$1/2$
Dry through, hours	—	4
Baking at 230°F (110°C)		
Dry through, hours	—	2
Dielectric strength (ASTM D 115)		
Dry, volts/mil	1,200	—
Wet, volts/mil	300	—

• MIL-V-173, *Varnish, Moisture-and-Fungus-Resistant (for Treatment of Communications, Electronic, and Associated Equipment)*. This specification is mandatory for use by all departments and agencies of the Department of Defense. The document covers one type of moisture-and-fungus-resistant varnish consisting of a para-phenyl phenol-formaldehyde resin in combination with tung oil and suitable solvents which has been made fungistatic by the addition of 7.0 ± 1.0 percent salicylanilide or one percent copper 8-quinolinolate. The resin shall conform to Federal Specification TT-R-271 and the tung oil shall conform to Federal Specification TT-T-775. The solvent shall not constitute a hazard to personnel applying the varnish. Methanol, benzene, or chlorinated hydrocarbons shall not be used. Nonvolatile content of the varnish shall be a minimum of 50 percent of the weight of the varnish excluding fungistatic agents. Of this, 55 ± 2 percent shall be tung oil with the balance resin.

• EPA-450/3-84-019, *Determination of Volatile Content Procedures for Certifying Quantity of Volatile Organic Compounds Emitted by Paint, Ink, and other Coatings*.

ASTM STANDARDS FOR ELECTRICAL INSULATING VARNISHES

The American Society for Testing and Materials (ASTM) has established these standards for electrical insulating varnishes:

D 115, *Testing Varnishes Used for Electrical Insulation*

D 374, *Tests for Thickness of Solid Electrical Insulation*

D 618, *Conditioning Plastics and Electrical Insulating Materials for Testing*

D 1932, *Test for Thermal Endurance of Flexible Electrical Insulating Varnishes (up to 180C)*

D 2369, *Standard Test Method for Volatile Content of Coatings*

D 2436, *Specifications for Forced-Convection Laboratory Ovens for Electrical Insulation*

D 2519, *Test for Bond Strength of Electrical Insulating Varnishes by the Helical Coil Test*

D 2756, *Weight Loss of Electrical Insulating Varnishes*

D 3056, *Gel Time for Solventless Varnishes*

D 3145, *Test for Thermal Degradation of Electrical Insulating Varnishes by Helical Coil Method*

D 3251, *Test Method for Thermal-Aging Characteristics of Electrical Insulating Varnishes Applied Over Film-Insulated Magnet Wire*

D 3312, *Percent Reactive Monomer in Solventless Varnishes*

D 3377, *Weight Loss of Solventless Varnishes*

D 4217, *Gel Time of Thermosetting Coating Powder*

D 4733, *Test Methods for Solventless Electrical Insulating Varnishes*

The Manual on the Determination of Volatile Organic Compounds in Paints, Inks, and Related Products, ASTM Manual Series, MNL-4, contains a review of the regulations, methods of measurement, and calculation of VOC compound in solvent-containing compositions.

UL STANDARD 1446, SYSTEMS OF INSULATING MATERIALS—GENERAL

Although varnishes are only one component of an insulation system, they come into contact with all other components, serving to bond the system into a solid unit. It is, therefore, essential that a varnish be compatible with other elements in the system. Otherwise, the system may be subject to premature failure. UL 1446 deals with this issue. Noteworthy sections of this standard include the following:

• Covered are test procedures to be used in the evaluation of Class 130 or higher electrical insulation systems intended for connection to branch circuits rated 600 volts or less.

• An insulation system is an intimate combination of insulating materials used in electrical equipment.

• Insulation system components include ground insulation, magnet wire coating, encapsulant, phase-to-phase insulation, varnish, wedges, tapes, lead wire, tie cord, sleeving, and all similar parts.

• Insulating materials having different assigned temperature classes may be combined to form an insulation system having a temperature class that may be higher or lower than that of any of the individual components.

• The compatibility of an insulating material with other materials in the same insulation system shall be investigated to determine whether thermal aging of such components makes the system vulnerable to unacceptable deterioration in normal service at the assigned temperature class of the system.

APPLICATION METHODS

The first method for applying insulating coatings and impregnants to electrical apparatus was by brushing or dipping followed by curing at room temperature or in ovens. This method is still used by service shops for repair and rehabilitation of rotary equipment and transformers. However, faster and more effective processing is needed where production volumes and energy consumption are high and compliance with environmental, health, and safety legislation is required. The most widely used modern methods are summarized in this section.

Vacuum Pressure Impregnation (VPI)

This technique, now in general use, features better impregnation of absorbent materials and elimination of voids. Suitable impregnants include solventless epoxies and polyesters. Key steps in this process include:

1. Loading the preheated part in the treatment tank
2. Evacuating the air from the tank to the desired pressure (1 mm of mercury for some solventless varnishes)
3. Introducing the varnish and holding low pressure
4. Pressurizing the tank and holding under pressure with inert gas
5. Relieving pressure
6. Repeating the vacuum/pressure cycle for the appropriate number of cycles
7. Removing the part from the treatment tank
8. Curing the varnish by oven baking, infrared, or resistance heating.

Trickle Impregnation

This method requires preheating the parts, dispensing on them a measured amount of varnish, and heat-curing the coating. In this method, 100 percent of the varnish is utilized, thereby minimizing material losses. Solventless two-part types of both polyester and epoxy varnishes are widely used, with polyesters predominating in the U.S. and epoxies in Europe. Polyesters are lower cost and are easier to process, but have higher shrinkage. Epoxies have better adhesion. Varnishes are cured with a very active catalyst such as tertiary butyl perbenzoate.

In this process, varnishes are trickled on rotating parts preheated at 180° to 250°F (82°–121°C). Volume and rate of application of varnish and rotation speed must be carefully controlled for best results.

Several types of equipment are used for the trickle impregnation method, ranging from single-station machines for large parts to continuous flow-through machines utilizing conveyor belts for high-volume parts as they pass through the preheat, dispensing, and post-curing zones. Final cure at 250° to 350°F (121°–171°C) is accomplished by one of these means:

- Convection oven with gas or electric heat
- Radiant heat using infrared or ceramic heaters
- Resistance heat generated by passing a current through the wound coils of the part
- Heat generated by the high exotherm of the catalyzed varnish.

Fluidized Bed

Dry, powdered resin may be applied to parts by the fluidized bed process where film thickness of 5 mils and over is permissible and impregnation is not required. This method is suitable for coating slots in electrical apparatus, where it competes with polyamide paper (Nomex™) and polyester ribbon (Mylar™). To be effective, powder must be applied at a rate to produce a 10–15 mil coating to the face of the part so that slots are coated with at least 6–10 mils and corners have a minimum of 5–6 mils coating, as required by Underwriters Laboratories.

In this process, dry, powdered resin is suspended in a chamber by controlled velocity hot air or inert gas introduced from the bottom of the chamber through a screen and porous plats. The suspended powder is in constant motion and acts as a low-viscosity, low-surface-tension fluid. Before passing through the chamber, the part is treated to facilitate adhesion of powder by heating or application of an adhesive primer. After coating, the resin is cured by a conventional method. The complete process may be a continuous operation.

To be effective, this method requires precise control of several process variables:

- Polymer type
- Powder particle size
- Pretreatment techniques and temperature
- Chamber zone temperatures

- Time of exposure
- Curing method, temperature, and time.

Electrostatic techniques may be incorporated with this process by applying a high-voltage dc field to give resin particles a negative charge. The part to be coated is electrically grounded so that particles are electrostatically attracted to it as it passes through or is suspended above the fluidized bed. This eliminates the need to heat the part or coat it with an adhesive, and permits powder to be easily removed by air knife or other method from areas where coating is not desired. Due to their excellent adhesive and other physical properties, epoxies are the most commonly used resins, although other polymers are also used, including silicones, polyesters, phenolics, nylon, and polyvinyl chloride.

Radiation Curing

This method employs ultraviolet light (UV) or electron beam (EB) technology and cures only resins reached by the light or beam. A baking step is subsequently required to cure resins not exposed to radiation.

With UV or EB radiation curing there is no energy wasted in heating the process equipment, atmosphere, or apparatus; or in evaporating solvents. For example, a UV curable class H (180°C) solventless varnish requires for curing 4 hours at 150°C from the time the part is at temperature compared with 5–7 seconds when exposed to UV light at 200 watts per linear inch supplied by mercury vapor lamps. The arc between lamp electrodes generates electromagnetic energy of varying wavelengths, giving off infrared and visible light as well as the desired ultraviolet radiation.

Conformal Coating

Circuit materials and delicate electrical/electronic components may be protected dielectrically and from the environment by a coating which conforms closely to the shape of the part. Commonly used coatings, such as epoxy, acrylic, urethane, and silicone, may be applied in several ways, including:

- Dip coating
- Spraying with both pressure and atomizing sprayers
- Flow-coating and brushing

A unique conformal coating and method of application, the parylene process, has been developed by Union Carbide. Parylene, a polymer based on para-xylylene or its substituted derivatives, cannot effectively be melted, extruded, molded, calendered, or dissolved to form solvent systems. In this process, a solid dimer is converted to a reactive vapor of the monomer, which when passed over parts at room temperature coats them with polymer in thicknesses of 0.25–1.5 mils. No catalysts are employed. Although the process requires high heat at one point, vapor-phase deposition occurs at room temperature; thus, temperature-sensitive parts are not damaged. The process basically consists of these steps:

1. Vaporizing the para-xylene dimer at 175°C (347°F) and 1 torr (the pressure to support 1 mm of mercury at 0°C (32°F)

2. Pyrolizing the dimer at 680°C (1256°F), 0.5 torr, to form the monomer

3. Formation of the polymer as it deposits on the part at 25°C (77°F), 0.1 torr

4. Treatment in a cold trap at –70°C (–94°F) and reduction of pressure to 0.001 torr.

DEVELOPMENT PROGRAMS

The emphasis in development programs is on replacing varnishes containing hazardous volatile organic compounds (VOCs) in the following ways:

• *High solids* (up to 100 percent), low volatile liquid varnishes which reduce or eliminate emissions to levels in compliance with Environmental Protection Agency and Occupational Safety and Health Administration standards and regulations.

• *Water-based varnishes* are now produced by most broadline manufacturers of insulating coatings and impregnants. Earlier versions had poor moisture resistance, inadequate physical properties, and were lacking in tank stability. While formulations are proprietary, a distinction may be made between water-soluble systems and water-borne systems. The former may be diluted 100 percent with water, while the latter requires a cosolvent for extreme dilution with water.

One commercially available product is an alkyd-phenolic modified polyester in which carboxyl terminated backbones are reacted with an amine to form amine salts. The amine salt and coupling agent, butyl cellosolve, blend in water at pH 8.2–8.5 (controlled by triethylamine) to form the resin system. On baking at 275° to 325°F (135°–163°C), the amine is driven from the film, permitting the carboxyl group to react with the other ingredients, forming a film which is resistant to moisture and chemicals, and possesses good mechanical strength. As supplied, the varnish is 40 percent minimum nonvolatile. Weight is approximately 8.3 pounds per gallon. Flash point is over 200°F (93°C). Thermal rating is Class 180. Thinning to the desired application consistency is done by adding water.

Another product, a water-borne system, contains by weight 50 percent resins such as phenolic, melamine, and solvent-based modified polyester. The volatile portion is a mixture of 80 percent water and 20 percent cosolvent 2-butoxyethanol. The pH is maintained at 7.5–8.0 for optimum water solubility by t-alkanolamine. As this varnish is cured, all water is evaporated at 225°F (107°C), well below the polymerizing temperature of the resin system. Thus, the likelihood of premature surface curing is greatly reduced compared with solvent-based varnishes containing solvents evaporating at much higher temperatures. After the water is driven off, curing is completed at 300° to 325°F (147°–163°C). It is claimed that this varnish has excellent humidity resistance and chemical resistance superior to traditional polyester varnishes. It may be formulated to offer very high bond strength, especially at elevated temperatures, where it competes with epoxies. As supplied, specific gravity is 1.040–1.060. Flash point is over 200°F (93°C). Thermal endurance for 20,000 hours is over 180°C. Dilution should be with a blend of 80 parts water, 20 parts butoxyethanol.

Both of the above varnishes may be used in vacuum pressure impregnation processing.

Although the specific heat and heat of vaporization of water are significantly higher than for varnish solvents, some authorities claim that water-based varnishes take less energy to cure for these reasons:

- Water has less affinity than solvents have for resins (lower hydrogen bond strength).
- The film formed in curing water-based varnishes stays open during curing. With solvent-based varnishes, solvents have to fight their way through the film that forms early in the curing process.
- Lower air volumes are required in curing waterbased varnishes.

• *Powder coating* employing fluidized-bed technology is practical where there is high-volume production for a part and its configuration is open to powder deposition. Typically, the powder is a one-component epoxy. Polyester powders are also used, but since they are made from solvents, they may contain residual solvent which could cause pinholing during the curing process. Uniformity of film thickness, especially at edges, is often a problem with powder coatings. To obtain sufficient coverage at critical points, it is often necessary to coat other surfaces with thicker-than-required film. The powder coating process is discussed in the "Fluidized Bed" section of this chapter and in "Development Programs," p. 206.

MARKET TRENDS

There is increasing emphasis on compatibility of coatings and impregnants with other components of insulation systems. It is especially important that the insulating varnish is compatible with the magnet wire enamel in the same system. Otherwise, bonding may be inadequate, leading to system failure. Both magnet wire and varnish manufacturers are concerned with system compatibility, and many will furnish helpful information on this subject.

Varnish producers and users must now comply with increasingly more stringent Environmental Protection Agency and Occupational Safety and Health Administration standards and regulations, especially with regard to solvents. The ideal is to eliminate solvents entirely, but because of the large investment in production equipment designed for solvent systems, it is often more practical to use varnishes made with exempt (acceptable) solvents or to provide for recycling or disposal of nonexempt vapors safely and effectively. There is little doubt, however, that most new systems will be designed to use solventless, water-based, or high-solids varnishes.

INHERENT CHARACTERISTICS OF COATING
AND IMPREGNATING VARNISHES

• *Oleoresinous* varnishes are lowest cost with adequate properties for many applications.

• *Alkyd polyesters* are moderate cost and the most widely used solvent-type varnishes, with good overall properties over a wide useful temperature range.

• *Unsaturated polyesters* are the most commonly used solventless VPI varnishes, and are lower cost than solventless epoxies.

• *Epoxies* are noted for their good dielectric properties strong bond strength, and resistance to chemicals and high humidity. They are higher cost than polyesters.

• *Parylene* has excellent dielectric and barrier properties, and is highly resistant to attack by most solvents and chemicals. It may be applied only by a special process and is high cost.

• *Polyurethanes* have good film adhesion and flexibility, but are not as resistant to moisture and chemicals as other varnishes.

• *Silicones* have the best thermal properties and water resistance, but have relatively low mechanical properties and adhesion, and are high cost.

PRINCIPAL TYPES OF COATINGS AND
IMPREGNANTS

The chemistry, grades, properties, and electrical/electronic uses of principal types of coatings and impregnants are described in the sections which follow in this order:

• Early Oleoresinous Insulating Varnishes
• Alkyd Polyesters
• Unsaturated Polyesters
• Epoxies
• Parylene
• Phenolics
• Polyurethanes (PURs)
• Silicones

Early Oleoresinous Insulating Varnishes

Electrical apparatus will not function without adequate insulation. At first, modified wood finishing varnishes made in open kettles sufficed. Those varnishes were made with drying vegetable oils, such as linseed and China-wood (tung) oils, cooked at temperatures up to 600°F (316°C) together with rosin, ester gum, copal, and other resins, and reduced on cooling to usable consistency with compatible solvents. The drying, or oxidizing, process was facilitated by addition to formulations of cobalt and manganese compounds, such as acetates and napththenates.

It was soon found that incorporation into varnishes of materials such as gilsonite, pitches, coal tars, and petroleum asphalts resulted in improved dielectric properties and resistance to moisture, acids, and alkalis. Yellow varnishes were still superior in oil resistance and physical properties. Further improvement resulted when the equipment was baked after treating it with varnish. This led to the development of varnishes with lower oil content (short-oil varnishes) specifically designed for oven curing. Some of these varnishes are still used.

Alkyd Polyesters
(Varnish Resins)

In the coatings field, the generic term *polyester* applies to reaction products of a polyhydric alcohol with a polybasic acid. There are two types of polyesters used in the manufacture of varnishes. This section covers alkyd polyesters, also called *glyptals*, the most widely used, which are polyesters modified with fatty acids of vegetable oils. The section following this discusses unsaturated polyesters, which may be crosslinked by reaction with a monomer such as styrene, vinyl toluene, or diallyl phthalate.

Alkyd polyesters became available in the mid-1920s, and soon became the predominant polymers for coatings and impregnants for application to all types of electrical equipment.

Chemistry

The sequence of steps in producing an alkyd resin is, first, the partial esterification of a polyhydric alcohol with a fatty acid, followed by reaction with a dibasic acid:

(1) $HO-CH_2-CH_2-OH + R-\underset{\underset{O}{\|}}{C}-OH \longrightarrow$

     glycol   fatty acid

$$HO-CH_2-CH_2-O-\underset{\underset{O}{\|}}{C}-R + H_2O$$

            monoester of glycol

(2) phthalic acid $+ nHO-CH_2-CH_2-O-\underset{\underset{O}{\|}}{C}-R \longrightarrow$

     phthalic acid   monoester of glycol

$$\left[\begin{array}{c} \underset{\text{C}}{\overset{\overset{\displaystyle O}{\|}}{}} \\ \underset{\overset{\displaystyle O}{\|}}{\text{C}}-\text{O}-\text{CH}-\text{O} \\ \overset{\displaystyle \text{CH}_2}{} \\ \overset{\displaystyle \text{O}}{} \\ \overset{\displaystyle \text{C}=\text{O}}{} \\ \text{R} \end{array} \right]_{2n} + 2n\text{H}_2\text{O}$$

modified alkyd polymer

Commonly used polyols, in addition to glycol, are glycerol, polyglycerol, and pentaerythritol (a key ingredient in insulating varnishes). Polybasic acids, other than phthalic, include isophthalic, maleic, adipic, and sebasic, with longer chain acids yielding softer resins that made good polymeric plasticizers for harder resins. Fatty acids are derived from vegetable oils such as linseed, tung (China-wood), soya, safflower, perilla, and castor. The resin can be prepared by either bulk or solution polymerization methods with cooking and refluxing continued until the viscosity and acid number of the reaction mixture reach predetermined values. The resin may be supplied in solution which the user may further reduce. Long-oil alkyds are soluble in mineral spirits, but short-oil alkyds require stronger solvents such as xylene or toluene.

Alkyds may be further modified to achieve certain properties by blending with other varnishes or by reacting with other resins, including phenolics for improved water and alkali resistance, epoxies for better adhesion, silicones for better thermal and dielectric properties, and aminos for greater durability. Alkyds may also be combined with reactive monomers, such as styrene, vinyl toluene, and diallyl phthalate for fast curing varnishes.

Grades

A wide variety of alkyd grades is marketed by major varnish manufacturers. Representative baking grades are listed in Table 8-2. Air drying/baking grades are also available, but these usually have lower properties.

Processing

See "Application Methods" in this chapter.

Properties

As is shown in Table 8-2, alkyds may be formulated with excellent dielectric, mechanical, and chemical resistance properties covering a wide useful temperature range.

TABLE 8-2 Key Properties of Representative Baking-Type Alkyd Varnishes

FEATURE	MAXIMUM USEFUL TEMPERATURE (°C)	CURING (hours)	CURING (°C)	DIELECTRIC STRENGTH ASTM D 115 (volts/mil) dry	wet	BOND STRENGTH ASTM D 2519 (pounds) 23°C	155°C	CHEMICAL RESISTANT WATER	(10% H_2SO_4)	(1% NaOH)	OIL ASTM D 115
Water soluble	180	1½–6	135–163	4,000	2,900	30	5	na	na	na	na
Alternative to silicone	180	2–8	135–177	4,100	3,000	15	2.5	Ex	Ex	Ex	Passed
Phenolic modified flexible	195	½–4	135–163	4,100	2,900	35	5	Ex	Ex	Ex	Passed
Phenolic modified with alkyd Rule 66 compliance	200	½–4	121–163	3,400	2,800	35	4.5	Ex	Ex	Ex	Passed
Flash point 38°C+ general purpose	215	1–5	135–163	4,000	2,900	28	2.5	Ex	Ex	Ex	Passed

Note: All of the above except water soluble comply with Federal Specification MIL-I-24092.

na—not available.

Ex—excellent.

Electrical/Electronic Uses

Alkyds are the most widely used coatings and impregnants for application to all types of electrical/electronic items, including rotating equipment, transformers, and electronic components.

Unsaturated Polyesters
(Varnish Resins)

The trend in insulating varnishes is toward greater use of solventless types suitable for vacuum pressure impregnation (VPI). Unlike films formed by solvent-type varnishes, solventless varnish films do not contain voids formed by solvent evaporation during curing. Solventless polyester varnishes have been available since the late 1940s, and are the most commonly used in the VPI process. Solventless epoxies are also used, but to a lesser extent.

Chemistry

Unsaturated polyesters are formed by reacting an unsaturated polybasic acid, most frequently maleic or fumaric, with a dihydroxy alcohol, such as ethylene, propylene, diethylene, or dipropylene glycol. Saturated acids, for example, phthalic and adipic, may also be included in the formulation to control unsaturation and make the final resin tougher and more flexible.

$$n\text{HO}-\overset{O}{\overset{\|}{C}}-\overset{H}{\overset{|}{C}}=\overset{H}{\overset{|}{C}}-\overset{O}{\overset{\|}{C}}-\boxed{\text{OH} + n\text{H}}\,\text{O}-\overset{H}{\overset{|}{C}}=\overset{H}{\overset{|}{C}}-\text{OH} \xrightarrow{-\text{H}_2\text{O}}$$

maleic acid   ethylene
                               glycol

$$\left[\!\!-\text{O}-\overset{O}{\overset{\|}{C}}-\overset{H}{\overset{|}{C}}=\overset{H}{\overset{|}{C}}-\overset{O}{\overset{\|}{C}}-\text{O}-\overset{H}{\overset{|}{C}}=\overset{H}{\overset{|}{C}}-\text{O}\!-\!\right]_n$$

unsaturated polyester
resin

To this resin is added a vinyl-type monomer such as styrene, alphamethylstyrene, vinyl toluene, diallyl phthalate, divinylbenzene, and triallylcyanurate, or a mixture of these. Curing proceeds by free-radical addition across double bonds, and is initiated by organic peroxides, for example, dicumyl peroxide, heat, or ultraviolet light. The base resin formulation and monomer selection each significantly affect all varnish properties.

Grades

Several manufacturers offer a selection of one- and two-component solventless, unsaturated polyesters with varying thermal classifications, gel time, and properties. Representative grades are shown in Table 8-3.

TABLE 8-3   Key Properties of Representative VPI Baking-Type Solventless Polyester Varnishes

FEATURE	THERMAL CLASS (°C)	GEL TIME @ 100°C (minutes)	DIELECTRIC STRENGTH ASTM D-1346 @ 23°C (volts/mil)	DIELECTRIC CONSTANT** ASTM D-150		DISSIPATION FACTOR** ASTM D-150		INSULATION RESISTANCE ASTM D-257 (megohms)		BOND STRENGTH ASTM D-2519 (pounds)	
				23°C	155°C	23°C	155°C	23°C	155°C	23°C	155°C
Rigid 2-part	180	40*	1,800	5.2	8.7	0.046	0.082	1×10^6	1×10^4	32	7.5
Semi-flexible high bond strength thixotropic 2-part	180	90*	2,000	4.5	7.4	0.017	0.076	1×10^6	1×10^5	45	8.5
Rigid high bond strength high flash (165°C) Freon resistant 1-part precatalyzed	205	125	1,750	3.1	na	na		na		42	20

*With 1% dicumyl peroxide.

**At 100 Hz.

na—not available.

Processing

See "Application Methods," p. 239 in this chapter.

Properties

Solventless polyesters are significantly lower cost than solventless epoxies, and have dielectric, mechanical, and chemical resistance properties adequate for most applications. However, where a part or unit must withstand submersion or prolonged exposure to high humidity, or where the highest bond strength is required, epoxies are recommended. Because of the virtual elimination of voids on curing, solventless varnishes provide better heat dissipation and reduce the possibility of corona damage.

Table 8-3 lists key properties of representative baking-type solventless polyester varnishes.

Electrical/Electronic Uses

Unsaturated solventless polyesters are the most widely used vacuum pressure impregnation varnishes in the manufacture of high-volume electrical equipment.

Epoxies
(Varnish Resins)

Epoxy resins, available commercially since 1947, are second in usage in insulation varnishes only to polyesters, and are favored for critical applications requiring high bond strength, toughness, and chemical resistance. Epoxies may be solventless, with solvent, or in the form of powder.

Chemistry

While the basic epoxy resin is the reaction product of epichlorohydrin, bisphenol A, and sodium hydroxide, countless variations are available with a wide range of properties and applications in the coatings field. For convenience in discussion, epoxies may be classified as follows:

• *Unmodified epoxy resins* useful in making coatings include the basic resin and cycloaliphatic epoxies produced by epoxidation of cyclic olefinic compounds with peracidic acid. Formulas for these resins are shown in the "Epoxies" section of Chapter 6, *Embedding Compounds*. Widely used resins have an excess of epichlorohydrin over bisphenol A (typically a ten–one ratio) and are cured with amine hardeners. Cycloaliphatics are unreactive with amines and are usually cured with anhydride hardeners. A common practice is to blend these resins with urea–formaldehyde or phenol–formaldehyde varnishes.

• *Epoxy novolac resins* are formed by reacting epichlorohydrin with novolac resin obtained by condensing phenol or ortho-cresol with formaldehyde under acidic conditions. For additional information on novolacs, including formula, see "Epoxies" in Chapter 4, *Thermosetting Molding Compounds*.

Hardeners for the above epoxy types are discussed in the "Epoxies" section in Chapter 6, *Embedding Compounds*.

• *Epoxy ester resins* are synthesized by esterification of epoxy resins with fatty acids derived from tall oil, dehydrated castor oil, and other drying oils at 250°C under a nitrogen blanket to prevent oxidation. The ratio of acid to epoxy varies from 60/40 for long-oil esters to 40/60 for short-oil types. The addition of saturated fatty acids increases film flexibility and slows drying. Epoxy esters are usually supplied at 50 percent solids in xylol. Drying properties are enhanced by the addition of small percentages of metallic naphthenates.

Grades

Epoxies are available in solvent-type air-drying/baking grades, typically 50 ± 5 percent solids. Solventless grades for VPI processing are widely used for critical applications. Representative grades are shown in Table 8-4.

Processing

See "Application Methods," p. 239 in this chapter.

Properties

Used mostly for Class 130 and Class 155 applications, epoxies possess outstanding bond strength, along with resistance to high humidity and chemicals. Dielectric properties, however, tend to drop off sharply as temperatures approach the thermal class ratings of epoxy varnishes.

Table 8-4 lists key properties of representative baking-type varnishes.

Electrical/Electronic Uses

Because of their higher cost, epoxies are generally used for critical applications not well suited to polyesters, such as in hermetic motors and heavy-duty rotating equipment. They are also used in conformal coatings for circuit boards.

Parylene
(Varnish Resin)

A thin-layer protection of electrical/electronic components and assemblies is provided by parylene in a process and equipment developed by Union Carbide and commercialized in 1965.

Chemistry and Processing

Di-para-xylylene is produced by heating para-xylene to about 900°C in the presence of steam. This dimer, when vaporized at about 175°C, forms the monomer, para-xylylene. When heated to about 680°C at reduced pressure, the monomer pyrolyses to the di-radical, which then forms a polymeric coating on parts at room temperature and further reduced pressure. See also the description of the process in "Conformal Coating" in this chapter.

TABLE 8-4 Key Properties of Representative Baking-Type Epoxy Varnishes

FEATURE	THERMAL CLASS (°C)	GEL TIME @ 140°C (minutes)	DIELECTRIC STRENGTH ASTM D-115 @ 23°C (volts/mil)	DIELECTRIC CONSTANT** ASTM D-150		DISSIPATION FACTOR** ASTM D-150		INSULATION RESISTANCE ASTM D-257 (megohms)		BOND STRENGTH ASTM D-2519 (pounds)	
				23°C	155°C	23°C	155°C	23°C	155°C	23°C	155°C
Solventless rigid low viscosity 1-part	155	20–35	2,300	3.9	>5.0	0.004	>0.20	1×10^6	6×10^3	98	14
Solventless semi-flexible excellent H_2O resistance thixotropic 1-part	155	12–16	2,100	4.2	>6.0	0.004	>0.20	1×10^6	4×10^2	107	10
Solventless semi-rigid high bond strength, dielectric properties, chemical resistance thixotropic 1-part	155	7–10	2,700	2.8	3.8	0.0024	0.080	1×10^6	3×10^5	144	63
Solvent type for hermetic motors	210	60**	4,600	3.3	na	0.003	na		na	55	18

* At 100 Hz.
** Baking time @ 149°C (minutes).
na—not available.

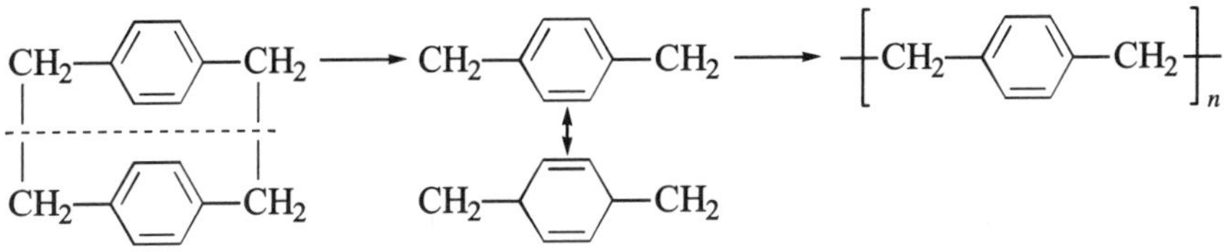

di-para-xylylene   para-xylylene   poly(para-xylylene)
"DPX"   (Parylene N)

Grades

Commercially available versions of parylene include:

• Parylene N, poly-para-xylylene, shown in the preceding diagram. This polymer exhibits superior dielectric strength and exceptionally high surface and volume resistivities, that are virtually unchanged over its useful temperature range exceeding 220°C in the absence of oxygen.

• Parylene C, poly-monochloro-para-xylylene:

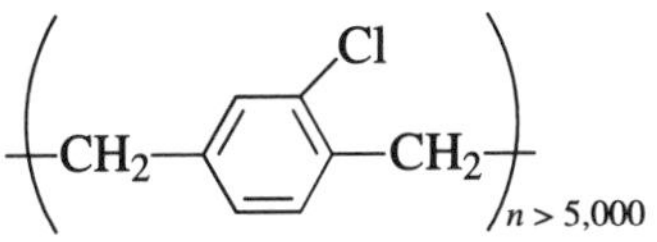

This version is characterized by significantly lower permeability to moisture and gases, while retaining excellent dielectric properties.

• Parylene D, poly-dichloro-para-xylylene:

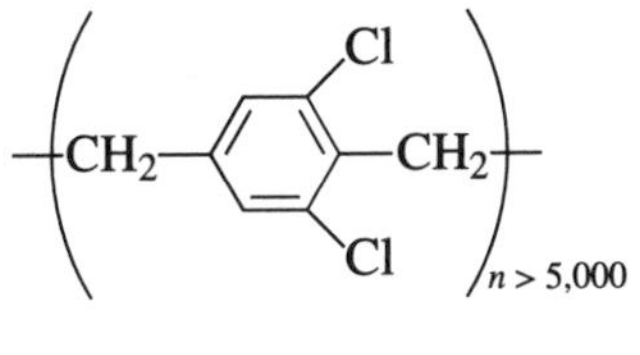

This polymer withstands continuous exposure to air at temperatures up to 150°C.

Parylene coatings are approved under Military Specification MIL-I-46058, *Insulating Compound, Electrical (for Coating Printed Circuit Assembles)*.

Properties

Noteworthy properties of parylene include:

• Uniform coating thickness over edges, points, and internal areas of parts
• Precise control of pinhole-free coating thickness from 0.004 to 1.25 mils
• Excellent dielectric values up to 150°C in air or to over 220°C in oxygen-free atmosphere

- Barrier properties superior to most other coatings
- Good mechanical properties over the temperature range from $-200°C$ to $275°C$
- High resistance to attack by most solvents and corrosive chemicals.

Table 8-5 shows key property values for parylene N, most widely used for electrical/electronic applications.

Electrical/Electronic Uses

A principal application of parylene is to protect circuit boards or modules from hostile environments, such as airborne contaminants, moisture, salt spray, and corrosive vapors. Other uses include protection of hybrid circuits, ferrites, resistors, thermistors, thermocouples, sensing probes, and photocells. A parylene coating on metal foil is suitable for making capacitors.

TABLE 8-5   Key Properties of Parylene N Conformal Coating Resin

	ASTM TEST METHOD	VALUE
Specific gravity*	D 792	1.11
Dielectric strength, 1 mil, short time, V/mil*	D 149	7,000
Dielectric constant*	D 150	
60 Hz		2.65
10^6 Hz		2.65
Dissipation factor*	D 150	
60 Hz		0.0002
10^6 Hz		0.0006
Volume resistivity, $\Omega \cdot cm$*	D 257	1×10^{17}
Arc resistance*	D 495	na
Water absorption, 0.029 in thick specimen, %	D 570	
24 h @ 73°F (23°C)		0.06
Deflection temperature, 264 psi, °F (°C)	D 648	405 (207)
Maximum recommended service temperature °F (°C)		428 (220)
Tensile strength at break, psi*	D 638	6,500
Izod impact strength	D 256	
Foot pounds per inch notched $^1/_8$ in specimen*		na
Flammability ratings		
UL Standard 94, $^1/_{16}$ in specimen		
Standard grade		na
Oxygen index, %	D 2863	
Standard grade		na

*At room temperature.

na = not available.

Phenolics
(Varnish Resins)

Phenolic resins were developed by Belgian chemist L. H. Baekeland in 1907, and became commercially available a few years later. Today, phenolic resins are widely used as modifiers to improve temperature resistance of other resins, permitting some formulations to qualify as Class 180 insulations where they offer a low-cost alternative to silicone varnishes.

Chemistry

Phenolic resins are the condensation reaction products of a phenol with an aldehyde. The basic resin is formed by reacting phenol with aqueous 37–50 percent formaldehyde at a temperature of 100°C or under in the presence of an alkaline catalyst. While formaldehyde is the commonly used aldehyde because of its relatively low price, many substituted phenols and acids as well as alkaline catalysts are employed to produce resins of different types and with different characteristics. Alkaline catalysts and high proportions of formaldehyde encourage formation of heat-reactive, highly crosslinked, insoluble resins. Acid catalysts yield thermoplastic resins, called novolacs, that are fusible and miscible in certain solvents. See also "Phenolics/Furfurals (Molding Resins)," p. 127.

Grades (Types)

The principal types of phenolics used in coatings and impregnants are as follows:

- Modified, oil-soluble resins
- Pure (100 percent) phenolic resins
    - oil-soluble, thermoplastic
    - oil-soluble, heat-reactive
    - alcohol- and ketone-soluble, oil-insoluble, thermosetting
- Phenolic dispersion resins

Modified, oil-soluble phenolics were the first phenolic resins used in the coatings field, and were usually modified with rosin or its glyceryl ester, ester gum, to make the resin soluble in drying oils. These resins replaced the natural resins of oleoresinous varnishes, improving drying properties, water and alkali resistance, and durability. Other phenolics, rendered oil-soluble, are based on m-cresol and 3,5-xylenol. These resins, although not as highly resistant and durable as the pure phenolics developed later, are lower in cost, have better adhesion properties, and are well-suited for spar varnishes and porch and deck enamels. Other types, however, are better suited for insulating varnishes.

Oil-soluble, thermoplastics, pure (100 percent) phenolics are made by condensing approximately equimolar proportions of a para-substituted phenol with formaldehyde in the presence of an acid catalyst. This resin is covered by Federal Specification

TT-R-271, *Resin, Phenol-Formaldehyde, Para-Phenyl*, and is required under Federal Specification MIL-V-173, *Varnish, Moisture-and-Fungus-Resistant* described in the "Federal Specifications and Practices" in this chapter.

Oil-soluble, heat-reactive, pure (100 percent) phenolics are the reaction products of a para-substituted phenol with formaldehyde in the presence of an alkaline catalyst. About 2 moles of formaldehyde are used to 1 mole of substituted phenol. The reaction is interrupted at an intermediate stage at which the resinous condensate is still oil-soluble, although reactive. The partly polymerized resin is subsequently heat-reacted with drying oils to produce varnishes with very rapid bodying properties. Short-oil industrial banking varnishes of this type make excellent insulating and impregnating varnishes for electrical equipment because of their fast, deep-curing characteristics and their good dielectric properties. These resins are often incorporated into formulations with lower grade resins to improve drying speed, water resistance, hardness, and durability.

Alcohol- and ketone-soluble, oil-insoluble, thermosetting, pure (100 percent) phenolics are produced by reacting oil-insoluble phenol or meta-substituted, phenol with formaldehyde in alkaline solution. A high ratio of formaldehyde to phenol imparts greater reactivity to these resins. The reaction is stopped at an intermediate stage of polymerization while the resin is still alcohol-soluble and reactive. The partially polymerized resin is supplied as a solution in an appropriate solvent. The applied coating, when heated, converts to a completely polymerized thermoset, insoluble film that is highly resistant to corrosion, abrasion, and chemical attack. In these respects, this type is superior to other types of pure phenolic resins. Room temperature curing is facilitated by adding amine-type or other alkaline catalysts, although heat curing imparts maximum chemical and solvent resistance.

Phenolic dispersion resins are highly polymerized phenolic resin-oil complexes dispersed in volatile organic solvents. They are nonoxidizing and dry quickly by evaporation of solvent to form weather-resistant films with a high degree of impermeability to moisture, and good resistance to alcohols, ketones, esters, ethers, and dilute alkalis. These resins are often blended with other resins for property enhancement.

Processing

See "Application Methods," p. 239 in this chapter.

Properties

See discussion in this section under "Grades" of properties of various types of phenolic varnishes. A table of specific properties is not meaningful because of the limitless number of possible formulations for the various types of phenolic varnishes.

Electrical/Electronic Uses

Principal uses for phenolic varnishes are for moisture-and-fungus-resistant varnish, and to blend with other varnishes for improved temperature and chemical resistance.

Polyurethanes (PURs)
(Varnish Resins)

Polyurethane chemistry originated in Germany in the late 1930s, but commercial production by U.S. manufacturers did not start until after World War II. Although not as widely used in insulating coatings as polyesters or epoxies, polyurethanes are available for this purpose.

Chemistry

Polyurethanes used in embedding compounds and wire enamels are covered in other chapters. Principal types of polyurethanes are discussed in "Polyurethanes" in Chapter 6, *Embedding Compounds*.

Type 1 one-component urethane alkyds are used in insulating varnishes, enamels, and conformal coatings. The manufacturing process involves partial hydrolysis of drying oils to produce a mixed hydroxy ester with free hydroxyl groups which can then react with a diisocyanate to form a prepolymer resin. The reaction is controlled so that isocyanate groups are fully reacted while leaching the unsaturated fatty acid portion of the polymer free to combine with oxygen, or cure when catalyzed by metallic naphthenates. Curing can be rapid (a few minutes), especially at low baking temperatures.

Type 2 one-component moisture-cured systems, 50 percent solids content, are used for conformal coatings. Curing is dependent on evaporation of solvents followed by reaction with ambient moisture where relative humidity is 30 percent or higher. To extend shelf life, precautions should be taken to ensure that the atmosphere in containers is moisture-free.

Type 5 two-component resins are used in both solventless and solvent-type coatings (30–60 percent solids). Pot life of mixed components is limited. Coatings may emit toxic fumes and cause skin problems unless precautions are taken.

Grades

Table 8-6 lists some representative grades of polyurethane coatings.

Processing

See "Application Methods," p. 239 in this chapter.

Properties

As shown in Table 8-6, polyurethanes serve Class 155 applications, with abrasion resistance being their outstanding feature.

Electrical/Electronic Uses

Principal uses of polyurethane coatings are for conformal coatings on circuit boards and for enamel coatings on electrical equipment.

TABLE 8-6 Key Properties of Representative Polyurethane Coatings

FEATURE	THERMAL CLASS (°C)	CURING (hours)	(°C)	DIELECTRIC STRENGTH ASTM D-115 (volts/mil) DRY	WET	WATER	CHEMICAL RESISTANCE (10% H$_2$SO$_4$)	(1% NaOH)	OIL ASTM D-115
Clear conformal coating superior abrasion resistance fast cure Meets MIL-I-24092	155	$^1/_4$	23	2,000	1,200	Ex	Ex	Ex	Passed
Red enamel superior abrasion resistance fast cure	155	$^1/_2$	23	2,000	1,200	Ex	Ex	Ex	Passed
Black enamel same features as red enamel	155	$^1/_2$	23	1,800	1,200	Ex	Ex	Ex	Passed

Ex—excellent.

Silicones
(Varnish Resins)

Silicone resins have been commercially available since about 1944. Varnishes made from these resins are characterized by excellent thermal stability and water resistance, but because of their high cost, they are used only where other varnishes are not suitable.

Chemistry

The basic reaction in making silicones is a condensation of hydroxy silanes (silanols) derived from mono-, di-, and trichlorosilanes. Commercial processes are either direct or employ a Grignard reagent. Pure silicones have the following structure:

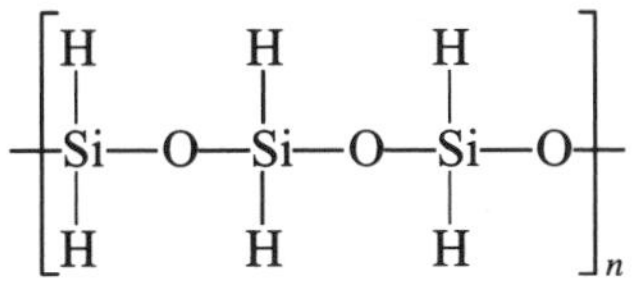

$$\begin{bmatrix} H & H & H \\ | & | & | \\ -Si-O-Si-O-Si-O- \\ | & | & | \\ H & H & H \end{bmatrix}_n$$

Practical resins, however, are organosilicon compounds of high molecular weight in which some of the hydrogen atoms are replaced with various alkyl and aryl groups capable of both linear and crosslinked polymerization. Linear polymers are used in embedding resins and elastomers, discussed in Chapter 6, *Embedding Compounds* and in Chapter 5, *Extrusion Compounds*.

Crosslinked types are formulated into varnishes. Crosslinking is accomplished by introducing at intervals along the siloxane chain a third oxygen atom in place of one of the organic groups. This leads to oxygen bridges between trifunctional atoms on different chains. There is also crosslinking through polyfunctional organic side groups. Thus, the useful resin has a structure represented by the following, where R is an alkyl or aryl group:

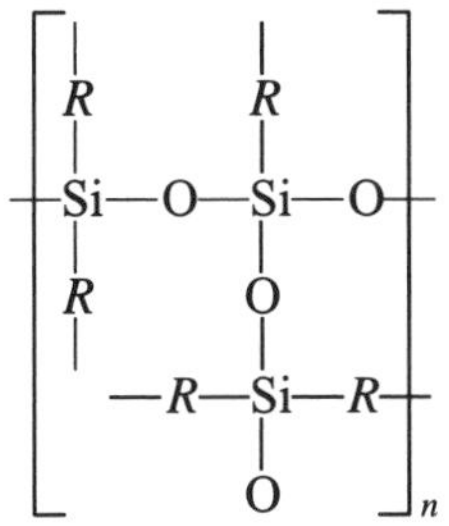

$$\begin{bmatrix} & | & & | \\ & R & & R \\ & | & & | \\ -Si&-O-Si&-O- \\ & | & & | \\ & R & & O \\ & & & | \\ & & -R-Si-R- \\ & & & | \\ & & & O \end{bmatrix}_n$$

silicone resin

Silicone resins are often modified with lower cost resins, resulting in economical formulations with better thermal properties than the lower cost resin alone and improved physical properties over pure silicones. Modifications with alkyds and epoxies are especially useful in making insulating varnishes. The introduction of halogen atoms or groups containing halogens into recipes renders resins nonflammable or less flammable.

Grades

A variety of grades used in insulating coatings and impregnants is commercially available, including room temperature curing and baking versions of both solvent and solventless types. Representative grades are listed in Table 8-7.

Processing

See "Application Methods," p. 239 in this chapter.

Properties

Outstanding features of silicones include:

- Excellent dielectric properties over a wide temperature range
- Long term service at temperatures up to 482°F (250°C)
- Greater resistance to corona than other coatings and impregnants
- Excellent water and moisture resistance.

Mechanical properties and adhesion are inherently lower than epoxies, polyesters, and polyurethanes. Solubility, chemical resistance, and compatibility with other materials vary with the nature of the side chains and molecular weight.

Table 8-7 lists key properties of representative silicones.

TABLE 8-7 Key Properties of Representative Silicone Varnishes and Conformal Coatings

	ASTM TEST METHOD*	BAKING VARNISH	ROOM TEMP. CURING NONCORROSIVE 1-PART CONFORMAL COATING
Dielectric strength	D 115	2,000	500
Dielectric constant	D 150		
100 Hz		3.10	2.64
100 kHz		3.00	2.63
Dissipation factor	D 150		
100 Hz		0.010	0.0016
100 kHz		0.007	0.0006
Volume resistivity	D 257	1×10^{14}	5×10^{14}
Maximum recommended service			
Temperature °F (°C)		356–428 (180–220)	356 (180)
Meets federal specifications		MIL-I-24092	MIL-I-46058 MIL-A-46146

*At room temperature.

Electrical/Electronic Uses

The high cost of silicones limits their use to applications not suitable for other coatings and impregnants, including high-temperature rotating equipment and conformal coatings for electronic parts exposed to harsh environments.

Unclad and Clad Structures

UNCLAD STRUCTURES

INTRODUCTION

Unclad structures have a long history of use as dielectrics, dating back to the earliest days of the electrical industry. As used in this handbook, unclad structures include high-pressure laminates, polyester glass laminates, sheet and bulk molding compounds, and pultrusion materials. In general, these materials are characterized by exceptionally high strength-to-weight ratios as well as good formability and machineability, under prescribed conditions, which dictate their use as dielectrics where these properties are required.

Since reinforcing materials are also used in thermoplastic and thermosetting compounds, this chapter is oriented toward production methods other than those covered in preceding chapters.

REINFORCEMENTS

The most common reinforcement used for resins is known as "E" glass. This glass was designed especially for dielectric applications, although it is now more widely used. E glass has 50–55 percent SiO_2, which is less than in most other glasses. Boric oxide (B_2O_3) content for E glass is 8–13 percent, among the highest of all glasses. Other major ingredients are Al_2O_3, Fe_2O_3, and CaO. Another glass used for dielectric applications is "S" glass. This glass, about 65 percent SiO_2, 0.25 percent B_2O_3, and 35 percent Al_2O_3 and MgO, has higher tensile strength than E glass, and is used in fabrics for laminates in high-performance parts where its strength-to-weight ratio is greater than that of most metals.

Glass reinforcements are supplied in the following forms:

- Continuous roving is made from 50 or more filaments to form a continuous strand. Spun roving is prepared by spinning staple fibers 8–15 inches long to form a continuous strand held together by a slight twist and sizing. Rovings are used in the pultrusion process.

- Chopped strands are made from continuous or spun roving cut in lengths of $^1/_8$–2 inches. Longer strands are mixed with resins to make compounds for compression and transfer molding, and for bulk molding compound (BMC). Shorter strands are more suitable for injection molding compounds.

- Cloth woven on textile looms is available in a variety of textures, weights, and thicknesses, and in plain, twill, and satin weaves. Unidirectional properties are obtained by emphasizing warp (lengthwise) yarns over weft (filling) yarns. Plain weave is used for most laminates. Satin weaves are better for parts with compound curves.

- Mats are composed of either continuous or chopped strands. Chopped strand mats have strands about 2 inches long, randomly distributed and bound together with resin to form flat sheets used primarily in hand lay-up, low-pressure laminating. Continuous strand mats are made from multiple layers of continuous filaments laid down in a swirl pattern and bonded with resin. Continuous strand mats are used as substrates for complex shapes.

All glass fibers must have a surface treatment, or sizing, that serves as a lubricant and bonding agent. Untreated filaments are prone to abrasion by one another and, unless bonded together, they would be difficult to handle in later operations. A temporary size is applied to warp (lengthwise) yarns preparatory to weaving into glass cloth. This size is usually composed of one or a combination of these materials: gelatin, hydrogenated vegetable oil, polyvinyl alcohol, dextrin, nonionic emulsifier, and cationic lubricant. This sizing is temporary since it must be burned off prior to resin treatment. Otherwise, satisfactory bonding of resin to glass would not be achieved.

After burning, another coating, called a compatible sizing, is applied to the glass cloth to promote bonding with resin and as a lubricant. A widely used product with polyester resins is trademarked Volan (methacrylatochromic chloride) by Du Pont. Organofunctional silanes and organo-titanates are also used.

Other reinforcing fibers used infrequently or not at all in dielectric applications include:

- Carbon (graphite) fibers, used in aerospace and sports equipment
- Aramid fibers, trademarked "Kevlar" by Du Pont, are also used primarily for aerospace and sports equipment
- Boron fibers, the strongest and most expensive fibers, used exclusively for military applications
- Sisal, low-cost vegetable fibers, are used alone or with glass in applications such as air conditioner and heater housings, and electrical outlet boxes.

RESINS

The principal resins employed in the manufacture of unclad and clad structures are the following:

- Epoxies
- Polyesters
- Phenolics
- Melamines
- Silicones
- Diallyl phthalate
- Polyimides

The standards, properties, and features of these resins are detailed in the sections which follow.

DEVELOPMENT PROGRAMS

The focus of development programs continues to be on achieving stronger, lighter, low-profile (low-shrink) composites through improved resin technology, new reinforcements, and more sophisticated fabrication methods. Lower cost materials and production methods are under continuous evaluation, especially for large, high-volume parts.

MARKET TRENDS

The principal impetus for above normal growth in the use of reinforced resins was the oil embargo of 1972–74. These materials, with their high strength-to-weight ratios, corrosion resistance, and versatility in fabrication techniques, were logical replacements for heavier metal parts, especially in the motor vehicle industry, where manufacturers were mandated to increase dramatically the fuel efficiency of vehicles according to a Department of Transportation time schedule.

Because of technological advances, these materials have become more attractive for a wide spectrum of uses in other industries, most notably in structural applications for calculators, computers, and business equipment such as copiers, typewriters, and electronic financial and retail terminals.

Unsaturated polyester resins with glass reinforcements will continue to predominate, although there will be greater use of other fibers, such as carbon (graphite), aramid, and boron, for specialized applications such as in sports and military/aerospace equipment.

PRINCIPAL UNCLAD STRUCTURES
FOR ELECTRICAL/ELECTRONIC USES

The chemistry, grades, processing, properties, and electrical/electronic uses of principal unclad structures are described in the sections which follow.

Unclad High-Pressure Thermoset Laminates

Usually referred to as industrial laminates, these materials are fabricated by machine tools into countless parts for electrical and electronic applications. Principal producers include Haysite Reinforced Plastics, Iten, Norplex, NVF, Spaulding, and Westinghouse.

Chemistry

The principal laminating resins, usually captively produced, are phenolics, epoxies, melamines, and silicones:

Phenolics are typically one-step liquid resins obtained by alkali-catalyzed condensation of phenol or cresol with excess formaldehyde.

Epoxies used as laminating resins are based principally on the polycondensation of bisphenol-A and epichlorohydrin, as described in "Epoxies" (p. 183) in Chapter 6, *Embed-*

ding Compounds. A widely used hardener is dicyandiamide (dicy), which cures through its four nitrogen-containing functional groups, consuming both epoxide and hydroxyl groups. It is a latent curing agent, relatively stable at room temperature, but activated by heat. Formulations may also contain 5–10 percent epoxy novolacs. Flame-retardant grades containing 20 percent bromine by weight are made by replacing bisphenol-A with tetra-bromobisphenol-A. Epoxy glyoxyl resin is added to brominated resin for FR-5 grade.

Melamine laminating resins are based normally on the reaction of 1 mole of melamine with 2 moles of formaldehyde. The nature and degree of polymerization depend on the pH of the reaction, usually in the eight–ten range.

Silicone laminating resins of the crosslinking type are discussed in "Silicones" in Chapter 8, *Insulating Coatings and Impregnants.*

Polyimide laminating resins may be prepared as follows:

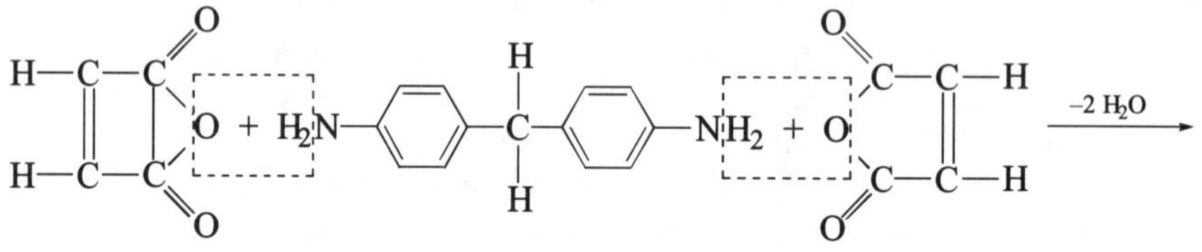

(A)	(B)	(A)
maleic anhydride	methylene dianiline	maleic anhydride

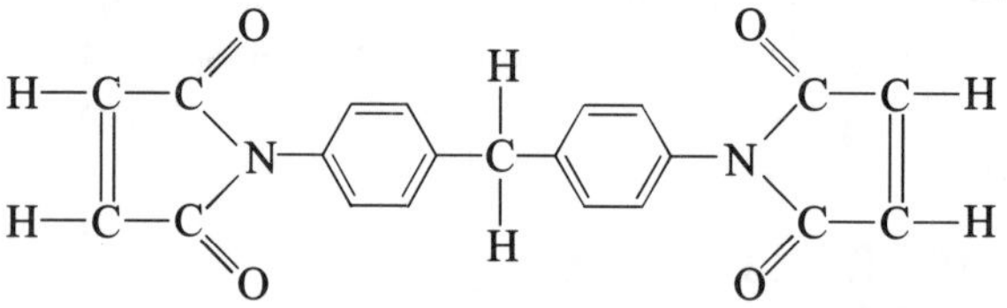

(C)
bismalemide

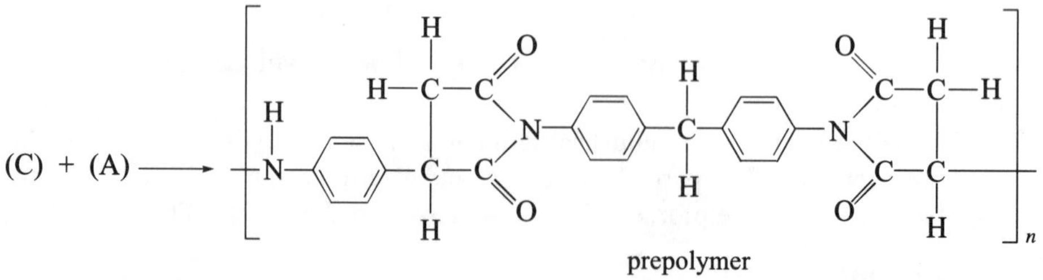

prepolymer

The prepolymer is finally cured in the laminating operation with pressure and at elevated temperature.

Another process is described in "Polyimide," p. 228.

Federal Specifications for Unclad Laminates

The following Federal specifications have been developed for unclad laminates:

- L-P-509, *Plastic Sheet, Rod and Tube, Laminated, Thermosetting*
- MIL-P-997, *Plastic Material, Laminated, Thermosetting, Electric Insulation, Sheets, Glass Cloth, Silicone Resin*
- MIL-P-13949/11, *Plastic Sheet, Laminated, Materials (for Printed Wiring Boards), GE Base Material, Glass Cloth, Resin Preimpregnated (B-Stage)*
- MIL-P-13949/12, *Plastic Sheet, Laminated, Materials (for Printed Wiring Boards), GF Base Material, Glass Cloth, Resin Preimpregnated (B-Stage)*
- MIL-P-13949/13, *Plastic Sheet, Laminated, Materials (for Printed Wiring Boards), GI Base Material, Glass Cloth, Resin Preimpregnated (B-Stage)*
- MIL-P-15035, *Plastic Sheet, Laminated, Thermosetting, Cotton Fabric Base, Phenolic Resin*
- MIL-P-15037, *Plastic Sheet, Laminated, Thermosetting, Glass Cloth, Melamine Base*
- MIL-P-15047, *Plastic Material, Laminated Thermosetting Sheets, Nylon Fabric Base, Phenolic Resin*
- MIL-P-18177, *Plastic Sheet, Laminated, Thermosetting, Glass Fiber Base, Epoxy Resin*
- MIL-P-19161, *Plastic Sheet, Laminated, Glass Cloth, Polytetrafluoroethylene Resin*
- MIL-I-24204, *Insulation, Electrical, High Temperature, Bonded, Synthetic Fiber Paper*
- MIL-P-25525, *Plastic Material, Phenolic Resin, Glass-Fiber Base, Laminated*

ASTM Standards for Laminated Thermosetting Materials

The American Society for Testing and Materials (ASTM) has established these standards for laminated thermosetting materials:

D 229, *Testing Rigid Sheet and Plate Materials Used for Electrical Insulation*

D 257, *Tests for D-C Resistance or Conductance of Insulating Materials*

D 348, *Testing Rigid Tubes Used for Electrical Insulation*

D 349, *Testing Laminated Round Rods Used for Electrical Insulation*

D 495, *Test for High-Voltage, Low-Current Dry Arc Resistance of Solid Electrical Insulation*

D 618, *Conditioning Plastics and Electrical Insulating Materials for Testing*

D 621, *Tests for Deformation of Plastics Under Load*

D 668, *Measuring Dimensions of Rigid Rods and Tubes Used for Electrical Insulation*

D 709, *Standard Specification for Laminated Thermosetting Materials*

D 785, *Test for Rockwell Hardness of Plastics and Electrical Insulating Materials*

D 883, *Definitions of Terms Relating to Plastics*

D 2304, *Thermal Evaluation of Rigid Electrical Insulating Materials*

Underwriters Laboratories has developed the following standards:

- UL 746D, *Polymeric Materials—Fabricated Parts*
- UL 746E, *Polymeric Materials—Industrial Laminates*

NEMA Grades for Unclad Laminates

The National Manufacturers Association (NEMA) in Standard LI 1 has established the following grade designations for industrial laminates:

- Grade X: phenolic resin, paper reinforced. Good mechanical properties, fair dielectric values, low resistance to moisture.
- Grade XX: phenolic resin, paper reinforced. Good mechanical properties, better dielectric than Grade X, and more resistant to moisture.
- Grade XXX: phenolic resin, paper reinforced. Mechanical properties about the same as Grade XX, but better dielectric because of high resin content. Suitable for use at radio frequencies and in high-humidity environments.
- Grade XP: phenolic resin, paper reinforced. May be punched up to $^1/_{16}$ inch at room temperature and punched hot up to $^1/_8$ inch. Dielectric properties and moisture resistance intermediate between Grades X and XX.
- Grade XPC: phenolic resin, paper reinforced. More flexible than XP with better cold punching and shearing properties.
- Grade XXP: phenolic resin, paper reinforced. Similar to Grade XP in machinability and to Grade XX in dielectric and moisture-resistant properties.
- Grade XXXP: phenolic resin, paper reinforced. Better than Grade XXX in dielectric properties and easier to hot punch. Highly moisture resistant.
- Grade XXXPC: phenolic resin, paper reinforced. Similar to XXXP, but easier to punch at lower temperatures.
- Grade FR-1: phenolic resin, paper reinforced. Similar to Grade XP, but less flammable.
- Grade FR-2: phenolic resin, paper reinforced. Similar to Grade XXXPC, but less flammable.
- Grade C: phenolic resin, cotton fabric reinforced. Good mechanical properties, but not recommended for dielectric use.
- Grade CE: phenolic resin, cotton fabric reinforced. Tougher than Grade XX and more resistant to moisture than Grade C, with dielectric properties suitable for low-voltage applications.
- Grade L: phenolic resin, fine weave cotton fabric reinforced. Not as tough as Grade C, but machines more easily. Not recommended for dielectric use.
- Grade LE: phenolic resin, fine weave cotton fabric reinforced. Similar to Grade L in mechanical and machining properties, but better in moisture resistance. Dielectric properties suitable for low-voltage applications.
- Grade N-1: phenolic resin, nylon fabric reinforced. Excellent dielectric properties, even under high-humidity conditions. Fungus resistant.
- Grade G-3: phenolic resin, glass fabric reinforced. High impact and flexural strength, but bond strength lowest of glass-base grades.

- Grade G-5: melamine resin, glass fabric reinforced. High mechanical strength and hardest laminated grade. Low burning rate and second only to silicone laminates in heat and arc resistance. Excellent dielectric properties under dry conditions.
- Grade G-9: melamine resin, glass fabric reinforced. Higher mechanical strength than Grade G-5. Dielectric properties similar to Grade G-5 with better moisture resistance.
- Grade G-7: silicone resin, glass fabric reinforced. Glass H insulation with excellent dielectric properties. Second only to Grade G-5 in low rate of burning.
- Grade G-10: epoxy resin, glass fabric reinforced. Excellent mechanical and good dielectric properties under both dry and humid conditions.
- Grade G-11: epoxy resin, glass fabric reinforced. Similar to Grade G-10, but with improved flexural strength.
- Grade FR-3: epoxy resin, paper reinforced. Superior in dielectric characteristics to Grade XXPC. Good mechanical properties. Suitable for punching at room temperature. Flame retardant. Used extensively as printed circuit substrate.
- Grade FR-4: epoxy resin, glass fabric reinforced. Similar to Grade G-10, but with reduced burning rate.
- Grade FR-5: epoxy resin, glass fabric reinforced. Similar to Grade G-11, but with reduced burning rate.
- Grade CEM-1: epoxy resin, glass fabric surfaces, cellulose paper core. Good punching grade. Low burning rate.
- Grade CEM-3: epoxy resin, glass fabric surfaces, glass paper core. Properties approach those of Grade FR-4.

Prepreg Manufacture

The basic materials for making industrial laminates are called *prepregs*. A prepreg is composed of a substrate, usually paper, cotton, or glass fabric, impregnated with a thermosetting resin, generally phenolic, epoxy, or melamine. Resin-to-substrate ratios typically range on a weight basis from 40/60 to 60/40.

The continuous manufacturing process consists of the following steps:

- Impregnation of substrate with resin is accomplished as it is fed from a roll by a series of rollers into a resin tank.
- After impregnation of the substrate, excess resin is removed by doctor blades, wire-wound rod, or reverse gravure roller.
- The material then passes through a single- or multizone oven in which temperature, dwell time, and air flow are critical factors and are tailored to each resin system.
- The resin is partially cured to make it relatively tack free.
- The B-stage (partially cured) prepreg is then rolled, sometimes with a polyethylene interlayer to prevent blocking.
- The prepreg roll is then stored in a refrigerated room at 40° to 60°F (4°–16°C), 50 percent relative humidity, to minimize further curing.

Manufacturers of industrial laminates produce virtually all the prepregs they require internally.

Unclad Laminates Manufacture

To make an industrial laminate, the prepreg is cut to size and laid up to the desired thickness, with allowance made for compression. A number of plies are thus required to form a "book." As many as ten such books separated by polished stainless steel plates are stacked in a press opening. Press pressure is equalized over the book surface by cushion sheets and additional steel plates on the top and bottom of the book stack. Books of different substrate styles, resins, and resin content may be cured in the same stack. A press cycle is about 3 hours and consists of these steps:

• Loading laminate books with cushions and plates onto a platen with temperature under 150°F (66°C).

• Closing the press and applying the required pressure (100–600 psi).

• Increasing temperature to intermediate range of 200° to 250°F (93°–121°C) and holding to allow the resin to flow, binding the layers of the laminate together at the desired thickness.

• After the resin has gelled, increasing the temperature to 325° to 400°F (162°–204°C) for final cure.

• Holding the laminate at the cure temperature for the required period of time.

• Cooling the press hydraulically and removing the laminate.

Compression presses may have five–ten openings, and may be heated by steam, oil, or electricity. Standard sizes (in inches) for finished laminates are 36 × 36, 36 × 48, 36 × 72, 48 × 60, 48 × 72, 48 × 96, and 48 × 120. The prepreg is cut 1 inch larger than the finished size to allow for trimming.

Thermoset laminates may also be made into convolutely wound tubing. After curing under high pressure and heat, an easily machineable material is formed with dielectric properties similar to those of high-pressure laminated sheets.

Processing Polyimide Laminates

The processing of polyimide/glass laminates is more critical than for other laminates. Powdered resin is dissolved in N-methylpyrrolidone (NMP) to form a 45–55 percent impregnating bath. After impregnation and passing through the oven, the prepreg should have a resin content of approximately 35 percent, with a flow rate of between 20 and 40 percent, and a solvent content between 7 and 9 percent for high-pressure laminating. This requires oven temperatures and dwell times between 250°F (120°C) for 75 minutes and 320°F (160°C) for 10 minutes. Prepregs packaged in polyethylene bags may be stored for two–three months at 32°F (0°C) and for longer periods at lower temperatures.

In the laminating operation, the following steps are recommended:

• The stack of prepreg sheets is placed between foils of aluminum on the press platen.

• Through press plates at 250°F (120°C), a pressure of 210–800 psi (depending on the stack thickness) is applied.

- The temperature is then increased to 360°F (180°C) at a predetermined rate for a stack no thicker than $^3/_8$ inch (a longer period is required for thicker stacks).
- Cure for 1 hour at 360°F (180°C).
- The laminate may be removed hot from the press.
- Post-cure for 24 hours at 480°F (250°C) or for 48 hours at 390°F (200°C).

Properties and Uses

There are numerous characteristics of laminated thermosetting materials that make them desirable for a wide spectrum of applications:

- They possess an exceptionally high strength-to-weight ratio, with specific gravity about half that of aluminum and one-sixth that of steel, making laminates desirable for uses where high strength and minimum weight are requirements.
- Gears cut from cotton cloth based laminates run more quietly than metal gears, require no lubrication, and are durable. Applications range from clock gears to gears for rolling mills.
- Industrial laminates are highly resistant to corrosion, and to attack by organic acids, most solvents, and dilute inorganic acids (but not to alkalis). Combined with easy machineability and good mechanical properties, these materials are suitable for use, where metals are not, in the chemical and allied industries.
- Excellent dielectric values, good structural properties, light weight, and ease of fabrication make laminates favorites for a multitude of applications in the manufacture of electrical and electronic equipment. Laminate types include:
  - Phenolic resin glass-reinforced laminates which retain good dielectric properties at moderately high temperatures and are stronger than paper-reinforced phenolics.
  - Epoxy resin paper-reinforced laminates which have good mechanical properties with low burning rate. Dielectric properties remain stable in humid environments.
  - Epoxy resin glass-reinforced laminates which have higher mechanical strength than paper-reinforced epoxies. Dielectric values are excellent. These laminates are used extensively as substrates for printed circuits.
  - Diallyl phthalate glass-reinforced laminates which have excellent dielectric properties, are easy to process, and are useful up to 450°F (230°C). They are suitable for multilayer printed circuits.
  - Melamine resin glass-reinforced laminates which have outstanding resistance to arcing and burning, and have high mechanical strength.
  - Silicone resin glass-reinforced laminates which have excellent thermal properties and arc resistance, and are also excellent dielectrics.
  - Fluorocarbon resin glass-reinforced laminates which have stable dielectric properties, low moisture absorption, are chemically inert and flame resistant, and remain useful up to 500°F (260°C).
  - Polyimide resin glass-reinforced laminates which have exceptional thermal properties and remain useful over 500°F (260°C). They have high

TABLE 9-1   Typical Values for High-Pressure Electrical Grade Thermoset Laminates

	TEST METHOD	GRADE										
NEMA/ASTM Grade	X		XX	XXX	XP	XPC	XXP	XXXP	XXXPC	FR-1	FR-2	FR-3
MIL-P Grade			PBG	PBE-E				PBE-P	PBE-P	—	—	PEE
Federal L-P-509/513 Grade	X		PBG	PBE	XP	XPC	XXP	PBE-P	PBE-P	—	FR-2	—
PROPERTY												
Tensile strength, 10^3 psi	ASTM D 638											
Lengthwise		20	16	15	12.4	10.5	11	12	12.4	12	12.5	14
Crosswise		16	13	12	9	8.5	8.5	9.5	9.5	9	9.5	12
Compressive strength, 10^3 psi	ASTM D 695											
Flatwise		36	34	32	25	22	25	25	25	25	25	29
Edgewise		19	23	25.5	—	—	—	—	—	—	—	—
Flexural strength, 10^3 psi, $1/8$ in	ASTM D 790											
Lengthwise		25	15	13	14	12	14	16	12	16	17	20
Crosswise		22	14	12	12	10	12	13.2	10	13	15	16
Modulus of elasticity in flexure, 10^6 psi	ASTM D 229											
Lengthwise		1.8	1.4	1.3	1.2	1.0	0.9	1.0	1.0	1.2	1.0	1.3
Crosswise		1.3	1.1	1.0	0.9	0.8	0.7	0.7	0.7	0.9	0.9	1.0
Shear strength, 10^3 psi	ASTM D 732	12	11	10	8	9.5	11	1.2	11	12	10.8	11
Izod impact strength, ft·lb/in of notch	ASTM D 256											
Flatwise		4.0	1.3	1.0	—	—	—	—	—	—	—	—
Edgewise		0.5	0.35	0.35	—	—	—	—	—	—	—	—
Rockwell hardness, M scale	ASTM D 785	110	105	110	95	75	100	100	95	95	97	100
Bond strength, lb	ASTM D 229	700	800	950	1,100	1,400	—	1,100	1,000	1,200	1,200	950
Coeff. of thermal expansion, cm/cm/°C $\times 10^{-5}$, x-axis	IPC-TM 650-2.4.24	1.3	—	1.5	1.3	1.5	—	1.6	1.7	1.6	1.6	2
Water absorption, % 24 h, $1/16$ in	ASTM D 229	6	2	1.4	3.6	5.5	1.8	—	0.75	—	0.75	0.65
Dielectric strength, volts/mil, short time, $1/16$ in	ASTM D 229	700	700	650	650	625	700	650	650	650	650	600
Dissipation factor, 10^6 cycles	ASTM D 150	0.06	0.05	0.04	0.06	0.06	0.04	0.04	0.04	0.065	0.035	0.035
Thermal Class, electrical, $1/16$ in and over		130	130	130	130	130	130	130	130	105	105	130
Flammability	UL 94	HB	HB	HB	HB	HB	HB	HB	HB	V-0	V-0	V-0

TABLE 9-1   (*Continued*)

	TEST METHOD	GRADE									
NEMA/ASTM Grade		FR-4 FR-5	FR-6	CE	LE	N-1	G-7	G-9	G-10 G-11	CEM-1	CEM-3
MIL-P Grade		–4 GEE	—	FBG	FBE	NPG	GSG	GME	GEE	—	—
Federal L-P-509/513 Grade		–5 GEB	—	CE	LE	N-1	—	—	GEB	—	—
PROPERTY											
Tensile strength, 10^3 psi	ASTM D 638										
Lengthwise		40	9	12	12	8.5	23	37	40	45	—
Crosswise		35	9	9	8.5	8	18.5	30	35	34	—
Compressive strength, 10^3 psi	ASTM D 695										
Flatwise		60	29	39	37	20	45	70	60	50	—
Edgewise		35	29	24.5	25	—	14	25	35	—	—
Flexural strength, 10^3 psi, $^1/_8$ in	ASTM D 790										
Lengthwise		55	20	17	15	10	23	55	55	35	40
Crosswise		45	20	14	13	9	20	35	45	28	32
Modulus of elasticity in flexure, 10^6 psi	ASTM D 229										
Lengthwise		2.7	0.7	0.9	1.0	0.6	1.4	2.5	2.7	2.5	—
Crosswise		2.2	0.7	0.8	0.9	0.5	1.2	2.0	2.2	2.1	—
Shear strength, 10^3 psi	ASTM D 732	19	14	11	11	14	17	20	19	15	14
Izod impact strength, ft·lb/in of notch	ASTM D 256										
Flatwise		7.0	3.5	2.3	1.8	4.0	8.5	12.0	7.0	1.6	1.6
Edgewise		5.5	3.5	1.4	1.0	3.5	1.5	8.0	5.5	1.2	1.4
Rockwell hardness, M scale	ASTM D 785	111	114	105	105	105	100	120	111	102	—
Bond strength, lb	ASTM D 229	—	—	1,800	1,600	1,000	650	1,700	2,200	1,400	—
Coeff. of thermal expansion, cm/cm/°C $\times 10^{-5}$, x-axis	IPC-TM 650-2.4.24	0.9	1.0	2.0	2.0	8.0	1.0	1.0	0.9	1.3	—
Water absorption, % 24 h, $^1/_{16}$ in	ASTM D 229	0.25	0.18	2.2	1.95	0.6	0.3	0.8	0.25	0.3	0.25
Dielectric strength, volts/mil, short time, $^1/_{16}$ in	ASTM D 229	500	500	500	500	600	400	400	500	675	—
Dissipation factor, 10^6 cycles	ASTM D 150	0.025	0.017	0.055	0.055	0.038	0.003	0.017	0.015	0.035	0.025
Thermal Class, electrical, $^1/_{16}$ in and over		130	130	130	130	105	180	—	130	130	—
Flammability	UL 94	V-0	V-0	HB	HB	HB	V-0	V-0	HB	V-0	

resistance to radiation, outstanding dimensional stability, chemical inertness, and are inherently nonflammable.

– Nylon-reinforced laminates which retain high insulation resistance under humid conditions. They have exceptionally low dielectric constant and dissipation factor, making them preferred for high-frequency electronic applications.

– Composite laminates which are easy to punch and are suitable for substrates for printed circuits.

Use of high-price silicone, fluorocarbon, and polyimide resin laminates is limited to those applications where their superior properties are required.

Table 9-1 shows key property values for NEMA grade high-pressure thermoset laminates.

Polyester Glass Mat Laminates

These materials, some of which are made to NEMA standards, may also easily be formulated to meet individual customer specifications at relatively low cost compared with other types of reinforced resins. Major suppliers are Cincinnati Development, Glastic, and Haysite Reinforced Plastics.

Chemistry

Unsaturated thermosetting polyesters used with reinforcing substrates are condensation polymers formed by reacting an anhydride or dibasic acid with a dihydroxy alcohol. At least one of the acidic components must contain an olefinic bond. Saturated acids may be included to control unsaturation and improve physical properties. A reactive monomer, such as styrene or vinyl toluene, is added to control viscosity and to promote crosslinking. A small amount of inhibitor, for example hydroquinone, is added to retard the reaction prior to use. Thus, a wide range of resin properties is obtainable by varying the ratio of ingredients. For a general purpose resin, a typical mole ratio would be:

RAW MATERIAL	MOLES
Propylene glycol	2.1
Phthalic anhydride	1.0
Maleic anhydride	1.0
Styrene	1.8

See also the "Unsaturated Polyesters" (p. 248) in Chapter 8, *Insulating Coatings and Impregnants*.

Federal Specifications for Polyester Glass Mat Laminates

The following Federal specifications have been developed for polyester glass mat laminates:

- L-P-383, *Plastic Material, Polyester Resin, Glass Fiber Base, Low Pressure Laminate*
- MIL-P-24364, *Plastic Sheet, Laminated, Thermosetting Electrical Insulating Sheet, Glass Mat*
- MIL-P-24364/1, *Plastic Sheet, Laminated, Thermosetting Electrical Insulating Sheet, Polyester Glass Mat Grade GPO-N1 (Classes 130, 155, and 180)*
- MIL-P-24364/2, *Plastic Sheet, Laminated, Thermosetting Electrical Insulating Sheet, Polyester Glass Mat Grade GPO-N2 (Class 130)*
- MIL-P-24364/3, *Plastic Sheet, Laminated, Thermosetting Electrical Insulating Sheet, Polyester Glass Mat, Grade GPO-N3*
- MIL-P-25395, *Plastic Material, Heat Resistant, Low Pressure Laminated Glass Fiber Base, Polyester Resin*

NEMA and ASTM Standards for Polyester Glass Mat Laminates

The following standards and specifications cover polyester glass laminates:

- NEMA Standard LI 1, *Industrial Laminated Thermosetting Products*
- ASTM D 1532, *Polyester Glass-Mat Sheet Laminate*

Grades

NEMA Standard LI 1 and ASTM Standard D 1532 list these grades:

- Grade GPO-1: general purpose
- Grade GPO-2: general purpose that requires resistance to flame
- Grade GPO-3: general purpose that requires resistance to flame and tracking.

Punching grades of the above are also available.
Standard sheet sizes: 24–48 inches in width, 36–90 inches in length.
Suppliers have their own proprietary grades, and will also make laminates to meet customer specifications.

Processing

A method, often called the hand lay-up process, does not require prepreg or high-pressure techniques. It is employed to make polyester sheets or shapes with glass mat or glass fabric reinforcement. The following steps are involved in making sheets (shapes are made in molds, but the process is essentially the same):

- Substrates are stacked by hand or by machine to the desired thickness on a platen in a low-pressure press.
- A predetermined amount of catalyzed polyester resin is poured on the plies of substrate.
- The material is "worked" by roller, squeegee, or other means to eliminate entrapped air and to ensure uniform distribution of resin.

TABLE 9-2   Typical Property Values for NEMA Grade Polyester Glass Mat Laminates

PROPERTY	TEST NUMBER		GRADE		
	ASTM	UL	GPO-1	GPO-2	GPO-3
Tensile strength, 10^3 psi	D 638		9.4	8.9	7.8
Compressive strength, 10^3 psi	D 695		38.9	39.0	33.1
Flexural strength, $^1/_{16}$ in, 10^3 psi	D 790		22.3	24.6	22.1
Shear strength, 10^3 psi	D 732		13.4	13.4	11.6
Izod impact, ft·lb/inch of notch	D 256		8.0	10.7	8.9
Coeff. of thermal expansion in/in/°C × 10^{-5}	D 696		2.2	2.0	2.0
Water absorption, % 24 h, $^1/_{16}$ in	D 570		0.3	0.6	0.4
Dielectric strength, volts/mil short time, $^1/_{16}$ in	D 149		420	430	450
Dissipation factor, 10^6 cycles	D 150		0.01	0.01	0.01
Dielectric constant, 10^6 cycles	D 150		4.2	3.9	4.1
Temperature index		746B			
Electrical			—	130	130
Mechanical			—	160	160
Flame resistance		94	HB	V-0	V-0
Oxygen index	D 2863		21.8	36.0	35.0
Thickness as supplied, inches					
Minimum			0.031	0.031	0.031
Maximum			2	2	2
Specific gravity	D 792		1.78	1.97	1.81

Note: Tests are at room temperature except as indicated.

• The remaining steps are similar to those described in "Unclad Laminates Manufacture" (p. 272) in this chapter, except that pressures and temperatures are much lower and tailored to the resin and catalyst used.

Properties

These low-cost laminates possess a combination of desirable properties, including ease of fabrication, high flexural and impact strength, flame, arc, and tracking resistance, and good dielectric values.

Table 9-2 shows key property values for NEMA grade laminates.

Electrical/Electronic Uses

These materials are used in switchgear, transformer, rotating equipment, air conditioner, and television applications, as well as for substrates for printed circuits.

Sheet Molding Compound (SMC)

Sheet molding compounds (SMC) are especially well suited for making large parts by matched-die molding techniques for a wide variety of end uses. There are several merchant suppliers of sheet molding compound including Armco Composites, Hatco Polyester Division of W. R. Grace, Haysite Reinforced Plastics, Plumb Chemical, PPG Industries, Premix, and Rockwell International. High-strength formulations are offered by Armco (USMC) and PPG (HMC and XMC).

Chemistry

Sheet molding compound is principally based on unsaturated polyesters to which are added vinyl-type monomers, such as styrene, vinyl toluene, diallyl phthalate, divinyl benzene, and triallylcyanurate. For more on this subject, see "Unsaturated Polyesters" (p. 248) in Chapter 8, *Insulating Coatings and Impregnants*.

The inherent tendency of polyester compounds to shrink as crosslinking occurs can be counteracted by addition of 10–15 percent by weight of a thermoplastic polymer which swells during the hot molding cycle. Acrylic copolymers, polyvinyl acetate, and polycaprolactones are commonly used "low-profile" additives which provide nearly zero shrinkage. A major product is acrylic-modified polyvinyl acetate in styrene solution, 40 percent solids. Low-profile additives are supplied as either two-part systems, to be mixed just prior to use, or as one-part premixed systems. Other compound ingredients include chopped glass fibers of $^1/_2$–2 inches in length, fillers, lubricants, thickeners, and catalysts.

Grades

Although sheet molding compound grades are commercially available from several major companies, most sheet molding compound molders produce their own sheet molding compound formulated to meet their specific requirements. Formulations cover a wide range of ratios of resin-to-glass-to-filler. Sheet molding compound is made by a continuous process, and is supplied in widths of 24, 48, and 60 inches.

Sheet molding compounds with exceptionally high physical properties are also available from a few suppliers. These products are known by designations such as HMC, XMC, and USMC (unidirectional sheet molding compound), and are characterized by glass content of 60–80 percent, compared with 30 percent for a typical sheet molding compound. Fire-retardant grades are made by incorporating alumina trihydrate into compound recipes.

Processing

A continuous process is used for making sheet molding compounds:

• A thoroughly mixed resin blend is continuously fed onto a polyethylene sheet on a moving conveyor belt.

• Glass fibers chopped to $^1/_2$–2 inches in length are deposited on the resin mix.

• A second application of resin mix is applied to coat the glass fibers.

• This composition is then covered with another polyethylene sheet and passed through kneading and compacting rollers to obtain a uniform dispersion.

• Finally, this "sandwich" is wound in rolls and stored for 24–72 hours or more until the sheet molding compound has reached through chemical thickening the right consistency (to dry sheet form) for molding.

To allow for poor distribution of resin and glass at the edges, the production sheet width is 4 inches wider than finished product width to permit trimming off 2 inches on each edge.

Prior to molding, the polyethylene films are removed, the sheet molding compound is cut to the desired size, and the charge is placed in the mold. To facilitate flow, the charge may be preheated at 100° to 120°F (38°–49°C). Mold temperature and cycle time are unique for each part design, and must be carefully determined by trial before production runs. Sheet molding compound, often processed with high pressure (1,000–1,200 psi) techniques, may also be molded at 300–700 psi using specially formulated resins. This permits use of larger platen areas and significantly lower cost presses.

Properties

Sheet molding compound molded parts are replacing metal in many applications where they have inherent advantages, including lighter weight, corrosion resistance, greater design latitude, and lower tooling cost. Dielectric properties are improved by increased resin content at the expense of reduced physical values. Products may be made flame retardant, but this may also lower other property values. Therefore, compounds should be formulated with consideration given to achieving the optimum balance of mechanical, dielectric, and flame-retardant properties to meet the requirements of the part to be molded.

Electrical/Electronic Uses

Housings and components for business machines and appliances are major consumers of sheet molding compound. The principal dielectric use is in industrial controls and switchgear where high arc resistance and flame-retardant properties are required.

Bulk Molding Compound (BMC)

Similar chemically to sheet molding compounds, bulk molding compounds (BMC) use shorter length fibers of several types and are putty-like mixtures suitable for molding in match-metal dies. Principal merchant producers include American Cyanamid, Armco, Cincinnati Development, Glastic, Haysite Reinforced Plastics, Plumb Chemical, Premix, and Rodgers Engineering.

Chemistry

Polyester resins used in bulk molding compounds are similar to those in sheet molding compounds, including blending styrenated polyesters with low-profile (low-shrink) additives to produce a thickened compound. Alternatively, the resin is modified with vinyl toluene in nonthickened compounds where low-profile additives are optional.

Reinforcing fibers shorter than $^1/_2$ inch include glass, sisal, asbestos, nylon and other synthetic fibers, with glass predominating.

Grades

As is the case with sheet molding compounds, most large bulk molding compound molders make their own compound. However, both thickened and nonthickened grades are commercially available. Typical grades commonly contain 15–35 percent fiber reinforcement. Bulk molding compound is available in bulk, extruded log, and close tolerance slug forms.

Processing

Bulk molding compounds may be made either in a batch process or in a continuous mixer. Batch equipment is essentially a sigma blade (shaped somewhat like the capital Greek letter sigma lying on its side) or spiral blade mixer with a wide clearance between the wall and the blades to avoid breaking of fibers during mixing. In this process, fibers are added last to minimize fiber damage.

For continuous mixing, special equipment such as a Baker-Perkins Ko-Kneader is required. This machine creates a kneading action by a combination of rotary and axial motion of its agitator.

Thickened systems are aged for 24 hours or more to reach the right consistency for molding. Bulk molding compounds may be molded by compression, transfer, and injection molding methods described in Chapter 4, *Thermosetting Molding Compounds*.

Properties

Bulk molding compound characteristics are similar to those of sheet molding compounds, although mechanical properties are somewhat lower because of shorter fiber lengths. The processing rate is one of the highest of any reinforced resin, and the process cost is one of the lowest for high-volume production. High mold cost makes the process uneconomical for short production runs.

Uses

The largest application for bulk molding compounds is in parts for motor vehicles using low-cost sisal- or glass-reinforced materials. Other major applications include power tool housings and parts for appliances and business machines. Parts made by injection molding include circuit breaker parts, standoff insulators, and electronic uses such as solenoids, transformer coils, switch bases, coil forms, and TV tuner parts.

Pultrusion

This rapidly growing technology is suitable for making profiled shapes in a continuous operation. Major pultruders serving the electrical/electronics industries include Creative Pultrusions, Glastic, Haysite Reinforced Plastics, and Stoughton Composites.

Chemistry

Unsaturated polyesters, similar to those described earlier in this chapter, are the principal resins used in pultrusion. Benzoyl peroxide is the preferred catalyst. Other resins used include epoxies, silicones, and vinyl esters. Formulations typically incorporate fillers, pigments, and flame retardants.

Grades

Resin systems are usually formulated to meet specific end use requirements.

Processing

The following steps are typical of the pultrusion process:

• Filaments, most often in the form of glass roving, are dispensed from center-pull packages or tangentially from rolls arranged to feed together at an impregnating tank.

• The filaments are saturated in the resin bath in the tank equipped with side guides, wet-out bars, orifices, and squeeze rolls.

• Profiles are preformed to approximate shape after leaving the resin bath and before entering the die.

• Final shaping and curing occur in a chrome-plated steel die designed to withstand the pressure of expansion during the curing process, which is exothermic. Die temperatures for polyesters range from 240° to 375°F (115°–190°C). Electrically heated dies permit heating zones within the die. Process speed is determined by the die length, resin characteristics, and type of preheating equipment (if used).

• The cured profile is pulled by devices of various designs positioned to permit the material to cool somewhat.

• Finally, a cutoff saw or shear operates automatically to cut the profile to the desired length.

Properties

Advantages (some of which are unique) of pultruded profiles include:

• Shapes are continuously produced
• Exceptional high strength in the longitudinal direction
• Adaptability to small cross-sectional areas
• Corrosion resistance (to replace metal)
• Adequate dielectric and low maintenance properties for high-strength structural parts.

Electrical/Electronic Uses

Typical applications include safety ladders, aerial booms for trucks, third rail covers for electric railroads, poleline hardware, and component parts for control and switchgear devices.

CLAD STRUCTURES

INTRODUCTION

Clad structures are used to make printed circuits, a technology that has evolved rapidly since the initial impetus provided by specialized military applications during World War II. For a time, it appeared that the manufacture of printed circuits was ideally suited to small businesses headed by technically competent, innovative engineers. However, certain inherent problems, which still exist to a lesser degree, blocked the road to profitability for these companies:

• Specifications and performance requirements for finished circuits were often difficult to meet with available basic materials and production processes. This resulted in a high percentage of rejects and low productivity.

• The manufacturing process involving photochemical, electrochemical, and chemical operations required continuous monitoring and adjustment throughout production runs.

• It is difficult and expensive to comply fully with increasingly stringent Environmental Protection Agency (EPA), Occupational Safety and Health Administration (OSHA), and state and local regulations and ordinances.

Most large users of printed circuits now etch their own circuits. Some companies also make resins, prepregs, and clad laminates, thereby controlling the entire manufacturing process. The circuits made by these large producers are often incorporated in devices of their own manufacture, such as computers, television sets, radios, telephones, calculators, and many motor vehicle applications.

SUPPLIERS

Rigid clad circuit materials are sold commercially by:

Cincinnati Milacron	NVF
General Electric	Polyclad Laminates
New England Laminates	Rogers
Norplex Oak	Spaulding Composites

Flexible clad circuit materials are sold commercially by:

Circuit Materials	Norplex Oak
Chase-Foster (Keane)	Rexham
Du Pont	Rogers
3M	Sheldahl
New England Laminates	TME

TECHNOLOGY

A printed circuit is a network of conducting metal (mostly copper) paths on, or imbedded in, a dielectric material. The circuit interconnects electronic components with significant savings in space, weight, and cost compared with circuits in which the components are interconnected by wires. Circuits may be single side, double side, or multilayer formed by laminating single side circuits. There are two methods in general use for producing printed circuits:

- The etched foil, or subtractive process, the most widely employed
- The additive process

Other, less commonly used methods include applying a die-stamped foil circuit to a dielectric, and forming a circuit with conductive ink.

The *etched foil* method employs a photochemical process to print the circuit pattern on a metal clad substrate. This pattern becomes chemically protected in the developing process, and all unprotected metal is etched away after the undeveloped chemicals are stripped with solvents. Etchants include ammonium persulfate, chromic and sulfuric acids, cupric chloride, and ferric chloride. Where circuits on both sides of a substrate or on layers in a multilayer laminate are to be connected, holes are drilled at appropriate locations. Circuits may then be connected mechanically using eyelets or tubelets, or by plated-through holes.

In the *additive* method, the circuit path is applied to the substrate by metal plating. This is accomplished by either the *fully electroless* process or the *semi-electroless* process. The term *electroless* means the metal is deposited while the part is in a bath by chemical reduction without application of an electric current. To facilitate the process, a noble metal plating catalyst is dispersed in the resin system. The cured substrate is coated with about 1 mil of adhesive consisting of an epoxy/phenolic resin blended with an elastomer, such as butadiene. To promote bonding, the adhesive surface is treated with a solution of chromic and sulfuric acids. Adhesives may be applied by curtain, roller, or dip coating. Curing is accomplished in a circulating air oven or by infrared radiation. Circuits are formed by screen printing, masking, and photoprinting techniques.

Additive circuit conductors are marginally less ductile than circuits formed by etching. Conductor paths are formed only where required, eliminating all etching and cleaning steps needed for the subtractive method. However, electroless copper is higher in cost than copper used in the subtractive method, which offsets savings in labor, waste copper, and process materials. The additive method should, in general, be considered when less than 35 percent of the surface area of the substrate is to be covered with copper, or when it is desired to eliminate the problem of undercutting of the sidewall inherent in the etched process—an advantage in extremely fine line circuits.

The semi-electroless process requires only a thin electroless deposition on the substrate surface, since desired conductor thickness is achieved by subsequent electroplating. Adhesion promotion and catalyzation or precatalyzation of the substrate is required for adequate bonding and metalization. Further processing parallels that of the subtractive method, although copper loss is about one-tenth as much, since the metal film is only one-tenth of a mil thick. Both additive processes require that the surface of the substrate be free of defects which would adversely affect adhesive coating and bonding operations.

A relatively new additive circuit technology involves the use of *polymer thick film inks* on organic resin substrates. This ink consists of a polymer, usually epoxy base, capable of crosslinking on heating, a conductive material, such as silver, dispersed in the polymer, solvents, and other ingredients necessary to adjust the physical characteristics to the consistency for proper application. The resulting circuit paths are of higher resistance than in the subtractive method, and their reliability and precision are inferior. However, inks for use on ceramic substrates requiring firing at relatively high temperatures do produce highly reliable conductor lines. These inks consist of mixtures of metal, oxide, and glass powders less than 10 microns in diameter, suspended in an organic vehicle. Application is by screen printing (the preferred technique), spraying, or brushing. After application, the ink is dried at 100°C to remove the solvent, and then fired at 300°–1,000°C to decompose the organic binder and bond the glass to the substrate in a film 0.0005–0.002 inch thick. Resistor thick film compositions may be applied in a similar way.

When production runs are long enough to absorb the cost of dies, *die stamping and embossing* techniques are often the most advantageous methods. A die is made to cut out from the conductive metal material a pattern which is then bonded to the substrate by a heat-activated adhesive. This method eliminates the chemical processes and number of production steps required by other methods, and ensures uniformity and good registration and line definition without undercutting. It also permits use of thicker conductors with greater current-carrying capability than possible with other printed circuit methods. However, this method is not as suitable for fine line or closely spaced circuit paths.

SUBSTRATES FOR PRINTED CIRCUITS

Widely used substrates for printed circuits include the following:

- High-pressure thermoset laminates described earlier in this chapter, with FR-4 grade accounting for about 80 percent of total.
- Alumina ceramics containing 75–99 percent Al_2O_3 which feature:
  - Permanent dimensional stability
  - Extreme hardness
  - Chemical inertness
  - Good impact strength
  - High operating temperature
  - Moderately high thermal conductivity.
- Sapphire (Al_2O_3), a crystalline variety of alumina. Synthetic sapphire is made by a crystal-growing process, including ribbon shapes suitable for substrates after polishing. This material has exceptionally high thermal stability and dielectric constant of 11.5 parallel to the C-axis. It is widely used for silicon-on-sapphire (SOS) hybrid IC and other microwave systems.

- Beryllia (BeO) porcelains, possessing exceptionally high thermal conductivity (30–100 Btu versus 16 Btu for alumina) and a dielectric constant of about 6.4. They are used where these properties are desired.

- Quartz and vitreous silica (SiO_2) substrates having dielectric constant of 3.2, lowest of any inorganic material.

- Polytetrafluoroethylene resin glass microfiber or alumina microfiber reinforced structures, noted for a uniform dielectric constant of 2.20–2.33 over a wide frequency range. These substrates are especially well suited for microwave circuit applications.

- Flexible substrates:
  - Polyester film
  - Polyester film on woven glass fabric
  - Polyimide film
  - Polyimide film on woven glass fabric
  - Polyamide-imide film
  - Epoxy coated polyester mat
  - Epoxy coated polyamide mat
  - Epoxy coated woven glass fabric
  - Fluorinated ethylene-propylene (FEP) film
  - Polytetrafluorethylene (PTFE) coated woven glass fabric
  - Polyamide film
  - Polyurethane film
  - Polyvinyl fluoride film
  - Polyethylene film
  - Polyvinyl chloride film
  - Polycarbonate film

- The advent of multichip modules (McM-D) has required novel dielectric insulations consisting of spin-deposited polymers, either polyimide (PI) or benzocyclobutene (BCB). These polymers form dielectric bases with low moisture absorption, low dielectric constant, high thermal stability, and excellent planarization.

For more information on ceramic substrates, see Chapter 15, *Ceramic and Glass Insulations*, p. 417.

SEMICONDUCTOR SILICON

The tremendous strides in dramatically reducing the size, while significantly improving the capabilities, capacities, and speed of electronic equipment such as computers, calculators, and communications equipment is attributable to the development of semiconductor technology based on the silicon chip, or integrated circuit (IC). Other semiconductor materials include gallium arsenide, germanium, silicon carbide, and selenium.

Ultrapure (99.97 percent pure) silicon is inherently a poor conductor of electricity. It is made semiconducting by diffusing controlled amounts of selective impurities, or dopants, such as phosphorous (*n*-type) or boron (*p*-type) into its surface. In the *n*-

type semiconductor, the semiconductor material has fewer external ring electrons than the dopant. The extra electrons do not fit the crystal structure and are available to carry current. The p-type semiconductors contain an impurity that has fewer electrons in the outer ring than the semiconductor has, thereby providing locations for electrons to visit in producing current.

The semiconductor grade is made by reduction of purified silicon tetrachloride or trichlorosilane with hydrogen. Silicon is deposited on hot (1,470°F, 800°C) filaments of tantalum or tungsten. In a newer, exothermic process, sodium fluorosilicate is reacted with sodium to produce silicon tetrafluoride. Further reaction with sodium yields high-purity silicon.

To make single-crystal silicon suitable for slicing into wafers, single-crystal rods are grown from the purified silicon by either the Czochralski or the floating-zone method. In the former process, a seed crystal is touched to the surface of a molten crucible charge of silicon. As the molten silicon solidifies on the seed crystal, the growing crystal is slowly pulled from the melt. This method is used to produce most commercial single-crystal silicon rods up to 100 millimeters in diameter.

The floating-zone method starts with a rod of polycrystalline silicon. A small zone at one end is melted using induction heating. Then, by moving the induction coil, the molten zone is made to move along the length of the rod. Under proper conditions, the material behind the molten zone will solidify as a single crystal. A major advantage of this process is that the molten material does not contact any other material from which it can pick up impurities. Both methods of growing single crystals also further refine the silicon because most impurities remain in the molten silicon rather than solidify in the crystal.

Circuit paths as narrow as $1/25,000$ inch are created by complex photochemical, etching, diffusion, and vacuum deposition processes. Using these techniques, digital watches are made with about 5,000 transistors on a single chip no larger than a child's fingernail; about 20,000 transistors on a chip are required for a pocket calculator; and 100,000 transistors are located on a chip for a small computer. Chips using $1/2$ micron geometries which will hold up to 4,000,000 transistors are under development.

COPPER FOILS

Copper foils used in printed circuits may be produced either in rolling mills or by electrodeposition. The rolling process foil has twice the tensile strength of the electrodeposited copper and tends to have fewer pinholes and better flex resistance, making it more suitable for flexible cables subject to frequent or severe bending. However, electrodeposited copper predominates for etched circuits. In this process, copper from a solution of copper sulfate or copper cyanide is plated on a revolving steel drum, from which it is continuously stripped, cleaned, and dried. The side next to the drum is smooth and shiny, while the other side has a grainy matte finish for bonding to the substrate. Adhesion is further improved by oxide treatment, known as Treatment "A," applied to the rough side of the copper, resulting in a tenfold increase in bond strength. More recently introduced processes include "XP," a light oxide treatment developed to reduce oxide staining in multilayer laminates, and "GT," a nickel flash treatment designed for use primarily with polyimides to increase bond strength at elevated temperatures. About 60 percent of the copper foil used for printed circuits is 0.0014 inch thick, or 1 ounce per square foot; 30 percent is 2 ounce copper; the remaining 10 percent is $1/2$ ounce or 3 ounce copper. Other metal foils are used for special purposes. These foils include nickel alloys such as Kovar, aluminum, and beryllium copper.

ADHESIVES

Foils are usually bonded directly to epoxy/glass prepregs, with the epoxy resin acting as the bonding agent. A polyvinyl butyral/phenolic adhesive is used to bond foils to phenolic substrates.

Adhesives for flexible clad circuits are predominantly polyesters of a saturated type containing hydroxyl groups and isocyanates for crosslinking upon application of heat and pressure. These adhesives may be applied in solution form or as cast film.

Copper foil may be bonded to fluorocarbon and polyimide type flexible substrates with a silicone pressure-sensitive adhesive. A high-molecular-weight polyacrylic adhesive is also used for polyimide flexible circuits.

COVERCOATS

The function of a covercoat for a printed circuit is to provide complete electrical insulation over the conductor paths, to seal out moisture, and to add mechanical strength to the laminate. The covercoat may be melt-bonded, adhesive-bonded, or applied by solution coating.

• In *melt-bonding* or *fusion bonding*, the overcoat film is adhered to the base in a platen press or roll laminator. Sufficient heat and pressure are applied to melt the film momentarily, after which the construction is cooled under pressure. To ensure good adhesion, the conductor foil surface requires special treatment.

• *Adhesive-bonding* permits use of polyester and polyimide covercoat films, which are not suitable for melt-bonding. Since the high heat required for melt-bonding is not needed, adhesive-bonding provides better dimensional control for the construction. It is standard practice to use the same covercoat film as used in the base laminate.

• *Solution-coating*, the lowest cost method, eliminates the need for lamination of a preformed, prepunched film by heat and pressure. Repairability of circuits is also facilitated where solution coating is used: the covercoat may be selectively stripped by solvents, or a low-melting-temperature covercoat may be used which will melt or vaporize under heat of a soldering iron. The disadvantages of this method are that coatings are more likely to contain pinholes, and solvents must comply with worker safety and environmental guidelines. A wide variety of coatings is available:
  − Silicones (thermal properties)
  − Epoxies (moisture and abrasion resistance)
  − Polyurethanes (low temperature flexibility, solderable)

- Butyl rubbers (low temperature flexibility, moisture resistance)
- Polyesters (solderable)
- Acrylics (strippable)

MULTILAYER CIRCUITS

Printed circuit materials, usually of epoxy or polyimide types, may be layered to form multilayer circuits. These high-density constructions, requiring precisely controlled production techniques, offer significant savings in size and weight.

The manufacturing process consists of these basic steps:

1. Circuits are etched on thin laminates (less than 0.032 inch thick).

2. The laminates are then laid up to the desired construction with two or three bonding layers of prepreg between circuit layers.

3. These multilayer books are cut to size and positioned in a molding frame by means of locating pins.

4. The frame with lay-up is placed in a press, making sure platens are parallel to within 0.001 inch (0.0025 centimeters).

5. The press is closed and heated to the desired temperature, usually in the range of 300° to 450°F (149°–232°C) at a pressure of about 500 pounds per square inch (35.2 kilograms per square centimeter).

6. After approximately 1 hour, the press contents are cooled under pressure to about 120°F (49°C).

7. The frame is unloaded from the press and the circuits removed.

Each step must be carefully monitored to ensure circuits of acceptable quality. Causes of rejects include voids, blisters, displaced layers, delamination, and board distortions.

Circuits may have up to 40 layers with a thickness of only 0.40 inch (1 centimeter) and contain plated-through holes 0.020 inch (0.05 centimeter) in diameter on 0.050 inch (0.13 centimeter) centers. Holes must be located and drilled with great accuracy. Interconnections from layer to layer are made by electroplating copper through holes. For multilayer circuits to be acceptable, it is essential that circuit paths be exactly aligned.

FLEXIBLE CIRCUITS

Flexible circuits provide the capability to work with complex configurations in three dimensions, and are ideal materials for applications where they are subject to continual flexing. A wide variety of substrates is suitable for making flexible circuits, including those listed in "Substrates" in this chapter and discussed in Chapter 10, *Dielectric Films*.

Adhesives required to bond the layers together are discussed in "Adhesives" in this chapter. The most commonly used method for applying adhesives is from solution. This process consists of the following steps:

1. The adhesive coating is applied to the substrate or foil by reverse or direct roll, gravure, or knife-over-roll technique.

2. Solvent is evaporated in a heated tunnel.

3. The web is passed around a hot roll to activate the adhesive.

4. The next layer is laminated to the composite by a rubber-covered nip roll.

5. The process is repeated until all the desired layers are laminated.

Epoxy prepregs contain no solvent, and may be bonded directly to the foil without the evaporation step. Pressure-sensitive silicone and acrylic adhesives require only pressure to bond at room temperature. In another process, adhesive melt is applied by extrusion, after which layers are laminated by a nip roll.

When a thin film is laminated to a plain weave fabric by a high-bond-strength, elastic adhesive, the resulting material exhibits a significant increase in bias stiffness compared with the combined bias stiffness of component layers. This is caused by the bonding of the fibers to each other and also to the film.

FEDERAL SPECIFICATIONS FOR PRINTED CIRCUITS

There are several Federal specifications for printed circuits, among which the following are significant:

• MIL-P-13949, *Plastic Sheet, Laminated, Metal Clad*. This specification is approved for use by all departments and agencies of the Department of Defense, and covers the requirements for fully cured, metal-clad laminated, plastic sheets (glass and paper base), and semi-cured (B-stage), resin impregnated, glass cloth (prepreg) to be used primarily for the fabrication of printed-wiring boards for electrical and electronic circuits. Substrates are identified as follows:

PX—Paper base, epoxy resin, flame resistant

GE—Glass (woven-fabric) base, epoxy resin, general purpose

GF—Glass (woven-fabric) base, epoxy resin, flame resistant

GB—Glass (woven-fabric) base, epoxy resin, hot strength retention

GH—Glass (woven-fabric) base, epoxy resin, hot strength retention and flame resistant

GP—Glass (nonwoven-fiber) base, polytetrafluoroethylene resin, flame resistant

GR—Glass (nonwoven-fiber) base, polytetrafluoroethylene resin, flame resistant, for microwave application

GT—Glass (woven-fabric) base, polytetrafluoroethylene resin, flame resistant

GX—Glass (woven-fabric) base, polytetrafluoroethylene resin, flame resistant, for microwave application

GI—Glass-fabric base, polyimide resin, high temperature

GY—Glass (woven-fabric) base, polytetrafluoroethylene resin, flame resistant for microwave application.

Among test requirements are tests for peel strength, peel strength after thermal stress, after temperature cycling, at elevated temperatures, and after exposure to process conditions.

- MIL-P-55110, *Printed-Wiring Boards*. This specification is approved for use by all departments and agencies of the Department of Defense, and covers the certification and performance requirements for rigid single-sided, rigid double-sided, and rigid multilayer printed-wiring boards with plated-through holes. Possible major defects are classified as follows:
  – Bonding of conductor and peeling of conductor: Any looseness of bond on any conductor length, any peeling of conductor (defect most prevalent at terminals and at ends of conductor contacts).
  – Broken eyelet: Part of eyelet missing. Circumferential splits.
  – Seating of eyelet: Eyelet not properly seated perpendicular to the board.
  – Plated-through hole: (1) In excess of 10 percent void of surface in hole rejectable; (2) voids at interface of hole rejectable; (3) circumferential separation rejectable.
  – Superfluous conductor: Potential cause of short. Clearance less than that specified in master drawing for electrical spacing.
  – Eyelets: Total voids in solder around eyelet exceeding 30 percent of the periphery in funnel flanged eyelets. Cracks in solder around eyelet.
  – Reduction in width of conductor at any point exceeding 20 percent of the minimum specified conductor width on the master drawing.
  – Conductor and lead spacing violating the requirements on the master drawing.
  – Plating on top surface of conductors and in plated-through holes: Lack of plating or less than minimum specified. Plating adhesion defect.
  – Cracks, chips, or bulges on board surfaces.
  – Unspecified removal of board material: Any visible unspecified removal of board material (such as removal about conductors to increase insulation resistance or indication of contamination). A minor change in surface appearance due to removal of unspecified material.
  – Warp or twist of board: Warp or twist in any board shall not exceed 1.5 percent.
  – Spacing of holes: Other than that specified on the master drawing.
  – Annular ring (at narrow point):
    a. The projecting flange or eyelet or standoff terminal extending beyond the annular ring.
    b. Measuring less than 0.015 inch (0.38 mm) beyond the edge of unsupported hole.
    c. Measuring less than 0.002 inch (0.05 mm) for internal terminal areas.
    d. Measuring less than 0.05 inch (0.13 mm) beyond the edge of plated-through hole.

- Terminal holes: Ragged holes with chipping or cracking in the wall of the holes, and bulging around the holes or reduction of the hole diameters with base laminate material such as fibers. Ragged metal foil edge. Metal foil deformed into a hole, torn or lifted. Size of holes not as specified in the master drawing.
- Delamination: Internal or external separation of base material (paper or glass).
- Spurs or whiskers: Presence of spurs or whiskers.
- Nodules: Cause reduction in hole diameter to less than specified on the master drawing.

Minor defects include:

- Cuts, cracks, or scratches in conductor: Board visible through copper (copper may be visible through overplating when overplating is specified).
- Cuts, cracks, or scratches completely across conductor or more than $\frac{1}{2}$ inch along conductor.
- Size of holes: Size of holes (other than terminal holes) not as specified in the master drawing.
- Loose standoff terminal: Any standoff terminal that can be turned or removed by hand before soldering.

IPC SPECIFICATIONS FOR PRINTED CIRCUITS

The Institute for Interconnecting and Packaging Electronic Circuits (IPC) has the following applicable specifications:

- IPC-L-108A, *Thin Laminates, Metal Clad Primarily for High-Temperature Multilayer Printed Boards*
- IPC-L-109A, *Glass Cloth, Resin Preimpregnated (B-Stage) for High-Temperature Multilayer Printed Boards*
- IPC-L-110A, *Preimpregnated, B-Stage Epoxy-Glass Cloth for Multilayer Printed Circuit Boards*
- IPC-L-112, *Foil Clad, Polymeric, Composite Laminate*
- IPC-L-115A, *Plastic Sheet, Laminated, Metal Clad for High-Temperature Performance Printed Boards*
- IPC-L-130, *Thin Laminates, Metal Clad, Primarily for General Purpose Multilayer Printed Boards*
- IPC-FC-220B, *Flat Cable, Flat Conductor, Flexible, Unshielded*
- IPC-FC-222, *Flat Cable, Round Conductor, Unshielded*
- IPC-FC-231, *Flexible Bare Dielectrics for Use in Flexible Printed Wiring*
- IPC-FC-232, *Adhesive Coated Dielectric Films for Use as Cover Sheets for Flexible Printed Wiring*
- IPC-FC-240C, *Flexible Printed Wiring*

- IPC-FC-241, *Metal-Clad Flexible Dielectrics for Use in Fabrication of Flexible Printed Wiring*
- IPC-FC-250, *Double Sided Flexible Wiring with Interconnections*
- IPC-D-300F, *Dimensions and Tolerances for Single and Two Sided Printed Wiring Boards*
- IPC-AM-361, *Rigid Substrates for Additive Process Printed Boards*
- IPC-ML-905B, *Performance Specifications for Multilayer Printed Wiring Boards*
- IPC-ML-990, *Performance Specification for Flexible Multilayer Wiring*

ASTM STANDARDS FOR COPPER-CLAD THERMOSET LAMINATES

The American Society for Testing and Materials (ASTM) has established these standards for copper-clad thermoset laminates:

- D 1867, *Copper-Clad Thermosetting Laminates for Printed Wiring*. This specification covers ten grades of copper-clad thermosetting laminates, with copper foil bonded to one or both surfaces. These grades (similar to NEMA grades described earlier in this chapter) and their key requirements are listed in Table 9-3. Table 9-4 lists thickness tolerances for copper-clad sheets.
- D 5109, *Test Methods for Copper-Clad Thermosetting Laminates for Printed Wiring Boards*

UL STANDARD FOR PRINTED WIRING BOARDS

Underwriters Laboratories has developed the following standard:

- UL 796, *Printed Wiring Boards*

DEVELOPMENT PROGRAMS

Solventless adhesives for bonding flexible circuits are being evaluated. These would greatly reduce or eliminate measures otherwise necessary to comply with Environmental Protection Agency (EPA) and Occupational Safety and Health Administration (OSHA) standards and regulations governing solvent emissions.

High-performance adhesives and improved processing systems and substrates for the additive process are under continuous development.

Multilayer circuit technology is being refined to improve production precision and efficiency.

TABLE 9-3 **Requirements for Copper-Clad Thermoset Laminates for Printed Circuits**
(Conditioning by ASTM Method D 618)
(ASTM D 1867)

PROPERTY	GRADE									
	XXXP	XXXPC	FR-2	FR-3	FR-4	FR-5	FR-6	CEM-1	G-10	G-11
Peel strength, min lb/in width 1-oz copper	6	6	6	8	8	8	7	7	8	8
Volume resistivity, min, $\Omega \cdot cms$ $1/16-1/8$ in	10^7	10^7	10^7	10^8	10^9	10^9	10^9	10^9	10^9	10^9
Surface resistance min, ohms	10^6	10^6	10^6	10^7	10^7	10^7	10^7	10^7	10^7	10^7
Dielectric breakdown parallel to laminations, min kV, step-by-step	15	15	15	30	40	40	30	40	40	40
Dielectric constant max @ 1,000 Hz	5.3	5.3	5.3	4.8	5.4	5.4	4.3	—	5.4	5.4
Dissipation factor max @ 1,000 Hz	0.05	0.05	0.05	0.040	0.035	0.035	0.03	0.040	0.035	0.035
Flexural strength min, psi										
Lengthwise	12,000	12,000	12,000	20,000	60,000	60,000	15,000	50,000	60,000	60,000
Crosswise	10,500	10,500	10,500	16,000	50,000	50,000	15,000	40,000	50,000	50,000
Water absorption max, %, $1/16$ in	1.00	0.75	0.75	0.65	0.25	0.25	0.40	0.30	0.25	0.25

TABLE 9-4 Thickness Tolerances for Copper-Clad Sheets
(Plus or Minus, Inches)
(From ASTM D 1867)

OVERALL THICKNESS INCLUDING COPPER, in[b]	CLASS I			CLASS II	
	GRADES XXXP, XXXPC, FR-2, FR-3		GRADES FR-4 FR-5, CEM-1[c], FR-6, G-10, G-11 ALL WEIGHTS OF COPPER— 1 AND 2 SIDES	GRADES XXXP, XXXPC, FR-2, AND FR-3 ALL WEIGHTS OF COPPER— 1 AND 2 SIDES	GRADES FR-4, FR-5 G-10, G-11 ALL WEIGHTS OF COPPER— 1 AND 2 SIDES
	1 oz, 1 SIDE	1 oz, 2 SIDES 2, 3, 4, AND 5 oz COPPER			
1/32 (0.031)	0.004	0.0045	0.0065	0.003	0.004
3/64 (0.046)	0.005	0.0055	0.0075	0.0035	0.005
1/16 (0.062)	0.0055	0.006	0.0075	0.004	0.005
3/32 (0.093)	0.007	0.0075	0.009	0.005	0.007
1/8 (0.0125)	0.0085	0.009	0.012	0.006	0.009
5/32 (0.156)	0.0095	0.010	0.015	0.007	0.010
3/16 (0.187)	0.010	0.011	0.019	0.008	0.011
7/32 (0.218)	0.011[a]	0.012[a]	0.021	0.009	0.012
1/4 (0.250)	0.012[a]	0.012[a]	0.022	0.009	0.012

[a]These values do not apply to grades XXXPC and FR-2.
[b]Conversion factor: 0.001 in = 0.0254 mm

  1 oz = 28 g.

[c]No standard has been developed for thicknesses greater than 3/32 in.

Note 1—At least 90% of the area of the sheet shall be within the tolerances given, and at no point shall the thickness vary from the nominal by a value greater than 125% of the specified tolerance.

Note 2—For a sheet having a nominal thickness not listed in the table, the tolerance shall be the same as for the next larger nominal thickness.

Note 3—Cut panels, less than 18 by 18 in, meet the applicable thickness tolerances in 100% of the area of the panel.

(Reprinted with permission from the *Annual Book of ASTM Standards*. Copyright, ASTM, 1916 Race Street, Philadelphia, PA 19103.)

The goal of integrated circuit engineers, said to be well within the realm of possibility, is to squeeze 10,000,000 transistors on a single chip.

MARKET TRENDS

The trend to miniaturize electronic circuits has made dramatic progress through the 70s, 80s, and 90s. This trend and progress is certain to continue with increasing use of integrated microcircuits on chips of silicon or gallium arsenide (much higher price, but six times faster).

Thin laminates for multilayer circuits are growing at almost double the rate for standard clad laminates.

There is increasing demand for flatter boards to avoid "popping" components once they are inserted.

Dielectric Films

INTRODUCTION

The term *films*, as used in this handbook, also includes thin sheets. Generally speaking, films have thicknesses of 0.003 inch or less and sheets are over 0.003 inch thick. However, Federal Specification MIL-I-631 sets 0.002 inch as the demarcation thickness and ASTM Standard D 2305 covers "polymer films not over 15 mils (0.0015 inch) thick." Thicker sheets and substrates used for flexible printed circuits and flexible cable are covered in Chapter 9, *Unclad and Clad Structures*. Tapes slit from films are covered in Chapter 12, *Tapes and Coated Fabrics*.

Although by far the largest volume use for films is in packaging, they have important applications as phase and layer insulation in transformers and rotating equipment and as electronic capacitor dielectrics.

TECHNOLOGY

Film manufacture is accomplished by several processes, but consistently high-quality products are as much the result of process design and control as of equipment type or capacity. Production techniques may be quite simple, as in the case of extruded film, or highly complex, as in the case of blown film. The major film forming processes are these:

• *Extrusion* is one of the most versatile and widely used methods. A single-screw extruder for films consists essentially of a hopper (for resin storage), a barrel provided with means to both add and extract heat, a metering screw (which picks up resin under the hopper, heats and carries it through the barrel, and develops sufficient pressure to force it through a die), and a die (which spreads the melt into film). There are many styles of screw design. Screw selection depends primarily on the resin to be processed. The ratio of the barrel length to the bore diameter, commonly referred to as the L/D ratio, is usually in the range of 20:1–90:1. Shear energy imparted by the action of the screw on some resins often requires heat to be extracted from the melt by cooling the barrel. With other resins, it is necessary to add heat to obtain the desired consistency in the melt. Generally, film in widths up to 120 inches is extruded onto three nonnipping cooling rolls of 12–36 inches in diameter. Alternatively, a water bath is sometimes used for cooling the film. Multiscrew extruders are available for specialized applications, such as the processing of heat-sensitive materials or materials that must leave the die at relatively low temperatures. Extruded film may be oriented by stretching as it passes between heated nip rolls, where a second set of rolls rotate slightly faster than the first. Transverse orientation is accomplished by tenter frames, similar to those used in stretching textiles, with the film first heated to increase its ductility. The tendency of the stretched film to relax is reduced by passing it through a heated oven. After cooling, the film is wound into master rolls.

• *Blown film* is made by vertical extrusion of a plastic tube through an annular die provided with a means of feeding compressed air into the center of the tube. At the start of the process, the open end of the tube is drawn together and sealed. The air line is opened and a bubble is formed, usually $1^{1}/_{2}$–3 times the tube diameter. The bubble is cooled by cold air, and then collapsed by wood slats or rollers, changing shape to a two-layer, flat film, which now becomes single-ply. Machine tension orients the film in the direction of bubble travel. The blowing action imparts transverse orientation. Thus, the film is biaxially oriented. The many variables in this process make precise control necessary to avoid variations in film thickness, low tensile strength, low impact strength, hazy film, blocking, and wrinkling. In comparing blown film with unoriented extruded flat film, the former has no losses in edge trimming (flat film edges are thickened because of neck-in), mechanical properties are better because of biaxial orientation, and equipment and die costs are lower. The extruded flat film process, however, yields superior optical clarity, has generally higher output, maintains closer thickness tolerances, and handles a wider range of melt viscosities.

• *Coextrusion* involves the simultaneous extrusion through a common die head of the same or different materials. Many variations of this technique are possible, including using a center layer as an adhesive, and foamed/film constructions (alternatively made in a separate laminating operation). Die design is crucial in this process where two or more extruders are coupled to a single die head. Most coextrusions bring the separate melts together prior to being extruded, although this can also be done outside the die. Precise control is necessary over screw speed, and melt pressures and temperatures. This process, although complex, makes possible the manufacture of a single film with superior properties to those of its component films taken separately.

• *Solvent casting* consists of forming a solution of the film resin particles and additives in a suitable solvent and thoroughly filtering this solution to remove all contaminants. The purified solution is then cast from a slot die onto a moving, endless, highly polished stainless steel belt. The solvent is removed by controlled heating as the belt passes through an oven. Finally, the finished film is stripped from the belt at the take-off station. Gauge control is achieved by adjusting the die opening across its width. The film produced by this method is free of strain and orientation with equal properties in all directions. If desired, the film may subsequently be oriented. This process is always used for cellulose triacetate, and is also used where wanted for fluoropolymers, polysulfones, polycarbonates, and polyimides (before imidization). Solvent casting is slower and much more expensive than extrusion or blown film methods, and measures must be taken to ensure solvent handling complies with worker safety and health and environmental guidelines and legislation.

ASTM STANDARD FOR TESTING FILMS

The American Society for Testing and Materials has developed the following standard:
• D 2305, *Standard Methods of Testing Polymeric Films Used for Electrical Insulation.*

This standard covers the testing of homogeneous organic polymer films not over 15 mils (0.4 mm) thick that are to be used for electrical insulation. Procedures include:

- Conditioning
- Dielectric breakdown voltage and dielectric strength
- Insulation resistance
- Permittivity (dielectric constant) and dissipation factor
- Resistance method for measuring the tendency to corrode metals
- Sampling
- Strain relief
- Tensile properties
- Thickness
- Volume resistivity

Applicable ASTM documents are cited for conducting these tests.

FEDERAL SPECIFICATION FOR FILMS

The governing Federal Specification for films is:

• MIL-I-631, *Insulation, Electrical, Synthetic-Resin Composition, Nonrigid.* This specification has been approved by the Department of Defense for use of the Departments of the Army, Navy, and the Air Force. It covers only the basic characteristics required by the synthetic insulating material, as such, and the methods suited to their satisfactory determination. *Film* is defined as material in thicknesses of 0.002 inch or less, in widths greater than 3 inches. Material 3 inches or less in width, slit from film, is defined as *film tape. Sheet* is defined as material in thicknesses greater than 0.002 inch and in widths greater than 3 inches. Material 3 inches or less in width slit from sheet is defined as *sheet tape.*

The following film types are identified:

Type A—Polyethylene

Type B—Cellulose acetate butyrate

Type C—Cellulose acetate or cellulose triacetate

Type D—Ethyl cellulose

Type E—Vinylidene chloride copolymer

Type F—Polyvinyl chloride and its copolymers

Type G—Polyethylene terephthalate

Requirements are listed for these properties for each type:

- Dielectric strength
- Dielectric constant
- Dissipation factor
- Volume resistivity
- Tensile strength
- Elongation
- Softening temperature
- Lengthwise shrinkage

DEVELOPMENT PROGRAMS

There is available technology for producing a virtually unlimited number of films and film composites to meet requirements for state-of-the-art electrical insulation applications. Development efforts focus on production techniques to make films free of contaminants and surface imperfections.

MARKET TRENDS

Use of dielectric films and composites will parallel trends in the electrical industry.

DIELECTRIC FILMS

These films are covered in the sections which follow, with typical property values shown in Table 10-1.

- Cellulose acetate (CA) and cellulose triacetate (CTA)
- Fluoropolymers (PTFE, PFA, FEP, ETFE, PVDF, ECTFE, PCTFE)
- Ionomer
- Polycarbonate (PC)
- Polyester (PET)
- Polyethylene (PE)
- Polyimide (PI)
- Polyparabanic acid (PPA)
- Polypropylene (PP)
- Polystyrene (PS)
- Polyvinyl chloride (PVC)
- Polysulfone
- Polyethersulfone

Cellulose Acetate (CA)
(Film)

Cellulose films were among the first films to be used as dielectrics. Cellulose acetate (CA) film is produced by ICI, 3M, and XCEL. Cellulose triacetate (CTA) film is made by Eastman Chemical and XCEL.

Chemistry

Cellulose acetate, also called *secondary acetate* or *diacetate*, is made by acetylating wood pulp or cotton linters with acetic acid or acetic anhydride, using a sulfuric acid catalyst. This ester is then partially hydrolyzed to form a thermoplastic product soluble in acetone and capable of being plasticized by lower alkyl phthalates and alkoxy-alkyl phthalates.

Cellulose triacetate is fully acetylated, purified cellulose, made in a similar reaction but omitting the hydrolysis step. Solvents include methylene chloride, chloroform, and tetrachloroethane. The film is formed by the solvent casting method.

Federal Specification for Cellulose Acetate Film

The following Federal specification has been developed for cellulose acetate film:

- L-P-504, *Plastic Sheet and Film, Cellulose Acetate*

Grades

Cellulose acetate and cellulose triacetate are available as glossy, clear films, or with a matte finish on one side to prevent slippage and increase visibility in coil winding applications.

Properties

Cellulose acetate and cellulose triacetate are designated as Class 105 insulations. Dielectric properties are good except for low moisture resistance. Key property values are shown in Table 10–1.

Electrical/Electronic Uses

Cellulose acetate is used to wrap coils, as layer and interphase insulation, as wire covering, and as capacitor insulation. A major use for cellulose triacetate is as a base for magnetic tape.

Fluoropolymers
(Film)

This family of high-price films is useful for applications requiring performance beyond the capabilities of other films. In general, fluoropolymer films are characterized by exceptionally wide useful temperature range, excellent dielectric properties, and outstanding resistance to attack by chemicals.

Chemistry

The basic chemistry and structures of the fluoropolymer resins used for films are covered in "Fluoropolymers" (p. 146) Chapter 5, *Extrusion Compounds*.

ASTM Standards for Fluoropolymer Films

The American Society for Testing and Materials has developed the following specifications for fluoropolymer films:

- D 3368, *FEP-Fluorocarbon Resin Sheet and Film*
- D 3369, *Polytetrafluoroethylene (PTFE) Resin Cast Film*
- D 3595, *Polytrifluoroethylene (PCTFE) Extruded Sheet and Film*

TABLE 10-1   Typical Property Values for Dielectric Films

	ASTM TEST	CELLULOSE ACETATE (CA)	CELLULOSE TRIACETATE (CTA)	FLUOROPOLYMERS PTFE	PFA	FEP	ETFE	PVDF	ECTFE	PCTFE	IONOMER
1. Tensile strength, psi $\times 10^3$	D 882	7–16	9–16	3.3	4–7	2.5–3.0	7–8	6.0–6.5	8.0–10	5–10	5
2. Elongation %	D 882	15–55	10–50	300	200–600	300	300	300–500	150–250	200–600	250–450
3. Bursting strength, 1 mil thick, Mullen points	D 774	30–60	50–70	—	—	11	—	—	—	28–31	—
4. Tearing strength (propagating), g/mil	D 1922	4–10	4–30	10–100	40–70	125	600–900	50	900–1,300	2.4–40	30–125
5. Water absorption, % 24 h 1 mil thick	D 570	3–8.5	2–4.5	nil	0.03	0.01	0.02	0.04	0.02	nil	0.4
6. Water vapor transmission rate, g/100 sq in 24 h	E 96E										
@ 73°F (23°C), 90% R.H.		10–40	30–40	—	—	—	—	2.6	—	—	—
@ 100°F (38°C), 90% R.H.		—	—	—	—	0.4	1.65	—	0.6	0.02–0.05	1.5–2.0
7. Permeability to gases cm^3/100 sq in/24 h @ atmospheric pressure, 73°F (23°C), 0% R.H.	D 1434										
O_2		117–150	150	—	—	750	100	14	25	7–15	2.6
CO_2		860–1,000	880	—	—	1,670	250	5.5	110	16–40	10–12
8. Dielectric strength, volts/mil, short time 50% R.H., 73°F (23°C), 1 mil film	D 149	4,000	4,000	2,200–4,400	4,000–5,000	5,000	5,000	—	5,000–6,000	3,000–3,900	—
9. Dielectric constant @ 73°F (23°C)	D 150										
10^3 cycles		3.6	3.2–4.5	2.0	2.0	2.0	2.6	8.4	2.6	2.6	2.4
10^6 cycles		3.2	3.3–3.8	2.0	2.0	2.0	2.6	6.6	2.6	2.3	2.4
10. Dissipation factor @ 73°F (23°C)	D 150										
10^3 cycles		0.013	0.016	0.0001	0.0002	0.0002	0.008	0.018	0.002	0.023	0.002
10^6 cycles		0.038	0.033	0.0001	0.0002	0.0003	0.0005	0.170	0.013	0.012	0.007
11. Volume resistivity, Ω cm, 73°F (23°C)	D 257	10^{15}	10^{15}	10^{18}	10^{18}	10^{18}	10^{16}	2×10^{14}	10^{16}	10^{18}	10^{16}
12. Chemical resistance (G = Good, F = Fair, P = Poor)											
Strong acids	D 543	P	F	G	G	G	G	G	G	G	G
Strong alkalies	D 543	P	P	G	G	G	G	G	G	G	G
Greases and oils	D 722	G	G	G	G	G	G	G	G	G	G
Organic solvents	D 543	P	F-P	G	G	G	G	G	G	G	G
13. Maximum use temperature °F (°C)		250 (120)	250 (120)	500 (260)	500 (260)	400 (205)	300 (150)	300 (150)	330 (165)	300 (150)	190 (90)
14. Thickness range, mils											
Minimum		0.88	2.0	0.25	0.5	0.5	—	0.5	0.5	0.75	1.5
Maximum		30	20	30	90	30	—	40	90	10	3.8
15. Maximum available width, inches		60	47	48	48	48	48	50	54	48	72
16. Specific gravity	D 1505	1.3	1.3	2.1	2.2	2.2	1.7	1.8	1.7	2.1	0.9

*MD = Machine direction.

TD = Transverse direction.

TABLE 10-1   (*Continued*)

POLYCARBONATE (PC)	POLYESTER (PET)	LOW DENSITY POLYETHLYENE (LDPE)	POLYIMIDE (PI)	POLYPARABANIC ACID (PPA)	POLYPROPYLENE		POLY-STYRENE (PS)	POLYVINYL CHLORIDE (PVC)	SULFONE POLYMERS	
					(PP) UNORIENTED	BIAXIALLY ORIENTED	BIAXIALLY ORIENTED	PLASTICIZED, EXTRUDED	POLYSULFONE EXTRUDED	POLYETHERSULFONE EXTRUDED
7.5–9.5	1.5–4	1.5–4	25	16	4–7	23–25	9–12	1.4–10	8.4–10	10–12
85–110	150	200–800	70	75	800–850	90	10–20	100–500	64–110	20–150
no brk	55–80	10–12	75	—	—	—	16–35	20	—	—
20–25	12–27	50–300	8	6	MD-20*	MD-3*	5	60–1000	9–12	7–16
					TB-450*	TD-10*				
0.35	0.5	0.01	2.9	2.8	0.005	0.005	0.04–0.10	neg.	0.3	2.1
—	—	—	—	—	—	—	—	—	—	340
11	1.6–1.8	1.0–1.5	5.4	—	0.7	0.25–0.4	7–10	5–30	18	—
300	3–6	500	25	—	150–240	160	300	300–1,100	230	90
1,075	15–25	2,700	45	—	500–800	540	900	30–6,000	950	405
6,300	7,000–7,500	5,000	7,000	6,400	—	5,000	5,000	1,800 (5 mil)	7,500	5,800
3.0	3.2	2.2	3.5	3.4	2.2	2.2	2.4–2.7	4–8	3.1	3.5
2.9	3.0	2.2	3.4	—	2.2	2.2	2.4–2.7	3.3–4.5	3.0	3.5
0.0015	0.005	0.0003	0.0025	0.003	0.0003	0.0003	0.0005	0.07–0.16	0.0008	0.00035
0.010	0.016	0.0003	0.010	—	0.0003	0.0003	0.0005	0.04–0.14	0.0034	0.006
10^{16}	10^{18}	10^{16}	10^{18}	—	3×10^{15}	5×10^{14}	10^{16}	10^{14}	10^{16}	10^{17}
G	G	G	G	G	G	G	G	G	G	G
P	P	G	P	P	G	G	G	G	G	G
G	G	P	G	G	G	G	G-P	G	G	G
G-P	G	F	G	G	G	G	G-P	P-F	P	G-P
270	300	200	465	340	270	270	160	220	300	356
(132)	(150)	(90)	(240)	(170)	(130)	(130)	(70)	(105)	(150)	(180)
0.25	0.08	0.3	0.3	1.0	0.75	0.45	0.25	0.5	0.1	0.1
30	14	10	5	5	10	3.5	20	40	30	30
45 cast 54 extr	120	240	60	—	69	120	76	80	52	52
1.2	1.4	0.9	1.4	—	0.9	0.9	1.05	1.2–1.4	1.2	1.4

Federal Specification for Fluoropolymer Film

The following Federal specification has been developed for fluoropolymer film:

• L-P-523, *Plastic Sheet and Film, FEP-Fluoropolymer, Extruded*

Grades and Characteristics

Suppliers, grades, advantages, and limitations of fluoropolymer films are shown in Table 10-2.

Processing

All fluoropolymer films except polytetrafluoroethylene are made by conventional extrusion techniques. Polytetrafluoroethylene film is skived (shaved continuously from the surface of a cylindrical block of molded resin), paste extruded and sintered, or cast from aqueous dispersion, after which the water is removed by heat and the resin fused at over 700°F (370°C). Polyvinylidene fluoride film is solvent-cast as well as extruded.

Properties

Key film property values are shown in Table 10-1.

Table 10-2   Summary of Characteristics of Fluoropolymer Films

FLUOROPOLYMER	SUPPLIERS	GRADES	ADVANTAGES	LIMITATIONS
Polytetrafluoro-ethylene (PTFE)	Carborundum Chemplast Dixon Garlock W. L. Gore Norplex Oak Polymer Corp. Carborundum	Cast Paste extruded followed by sintering Dispersion cast (free of pinholes)	Widest useful thermal range of all plastics (−267° to 260°C). Unaffected by virtually all chemicals. Dielectric constant is relatively unchanged (about 2.1) with time, frequency, or temperature. Arc resistance over 300 with no carboniza-tion. Oxygen index over 95%.	Cannot be melt processed. Poor resistance to corona and radiation.
Perfluoroalkoxy (PFA)	Du Pont	Standard grade Modified grade treated to adhesions on one side.	Dielectric, mechanical, thermal, and chemical properties equal to PTFE. Better mechanical properties than FEP above 150°C. Can be processed by conventional extrusion techniques.	Lower abrasion resistance and higher in price than PTFE.
Fluorinated ethylene-propylene copolymer (FEP)	Du Pont	Standard grade Modified grade treated to promote adhesion on one or both sides	Dielectric, mechanical, chemical properties equal to PTFE and PFA. Thermal range −73° to 205°C. Can be processed by conventional extrusion techniques. Oxygen index of 95%.	Narrower useful thermal range than PTFE or PFA. Not as tough as PTFE. Price between PFA and PTFE.

(continued)

Table 10-2   (Continued)

FLUOROPOLYMER	SUPPLIERS	GRADES	ADVANTAGES	LIMITATIONS
Ethylene-tetra-fluoroethylene (ETFE)	Du Pont Hoechst Celanese	Standard grade Modified grade treated to promote adhesion on one side	Mechanical properties better than other fluoroplastics except PCTFE. Can be processed by conventional extrusion techniques.	Dielectric properties not quite as good as PTFE, PFA, and FEP. Lower max operating temperature than PTFE, PFA, FEP, and ECTFE. OI of 28–32%. Price about the same as PFA.
Polyvinylidene fluoride (PVDF)	Ferro Rexham Westlake	Cast Cast Extruded	Can be solvated by organic esters and ketones as well as processed by conventional extrusion techniques. High abrasion resistance and resistance to UV and nuclear radiation. Oxygen index of 44%. Lowest price of fluropolymers.	Highest dielectric constant and dissipation factor of all fluoropolymers. Chemical resistance lowest. Useful thermal range ($-62°$ to $150°C$) narrowest.
Ethylene-chloro-trifluro ethylene copolymer (ECTFE)	AlliedSignal	Standard extrusion grade	Highest tensile strength abrasion resistance of all fluoropolymers. Good dielectric properties. Can be processed by conventional extrusion techniques. Oxygen index of 60. Lowest specific gravity (1.68) of all fluoropolymers. Outstanding permeability resistance to water vapor, gases, liquids.	Attacked by hot amines (otherwise good chemical resistance). Price about equal to FEP.
Polychlorotrifluoro-ethylene (PCTFE)	AlliedSignal	Good barrier property grade Good dimensional property grade	Excellent resistance to oxygen, ozone, fuming oxidizing acids, sunlight. Outstanding barrier to gases and liquids. Good abrasion resistance. Can be processed by conventional extrusion techniques. Good dielectric properties.	Precise control of temperature required in processing to prevent degradation. Attacked by some halogenated solvents. Price is highest of all fluoropolymers.

Electrical/Electronic Uses

Principal uses are for severe conditions of temperature or environment. Applications include:

- Wrapped insulation on high-temperature wire and cable
- Motor coil, phase, and ground insulation
- Capacitor dielectric
- Substrate for flexible printed circuits and flexible cables
- Pressure-sensitive tapes

Imide Polymers: Polyetherimide (PEI)
(Film)

A recent (1993) entry into the dielectric film market, polyetherimide film, produced by General Electric with the trademark Ultem 5000, competes with polyimide film for some Class 180 and above transformer and motor applications. The film is reported to have dielectric strength comparable to polyimide film with higher thermal conductivity and lower moisture absorption at significantly less cost. Minimum film thickness is 1 mil.

Compared with aramid paper, PEI film has much higher dielectric strength and lower moisture absorption, thus permitting smaller and more efficient equipment designs.

For information on the chemistry and key properties of PEI resin, see "Imide Polymers: Polyetherimide (PEI)," p. 83, Chapter 3, *Thermoplastic Molding Compounds*.

Imide Polymers: Polyimide (PI)
(Film)

Polyimide film, first commercially available in 1966, has the best thermal properties of all organic films. It is inherently flame resistant and remains useful in environments of intense radiation. The film is made by Du Pont under the trademark Kapton in the United States and by ICI with the trademark Upilex. There are also several Japanese producers of polyimide-type films.

Chemistry

Polyimide film is the condensation product of pyromellitie dianhydride (PMDA) with an aromatic diamine:

pyromellitic	aromatic diamine
dianhydride (PMDA)	(4,4'-diamino-
	diphenyl ether)

poly(amide-acid)

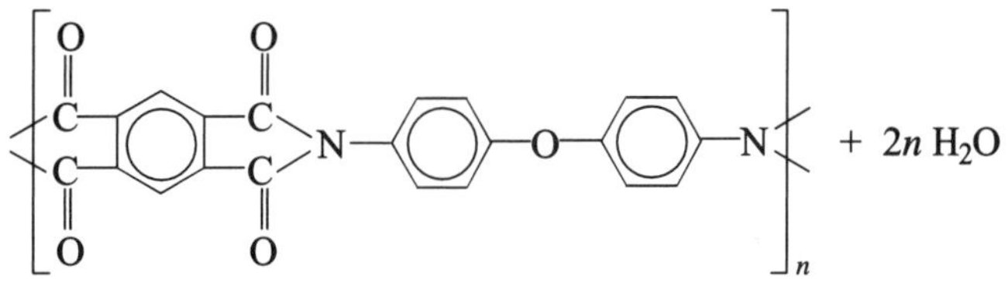

polyimide resin

The polymer is said to be cast into a film from solvent solution before the imidization step. After imidization, the film is usually biaxially oriented.

Federal Specification

The governing military specification for polyimide film is MIL-P-46112, *Plastic Sheet and Strip, Polyimide*. This specification covers polyimide sheet and strip with or without heat sealable FEP-fluorocarbon coatings. The following types and grades are designated:

> Type I—General purpose
> Type II—Heat sealable
>   Grade A—One side coated
>   Grade B—Two sides coated.

Requirements are listed for Type I sheet and strip, including tensile strength, elongation, shrinkage, moisture absorption, dielectric strength, volume resistivity, dielectric constant, and dissipation factor. Requirements for Type II sheet and strip include dielectric strength and minimum heat seal strength.

ASTM Standards for PI Films

The American Society for Testing and Materials (ASTM) has developed the following standards for polyimide films:

> • D 5213, *Specification for Polyimide Resin Film for Electrical Insulation and Dielectric Applications*
> • D 5214, *Test Method for Polyimide Resin Film for Electrical Insulation and Dielectric Applications*

Grades

There are three principal types of Kapton commercially available:

> • Type HN, an all-purpose film that can be laminated, metallized, and adhesive coated.
> • Type VN, similar to Type HN but with better dimensional stability.

• Type FN, a Type HN film coated on one or both sides with Teflon™ FEP fluorocarbon resin to impart heat sealability, provide a moisture barrier, and enhance chemical resistance.

Properties

Kapton film has no melting point and remains useful over the temperature range of –450 to 660°F (–269°–350°C). It will function continuously at 465°F (240°C). Physical attributes include high tensile strength, and high resistance to creep, cut-through, and abrasion. Dielectric properties are generally good, although except for dielectric strength, they decrease significantly with increasing temperature. Radiation resistance and resistance to ultraviolet light are outstanding. However, the film is vulnerable to attack by alkalis and strong inorganic acids, and moisture absorption rate is among the highest for all dielectric films.

Key property values are shown in Table 10-1.

Electrical/Electronic Uses

Because of its relatively high price, Kapton film is used principally where its unique properties make it the only suitable film, and also where its use permits economies in design which offset its high price. Typical applications include use in motors, where it replaces thicker dielectrics, permitting more powerful, cooler running motors with no increase in frame size; use in aerospace electrical equipment subject to temporary overloads; use in high-performance wire and cable constructions where it provides significant savings in space and weight; use as a substrate for flexible printed circuits and flat cables; and use as a capacitor dielectric.

Ionomer
(Film)

Ionomer resins were introduced in the middle 1960s with the trademark Surlyn by Du Pont, and have since found extensive use in molded products such as cut-resistant golf balls and automotive bumper components. Powdered resins are fused by heat or infrared radiation to no-return glass beverage bottles, making them shatter-resistant and permitting 25 percent savings in glass with quieter processing. By far the largest use for inomer, however, is in packaging applications, usually in combination with paper, foil, and other films.

Chemistry

The term ionomer, a contraction of ion and polymer, describes ethylene copolymers with methacrylic or acrylic acid, bonded together by ionic forces. Interchain bonding takes place through metallic ions, predominantly Zn^{++} formed as carboxylic groups on methacrylic or acrylic acid which are neutralized:

$$
\begin{array}{c}
CH_3 \\
| \\
-CH_2-C-(CH_2-CH_2)_n- \\
\end{array}
$$

This type of crosslinking is thermally reversible, making it possible to process ionomers by conventional thermoplastic techniques.

Grades

Ionomer film is available in extruded grades, but is generally supplied in combination with paper, foil, and other films for specific end uses. Military purchases for extrusion grades are made to MIL-P-46124, *Plastic Molding and Extrusion Material, Ionomer Resins*.

Properties

The oustanding property of ionomer is its toughness and abrasion resistance, as exemplified by its use as a golf ball cover and for automotive bumper parts. Its low specific gravity of 0.9–1.0 permits high yields. Maximum use temperature of 190°F (90°C) is relatively low.

Key property values are shown in Table 10-1.

Electrical/Electronic Uses

Although ionomer possesses good dielectric properties, its use is limited as electrical insulation since the polymers with which it competes (polyvinyl chloride, polyethylene) are much lower in price. However, this price disadvantage is offset in the case of polyvinyl chloride by its high yield and greater toughness, permitting thinner ionomer films to be substituted for thicker polyvinyl chloride films. Ionomer films also do not have plasticizer problems common to polyvinyl chloride films.

Polycarbonate (PC)
(Film)

One of the more recently developed resins for extruded and solvent cast film, polycarbonate is used for a variety of electrical/electronic applications. Producers of electrical grade film include General Electric (extruded) and Kimberly-Clark (solvent cast).

Chemistry

Polycarbonate is made by reacting bisphenol A with phosgene in the presence of a caustic:

bisphenol A   phosgene

polycarbonate

Polymerization may be in aqueous emulsion or in nonaqueous solution. Methylene dichloride is a preferred solvent in the solvent casting process.

Grades

Capacitor grades are made by solvent casting with Kimberly-Clark the sole producer. Grades for other electrical/electronic applications are made by General Electric, including a UV-stabilized grade. Films are usually biaxially oriented.

Properties

Polycarbonate film as good flexibility, heat resistance, dimensional stability, and thermoforming characteristics. Dielectric constant is virtually unaffected by temperature or frequency within the temperature range of $-75°$ to $250°F$ ($-60°$–$120°C$). The film resists attack by oils, fats, and dilute acids, but is affected by alkalis, amines, ketones, esters, and aromatic hydrocarbons. Water vapor and gas transmission rates are relatively high. The film can be heat sealed.

Key property values are shown in Table 10-1.

Electrical/Electronic Uses

Solution cast film is used almost exclusively in capacitors. Extruded film is used for coil insulation, slot insulation, conductor insulation, and as a substrate for flexible printed circuits.

Polyester (PET)
(Film)

Polyethylene terephthalate film is an exceptionally strong and durable film, used for a wide variety of industrial and packaging applications. Major producers for the merchant market are Hoechst Celanese (Celanar™), Du Pont (Mylar™), ICI (Melinex™), and 3M (Scotchpar™, Scotchpak™).

Chemistry

Polyester resin is produced by direct esterification of ethylene glycol and terephthalic acid, or by catalyzed ester exchange between glycol and dimethyl terephthalate, followed by polycondensation to a high-molecular-weight polymer:

$$HOOC\!-\!\bigcirc\!-\!COOH + HOCH_2CH_2OH \longrightarrow HO\!-\!(OC\!-\!\bigcirc\!-\!COOCH_2CH_2O)_nH + 2nH_2O$$

terephthalic acid   ethylene glycol   PET

$$CH_3OOC\!-\!\bigcirc\!-\!COOCH_3 + HOCH_2CH_2OH \longrightarrow HO\!-\!(OC\!-\!\bigcirc\!-\!COOCH_2CH_2O)_nH + 2nCH_3OH$$

dimethyl terephthalate   PET

Grades

Polyester film by itself is clear and slippery. It can be made heat-sealable by coating with polymers or laminating with polyethylene film. Friction-coated films are produced to eliminate slippage in wrapping operations. Film with a matte finish on one side is also available. Most polyester film is biaxially oriented.

Properties

Polyester film is characterized by outstanding tensile strength and flexibility, desirable dielectric properties, general chemical inertness, and useful temperature range of $-80°$ to over $300°F$ $(-20°–150°C)$. However, propagating tear strength is low and resistance to ultraviolet light is poor. Under humid conditions and high temperatures, the film has a tendency to hydrolyze. The film is designated Class 130 insulation.

ASTM Standards for PET Film

The American Society for Testing and Materials (ASTM) has developed the following standards for polyester film:

• D 3664, *Specification for Biaxially Oriented Polyethylene Terephthalate Film for Electrical Insulation and Dielectric Applications*. The material is biaxially oriented to improve the tensile properties in both the machine and transverse directions (MD and TD). For the purposes of this standard, film is defined as a piece of material not exceeding 0.250 mm (0.01 inch) in thickness and exceeding 100 mm (4 inches) in length and width. Tests for the properties covered are shown in Table 10–3.

Key property values are shown in Table 10-1.

• D 5047, *Specification for Polyethylene Terephthalate Film and Sheeting*

Electrical/Electronic Uses

Polyester film is used extensively for flexible printed circuits and flat cables, for wrapping wire and cable, for slot-wedge and phase insulation in motors, as barrier and layer insulation in transformers, and as the dielectric in electronic capacitors.

**Table 10-3   Property Tests and Test Methods for
Biaxially Oriented Polyethylene Terephthalate Film
(ASTM D 3664)**

PROPERTY MEASURED	ASTM METHOD
Dielectric breakdown	D 149
Dielectric constant (permittivity)	D 150
Sampling	D 202
DC resistance of conductance	D 257
Thickness	D 374
Chemical resistance	D 543
Water absorption	D 570
Conditioning	D 618
Folding endurance	D 643
Basis weight	D 646
Grease resistance	D 722
Bursting strength	D 774
Tensile properties	D 882
Tear resistance	D 1004
Gas transmission rate	D 1434
Outdoor weathering	D 1435
Density	D 1505
Melting point	D 2117
Folding endurance	D 2176
Combustibility	D 2863
Impact resistance	D 3420
Water vapor transmission	E 96

Polyester Film Composites

Polyester film is often combined with other materials, including polyester mat. These composites, used primarily as dielectric barriers with thermal rating 130°C, are covered by military specification MIL-I-22834, *Insulation, Electrical, Dielectric Barrier, Laminated, Plastic Film and Synthetic Fiber Mat*. The following compositions are designated:

   • Composition D100—One layer of polyethylene terephthalate fiber mat laminated to each side of a polyethylene terephthalate film or sheet (triplex). The mat is completely impregnated with epoxy resin.

   • Composition D70—One layer of polyethylene terephthalate fiber mat laminated to each side of a polyethylene terephthalate film or sheet (triplex). The mat is 70 percent impregnated with epoxy resin.

   • Composition D—One layer of polyethylene terephthalate fiber mat laminated to each side of a polyethylene film or sheet (triplex). The mat is heat bonded to the film.

   Minimum requirements are listed for dielectric strength, volume resistivity, tensile strength, tear resistance, and absorbency.

   See also MIL-I-19632 on paper-film composites, summarized (p. 324) in Chapter 11, *Dielectric Papers and Boards*.

Polyethylene (PE)
(Film)

Polyethylene film is made from a wide variety of polyethylene densities, molecular weights, and melt indexes as described in "Ethylene Polymers" (p. 154) of Chapter 5, *Extrusion Compounds*. Principal electrical/electronic applications for polyethylene is as an extruded dielectric in wire and cable (q.v.). There is some usage for films as splicing tape after it has been coated with pressure-sensitive adhesive, and as a capacitor dielectric. The relatively low maximum service temperature of 80° to 90°C (176°–194°F) (depending on grade) limits use for other applications.

Key film property values of low-density polyethylene (LDPE) are shown in Table 10–1.

Polyparabanic Acid (PPA)
(Film)

Polyparabanic acid (PPA) film was developed by Exxon Chemical, and became commercially available in 1979 with the trademark Tradlon. The film is no longer commercially available.

Chemistry

Polyparabanic acid film is made by solvent casting techniques. Known solvents include cyclic ketones, such as cyclohexanone. Polyparabanic acid has the following structure:

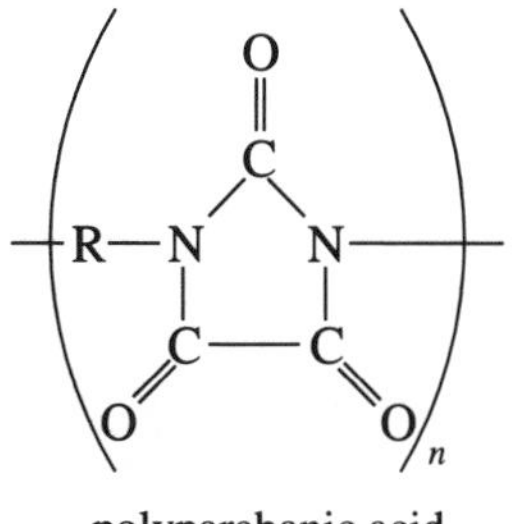

polyparabanic acid

Properties

Polyparabanic acid film has good dielectric properties, and is suitable for use in a temperature range from –200° to 170°C (–328°–338°F). It is resistant to attack by chlorinated hydrocarbons, alcohols, ethers, and acids, but is affected by strong bases.

Key film property values are shown in Table 10–1.

Electrical/Electronic Uses

Polyparabanic acid film is suitable for use in specialty applications as a wire and cable wrap, as motor insulation, and as a substrate for flexible printed circuits and flexible flat cable. When coated with suitable adhesive, it can be used in electrical grade pressure-sensitive tapes.

Polypropylene (PP)
(Film)

Polypropylene film is relatively low priced with physical and dielectric properties suitable for many electrical and electronic applications. Electrical grades of film are produced by Himont and General Electric. Electrical grade resin is made by Himont and Rexene Polymers, Division of Dart Industries.

Chemistry

Polypropylene is formed by polymerizing high-purity propylene in the presence of a Ziegler-type catalyst, such as aluminum alkyl. Principal processes are discussed in "Propylene Polymers" (p. 160) of Chapter 5, *Extrusion Compounds*.

Grades

Capacitor grades and film grades for other electrical/electronic applications are available from Himont and General Electric. Swelling of polypropylene films with impregnation by dielectric liquids is much lower with tenter process grades than with bubble process grades.

Properties

Oriented polypropylene copolymer film attributes include low price, high dielectric strength, low and stable dielectric constant and dissipation factor over a wide frequency range, good resistance to short-term voltage overloads, and excellent resistance to moisture, greases, and oils.

Key property values are shown in Table 10-1.

Electrical/Electronic Uses

A major use for propylene film is as a dielectric for capacitors. It is also extensively used as a cable wrap and for layer and phase separators in rotating electrical equipment and transformers.

Polystyrene (PS)
(Film)

Uniaxially oriented polystyrene film is used to some extent as a capacitor dielectric, which field is dominated by polypropylene and polyimide films. Low maximum service temperature (70°C) and high flammability limit its suitability for other electrical/electronic applications. The predominant producer of polystyrene film is Dow Chemical. HMC is the major supplier of capacitor grade film. See "Styrenics: Polystyrene (PS)" (p. 97) Chapter 3, *Thermoplastic Molding Compounds*.

Key film property values are shown in Table 10-1.

Polyvinyl Chloride (PVC)
(Film)

The principal use of polyvinyl chloride film is in the form of pressure-sensitive tapes. This supject is covered in Chapter 12, *Tapes and Coated Fabrics.*

For comparison with other films, key property values for polyvinyl chloride films are shown in Table 10-1.

Sulfone Polymers: Polysulfone
(Film)

Polysulfone resins became commercially available in 1965. They possess the highest radiation resistance of all organic polymers. Union Carbide is the sole producer of the resin, using the trademark Udel. Film was made by Kimberly-Clark (solvent cast), Rowland (extruded), and Westlake (extruded), but no longer is commercially available.

Chemistry

Polysulfone is made by reacting the sodium salt of bisphenol A with 4,4'-dichlorophenyl sulfone in dimethyl sulfoxide or a similar polar solvent:

sodium salt of bisphenol A   4,4'-dichlorophenyl sulfone

polysulfone

Properties

Resistance of the film to radiation is outstanding. It has good resistance to attack by mineral acids, alkalis, salt solutions, alcohols, oils, and greases, but it is affected by esters, ketones, chlorinated hydrocarbons, and aromatic hydrocarbons. Useful temperature range is $-150°$ to $300°$ ($-100°$–$150°C$). Dielectric properties are superior to those of polyethersulfone, but thermal, mechanical, and flammability values are lower.

Key property values are shown in Table 10-1.

Electrical/Electronic Uses

The film is used in flexible and rigid printed circuits, specialty laminates, and capacitors.

Sulfone Polymers: Polyethersulfone
(Film)

First produced in the United Kingdom, polyethersulfones became available in the United States in 1972–73. High-temperature properties are outstanding. Resins are produced by ICI Americas under the trademark Victrex. Film is made by Futurex, Kimberly-Clark (solvent-cast), and Westlake Plastics.

Chemistry

Polyethersulfone is made by Friedel-Crafts condensation of diphenyletherchlorosulfone using anhydrous aluminum chloride or similar metallic halide as a catalyst:

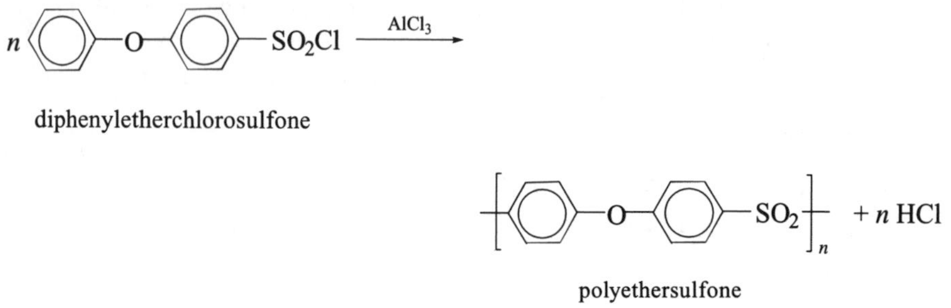

$$n \; \bigcirc - O - \bigcirc - SO_2Cl \quad \xrightarrow{\text{AlCl}_3}$$

diphenyletherchlorosulfone

$$\left[\bigcirc - O - \bigcirc - SO_2 \right]_n + n \text{ HCl}$$

polyethersulfone

Grades

Both solvent cast from methylene chloride (for capacitors) and extruded grade films are commercially available.

Properties

Dielectric and mechanical properties of this film are largely retained up to 356°F (180°C). Dielectric values are somewhat lower than those of polysulfone, but thermal, mechanical, and flammability values are higher. Chemical resistance is similar to polysulfone.

Key property values are shown in Table 10-1.

Electrical/Electronic Uses

High price limits uses to specialized flexible circuits, flat cables, and integrated circuits. Other uses are justified by the film's high-temperature attributes.

11

Dielectric Papers and Boards

INTRODUCTION

Dielectric papers and boards may be produced from a variety of materials, including wood, cotton, organic fibers, glass, ceramics, and mica. The distinction between paper and board is not specific, but paper is thinner gauge, generally below 0.030 inch and boards are 0.030 inch and over. Above 0.25 inch, boards are laminated with adhesive to any desired thickness. Electrical insulating boards are referred to as pressboard, transformer board, fuller board, and presspan.

This chapter covers cellulose, vulcanized fibre, aramid, mica, glass, alumina-silica, and ceramic papers and boards.

DEVELOPMENT PROGRAMS

There is continuing emphasis among manufacturers to increase dielectric paper purity by eliminating all contaminants.

MARKET TRENDS

Because of the designation of asbestos as a toxic substance under the Toxic Substances Control Act (see Chapter 19, *Government Activities*) and publicity regarding health problems of asbestos workers, asbestos is rarely used for paper and board products. Substitute materials include aramid paper, especially the mica composite types, mica paper and composites, alumina-silica paper, ceramic paper and composites, and glass paper.

DIELECTRIC PAPERS AND BOARDS

The technology, chemistry, grades, properties, and electrical/electronic uses of principal types of dielectric papers and boards are described in the sections which follow in this order:

- Cellulose papers and boards
- Vulcanized fibre and fish paper
- Aramid papers and boards
- Mica paper
- Other papers and boards

Cellulose Papers and Boards

Cellulose base papers and boards may be made from a variety of materials, including wood pulp, cotton, hemp, ramie, jute, flax, bagasse (from sugar cane), straw, papyrus, and bamboo. The principal producers of cellulose insulating papers and boards include

Cottrell Paper, Dennison (crepe grades), Manning Paper, Mosinee, Riegel, Spaulding, Stevens, and Union Camp.

Technology

Dielectric grade paper is most often made from sulfate (kraft) wood pulp derived from coniferous, or softwood trees, such as spruce, hemlock, and pine. The first step in this process is to remove undesirable components in the wood, principally pentosans and lignin. Pentosans are complex carbohydrates which hydrolyze to sugar. Lignin is a phenylpropane polymer of amorphous structure comprising seventeen to thirty percent of wood. These materials are removed by boiling under pressure wood particles with a solution of sodium sulfite and sodium hydroxide. Sodium sulfate is then added to the caustic pulp liquors, where it is reduced to the sulfide, which then acts as a digesting agent.

This slurry, about 5 percent solids, is agitated in a tank, or "beater," by a rotating drum equipped with closely spaced sharp knives. Another set of knives is secured to the bottom of the beater. As the drum circulates the slurry, the fibers are disintegrated by the cutting and tearing action of the knives, rather than by any beating action, as the term "beater" implies. This is the most critical step in the process, requiring careful control of ingredient concentrations and time of beating, which varies with the type and quality of the paper being made. The higher the degree of beating, the greater the ability of the finished paper to absorb moisture.

The continuous paper sheet is formed on a Fourdrinier machine, where the pulp is first diluted to contain over 98 percent water. The mixture is then filtered through sand traps to remove grit, over magnets to remove magnetic particles, and through strainers to extract coarse fibers. It next pours onto a moving screen, where felting of the fibers is induced by lateral shaking. The water is screened out, with the assistance of suction boxes beneath the screen. The damp pulp is carried by the screen between felt-covered rollers, where excess water is squeezed out. The newly formed web is next transferred to a felt conveyor, which takes it through a pair of press rolls to remove more water. It then passes to another felt, and is finally dried on a series of steam-heated drums which increase in temperature from first to last, until the moisture content of the paper is reduced to 4–6 percent.

High-grade insulating papers are also made by an analogous process from hemp and cotton rags or linters, and combinations of these with wood pulp.

Cellulose $(C_6H_{10}O_5)_n$ is a high-molecular-weight polymer consisting of crystalline areas embedded in amorphous areas. The polymer is essentially linear, but with some crosslinking. The polymer unit may be represented as follows:

Through a process termed *cyanoethylation*, the moisture-absorbing tendency is reduced and dielectric properties of cellulose are significantly improved. This process introduces the group —OCH_2CH_2CN into the cellulose molecule by reaction of acrylonitrile with a reactive hydrogeh of a hydroxyl group, thereby changing the molecular structure to one with fewer troublesome hydroxyl groups. Cyanoethylation is done before the beating step.

ASTM Standards for Dielectric Papers and Boards

The American Society for Testing and Materials (ASTM) has developed these standards covering dielectric papers and boards:

• D 202, *Sampling and Testing Untreated Paper Used for Electrical Insulation*. These methods cover the procedures for sampling and testing untreated paper to be used as an electrical insulator or as a constituent of a composite material for electrical insulating purposes.

• D 1305, *Electrical Insulating Paper and Paperboard—Sulfate or Kraft Layer Type*. This specification covers electrical grade unsized, unbleached sulfate paper and paperboard for use as layer insulation in coils, transformers, and similar apparatus.

• D 2413, *Preparation and Electrical Testing of Insulating Paper and Board Impregnated with a Liquid Dielectric*. These methods have been found practicable for papers having nominal thickness of 0.05 mm (2 mils) and above. They have been used successfully for insulating board as thick as 6 mm ($^1/_4$ inch) when care is taken to assure the specimen geometry necessary for valid measurement of dielectric properties.

• D 3394, *Sampling and Testing Electrical Insulating Board*. These boards are fibrous, porous sheets used for dielectric and structural purposes in electrical apparatus. These methods are not intended for testing vulcanized fibre or molded laminated sheets.

• D 4063, *Electrical Insulating Board*. This specification covers electrical insulating board (pressboard) manufactured from kraft, cotton, or kraft and cotton pulps. This board is intended for dielectrical or structural purposes in transformer and other electrical apparatus.

Federal Specification for Paper and Pressboard

Following is the governing military specification for paper and pressboard:

• MIL-I-545, *Insulation Sheet, Electrical, Paper, Pressboard*. This specification is approved for use by all departments and agencies of the Department of Defense. This specification covers the requirements for paper, pressboard, electrical insulation sheet. The insulation is class O insulation with a continuous-operating temperature limit of 90°C. There are two grades:

Grade B—Moderate dielectric strength, made from undyed and unbleached coniferous (kraft) wood pulp

Grade C—High dielectric strength, made up of cotton or linen rag stock

Paper-Film Composites

Paper is often laminated to plastic film, primarily for use as dielectric barriers. This insulation type is covered by military specification MIL-I-19632, *Insulation, Electrical, Dielectric Barrier, Laminated, Plastic Film and Rag Paper*. These composites have thermal rating of 105°C. The following compositions are designated:

Composition FR—One layer of polyethylene terephthalate film or sheet laminated to one layer of 100 percent rag paper (duplex).

Composition RFR—One layer of 100 percent rag paper laminated to each side of a polyethylene terephthalate film or sheet (triplex).

Composition RoFRo—One layer of rope paper laminated to each side of a polyethylene terephthalate film or sheet (triplex).

Minimum requirements are listed for dielectric strength, volume resistivity, tensile strength, and tear resistance.

Grades

Available grades include:

• Capacitor tissue made solely of unbleached sulfate paper without additives, in thicknesses from 0.20 to 2.0 mils

• Paper grades made from kraft, cotton, hemp, and combinations of these, in thicknesses under 30 mils

• Crepe grades with controlled stretch from 20 to 300 percent, in thicknesses from 1.9 to 32 mils

• Board grades made from kraft, cotton, or kraft and cotton, in thicknesses 30 mils and over

• Composite grades to meet specification MIL-I-19632 and custom-made composites made with a variety of papers and films.

Properties

Untreated cellulosic insulating papers and boards have dielectric strength little better than air, dielectric loss factors which increase sharply with increasing frequencies, poor mechanical strength, and perhaps their greatest weakness, a high affinity for moisture. Thus, by themselves, they are not useful for dielectric applications except as a spacer. However, when treated or combined with other insulating materials, paper and boards become widely useful. Treatments include impregnation and coating with resins and varnishes for low-voltage applications, and impregnation with liquid dielectrics for higher voltage equipment. As an example of the improvement through impregnation with mineral oil, for 5 mil thick untreated paper, the dielectric strength is 200 volts per mil, but when impregnated with oil, the dielectric strength becomes 1340 volts per mil, a 5.7-fold increase. Samples were first dried at 240°F (115°C) and an absolute pressure of 0.02 torr or less for 16 hours. (1 torr is the pressure required to support 1 millimeter of mercury at 0°C.) The impregnating oil is dried and degassed by the same process. Drying of paper and degassing of oil are essential procedures for good manufacturing practice as well. It has long been recognized that when these steps are taken, a significant improvement in all dielectric properties is obtained because of the elimination of moisture and air and their replacement with a dielectric liquid. Cellulose papers and boards have thermal index of 105.

Papers and boards may also be combined with dielectric films and resins, and even with mica flakes, to produce useful products with a wide range of properties.

Electrical/Electronic Uses

Typical applications include wire and cable wraps, capacitor dielectrics, coil wraps, separators and slot insulation for rotating equipment and transformers, panel boards, substrates for laminated plastics, fabricated and formed parts, and composites with other insulating materials. Creping reduces damage to paper in coil-winding operations and in wrapping structures involving curves, angles, and irregular contours.

Vulcanized Fibre and Fish Paper

Vulcanized fibre (the approved spelling), often referred to as fish paper in thicknesses under 0.10 inch, is one of the oldest types of electrical insulating materials. It is noted for its toughness and outstanding ability to absorb shock. The term *vulcanized* is applicable only to the extent that the process, which is unrelated chemically to the vulcanizing process for rubber, does enhance significantly the strength and stability of paper. Suppliers are NVF and Spaulding Composites.

Chemistry and Processing

Vulcanized fibre is made by saturating rag-base or wood pulp paper, layered to produce the desired thickness, in a bath of zinc chloride. This causes the paper to gelatinize and bond into a homogeneous sheet. The zinc chloride is then leached from the sheet with excess water. After the zinc chloride content has been reduced to a maximum of 0.1 percent (0.05 percent for electrical grades), the sheet is dried in a room with controlled temperature and humidity. Finally, the sheet is calendered to a smooth finish. Production time increases exponentially with sheet thickness so that several months are required to leach the zinc chloride from heavy-gauge sheets. Thin gauges have been produced continuously as the paper travels through a machine equipped with zinc chloride baths, leaching tanks with running water, and heated drying chambers.

Grades

There are three grades available as sheets, rods, tubes, and coils for dielectric applications. These grades are made to meet ASTM Standard D 710, *Specification for Vulcanized Fibre Sheets, Rods, and Tubes Used for Electrical Insulation:*

  • Commercial Grade, a general purpose electrical and mechanical grade with good machineability, available in thicknesses from 0.010 inch to 2 inches.

• Bone Grade, a harder, stiffer grade with good machineability, available in thicknesses from $1/32$ to $1/2$ inch.

• Electrical Insulation Grade, the premier grade for dielectric applications involving difficult bending or forming operations, available in thicknesses from 0.004 to $1/8$ inch.

Standard sheet sizes (in inches) for the above grades are 25 × 40, 48 × 72, and 52 × 84.

A built-up, or laminated grade is also available in thicknesses from $5/16$ to $1\,1/2$ inches, with sheet size (in inches) of 48 × 82. Because it requires much less time to fabricate with adhesives, this grade is significantly less expensive than the equivalent thickness in standard grades.

Properties

Noteworthy characteristics of vulcanized fibre include:

• High values for impact strength and resilience
• Good arc-resistant and arc-quenching properties
• Light weight (one-half that of aluminum)
• Easy machineability

The weakness of vulcanized fibre which severely restricts dielectric applications is its affinity for moisture. In only 2 hours immersion in water, a $1/16$ inch thick specimen will increase weight 30–50 percent. UL use temperatures are limited to 115°C (239°F) electrical and 110°C (230°F) mechanical. The MIL-F-1148 operating temperature limit is 105°C (221°F).

Electrical/Electronic Uses

While many of the major applications of vulcanized fibre take advantage of its high mechanical strength, there are many traditional dielectric uses, where moisture is not a problem, such as arc barriers in air circuit breakers, lightning arrestors for heavy-duty transformers, switch and appliance parts, and slot liners for rotating equipment. As corrugated strips, fibre is used for ducts in oil-filled distribution transformers. Fish paper is widely used to insulate electrical coils and windings, and in slots of electrical equipment.

Aramid Papers and Boards

Aramid paper is the generic term for a wholly aromatic polyamide paper, most commonly known by the trademark NOMEX of Du Pont, its developer and principal producer. This paper and its fiber counterpart have been commercially available since 1962, and are primarily designed to serve the thermal range of 155° to 220°C (220°–428°F)—above that of polyester film used for similar applications at lower temperatures.

Chemistry and Processing

Aramid fibers (floc) and small binder particles (fibrids) are made by reacting isophthaloyl chloride and m-phenyl diamine in the presence of acid acceptor, such as dimethylacetamide (DMAc) containing 5 percent lithium chloride. The polycondensation is carried out in this solution:

isophthaloyl   *m*-phenyl
chloride   diamine

poly (*m*-phenylene
isophthalamide)

Aramid paper is made from floc and fibrids in a Fourdrinier paper machine. No binders, sizes, or fillers are added in the process. The paper may then be calendered under high temperature and pressure to produce a dense, relatively nonporous sheet.

Aramid boards are made from the same materials as the papers, but are formed on commercial cylinder board-making machines, pressed and/or calendered to the required densities. In contrast to aramid paper, aramid pressboard is relatively porous and can be impregnated with mineral oil and various resins and varnishes.

Federal Specification

The governing Federal specification for aramid paper is MIL-I-24204, *Insulation, Electrical, High Temperature, Bonded, Synthetic Fiber Paper*. This specification is approved for use by the Naval Sea Systems Command and is available for use by all departments and agencies of the Department of Defense. Covered are calendered, flexible insulation used in the coils and slots of electrical machinery at temperatures up to 220°C (428°F). Required performance characteristics are listed (which are well below published values for commercially available products as shown in Table 11-1). Aramid papers also meet the requirements of NEMA, IEC, UL, and CSA standards.

Grades

Available commercial grades include:

• Standard, calendered NOMEX in thicknesses from 2 to 30 mils

• A higher tear strength calendered grade with greater conformability in thicknesses from 3.4 to 15 mils

• An uncalendered, porous grade in thicknesses from 5 to 23 mils

TABLE 11-1   Typical Property Values for Aramid Papers
5-Mil Specimens
(23°C, 50% R.H.)

	ASTM TEST METHOD	ARAMID		ARAMID AND MICA
		CALENDERED	UNCALENDERED	
Pounds per square yard		0.21	0.08	0.27
Tensile strength, lb/in	D 828			
Machine direction		80	10	34
Transverse direction		40	5	23
Elongation, %	D 828			
Machine direction		16	3.0	2.4
Transverse direction		13	5.0	2.4
Elmendorf tear, grams	D 689			
Machine direction		240	130	190
Transverse direction		520	180	290
Dielectric strength ac rapid rise V/mil	D 149	635	230	1,000
Dielectric constant	D 150			
60 Hz, 23°C, 50% RH		2.4		3.6
10^3 Hz			1.3	
Dissipation factor	D 150			
60 Hz, 23°C, 50% RH		0.006		0.12
10^3 Hz			0.003	

• A blend of aramid polymer and mica flakes, combining the strong mechanical properties of aramid with the high corona resistance of mica, in thicknesses from 3 to 14 mils

• Several types of aramid and aramid-blend pressboards of varying surfaces, properties, and densities, made of 100 percent aramid or blends of aramid materials with mineral or glass fibers, in thicknesses up to approximately 400 mils.

Laminated Products

In cases where very high strength and tear resistance are required, commercially available laminates with polyester film are often used with loss of some thermal capability to about 180°C (356°F). If necessary, higher price polyimide film laminates may be used to provide the required thermal and dielectric properties.

Properties

Aramid paper possesses high-temperature durability, flame retardancy, excellent mechanical strength, and good dielectric properties. It has limiting oxygen index (LOI) of about 27 and UL flammability rating of 94 V-0. Dielectric constant and dissipation factor increase slightly up to 220°C (428°F). The paper has good resistance to acids, alkalis, refrigerants, ketones, alcohols, and oils. It is compatible with most wire enamels and insulating varnishes. Aramid paper is susceptible to moisture absorption, and in fact requires 3–5 percent moisture content (based on dry weight) for the optimum balance of properties. Increasing the moisture content improves physical properties, but lowers insulating values, except dielectric strength, which remains essentially constant up to 95 percent relative humidity. This material is rated by Underwriters Laboratories for continuous use at 220°C (428°F).

Paper made with blends of aramid fibers with mica flakes has higher corona and arc resistance than unblended aramid paper, and is rated for continuous service at 220°C (428°F).

Typical property values are shown in Table 11-1.

Aramid boards exhibit the same thermal and electrical properties as aramid papers on a thickness basis.

Electrical/Electronic Uses

The first and still principal uses for aramid paper are in rotating electrical equipment for lead wire, coil, slot, phase, wedge, and end insulation; and in transformers for turn, layer, barrier, and tap insulation. Other applications include substrates for printed circuits, cable insulation, and spiral and convolute wound tubes and coil bobbins.

Mica Paper

Mica paper, or reconstituted mica, was first used for dielectric applications about 1950. Principal producers include Insulating Materials and Essex International.

For additional information on mica, see Chapter 14, *Mica and Mica Products*.

Technology

Mica paper is made from scrap mica. Mica muscovite splittings are broken up by heat at about 1470°F (800°C), followed by grinding particles into small platelets. Paper is formed in a modified Fourdrinier machine similar to that used in making kraft paper. The paper is then calendered to the desired thickness. The inherently low mechanical properties, moisture resistance, and dielectric strength of mica paper are improved by treatment with resins (shellac, epoxies, alkyds, polyesters, silicones) and by reinforcement (glass cloth, polyester mat and film, polyimide film).

Federal Specifications

The following Federal specifications cover mica paper products:

• MIL-I-19917, *Insulation Sheet, Electrical, Mica Paper, Silicone Bonded*. The following types are designated:

    Type MPF-S—Mica paper, flexible, silicone bonded
    Type MPR-S—Mica paper, rigid, silicone bonded

The nominal thickness range is 0.010–0.0625 inch for Type MPR-S and 0.010–0.030 inch for Type MPF-S.

• MIL-I-21070, *Insulating Sheet and Tape, Electrical, Reinforced Mica Paper*.
The following classes are designated:

Class 130—With organic resin binder and organic varnish coating (if used)
on glass fiber woven fabric reinforcement

Class 155—With epoxy or polyester resin binder and an epoxy or polyester
coating (if used) on glass fiber woven fabric reinforcement

Class 180—With silicone resin binder and silicone varnish coating (if used)
on glass fiber woven fabric reinforcement

The nominal thickness range for these composites is 0.005–0.025 inch. Requirements are specified for dielectric strength, volume resistivity, tensile strength, and tear
resistance for each construction and thickness.

TABLE 11-2   Typical Property Values for Mica Paper Composites

CONSTRUCTION	BINDER	% BINDER BY WEIGHT	FINISHED THICKNESS (inches)	THERMAL CLASS	SHORT TIME DIELECTRIC STRENGTH (volts/mil)	TENSILE STRENGTH (lb/inch)
0.0017 in glass cloth 0.004 in mica paper	epoxy B-staged	27–33	0.0065– 0.0075	155°C	600	70
0.0008 in polyester mat 0.004 in mica paper 0.002 in glass cloth	epoxy B-staged	36–42	0.009	155°C	450	80
0.0017 in glass cloth 0.002 in mica paper 0.00025 in polyester film	epoxy fully cured	22–28	0.005– 0.006	155°C	750	70
0.0026 in glass cloth 0.002 in mica paper 0.00025 in polyester film both sides	polyester fully cured	20–25	0.005– 0.006	155°C	1,000	70
0.0017 in glass cloth 0.004 in mica paper 0.00025 in polyester film	polyester fully cured	25–30	0.007– 0.008	155°C	650	70
0.002 in mica paper 0.002 in glass cloth	silicone fully cured	20–25	0.0045– 0.0055	200°C	500	80
0.00025 in polyester film 0.0018 in mica paper 0.002 in glass cloth	silicone fully cured	20–25	0.005	200°C	700–900	84
0.001 in KAPTON 0.002 in mica paper	silicone fully cured	10–12	0.0035	200°C	2,220	30

Note: The above constructions are available in full width rolls 36 in wide by 36 yards long and in tapes $1/2$–$1 1/4$ in wide
by 36 yd long.

NEMA Standard for Untreated Mica Paper

The National Electrical Manufacturers Association (NEMA) has developed the following standard for untreated mica paper:

• FI 2, *Untreated Mica Paper Used for Electrical Insulation.* Covered are procedures, conditions, and methods of sampling and testing uncalcined and partially calcined or chemically treated mica paper.

Grades

Because of its fragility, mica paper is usually supplied in composite structures with glass cloth, polyester film and mat, and aramid (NOMEX) reinforcements. Typical composites are listed in Table 11-2.

Processing

Mica paper is receptive to impregnation, including vacuum pressure impregnation with insulating varnishes, and is also easily combined with reinforcing fabrics, films, mats, and other papers.

Properties

The outstanding attributes of mica paper, which it contributes to composites, are its arc and corona resistance and its high-temperature properties. Weak physical properties often make use difficult or dependent on reinforcement with other materials.

Property values of typical mica paper composites are shown in Table 11-2.

Electrical/Electronic Uses

Mica paper composites are used extensively for wrapping coils in electrical equipment, as cable wraps, and in other applications requiring flexibility and high-temperature service.

Other Papers and Boards

Commercially available papers made primarily from inorganic materials include the following:

• *Glass paper* is composed of glass microfibers. Because adhesion of these fibers is much lower than for other paper fibers, for adequate bonding the pulp requires special processing in an acid medium or with an inorganic binder. Glass paper is not dense, since calendering would crush the fibers. The outstanding feature of glass paper is its thermal stability up to 1,000°F (538°C). Other attributes include high thermal conductivity, low moisture absorption, and good chemical resistance. Manning Paper and Crane are suppliers. Paper is available in thicknesses of 0.005 inch up to 0.070 inch or more, and in widths up to 100 inches.

• *Alumina-silica paper* contains 51 percent alumina (Al_2O_3) and 47 percent silica (SiO_2). From 2–5 percent organic binder may be used to improve handling characteristics. This paper can be used continuously at temperatures to 2,300°F (1,260°C). It is attacked by strong alkalis, hydrofluoric and phosphoric acids. It is furnished in thicknesses from 0.010 inch paper to $^1/_4$ inch board. Carborundum is a producer.

  • *Ceramic paper*, composed of a proprietary blend of "unfired ceramic-like material" with synthetic fibers and high-temperature binders, is produced by Quin-T Corporation as a substitute for asbestos-base papers in many applications. It is supplied in 0.005–0.060 inch thicknesses in rolls or sheets up to 48 inches wide. It is also furnished in laminates with polyester film and electrical grade fiberglass or improved mechanical strength. Dielectric properties depend on impregnation with many kinds of varnishes and resins. These materials are used in many applications at temperatures up to 428°F (220°C), depending on their organic binders.

12

Tapes and Coated Fabrics

PRESSURE-SENSITIVE TAPES

INTRODUCTION

The original U.S. patents for pressure-sensitive adhesives were issued about 1845. Around 1900, pressure-sensitive tapes became widely used for first aid purposes. The first pressure-sensitive electrical tape, friction tape, was marketed in the late 1920s. During World War II, the shortage of natural rubber encouraged the development of synthetic rubber adhesives, and since that time, both products and markets for pressure-sensitive tapes have grown phenomenally as a wide variety of adhesives applied to every conceivable type of backing has become available.

Adhesive formulations are usually complex and proprietary with tape manufacturers, although some of the secretiveness has diminished in recent years, due in large part to the marketing efforts of basic material suppliers.

Standards and specifications have been developed covering virtually all commonly used pressure-sensitive tapes. These documents are summarized in this section including product descriptions, nominal sizes, and basic property requirements.

SUPPLIERS OF PRESSURE-SENSITIVE TAPES

Producers of pressure-sensitive tapes include the following:

- Acme
- Anchor Continental
- CHR Industries
- DeWal Industries
- 3M
- Mystic
- Nashua
- Norplex Oak
- Permacel Division of Fasson
- Plymouth Rubber
- Polyken Division, Kendall
- Taconic
- Tuck Industries

TECHNOLOGY

Pressure-sensitive tapes comprise three or four layers, each with its own specific function:

• A backing that provides the required physical, chemical, and dielectric properties. Commonly used backings include:

FILMS	INSULATION CLASS
– Polyester	130°C
– Polyester acrylic	130°C
– Polypropylene	85°C
– Polyimide	155°C, 180°C
– Polytetrafluoroethylene	155°C, 180°C
– Vinyl	80°C, 105°C

COMPOSITES	
– Polyester film/mat	130°C
– Polyester film/polyester nonwoven tape	130°C
– Polyester film/paper tape	130°C
– Polyester film/glass filament tape	130°C

OTHER BACKINGS	
– Glass cloth	155°C
– Silicone resin coated glass cloth	180°C
– Cotton cloth	105°C
– Acetate taffeta cloth	105°C
– Kraft paper	105°C
– Creped kraft paper	105°C
– Aramid nonwoven fabric	155°C
– Self-bonding silicone rubber	not available

• An adhesive layer provides the means of adhering the backing to the unit or to the back of the tape when layered. Adhesives for electrical grade tapes are activated either by pressure or by heat. Adhesives activated with water are not acceptable for dielectric applications. Adhesives activated with organic solvents are not widely used because of their inconvenience and fire and health hazards. Commonly used adhesives for electrical grade tapes include those based on natural rubber, synthetic rubber (SBR, butyl, Buna-N), silicone rubber, polyacrylate esters, and block copolymers (styrene-butadiene-styrene, styrene-isoprene-styrene) and combinations of these polymers. Rubber and silicone adhesives which are thermosetting crosslink with heat, thereby increasing solvent resistance and adhesion to the part. Thermosetting should be accomplished before impregnation with solvent-type varnishes.

• A release coating permits the tape to be unwound easily, an important feature where automated equipment is employed. The release coating is usually very thin, permitting the tape to adhere well under pressure to itself and other surfaces.

• A primer coat is sometimes used to bond the adhesive layer to the backing so that delamination does not occur.

Natural and Synthetic Rubber Adhesives

Natural rubber (polyisoprene) was the first elastomer used in pressure-sensitive tapes, and it is still widely used. Natural rubber must be masticated to break down the gel content and to reduce the molecular weight before it is suitable for pressure-sensitive applications. This process also facilitates solution coating at reduced levels of solvent concentration.

Natural rubber becomes softer on oxidation aging because of chain scission, while styrene-butadiene rubber (SBR) rubber becomes harder with added crosslinking. Hence, by blending, aging properties of both materials can be significantly enhanced.

Improved dielectric, heat aging, solvent-resistant, and ozone-resistant characteristics are obtained in adhesives based on butyl and polyisobutylene elastomers. Butyl rubber is a copolymer of isobutylene with a small amount (3 percent) of isoprene which contributes double bonds to the macromolecule, thereby permitting the polymer to be crosslinked, or vulcanized:

$$(CH_3)_2C \!=\! CH_2 \qquad\qquad CH_2 \!=\! C(CH_3)CH \!=\! CH_2$$

$$\text{isobutylene} \qquad\qquad\qquad \text{isoprene}$$

Other ingredients in natural, synthetic, and blended rubber pressure-sensitive adhesive formulations include:

• Stabilizers and antioxidants, such as zinc dibutyl dithiocarbamate (0.20 or less percent by weight) and butylated hydroxytoluene (less than 0.1 percent by weight), are effective.

• Tackifiers, such as the polyterpenes, terpenephenolics, phenol-formal dehyde resins, modified rosins and rosin esters, and hydrocarbon resins, often in combination, impart tack in concentrations of 20–70 percent. Below 20 percent, addition of these resins has little effect on tack. Tack develops as concentration increases from 20 percent until it typically reaches a maximum at about 70 percent. Tack then falls precipitously to zero at about 80 percent concentration.

• Plasticizers, such as polybutene, paraffinic oils, petrolatum, and certain phthalates with long aliphatic side chains (e.g., ditridecyl phthalate) are used in combination with tackifiers to control cohesive strength.

• Fillers and pigments, such as zinc oxide (to increase tack and cohesive strength); clays, magnesium oxide, silica, hydrated silicas, calcium silicate, silicoaluminates (to increase stiffness); mica, graphite, and talc (to improve chemical resistance and gas permeability); aluminum hydrate, lithophone, whiting, and carbon black (to increase tack); and calcium carbonate (to lower cost), are among the multitude of materials used.

• Curing agents, such as active brominated phenolic resins, sulfur or sulfur-donor compounds, and p-quinone dioxime (QDO) or dibenzyl p-quinone dioxime (DBQDO), promote crosslinking, but reduce tack and solubility in hydrocarbon solvents, thereby limiting use. QDO and DBQDO should be combined with oxidizing agents such as MnO_2, PbO_2, Pb_3O_4, or benzothiazyl disulfide in curing butyl rubber adhesives. Zinc oxide or other metallic oxides are necessary to develop satisfactory cure in combination with sulfur or sulfur-donor compounds (not used in electrical tapes since they cause corrosion of copper wires).

Thus, the possible combinations of ingredients in a recipe are virtually limitless. The three critical parameters in formulating a pressure-sensitive adhesive are tack, peel adhesion, and shear creep resistance (cohesive strength). Obtaining maximum value for one of these parameters is often at the expense of values for the other parameters. Different resins will produce different ratios for the optimum balance of properties. In blends of two elastomers, a resin may have a selective effect and tackify only one. Although the science of adhesion has developed significantly in recent years, there is still a large element of art in formulating effective pressure-sensitive adhesives.

Silicone Rubber Adhesives

Silicone adhesives have thermal Class 180 rating, the highest of all pressure-sensitive adhesives. They are based on a silicone gum and a silicone resin with the following structures:

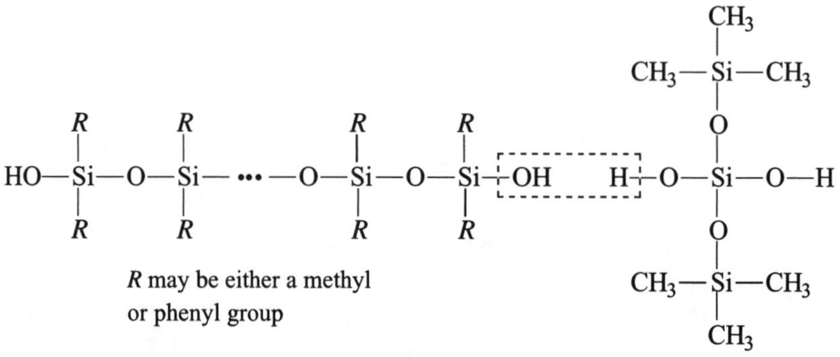

R may be either a methyl
or phenyl group

silicone gum   silicone resin

The adhesive is formed by condensation reaction in which water is removed. The resin acts as a tackifier, with the resin-to-gum ratio determining the tack and adhesion characteristics. The higher the resin content, the lower the tack. In high-resin-content adhesives, it is necessary to induce tack with heat and pressure. High-gum-content adhesives are naturally tacky at room temperature. Phenyl-modified silicones are noted for their unique combination of high viscosity, high peel strength, and high tack, and for their incompatibility with methyl-modified silicones. Adhesives may be made to cure through crosslinking by addition of benzoyl peroxide for curing at temperatures over 150°C (302°F), or by addition of 2, 4-dichlorobenzoyl peroxide for curing at 140°C (284°F) with higher cohesive strength resulting. Curing improves high-temperature shear properties at the expense of a slight loss of peel adhesion compared with adhesive that is noncuring.

Acrylic Adhesives

Acrylic adhesives have grown in use in recent years. They may be formulated from basic raw materials to be inherently pressure sensitive without the need for compounding, although incorporation of tackifying resins improves further tack and adhesion. Thermosetting acrylic systems may also be synthesized that are initially pressure sensitive, but still capable of a later thermosetting reaction. Adhesives are formed from alkyl acry-

**TABLE 12-1   Acrylic Monomers Used
in Pressure-Sensitive Adhesives**

Methylate acrylate	$CH_2\!=\!CH\!-\!COO\!-\!CH_3$
Ethyl acrylate	$CH_2\!=\!CH\!-\!COO\!-\!C_2H_5$
n-Butyl acrylate	$CH_2\!=\!CH\!-\!COO\!-\!C_4H_9$
2-Ethylhexyl acrylate	$CH_2\!=\!CH\!-\!COO\!-\!CH_2\!-\!\underset{\underset{\displaystyle C_2H_5}{\mid}}{CH}\!-\!C_4H_9$
Methyl methacrylate	$CH_2\!=\!\underset{\underset{\displaystyle CH_3}{\mid}}{CH}\!-\!COO\!-\!CH_3$
Acrylic acid	$CH_2\!=\!CH\!-\!COOH$

late and methacrylate monomers of 4–17 carbon atoms, as shown in Table 12-1. Commonly used are n-butyl acrylate and 2-ethylhexyl acrylate, which are copolymerized in emulsion or solution polymerization with acrylic acid and/or one or more of a multitude of other functional monomers capable of undergoing vinyl polymerization.

Although acrylic pressure-sensitive adhesives do not require compounding as do other elastomers, their adhesive properties may be varied by inclusion in formulations of other ingredients used with other elastomers.

Shear resistance of acrylic pressure-sensitive adhesives may be improved by introduction of divinyl benzene in formulas to promote crosslinking.

Acrylic pressure-sensitive tapes with polyester backing are rated thermal Class 130. With suitable backing, they are rated Class 155.

Block Copolymer Adhesives

Block copolymer thermoplastic rubber adhesives are of two principal classes:

1. Block copolymers in which the rubbery midblock is an unsaturated rubber. The two main types in this class are polystyrene-polybutadiene-polystyrene (S-B-S) and polystyrene-polyisoprene-polystyrene (S-I-S).

2. Block copolymers in which the elastomeric midblock is a saturated olefin rubber. Examples are polystyrene-poly(ethylene/butylene)-polystyrene (S-EB-S) and polystyrene-poly(ethylene/propylene)-polystyrene (S-EP-S).

In reality, there are many variations in molecular structure possible, and commercial products may have additional endblock and midblock resins. Formulations are usually proprietary.

Endblock compatible resins are used to:

• Improve the specific adhesion of the endblock phase

• Adjust melt viscosity

• Control the degree of softness or stiffness (stiffness increases with higher concentration of these resins)

Commonly used endblock compatible resins include polyaromatics (e.g., poly-alphamethylstyrene/vinyl toluene copolymer) and coumarone-indene polymers.

Midblock compatible resins are used:

- To impart pressure-sensitive tack
- To improve the specific adhesion of the midblock phase toward substrates
- To control the degree of softness or stiffness (softness increases with higher concentrations of these resins)
- As processing aids

Midblock compatible resins include rosin esters, polyterpenes, terpene phenolic resins, and aliphatic olefin derived resins.

Formulations may also include plasticizing oils, which perform the following functions:

- Decrease hardness
- Increase pressure-sensitive tack
- Improve low temperature flexibility
- Reduce melt and solution viscosity
- Lower cost

The plasticizer oil ideally should be insoluble in the endblock phase, but miscible with the midblock phase. Plasticizer oils include aromatic, naphthenic, and paraffinic hydrocarbons and mixtures of these. Polybutenes with molecular weight between 300 and 600 are effective plasticizers. Volatility loss by weight after 22 hours at 225°F (107°C) should be not over 2 percent.

S-I-S polymers are miscible with polyisoprene and natural rubber. S-B-S polymers are compatible with SBR and polybutadiene rubbers. It is, therefore, possible to blend these miscible elastomers to produce adhesives with enhanced properties. By incorporating low concentrations of thermoplastic rubber into unvulcanized natural rubber, polyisoprene, and SBR, cohesive strength will be improved, solution viscosity will be lowered, and adhesive strength may be higher. Including a small amount of a conventional rubber in a thermoplastic rubber may increase solution or melt viscosity somewhat without significant loss in other properties.

Other ingredients in thermoplastic rubber adhesive formulations may include:

- Fillers, such as clay and talc, to pigment the adhesive and to lower cost.
- Antioxidants, such as zinc dibutyl dithiocarbamate (R. T. Vanderbilt's Butyl Zimate™, Uniroyal's Butazate™, and Pennwalt's Butyl Ziram™), tri (nonylated phenyl) phosphite (Uniroyal's Polygard™), and 2-(4-hydroxy-3,5-tertiary-butyl anilino)4,6-bis(n-octyl thio)-1,3,5-triazine (Ciba-Geigy's Irganox 565™) which are effective stabilizers in concentrations of 5 percent or less against attack by oxygen.
- Antiozonants, such as nickel dibutyl dithiocarbamate (Du Pont's NBC™), dibutyl thiourea (Pennwalt's Pennzone B™), and Reichhold's Ozone Protector 80™, which are effective. Pennzone B should not be used in hot melt formulations as it accelerates crosslinking.

• Ultraviolet light inhibitors, such as 2,4-dihydroxy-benzophenone (GAF's Uvinul 400™), substituted benzotriazole (Ciba-Geigy's Tinuvin P™), octylphenyl salicylate (Eastman's Eastman OPS™), and octadecyl 3-(3,5-ditertiary-butyl-4-hydroxyphenyl) propionate (Ciba-Geigy's Irganox 1076™) which will provide adequate protection at a concentration of about 0.5 parts per hundred parts of elastomer.

PROCESSING

The traditional, although not the only, method for applying adhesives to backing is by solution or dispersion coating to a substrate. All natural and synthetic rubber adhesives are applied from solution in hydrocarbon solvents such as mineral spirits, hexane, heptane, naphtha, cyclohexane, toluene, and perchloroethylene.

Natural rubber is solvated[1] by granulating the rubber and slowly milling in the solvent. This process can be accelerated by addition on the mill of peptizers, such as pentachlorothiophenol. Before coating, it is customary to apply primers to some backings. On glass backings, epoxy silanes have been effective. Isocyanate primers are used for other textile substrates. Certain silicone solution coatings are suitable primers for silicone adhesives when cured for 15–30 seconds at 150°C (302°F) on polyester film.

Butyl polymers do not require milling, and may be solvated by gradual addition into a mixer with agitated solvent. Other ingredients are added after the butyl rubber is solvated. The blend may then be processed on mills or Bamburys if desired.

Unlike rubbers, acrylic adhesives may be dissolved in ethyl or butyl acetate, or in ketones, as well as in commonly used rubber solvents. Polymers exhibiting the best tack are n-butyl acrylate and 2-ethylhexyl acrylate.

In addition to solvent systems, acrylic aqueous dispersions are also widely used to make pressure-sensitive adhesives. Emulsion polymerization in water is effective in the presence of emulsifiers such as alkali salts of long-chain aliphatic, carboxylic or sulfonic acids, or sulfated ethylene oxide adducts. Alkali persulfates in small amounts (0.05 percent of monomer weight) may be used as initiators, for example, potassium peroxide disulfate. A typical reaction pH obtained with aqueous caustic is 7–8. The reaction is carried out by heat and pressure, staring at 70° to 80°C (158°–176°F) and ending at 80° to 90°C (176°–194°F). The viscosity of the solids content determines the completion of the reaction. Residual monomers are removed by passing a stream of nitrogen over the surface of the emulsion for 30 minutes before reducing the temperature. Following emulsion polymerization, the polymer is coagulated, washed, and typically solvated to 25–40 percent solids, although higher solids or solventless coatings are suitable for application by hot melt processing. Primers are not required before applying acrylic adhesives to substrates.

Silicone rubber pressure-sensitive adhesives are polymerized by condensation reaction in solution and processed with other ingredients in a manner similar to butyl rubber.

---

[1]*Solvation* is the adsorption of a microlayer or film of solvent on individual dispersed particles of a dispersion.

Block copolymer thermoplastic rubbers are prepared by polymerization in solvent using an alkyl lithium catalyst such as butyl lithium ($CH_3(CH_2)_3Li$). S-B-S and S-I-S types may be synthesized by sequential (linear) polymerization of all three blocks, or by sequential polymerization of two blocks followed by coupling with the third block, forming either a linear or multiarmed polymer, depending on the functionality of the coupling agent, usually an ester- or halogen-containing molecule. Crosslinking is physical, rather than chemical in nature as with conventional vulcanized rubber. This physical crosslinking can be "unlocked" by solvation or by heat and shearing, making the elastomer thermoplastic.

Typically, in block copolymer pressure-sensitive adhesives, the rubber midblock phase is the continuous phase and is present in higher concentration (70–85 percent). The higher the styrene content, the greater the hardness, the higher the tensile strength and modulus, and the shorter the elongation. As the styrene content is increased, the polymer becomes more like high-impact polystyrene.

Thermoplastic rubbers require compounding with certain resins, described earlier, to impact tack. Plasticizers and other ingredients may also be added to obtain the desired balance of properties. Premastication of the thermoplastic rubbers is not required since they dissolve readily in many solvents to make low-viscosity solutions.

Solvent selection, however, significantly affects the characteristics of the dried fiber. Two solubility parameters are involved: one for the endblock and one for the midblock. Then, with the addition of resin and plasticizer, third and fourth parameters are implicated. It is common practice, therefore, to employ solvent blends since it is unlikely that a single solvent would be equally effective on all ingredients. Natural and butyl rubber solvents are all candidates. Final film properties are usually determined by the solvent which evaporates last. It is preferable, therefore, that the solvent for the rubbery midblock be the last to evaporate for optimum adhesive properties. A typical solvent system (15 percent by weight of polymer) would contain 20 percent or more toluene blended with n-hexane for a low-viscosity matrix. Below 20 percent, toluene content viscosity increases sharply because n-hexane is a nonsolvent for the styrene endblock, while toluene is a solvent for both midblock and endblock. Thermoplastic rubbers require significantly less solvent than natural rubber. In fact, thermoplastic rubbers may be hot melt processed. Temperatures above 220°C (428°F) will decompose the rubber phase. The proper range for mixing and application is 120° to 180°C (248°–356°F). Mechanical mixing decreases melt viscosity significantly, permitting the adhesive to be more readily applied.

Many coating methods have been used successfully in the manufacture of pressure-sensitive tapes. The most widely used are reverse roll, calendering, and knife-over-roll. These methods vary in speed, and in viscosity and coating weight range capability. The function of all methods is to apply a uniform coating to a substrate and make it adhere. The *reverse roll coater*, in its simplest form, employs a metering roll, an applicator roll separated from the metering roll by a gap which determines the amount of coating to be applied, and a backing roll which carries the backing. The roll surfaces are going in opposite directions where they nearly contact each other and roll speeds are different. This method deposits a uniform coating thickness that is independent of backing thickness variations.

Calendering is best suited for large-volume manufacture of products which require a heavy adhesive coating. The prewarmed adhesive is fed onto the substrate as it is dragged through a narrow gap between heated rolls turning to oppose the direction of motion of the substrate. Alternatively, the adhesive sheet is first formed by the rolls, and then adhered in line to the substrate by passing between other rolls.

In *knife-over-roll* coating, the coating thickness is determined by the gap between the knife edge and the roll which passes the substrate under the knife. The adhesive is placed in a trough, or is contained as a free puddle by side dams only. There are many variations of this method and a similar system in which the knife contacts the substrate between two rolls (floating knife coater), and still another variation employing a driven "blanket" which carries the substrate (knife-over-blanket).

After coating, the solvent must be evaporated from the adhesive by convection or infrared drying, or a combination of both. Various methods and devices are employed to convey the web through the drying process. The simplest technique is the use of idler rolls which support the web as it is pulled through the dryer. It has been found that arranging the rolls (idler or driven) in an arch prevents edge curl. The web may also be conveyed by numerous smaller rollers, or on a belt. The newest technique uses heated air from nozzles both above and below the web to float the web as it is pulled through the dryer. Thus, the web is heated from both sides and no mechanical support is needed.

The design and control of the drying process are critical to the production of high-quality pressure-sensitive tape. Inadequate drying can produce blow holes or blisters in the adhesive film, and air with too high velocity can disturb the adhesive surface. Dryers usually have at least three temperature zones, with the lowest temperature zone first and the highest temperature zone last. Strict limits should be placed on the concentration of flammable vapor in any zone. Recycling of exhaust may be used to conserve energy provided these limits are not exceeded. Concentration limits depend on the flammability of the solvent or solvent mixtures used.

ADHESION TESTS FOR
PRESSURE-SENSITIVE TAPE

The property of tack lacks precise definition and has not yet been quantified by scientific test methods which give readily reproducible results. The thumb test is still regarded by some authorities as giving as good an indication of tack as any mechanical test. However, the thumb test applies only to tack between adhesive and skin. Mechanical tests in use include the following:

• A polished stainless steel ball 1.1 cm ($^7/_{16}$ inch) in diameter is rolled down an inclined track onto an upward-facing pressure-sensitive tape. The distance the ball rolls on the tape is an inverse measure of tack, but only between the adhesive and steel. Test conditions vary with the release height and how well the tape is held down. However, this method can be used for quality control in a plant where uniform conditions are maintained.

• Peel adhesion, commonly determined by applying the tape without pressure on a flat surface (which may be related to the end use for the tape). The tape is immediately peeled from the surface at 90 degrees at a rate of 30 centimeters per minute (12 inches per minute) employing a constant rate of extension tensile tester. A version of this method using a polished steel drum instead of a flat surface is described in ASTM D 1000, *Test Method of Testing Pressure-Sensitive Adhesive Coated Tapes Used for Electrical Insulation.*

• Peel adhesion, also determined by measuring the force necessary to strip a piece of tape adhered by a rubber covered steel roller at a 180 degree angle from a stainless steel plate at the same rate as in the above method. ASTM D 1000 describes this method also.

• Other tests for peel adhesion involve a rod of specified diameter and material which is brought into contact with the adhesive surface of a tape and then pulled away at a fixed rate using an Instron Tensile Tester or similar device.

• Shear creep resistance (cohesive strength), determined by applying a weight to the end of a strip of tape adhered to a specified vertical surface and measuring the amount of slippage in a given time, or the time for failure. In this test, it is essential that the tape make a 0 degree angle to vertical; otherwise, the tape will pull away from or toward the surface, or the weight will not apply an equal force across the tape.

These tests are suitable principally for comparing ostensibly identical tapes for a specific application. Variables which complicate adhesion tests include:

• Adhesive thickness
• Type and thickness of the backing
• Amount of pressure used to apply tape to test surface
• Type and condition of test surface
• Time of contact between adhesive and test surface before testing
• Test temperature
• Ambient humidity at time of test.

GRADES

Grades for all tapes covered by standards and specifications summarized in this section are commercially available.

ASTM STANDARDS FOR ELECTRICAL INSULATING PRESSURE-SENSITIVE TAPES

The American Society for Testing and Materials has developed the following standards for electrical insulating pressure-sensitive tapes:

SPECIFICATIONS

• Friction tape—D 69, D 4514
• Vinyl chloride backing—D 2301, D 3005

- Polyester film backing—D 2484
- Polytetrafluoroethylene backing—D 2686
- Glass cloth backing—D 2754
- Polyethylene backing—D 3006

TEST METHODS
- Pressure-sensitive tapes—D 1000
- Bondable silicone rubber—D 2148

The following pages provide a more detailed examination of these standards.

- D 69, *Friction Tape for General Use for Electrical Purposes*. This standard covers friction tape used for protecting and binding in place, insulation applied to joints of electrical wires and cables, and for other electrical and mechanical purposes. The fabric shall be thoroughly impregnated and evenly coated on both sides with the frictioning compound. This compound shall be a tacky adhesive compound containing practically no free sulfur or other substances that would have a deteriorating effect on copper or other metals, or on the fabric. The tape shall have the ability to stick to itself front to front, back to back, and back to front. The tape shall have a nominal thickness of 0.38 mm (0.015 inch) and shall be made in the following widths: 20, 25, 38, and 50 mm ($^3/_4$, 1, $1^1/_2$, and 2 inches). Each roll shall contain not less than 25 m (28 yards). Breakdown voltage shall not be less than 1,000 volts.

- D 1000, *Pressure-Sensitive Adhesive Coated Tapes for Electrical Insulation*. This standard describes methods of testing pressure-sensitive adhesive coated tapes used for electrical insulation. Two classes of tapes are identified:

    Class 1—Nonelastic backings such as:
    Paper, flat and creped
    Fabric, uncoated and coated
    Cellulose ester films
    Polyethylene terephthalate films
    Fluorocarbon polymer films
    Combinations of the above

    Class 2—Elastoplastic backings that are characterized by both high stretch and substantial recovery such as:
    Vinyl chloride and copolymers
    Vinylidiene chloride and copolymers
    Polyethylene

Table 12-2 lists the test methods.

- D 2148, *Bondable Silicone Rubber Tapes Used for Electrical Insulation*. This standard covers test methods for adhesion, bond strength, breakdown voltage, hardness, length, thickness, and width.

- D 2301, *Vinyl Chloride Plastic Pressure-Sensitive Electrical Insulating Tape*. This standard covers electrical insulating tape consisting of an elastomeric backing

**TABLE 12-2   Property Tests for Pressure-Sensitive
Adhesive Coated Tapes for Electrical Insulation
(ASTM D 1000)**

PROPERTY MEASURED
Accelerated aging (high-temperature tape only)
Adhesion strength
Breaking strength and elongation
Conditioning
Dielectric breakdown
Flagging
Flammability
High-humidity insulation resistance
Length of tape in a roll
Low-temperature elongation for Class 2 tapes
Low-temperature testing Class 2 tapes only
Oil resistance
Precautions
Resistance to accelerated aging (heat and moisture)
Resistance to penetration at elevated temperatures
Sampling
Shear strength after solvent immersion (Class 1 only)
Thermosetting properties

made from vinyl chloride plastic coated on one side with a pressure-sensitive adhesive. Two types of material are identified:

Type I—General-purpose tape, nominal thickness 0.007 inch (0.18 mm)

Type II—General-purpose tape, nominal thickness 0.010 inch (90.25 mm). Standard widths range from $^1/_4$ inch (6 mm) to 2 inches (50 mm). Standard lengths are 20 feet (6 m), 66 feet (20 m), and 108 feet (33 m).

The requirements for the tapes are shown in Table 12-3.

**TABLE 12-3   Requirements for Vinyl Chloride Plastic
Electrical Tapes
(ASTM D 2301)**

	TYPE I	TYPE II
Breaking strength, min, lb/in of width	18	26
(kg/25 mm of width)	(8.0)	(11.6)
Elongation, min, percent, average	125	150
Dielectric breakdown, min, volts, average		
Standard conditions	7,000	9,000
After conditioning 96 h at 23°C and 96 percent humidity	90% of original	90% of original
Adhesion, min, oz/in (g/25 mm) average		
On steel plate	22 (610)	22 (610)
On its own backing	18 (500)	18 (500)

Note: Tests are conducted in accordance with ASTM D 1000 procedures.

• D 2484, *Polyester Film Pressure-Sensitive Electrical Insulating Tape*. This standard covers electrical insulating tape consisting of polyethylene terephthalate film coated on one side with a pressure-sensitive adhesive. Five types of tape are identified:

   Type 1—General purpose tape, nominal thickness 0.0025 inch (0.06 mm) and coated with a thermoplastic adhesive that is not chemically heat reactive

   Type 2—General purpose tape, nominal thickness 0.0025 inch (0.06 mm) and coated with a heat-reactive thermosetting adhesive

   Type 3—General purpose tape, nominal thickness 0.0035 inch (0.09 mm) and coated with a heat-reactive thermosetting adhesive

   Type 4—General purpose thin tape, nominal thickness 0.0010 inch (0.025 mm) and coated with a heat-reactive thermosetting adhesive

   Type 5—Oil-resistant tape, nominal thickness 0.0025 inch (0.06 mm) and coated with a heat-reactive thermosetting adhesive. Standard widths range from $^1/_4$ inch (6 mm) to 2 inches (50 mm). Standard lengths are 108 feet (33 m), 164 feet (50 m), 180 feet (55 m), and 216 feet (66 m). Requirements for the tapes are shown in Table 12-4.

• D 2686, *Tetrafluoroethylene-Backed Pressure-Sensitive Electrical Insulating Tape*. This standard covers electrical insulating tape consisting of a polytetrafluoroethylene (PTFE) film coated on one side with a pressure-sensitive adhesive. Three types of tape are identified:

   Type 1—Standard backing coated with a silicone adhesive:
   Class A—0.002 inch (0.050 mm) backing thickness
   Class B—0.005 inch (0.125 mm) backing thickness

   Type II—Bondable backing tape coated with a silicone adhesive:
   Class A—0.002 inch (0.050 mm) backing thickness
   Class B—0.005 inch (0.125 mm) backing thickness

TABLE 12-4   Requirements for Polyester Plastic Pressure-Sensitive Tapes (ASTM D 2484)

	TYPE 1	TYPE 2	TYPE 3	TYPE 4	TYPE 5
Nominal thickness, inches (mm)	0.0025	0.0025	0.0035	0.0010	0.0025
	(0.060)	(0.060)	(0.090)	(0.025)	(0.060)
Breaking strength, min, lb/in of width	20	20	35	10	20
(N/m)	(3.5)	(3.5)	(6.1)	(1.7)	(3.5)
Dielectric breakdown, min, volts, average					
Standard conditions	4,500	4,500	6,000	3,000	4,500
After immersion in water	4,000	4,000	5,500	2,500	4,000
Adhesion, min, oz/in (N/m)					
On steel plate	20 (219)	35 (383)	35 (383)	20 (219)	20 (219)
On its own backing	15 (164)	15 (164)	15 (164)	10 (109)	15 (164)
On backing after thermosetting	—	35 (383)	35 (383)	35 (383)	30 (328)

Note: Tests are conducted in accordance with ASTM D 1000 and ASTM D 2305 procedures.

**TABLE 12-5   Requirements for PTFE Pressure-Sensitive
Adhesive Tapes
(ASTM D 2686)**

	CLASS A	CLASS B
Breaking strength, min, lb/in of width	10	20
(kN/m)	(1.7)	(3.5)
Elongation, min, percent	75	150
Adhesion, min, oz/in (N/m)		
On steel plate	15 (164)	20 (219)
On its own backing (Type II only)	15 (164)	15 (164)
Dielectric breakdown, min, volts, average		
Standard conditions	6,000	10,000
After immersion in water	6,000	10,000

Note: Tests are conducted in accordance with ASTM D 1000 procedures.

Type III—Standard backing coated with a special oil-resistant adhesive: Class A only—0.002 inch (0.050 mm) backing thickness.

Standard widths range from $^1/_4$ inch (6 mm) to 2 inches (50 mm). Standard lengths are 108 feet (33 m), 164 feet (50 m), 180 feet (55 m), and 216 feet (66 m). Requirements for the tapes are shown in Table 12-5.

• D 2754, *High-Temperature Glass Cloth Pressure-Sensitive Electrical Tape*. This standard covers high-temperature electrical insulating tape consisting of glass cloth coated on one side with a pressure-sensitive adhesive. Two types of tape are identified:

Type I—Woven glass cloth, silicone adhesive
Class A—nominal thickness 0.007 in (0.18 mm)
Class B—nominal thickness 0.010 in (0.25 mm)
Type II—Woven glass cloth impregnated with polytetrafluoroethylene, silicone adhesive, nominal thickness 0.005 in (0.13 mm).

Standard widths range from $^1/_4$ inch (6 mm) to 2 inches (50 mm). Standard lengths are 20 feet (6 m), 66 feet (20 m), and 108 feet (33 m). Requirements for the tapes are shown in Table 12-6.

• D 3005, *Low-Temperature Resistant Vinyl Chloride Plastic Pressure-Sensitive Electrical Insulating Tape*. This standard covers an electrical insulating tape for use at low temperature down to approximately 0°F (−18°C). The tape consists of a backing of vinyl chloride plastic, coated on one side with pressure-sensitive adhesive. Two types are identified:

Type I—nominal thickness 0.007 inch (0.178 mm)
Type II—nominal thickness 0.085 inch (0.216 mm)

Standard widths range from $^1/_4$ inch (6 mm) to 2 inches (50 mm). Standard lengths are 20 feet (6 m), 66 feet (20 m), and 108 feet (33 m). Requirements for the tapes are shown in Table 12-7.

TABLE 12-6   Requirements for High-Temperature Glass Cloth Pressure-Sensitive Tapes (ASTM D 2754)

	TYPE I		TYPE II
	CLASS A	CLASS B	
Breaking strength, min, lb/in	120	120	50
(kg/25 mm)	55	55	23
Dielectric breakdown, min, volts	2,000	2,500	3,000
Adhesion min, oz/in (g/25 mm)			
On steel plate	25 (710)	25 (710)	20 (570)
On its own backing	15 (425)	15 (425)	—

Note: Tests are conducted in accordance with ASTM D 1000 procedures.

TABLE 12-7   Requirements for Low-Temperature Vinyl Chloride Plastic Electrical Tape (ASTM D 3005)

	TYPE I	TYPE II
Breaking strength, min, lb/in	15	17
(kN/m)	(2.6)	(2.9)
Elongation, min, percent	150	150
Dielectric breakdown, min, volts		
Standard conditions	7,000	8,500
After conditioning 96 h at 23°C and 96 percent humidity	90% of dry	90% of dry
Adhesion, min, oz/inch (N/m)		
On steel plate	16 (175)	16 (175)
On its own backing	16 (175)	16 (175)
On its own backing at –7°C (19°F)	16 (175)	16 (175)

Note: Tests are conducted in accordance with ASTM D 1000 procedures.

TABLE 12-8   Requirements for Polyethylene Plastic Electrical Tape (ASTM D 3006)

Breaking strength, min, lb/in (kN/m)	16 (2.8)
Elongation, min, percent	150
Dielectric breakdown, min, volts	
Standard conditions	10,000
After conditioning 96 h at 23°C and 96 percent humidity	90% of standard
Adhesion, min, oz/in (N/m)	
On steel plate	16 (175)
On its own backing	16 (175)
On its own backing at –7°C (19°F)	16 (175)

Note: Tests are conducted in accordance with ASTM D 1000 procedures.

• D 3006, *Polyethylene Plastic Pressure-Sensitive Electrical Insulating Tape*. This standard covers electrical insulating tape consisting of a backing made from polyethylene plastic, coated on one side with a pressure-sensitive adhesive. Standard widths range from $^1/_4$ inch (6 mm) to 2 inches (50 mm). Standard lengths are 20 feet (6 m), 66 feet (20 m), and 108 feet (33 m). Thickness is 0.0095 inch (0.241 mm). Requirements for this tape are shown in Table 12-8.

• D 4514, *Friction Tape*. This specification presents requirements for adhesion, aged adhesion, breaking strength, breakdown voltage, parallelism, pinholes, tackiness, and thickness.

FEDERAL SPECIFICATIONS FOR ELECTRICAL INSULATING PRESSURE-SENSITIVE TAPES

The following Federal specifications have been developed for pressure-sensitive tapes:

• Nonelastic backings—MIL-I-15126
• Glass fiber backing—MIL-I-19166, MIL-I-22444
• Vinyl chloride—HH-I-595
• Polyethylene backing—MIL-I-15126
• Polyester film backing—MIL-I-15126
• Polytetrafluoroethylene backing—MIL-I-23594
• Silicone rubber tape—MIL-I-46852
• Unspecified backing—MIL-I-24391

The following pages provide a more detailed examination of these standards.

• HH-I-595, *Insulation Tape, Electrical, Pressure-Sensitive Adhesive, Plastic*. This specification was approved by the Commissioner, Federal Supply Service, General Services Administration, for the use of all Federal agencies. Covered are requirements of 80°C (176°F) polyvinyl chloride or copolymer of vinyl chloride and vinyl acetate, flame-retardant, cold- and weather-resistant, pressure-sensitive, electrical insulating tape. Standard widths of the tape are $^1/_2$, $^3/_4$, or 1 inch. The thickness of the tape shall be 0.007 inch. The length of tape in a roll shall be either 66 feet or 108 feet.

• Test procedures are described with reference, where appropriate, to applicable ASTM and UL standards. Property requirements include the following:

       Flame retardant, cold and weather resistant
       Low temperature
       Adhesion strength to backing at 19.5°F (−7.0°C)
       Unwind force at 19.5°F (−7.0°C)
       Static elongation at 19.5°F (−7.0°C)
       Breaking strength

• MIL-I-15126, *Insulation Tape, Electrical, Pressure-Sensitive Adhesive and Pressure-Sensitive Thermosetting Adhesive*. This specification has been approved by the Department of Defense and is mandatory for use by the Departments of the Army, the Navy, and the Air Force. Covered are requirements for pressure-sensitive adhesive and pressure-sensitive thermosetting adhesive electrical insulation tapes intended for use in construction and repair of electrical and electronic equipment. The following types of tape are identified:

> Type AFT—Acetate film backing, thermosetting adhesive
> Type ACT—Acetate cloth backing, thermosetting adhesive
> Type CFT—Cotton fabric backing, thermosetting adhesive
> Type GFT—Glass fabric backing, thermosetting adhesive
> Type PCT—Paper backing-crepe, thermosetting adhesive
> Type PFT—Paper backing-flat, thermosetting adhesive
> Type EF-7—Polyethylene backing—7 mils thick tape
> Type EF-9—Polyethylene backing—9 mils thick tape
> Type EF-20—Polyethylene backing—20 mils thick tape
> Type MFT-2.5—Polyethylene terephthalate film backing, thermosetting adhesive—2.5 mils thick tape
> Type MF-2.5—Polyethylene terephthalate film backing—2.5 mils thick tape
> Type MFT-3.5—Polyethylene terephthalate film backing, thermosetting adhesive—3.5 mils thick tape

The adhesive used in types EF-7, EF-9, EF-20, and MF-2.5 shall be pressure sensitive, and shall not require heat, moisture, or other special manner of preparation. The adhesive employed in other types shall be a pressure-sensitive thermosetting, or heat-curing adhesive. Nominal widths shall be in the range of $^1/_4$–$1^1/_2$ inches, in $^1/_8$ inch increments. Rolls shall contain up to 72 yards. Thickness and other requirements are shown in Table 12-9.

• MIL-I-19166, *Insulation Tape, Electrical, High-Temperature, Glass Fiber, Pressure-Sensitive.* This specification has been approved by the Department of Defense and is mandatory for use by the Departments of the Army, the Navy, and the Air Force. Covered is glass fiber, silicone adhesive, pressure-sensitive tape for construction and repair of electrical equipment operating at high temperatures. The tape shall consist of glass fiber textile fabricated from continuous filament yarn, coated on one side with a silicone pressure-sensitive adhesive. The tape shall require no heat, moisture, or other preparation prior to or subsequent to application. The tape shall be natural in color, 0.007 or 0.010 inch thick, and in widths of 0.50, 0.75, and 1.00 inch. Rolls shall be 36 yards in length. Dielectric breakdown shall be as follows:

	NOMINAL THICKNESS	
CONDITIONING	0.010 in (volts, min)	0.007 in (volts, min)
96 h at 23°C and 0% humidity	4,000	2,000
96 h at 23°C and 96% humidity	3,000	1,500

**TABLE 12-9   Requirements for Pressure-Sensitive
Adhesive and Pressure-Sensitive Thermosetting
Adhesive Tapes
(MIL-I-15126)**

TYPE	NOMINAL THICKNESS (inches)	TENSILE STRENGTH CONDITION 96/23/50* (lb/in width, min)	DIELECTRIC BREAKDOWN CONDITION 96/23/50* (volts, min)	CONDITION 96/23/96** (volts, min)
AFT	0.0035	15	4,500	3,500
ACT	0.0080	30	1,500	1,000
CFT	0.0105	40	1,000	500
GFT	0.0070	120	1,000	500
PCT	0.0105	15	1,000	500
PFT	0.0060	30	1,000	500
EF-7	0.0070	12	9,500	9,500
EF-9	0.0090	15	12,000	11,000
EF-20	0.020	25	15,000	12,000
MFT-2.5	0.0025	20	4,500	4,500
MF-2.5	0.0025	20	4,500	4,500
MFT-3.5	0.0035	30	7,000	7,000

*Conditioning for 96 h at 23°C and relative humidity 50 percent.
**Conditioning for 96 h at 23°C and relative humidity 96 percent.

• MIL-I-22444, *Insulation Tape, Electrical, Self-Bonding, Silicone Rubber Treated Bias Weave or Sinusoidal Weave Glass, Cable Splicing, Naval Shipboard.* The silicone rubber compound may be applied to the glass backing tape by an extrusion process and sealing the tape edge by extending the rubber compound $^1/_{32}$ of an inch beyond each edge of the glass tape. A separator tape shall be used between adjacent layers on the roll of tape to prevent bonding. Standard widths shall be $^1/_2$, $^3/_4$, 1, and $1^1/_2$ inches. Thickness shall be in the range of 0.015–0.050 inch. Rolls shall contain 20 yards of tape. Average dielectric strength after being conditioned for 96 hours at 23°C (73°F) and 96 percent relative humidity shall be 500 volts per mil.

• MIL-I-23594, *Insulation Tape, Electrical, High Temperature, Polytetrafluoroethylene, Pressure-Sensitive.* This specification is mandatory for use by all departments and agencies of the Department of Defense. Two types of tape are identified:

Type I—Smooth backing

Type II—Treated backing (for insulation varnish)

Backing shall be virgin fused polytetrafluoroethylene resin. Type II backing is treated to accommodate an insulation varnish. Backing thickness shall be at least 1.5 mil. The adhesive shall be a pressure-sensitive silicone polymer and shall not require heat treatment or other special means of preparation prior to application. Standard widths are $^1/_2$, 1, $1^1/_2$, and 2 inches. Standard roll length is 36 yards. This material is intended for use as splicing tape for electrical systems of underwater ordnance equipment, aircraft, missile, and other weapons systems. Tensile strength and dielectric breakdown values shall be in accordance with Table 12-10.

**TABLE 12-10   Tensile Strength and Dielectric Breakdown
Values for Polytetrafluoroethylene Pressure-Sensitive Tape
(MIL-I-23594)**

BACKING THICKNESS (mils)	TENSILE STRENGTH (lb/in width, minimum) TYPES I AND II	DIELECTRIC BREAKDOWN (volts, minimum) TYPE I	TYPE II
1.5–2.5	7.0	4,200	3,800
2.6–3.5	12.0	7,200	6,500
3.6–4.5	17.0	10,300	9,000
4.6–5.5	22.0	12,700	11,500
5.6–6.5	27.0	15,400	14,000
6.6–7.5	32.0	18,000	16,500

• MIL-I-24391, *Insulation Tape, Electrical, Pressure-Sensitive.* This specification is mandatory for use by all departments and agencies of the Department of Defense. Covered are requirements for one type of plastic, electric, pressure-sensitive, self-extinguishing insulating tape for use at any ambient temperature condition between 10°F (–12.2°C) and 150°F (65.6°C). The adhesive shall be water insoluble and shall require no solvent, heat, or any other preparation prior to application. Standard widths are $^1/_2$, $^3/_4$, or 1 inch. Tape thickness shall be 0.0085 inch and length of tape in a roll shall be a minimum of 36 yards. Minimum average dielectric breakdown after water immersion shall be 7,000 volts and after exposure for 4 hours at 158°F (70°C) shall be 6,000 volts. After aging for seven days at 150°F (65.6°C) and 90 percent relative humidity, adhesion shall be at least 75 percent of original value.

• MIL-I-46852, *Insulation Tape, Electrical, Self-Adhering Unsupported Silicone Rubber.* This specification is approved for use by all departments and agencies of the Department of Defense. Covered are requirements for an unsupported silicone rubber tape which will adhere to itself at room temperature. The tape is suitable for use at temperatures from –130° to +500°F (–90° to +260°C). The tape is intended to be used as electrical insulation and as a thermal protection barrier for components subjected to missile blast environments. The tape shall be furnished in two types:

  Type I—Rectangular cross-section
  Type II—Triangular cross-section with colored guidelines.

Dimensions for Type I shall be as follows:

  Width—0.50 inch 1.5 inches in 0.25 inch increments
  Thickness—0.010, 0.020, 0.030 inch
  Roll length—12 yards

Dimensions for Type II shall be as follows:

  Width—1, 1.25, 1.5 inches
  Thickness—0.020–0.080 inch in 0.010 inch increments
  Roll length—12 yards

Property requirements shall be in accordance with Table 12-11.

**TABLE 12-11   Requirements for Unsupported,
Self-Adhering Silicone Rubber Tape
(MIL-I-46852)**

PROPERTY	VALUE	TEST METHOD
Tensile strength, minimum lb/in^2	700	ASTM D 119
Elongation percent, minimum	300	ASTM D 119
Water absorption percent by weight, maximum	3	Fed. Test 601 Method 6215
Dielectric strength, volt/mil, minimum		
0.010, 0.020, 0.030 in thick	400	ASTM D 119
0.040 in thick	300	
0.050 in thick	275	
0.060 in thick	250	
0.080 in thick	200	
Volume resistivity $\Omega \cdot$ cm minimum	3×10^{14}	ASTM D 1458
Bond strength, 1 in width, minimum pounds	2	MIL-I-46852

UL STANDARD FOR INSULATING TAPES

Underwriters Laboratories has developed the following standard for insulating tapes:

- UL 510, *Electrical Insulating Tapes*

PROPERTIES

The minimum requirements for all types of pressure-sensitive tapes are shown in the accompanying ASTM Standards and Federal specifications. Features of tapes with different elastomer bases are as follows:

- Natural rubber provides excellent adhesion and can be made to crosslink readily at varnish curing temperatures with improved chemical resistance. Premastication is required.
- Butyl and polyisoprene elastomers have inherently better dielectric, heat aging, solvent-resistant, and ozone-resistant properties than natural rubber. These adhesives may also be made to crosslink. Premastication is not required.
- Acrylic adhesives do not require premastication, tackifier resins, or fillers. These adhesives are rated thermal Class 155, versus Class 130 for natural and butyl rubbers. Oil and oxidation resistance is superior to rubbers, but adhesion is lower. Thermosetting versions can be made by introduction of functional groups.
- Silicone adhesives remain effective at 180°C (356°F), the highest of all commonly used adhesives. They are inorganic, and if burned, leave a nonconducting residue. Low-temperature flexibility is superior to all other adhesives, but adhesive properties are somewhat lower. Silicone adhesives may be applied to etched Teflon

films and unetched polyolefin films to which organic elastomers will not adhere adequately.

    • Block copolymer adhesives do not require premastication. They dissolve readily, and because they are thermoplastic, they can be coated at 100 percent or high solids content using hot melt techniques.

TAPE SELECTION

In selecting a pressure-sensitive tape for a specific application, design engineers should consider the operating conditions to which the tape will be exposed:

- The required dielectric breakdown strength
- The operating and environmental temperature range
- Space limitations
- Exposure to solvents and other chemicals
- Required physical properties
- Conformance to customer, industry, or military specifications or standards (there is no UL 94 flammability rating for dielectric tapes)
- The surface to which the tape must adhere
- Esthetic requirements (color, printability)

ELECTRICAL/ELECTRONIC USES

Pressure-sensitive tapes are used to wrap and to hold down wires and wire harnesses, and for terminating and splicing wires and cables. Bus bars are wrapped with pressure-sensitive tapes for insulation. High-temperature applications are served by silicone rubber pressure-sensitive tapes on fluoropolymer or polyimide film backing.

    Thermosetting pressure-sensitive tapes are used in the manufacture of electrical equipment subsequently treated with varnishes and baked at varnish-curing temperatures.

    Polyester-backed pressure-sensitive tapes are used to mask printed circuit boards during plating operations.

DEVELOPMENT PROGRAMS

Because solvent systems are coming under increasing regulation aimed at protecting the environment and worker health and safety, development programs are principally focused on emulsion-based and hot melt adhesives. Acrylic and block copolymer adhesives are inherently better suited for these new systems than are rubbers, but rubbers are still unsurpassed in tack and low cost, even though they are the most energy intensive to process.

MARKET TRENDS

Pressure-sensitive tapes are firmly entrenched in electrical equipment manufacturing processes and in wire and cable wrapping applications. A wide range of products is offered to users who are cost and quality conscious, and are less concerned with adhesive recipes. While it is likely that rubber-base adhesives will lose market share to newer systems, the trend will be gradual.

NONADHESIVE TAPES AND COATED FABRICS

INTRODUCTION

Yellow and black varnished cambrics based on oleoresinous varnishes were among the earliest materials used to insulate electrical equipment. Although the largest volume use, as a tape wrap for power cables, has largely yielded most of its market share to extruded insulations, there has been an expansion of coating and substrate types for other electrical and electronic applications. Newer materials covered in this chapter include polyester, silicone, and polytetrafluoroethylene coatings; glass and dacron substrates; and synthetic and silicone rubber tapes.

SUPPLIERS OF NONADHESIVE TAPES AND COATED FABRICS

Producers of nonadhesive tapes and coated fabrics include the following:

- Acme
- CHR Industries
- Dewal Industries
- 3M
- Norplex Oak
- Permacel Division of Fasson

TECHNOLOGY

Electrical insulating coated fabrics and tapes, and unsupported tapes, come in a variety of types, among which the following are in common use:

- Black and yellow, straight-cut and bias-cut varnished cotton fabric and tape (ASTM D 373, MIL-I-3374)
- Silicone-varnished glass cloth and tape (ASTM D 1495, MIL-I-17205)
- Fully cured silicone rubber-coated glass fabric and tapes (ASTM D 1931)
- Varnished glass-polyester cloth and tapes (ASTM D 2400)

- Oleoresinous varnish-coated glass fabrics and tapes (ASTM D 3949, MIL-I-17205)
   - Nonmetallic conducting and electrically insulating rubber tapes (ASTM D 4325)
   - Polytetrafluoroethylene-coated glass fabric and tapes (ASTM D 4969, MIL-I-18746)
   - Natural and synthetic rubber tapes (HH-I-553)
   - Filler-type, flameproof insulation tape (MIL-I-17695)
   - Semi-cured thermosetting resin-treated glass tape (MIL-I-24178).

The above ASTM standards and Federal specifications are summarized later in this chapter.

PROCESSING

Varnished fabrics are made by passing fabric through a tank containing varnish or resin, after which the film is formed and cured in a vertical tower with temperature controlled to suit the varnish or resin system employed. The desired film thickness is built up in several passes (usually two–four) through the tower. Typically, a machine consists of three or four adjacent units to permit the fabric to travel continuously through successive tanks and heating chambers.

Alternatively, the oven may be horizontal with the fabric passing through it on edge with edges held by tenters. With this method, much larger ovens may be used and significantly faster processing is made possible. Varnish or resin must be applied in such a way that it does not flow excessively or form droplets which would cure to make film imperfections. Top and bottom of rolls of material are reversed after each pass to prevent film from becoming wedge shaped. This process is best suited for long runs, and was used extensively when varnished cambric tape was the principal insulation for power cables.

Varnishes and resin systems are similar to those discussed in Chapter 8, *Insulating Coatings and Impregnants*. Substrates include cotton cloth, glass cloth, dacron-glass cloth, canvas, polyester mat, aramid mat, and composites of these. Kraft paper is also varnished for electrical applications.

In processing solvent systems, vapors must be burned or recycled so that emissions are in compliance with EPA, OSHA, and local government standards and regulations. See "Federal Air Pollution Program" (p. 517) of Chapter 19, *Government Activities*.

Rubber-coated glass tapes are made by methods described in "Pressure-Sensitive Tapes" section of this chapter. Extruded rubber processing is discussed in Chapter 5, *Extrusion Compounds*.

GRADES

Grades for all materials covered by standards and specifications summarized in this section are commercially available.

ASTM STANDARDS FOR NONADHESIVE TAPES AND COATED FABRICS

The American Society for Testing and Materials has developed the following standards for nonadhesive tapes and coated fabrics. For a listing of product types and related standards, see "Technology" (p. 356).

• D 295, *Varnished Cotton Fabrics Used for Electrical Insulation.* This standard describes methods for testing varnished cotton fabrics and varnished cotton fabric tapes to be used as electrical insulation. Table 12-12 lists the test methods applicable to both straight-cut and bias-cut materials.

TABLE 12-12   Property Tests for Varnished Cotton Fabrics (ASTM D 295)

PROPERTY MEASURED	ASTM METHOD
Breaking strength	D 295
Conditioning	D 295
Dielectric breakdown voltage	D 149
Dielectric breakdown voltage under elongation	D 149
Dissipation factor and permittivity	D 150
Elongation	D 295
Resistance to oil	D 92
Selection of test specimens	D 295
Tear resistance	D 689
Thickness	D 374
Thread count	D 295
Volume resistance	D 257
Weight	D 295

• D 373, *Black and Yellow Straight-Cut and Bias-Cut Varnished Cotton Cloth and Tape for Electrical Insulation.* Breaking strength and elongation requirements are specified. Dielectric breakdown requirements are shown in Table 12-13.

• D 902, *Flexible Resin-Coated Glass Fabrics and Glass Fabric Tapes Used for Electrical Insulation.* This standard describes methods of testing resin-coated glass fabrics and glass fabric tapes to be used as electrical insulation. Table 12-14 lists the following test methods.

• D 1458, *Fully Cured Silicone Rubber-Coated Glass Fabric and Tapes for Electrical Insulation.* This standard describes methods of testing fully cured silicone rubber-coated glass fabric and tapes to be used for electrical insulation. Table 12-15 lists the following test methods.

• D 1459, *Silicone Varnished Glass Cloth and Tape for Electrical Insulation.* This standard covers clear silicone varnished glass fabrics in the form of sheets, rolls, and tapes for use as electrical insulation. These materials shall conform to Class H (180) requirements. Two types of materials are identified:

TABLE 12-13   Dielectric Breakdown Requirements for
Varnished Cotton Cloth
(ASTM D 373)

NOMINAL THICKNESS, mm (mils)	CUT	BLACK			YELLOW		
		UNSTRESSED	UNDER 6% ELONGATION	AFTER HOT OIL	UNSTRESSED	UNDER 6% ELONGATION	AFTER HOT OIL
0.18 (7)	bias	7.7	4.2	7.7	7.0	3.2	7.0
	straight	7.7	—	—	6.3	—	—
0.25 (10)	bias	11.0	9.0	11.0	11.0	7.0	11.0
	straight	11.0	—	—	10.0	—	—
0.30 (12)	bias	13.8	11.0	13.8	13.2	10.0	13.2
	straight	13.8	—	—	12.0	—	—

Note: Tests are conducted in accordance with ASTM D 295 procedures.

(Reprinted with permission from the *Annual Book of ASTM Standards.* Copyright, ASTM, 1916 Race Street, Philadelphia, PA 19103.)

Type A—For use where maximum electrical properties are required
Type B—For use where maximum physical properties are required.

Nominal thicknesses range from 3 to 25 mils (0.076–0.635 mm).

Average breaking strength requirements range from 70 pounds per inch of width for 3 mil material to 300 pounds per inch of width for 20 mil Type A material (12.25 newtons per mm of width for 0.076 mm thick material to 52.54 newtons per mm of width for 0.508 mm thick material). For Type B material, the range is 100 pounds per inch of width for 5 mil thick material to 350 pounds per inch of width for 25 mil thick material (17.51 newtons per mm of width for 0.127 mm thick material to 61.29 newtons per mm of width for 0.635 mm thick material). Short-time dielectric strength val-

TABLE 12-14   Property Tests for Varnished Glass Fabrics
(ASTM D 902)

PROPERTY MEASURED	TEST METHOD
Breaking strength	D 828
Conditioning	D 902
Dielectric breakdown voltage and dielectric strength	D 149
	D 295
Dissipation factor and relative permittivity	D 150
	E 104
Effect of elevated temperature	D 902
Resistance to oil	D 902
Sampling	D 902
Thickness	D 374
Thread count	D 902
Weight	D 902
Weight loss at elevated temperature	D 902

TABLE 12-15   Property Tests for Fully Cured Silicone
Rubber-Coated Glass Fabrics
(ASTM D 1458)

PROPERTY MEASURED	ASTM METHOD
Sampling	D 1458
Conditioning	D 1458
Thickness	D 1458
Breaking strength	D 1458
Breaking strength after creasing	D 1458
Dielectric breakdown	D 149
	D 295
	D 618
Dielectric proof-voltage	D 1389
Dissipation factor and relative permittivity	D 150
Volume resistivity	D 257
Weight	D 1458
Thread count	D 1458

ues using $^1/_4$ inch (6.4 mm) electrodes shall be in the range of 1,300 volts per mil for 3 mil thick material to 400 volts per mil for 20 mil thick material (51.2 kilovolts per mm for 0.508 mm thick material). The range for Type B material shall be 1,000 volts per mil for 5 mil thick material to 300 volts per mil for 25 mil thick material (39.4 kilovolts per mm for 0.127 mm thick material to 11.8 kilovolts per mm for 0.635 mm thick material). Tests are conducted in accordance with ASTM D 902 procedures.

• D 1931, *Fully Cured Silicone Rubber-Coated Glass Fabric and Tapes for Electrical Insulation.* This standard covers fully cured silicone rubber-coated glass fabric in sheet form, full-width rolls, and tapes to be used as electrical insulation. Physical and electrical requirements are listed in Table 12-16.

• D 2400, *Varnished Glass-Polyester Cloth Used for Electrical Insulation.* This standard applies to black and yellow varnished woven cloth and tape having as a base fabric poly(ethylene terephthalate) yarns in the warp direction and glass yarns in the filler direction. The base fabric shall have a nominal thickness of 0.004 inch (0.10 mm). The varnish shall be of the oleoresinous type. Dielectric breakdown strength values are shown in Table 12-17.

• D 3949, *Coated Glass Fabrics Used for Electrical Insulation.* This standard covers woven glass fabric coated with fully cured oleoresinous varnish, epoxy resin, polyurethane resin, or polyester resin. This standard has been developed for materials using fabrics woven from relatively coarse-filament yarns, and in some cases from relatively open weave constructions. The following types of material are identified:

Type 1—Black oleoresinous varnish
Type 2—Yellow oleoresinous varnish
Type 3—Epoxy resin
Type 4—Polyurethane (isocyanate) resin
Type 5—Polyester resin.

Nominal thicknesses and dielectric requirements are shown in Table 12-18.

TABLE 12-16   Physical and Electrical Requirements for Silicone Rubber-Coated Glass Fabric and Tapes (ASTM D 1931)

Physical:				
Nominal thickness, mm	0.12	0.18	0.25	0.38
in	0.005	0.007	0.010	0.015
Thickness tolerance, mm	±0.01	±0.03	±0.03	±0.03
in	±0.0005	±0.001	±0.001	±0.001
Nominal fabric thickness, mm	0.08	0.10	0.10	0.18
in	0.003	0.004	0.004	0.007
Breaking strength (warp), min, avg:				
kN/m	12.3	21.9	21.9	43.8
lbf/in	70	125	125	250
Breaking strength (creased), min, avg:				
kN/m	3.5	15.8	15.8	35.0
lbf/in	20	90	90	200
Weight, kg/100 m^2	16 ± 1.6	26 ± 2.6	37 ± 3.7	49 ± 4.9
lb/100 yd^2	30 ± 3	48 ± 4.8	68 ± 6.8	90 ± 9
Electrical:				
Dielectric strength, kV/mm (V/mil), min, avg:				
Condition 96/23/50*	36 (900)	30 (750)	34 (850)	26 (650)
Condition 24/23/96**	—	19 (475)	23 (575)	—
Dissipation factor, 2 kV/mm (50 V/mil) stress, max, avg:				
at 23°C (73°F)	0.015	0.015	0.015	0.015
at 100°C (212°F)	0.020	0.020	0.020	0.020
Relative permittivity, 2 kV/mm (50 V/mil) stress, max, avg, at 23°C (73°F)	4	4	4	4

*96 h at 23°C and 50 percent humidity.

**24 h at 23°C and 96 percent humidity.

Note: Tests are conducted in accordance with ASTM D 1458 procedures.

(Reprinted with permission from the *Annual Book of ASTM Standards*. Copyright, ASTM, 1916 Race Street, Philadelphia, PA 19103.)

TABLE 12-17   Electrical Requirements for Varnished Glass-Polyester Cloth (ASTM D 2400)

	BLACK			YELLOW	
Nominal thickness inches (mm)	0.008 (0.20)	0.010 (0.25)	0.012 (0.31)	0.008 (0.20)	0.010 (0.25)
Dielectric breakdown voltage, min ave kV					
Unstressed	10.5	12.5	15.0	9.5	11.0
Under 6% elongation	8.5	10.0	12.5	8.0	9.5
After hot oil	10.5	12.5	15.0	9.5	11.0

TABLE 12-18 Dielectric Strength Requirements for Coated Glass Fabrics (ASTM D 3949)

NOMINAL THICKNESS, in (mm)		CONDITION 48/23/50,* AVERAGE MIN. V/mil[a]					CONDITION 90/23/90,* AVERAGE MIN. V/mil[a]				
TOTAL	FABRIC	TYPE 1	TYPE 2	TYPE 3	TYPE 4	TYPE 5	TYPE 1	TYPE 2	TYPE 3	TYPE 4	TYPE 5
0.003 (0.08)	0.002 (0.05)	1,950	1,900	1,300	1,750	1,900	1,050	1,000	800	750	1,200
0.005 (0.08)	0.002 (0.05)	1,900	1,850	1,350	1,700	1,800	950	900	750	700	1,200
0.005 (0.13)	0.003 (0.08)	1,850	1,750	1,400	1,500	1,600	950	900	800	700	900
0.007 (0.18)	0.003 (0.08)	1,800	1,650	1,300	1,350	1,500	900	850	800	750	900
0.007 (0.18)	0.004 (0.10)	1,650	1,450	1,250	1,250	1,400	850	800	650	600	700
0.010 (0.25)	0.004 (0.10)	1,600	1,350	1,250	1,200	1,400	800	700	650	600	700
0.010 (0.25)	0.005 (0.13)	1,100	1,000	1,200	1,150	1,300	600	550	600	550	600
0.010 (0.25)	0.005 (0.13)	1,000	800	1,100	1,150	1,150	550	450	600	500	600
0.012 (0.30)	0.007 (0.18)	950	750	1,000	1,150	1,150	500	400	450	400	550
0.015 (0.38)	0.007 (0.18)	900	700	1,000	1,000	1,150	500	400	400	350	550
0.015 (0.38)	0.007 (0.18)	850	750	900	750	1,050	450	300	350	300	550
0.015 (0.38)	0.010 (0.25)	800	700	—	—	—	450	300	—	—	500
0.020 (0.51)	0.010 (0.25)	750	600	—	—	—	400	300	—	—	—
0.020 (0.51)	0.015 (0.38)	750	600	—	—	—	400	300	—	—	—
0.025 (0.64)	0.015 (0.38)	700	550	—	—	—	350	—	—	—	—
0.030 (0.76)	0.015 (0.38)	—	—	—	—	—	—	—	—	—	—

NOMINAL THICKNESS, in (mm) | | DIELECTRIC STRENGTH AT ROOM TEMPERATURE AFTER BENDING 180° AVERAGE MIN. V/mil[a] | | | | | PERCENTAGE RETENTION OF BREAKDOWN VOLTAGE AFTER AGING 168 h AT TEMPERATURE[b] AND BENT 180° MIN. %

TOTAL	FABRIC	TYPE 1	TYPE 2	TYPE 3	TYPE 4	TYPE 5	ALL TYPES
0.003 (0.08)	0.002 (0.05)	1,650	1,550	1,300	1,550	1,550	75
0.005 (0.13)	0.002 (0.05)	1,650	1,500	1,250	1,300	1,400	75
0.005 (0.13)	0.003 (0.08)	1,600	1,500	1,250	1,250	1,300	75
0.007 (0.18)	0.003 (0.08)	1,600	1,500	1,200	1,200	1,300	75
0.007 (0.18)	0.004 (0.10)	1,100	1,050	1,050	1,150	1,300	65
0.010 (0.25)	0.004 (0.10)	1,400	1,300	1,050	1,100	1,600	70
0.010 (0.25)	0.005 (0.13)	500	450	1,000	1,050	1,100	70
0.010 (0.25)	0.007 (0.18)	600	500	1,000	900	950	70
0.012 (0.30)	0.007 (0.18)	500	450	1,000	800	800	70
0.015 (0.38)	0.007 (0.18)	500	450	700	750	800	65
0.015 (0.38)	0.010 (0.25)	500	450	700	550	750	65
0.020 (0.51)	0.010 (0.25)	500	400	—	—	—	60
0.020 (0.51)	0.015 (0.38)	400	300	—	—	—	55
0.025 (0.64)	0.015 (0.38)	400	300	—	—	—	55
0.030 (0.76)	0.015 (0.38)	350	300	—	—	—	50

[a]Divide V/mil by 25.4 to obtain kV/mm.

[b]Temperature of aging: Types 1 and 2 = 130°C

Types 3 and 4 = 180°C

Type 5 = 200°C.

*Hours/temperature/percent relative humidity.

(Reprinted with permission from the *Annual Book of ASTM Standards*. Copyright, ASTM, 1916 Race Street, Philadelphia, PA 19103.)

• D 4325, *Nonmetallic Conducting and Electrically Insulating Rubber Tapes*. This standard describes methods of testing electrically insulating and nonmetallic conducting rubber tapes designed for splicing, terminating, and sheath repair of electrical wire and cable. Table 12-19 lists the test methods.

TABLE 12-19   Property Tests for Nonmetallic Conducting and Electrically Insulating Rubber Tapes (ASTM D 4325)

PROPERTY MEASURED	ASTM METHOD
Conditioning	D 4325
Dielectric strength	D 149
Dimensions	D 4325
Dissipation factor	D 150
Elongation	D 412
Heat resistance	D 4325
Fusion	D 4325
Ozone resistance	D 470
Permittivity	D 150
Sample requirements	D 4325
Tensile strength	D 412
Volume resistivity	D 257
Ultraviolet and weather resistance	D 750

• D 4388, *Nonmetallic Conducting and Electrically Insulating Rubber Tapes*. This standard covers nonmetallic conducting and electrical insulating tapes designed for the splicing and repair of electrical wire and cables operating at voltages up to 138,000 volts, phase to phase. Four types of tapes are identified:

Type I—A low-voltage rubber insulating tape designed for use on wires and cables operating up to 2,000 volts phase to phase in dry locations with temperatures up to 80°C (176°F) for continuous operation, to 95°C (203°F) for emergency overload conditions, and up to 150°C (302°F) for short-circuit conditions. This tape is not ozone resistant.

Type II—A medium-voltage, ozone-resistant, rubber insulating tape designed for use on wires and cables operating up to 35,000 volts phase to phase in either wet or dry locations with conductor temperatures up to 90°C (194°F) for continuous operation, to 130°C (266°F) for emergency overload conditions, and up to 250°C (482°F) for short-circuit conditions. This tape is not resistant to ultraviolet radiation.

Type III—A high-voltage, ozone-resistant, rubber insulating tape designed for use on wires and cables operating up to 138,000 volts phase to phase in either wet or dry locations with conductor temperatures up to 90°C (194°F) for continuous operation, to 130°C (266°F) for emergency overload conditions, and up to 250°C (482°F) for short-circuit conditions.

Type IV—A nonmetallic conducting tape designed for shielding various portions of joints and terminations in electrical wires and cables operating at any voltage under wet or dry conditions.

The tape shall consist of a noncrosslinked or partially crosslinked rubber nonmetallic conducting or insulating compound. Standard widths range from 0.75 inch (19 mm) to 2 inches (50 mm). Standard lengths range between 3 yards (2.7 m) and 10 yards (9.1 m). Thickness range is 0.020 inch (0.508 mm) to 0.040 inch (1.016 mm). Requirements for these tapes are shown in Table 12-20.

TABLE 12-20   Requirements for Nonmetallic Conducting and Electrically Insulating Rubber Tapes (ASTM D 4388)

REQUIREMENTS	TYPE I	TYPE II	TYPE III	TYPE IV
Tensile strength, min avg, psi (MPa)	250 (1.7)	250 (1.7)	250 (1.7)	>100 (0.69)
Elongation at break, min avg, %	300	500	700	300
Dielectric strength, min avg, V/mil (kV/mm):				
0.020 in (0.51 mm) thickness	—	—	700 (28)	—
0.030 in (0.76 mm) thickness	350 (14)	500 (20)	600 (24)	—
0.040 in (1.016 mm) thickness	—	400 (16)	—	—
Dissipation factor, max:				
24 h at 23°C[a]	—	0.05	0.05	—
168 h at 70°C[a]	—	0.05	0.05	—
Permittivity, max:				
24 h at 23°C[a]	—	4.5	4.0	—
168 h at 70°C[a]	—	4.5	4.0	—
Volume resistivity, ($\Omega \cdot$ cm):				
96 h at 23°C and 50% RH	—	10^{14} min avg	10^{14} min avg	10^{3} max
96 h at 23°C and 96% RH	—	10^{13} min avg	10^{13} min avg	—
168 h at 90°C	—	—	—	5×10^{4} max
Fusion—Flags $5/64$ in (0.2 mm) max	200% elongation	300% elongation	300% elongation	300% elongation
Ozone resistance	–	No visible cracks or checking	No visible cracks or checking	—
Heat resistance	Pass at 95°C	Pass at 110°C	Pass at 130°C	—
UV resistance	—	—	Pass	Pass

[a]After immersion in distilled water prior to testing at 23°C in accordance with Test Methods D 4325.

Note: Tests are conducted in accordance with ASTM D 4325 procedures.

(Reprinted with permission from the *Annual Book of ASTM Standards*. Copyright, ASTM, 1916 Race Street, Philadelphia, PA 19103.)

• D 4969, *Specification for PTFE Coated Glass Fabric*. Type 1 of this standard covers glass fabric coated on both sides with polytetrafluoroethylene polymer. Table 12-21 lists the minimum dielectric requirements for this material.

TABLE 12-21 Dielectric Requirements for PTFE-Coated Glass Fabric (ASTM D 4969)

	COATED THICKNESS inches*	DIELECTRIC BREAKDOWN volts (min)
Type 1	0.002	1,000
	0.003	2,500
	0.004	3,300
	0.005	3,900
	0.006	4,300
	0.008	4,700
	0.010	5,200
	0.014	7,500

*There is no specification for coating thickness.

(Reprinted with permission from the *Annual Book of ASTM Standards*. Copyright, ASTM, 1916 Race Street, Philadelphia, PA 19103.)

The key requirement of this specification is that, after exposure to a temperature of 150°C (302°F) for 12 hours, the average dielectric breakdown value of five samples shall not be less than 80 percent of the value determined before exposure.

FEDERAL SPECIFICATIONS FOR NONADHESIVE TAPES AND COATED FABRICS

The following Federal specifications have been developed for nonadhesive tapes and coated fabrics. For a listing of product types and related specifications, see "Technology" (p. 356).

• HH-I-553, *Insulation Tape, Electrical* (*Rubber, Natural and Synthetic*). This specification is for the use of all Federal agencies. Covered is rubber insulating tape for splicing wires and cables at both normal and high voltages. Two grades of tape are designated:

Grade A—Ozone resistant, 130°C (266°F)

Grade B—Regular, 600 volts maximum, 80°C (176°F)

Grade A tape shall consist of an unvulcanized self-amalgamating compound of natural rubber, synthetic rubber, or a mixture of the two. Grade B tape shall consist of an unvulcanized or partially vulcanized compound of natural rubber, synthetic rubber, or a mixture of the two. Nominal thickness of the tape shall be 0.030 inch with width according to ASTM D 119. Grade A tape shall not delaminate without rupture of the material after spiral wrapping and conditioning at 20° to 30°C (68°–86°F) for 24 hours. The fusion between layers of Grade B tape shall be in accordance with ASTM D 119.[1] Dielectric strength of both grades shall be in accordance with ASTM D 119.[1]

[1]Now discontinued by ASTM.

• MIL-I-3374, *Insulation Cloth and Tape, Electrical, Varnished Cambric*. This specification has been approved by the Department of Defense for use of the Departments of the Army, the Navy, and the Air Force. Covered is insulation intended for use as cable insulation, coil wrappers, and for insulating bus bars, joints and terminals of electric wires and cables. Insulation shall be of the following types and classes:

    Type I—Cloth
        Class 1—Bias cut
        Class 2—Straight cut
    Type II—Tape
        Class 1—Bias cut
        Class 2—Straight cut

The fabric shall be a closely knit cotton cloth with minimum thread count per inch of 60 for the warp and 55 for the filler. The coating shall be a high-grade yellow baking varnish, and the number of coats used shall be such that the finished material meets all dimensional and performance requirements of this specification. The varnish coating shall not crack when the cloth or tape is doubled upon itself, or when bent through 180 degrees over a $^1/_8$ inch mandrel after conditioning in an air oven at 125°C (257°F) for 150 hours. Requirements for these materials are shown in Table 12-22.

• MIL-I-17205, *Insulation Cloth and Tape, Electrical, Glass Fiber, Varnished*. This specification has been approved by the Department of Defense and is mandatory for use by the Departments of the Army, the Navy, and the Air Force. Insulation is classified as follows:

    Form C—Cloth
        Grade O—Varnished with yellow organic varnish
        Grade S—Varnished with silicone varnish for use in motors, generators, and transformers.
    Form T—Tape
        Grade O—Varnished with yellow organic varnish
        Grade S—Varnished with silicone varnish for use in motors, generators, and transformers.

Requirements for these materials are shown in Table 12-23.

• MIL-I-17695, *Insulation Tape, Electrical, Filler Type, Flameproof, Synthetic*. This specification covers the requirements for a filler type insulation in tape form intended for use as a filler material for splicing power and control cables. The unsupported material shall have application characteristics which are putty-like and capable of being easily moldable with reasonable finger pressure or applied winding pressure. It shall be able to maintain a continuity of material that will not crumble, flake, thin apart into stringiness or otherwise react to cause an unintended break in its normal application. The tape shall have maximum thickness of 0.125 inch and be $1^1/_2$ inches wide with a length of 5 feet. The material shall be flameproof in its stored form, in its application, and in use.

• MIL-I-18746, *Insulation Tape, Nonadhering, Glass Fabric Polytetrafluoroethylene Coated*. This specification covers one type of polytetrafluoroethylene-coated glass fabric for electrical insulation use at temperatures up to 260°C (500°F). The glass fabric shall be coated on both sides with polytetrafluoroethylene resin. Nominal widths

TABLE 12-22   Requirements for Varnished Cambric Cloth and Tape (MIL-I-3374)

			BREAKING STRENGTH CONDITION[1] C-96/23/50		DIELECTRIC STRENGTH		TEAR STRENGTH ACROSS WRAP
			WARP OR LENGTH-WISE	FILLER	CONDITION[1] C-96/23/0	CONDITION[1] C-96/23/96	CONDITION[1] C-96/23/50
		NOMINAL THICKNESS	lb/in width	lb/in width	volts/mil	volts/mil	grams
TYPE	CLASS	inch	(min. avg.)	(min. avg.)	(min avg.)	(min. avg.)	(min. avg.)
I—Cloth	1—Bias cut	0.007	38	—	1,000	250	—
		0.010	38	—	1,100	450	—
		0.012	38	—	1,000	500	—
I—Cloth	2—Straight cut	0.007	43	25	1,100	450	250
		0.010	43	25	1,100	600	250
		0.012	43	25	1,100	600	250
II—Tape	1—Bias cut	0.007	38	—	1,000	250	—
		0.010	38	—	1,100	450	—
		0.012	38	—	1,000	500	—
II—Tape	2—Straight cut	0.007	43	—	1,100	450	—
		0.010	43	—	1,100	600	—
		0.012	43	—	1,100	600	—

[1]Hours/°C/percent relative humidity.

are $\frac{1}{2}$, $\frac{3}{4}$, 1, and $1\frac{1}{2}$ inches. Nominal thicknesses, mechanical and electrical properties are shown in Table 12-24.

After exposure to a temperature of 150°C (302°F) for 12 hours, the average dielectric strength shall not be less than 80 percent of the value before exposure.

• MIL-I-24178, *Insulation Tape, Electrical, Semi-Cured Thermosetting Resin Treated Glass, Armature Banding, Naval Shipboard*. This specification covers semi-cured thermosetting resin treated, unidirectional glass electrical insulating tape intended for banding armatures of rotating electrical machines for shipboard use. Other applications may be banding of wound rotor coils, securing stator coils, surge rings, commutator string bands. There are two types of tape:

Type B—General Purpose, 130°C for continuous operation

Type H—High temperature, 180°C for continuous use

The insulating tape shall consist of nonwoven parallel glass yarns impregnated and bonded with a thermosetting resin. The tape shall be furnished in the semi-cured ("B" stage) state, and shall be flat, dry, and nonsticky. The surface may be slightly tacky, but shall unwind freely from the spool without transfer of resin or glass strands. The thermosetting resin shall be such that the treated tape may be completely cured

TABLE 12-23   Requirements for Varnished Glass Cloth and Tape (MIL-I-17205)

GRADE	FORM	NOMINAL THICKNESS OF VARNISHED MATERIAL	NOMINAL GLASS BASE FABRIC THICKNESS	TENSILE STRENGTH CONDITION C-96/23/50[1] WARP lb/in WIDTH (min. avg.)	FILL lb/in WIDTH (min. avg.)	DIELECTRIC STRENGTH CONDITION[1] C-96/23/0 volts/mil (min. avg.)	CONDITION[1] C-96/23/96 volts/mil (min. avg.)
		inch					
O	C	0.005	—	100	70	1,200	600
		0.007	—	100	70	1,200	600
		0.010	—	150	100	1,100	500
O	T	0.005	—	100	—	1,200	600
		0.007	—	100	—	1,200	600
		0.010	—	150	—	1,100	500
S	C	0.002	0.001	22	18	1,800	600
		0.004	0.002	70	40	1,400	1000
		0.007	0.003	100	70	1,300	900
		0.010	0.004	150	100	1,200	800
S	T	0.002	0.001	22	—	1,800	600
		0.004	0.002	70	—	1,400	1000
		0.007	0.003	100	—	1,300	900
		0.010	0.004	150	—	1,200	800

EFFECT OF ELEVATED TEMPERATURES

GRADE	FORM	POWER FACTOR AT 60 CYCLES Condition[1] C-96/23/0 (max. avg.)	Condition[1] C-96/23/96 (max. avg.)	DIELECTRIC CONSTANT AT 60 CYCLES Condition[1] C-96/23/0 (max. avg.)	Condition[1] C-96/23/96 (max. avg.)	CONDITION[1] C-168/23/50 Initial Bent Dielectric Strength volts/mil (min. avg.)	Condition[1] E-168/130 Percent Retention (min.)	Condition[1] E-168/250 Percent Retention (min.)
O	C	0.050	0.30	6.0	18.0	1,050	65	—
		0.050	0.30	6.0	18.0	1,050	65	—
		0.050	0.30	6.0	18.0	1,000	65	—
O	T	—	—	—	—	1,050	65	—
		—	—	—	—	1,050	65	—
		—	—	—	—	1,000	65	—
S	C	0.020	0.20	3.5	4.5	1,700	—	75
		0.020	0.20	4.5	6.0	1,300	—	75
		0.020	0.15	4.5	6.0	1,200	—	75
		0.020	0.20	4.5	6.0	1,100	—	75
S	T	—	—	—	—	1,700	—	75
		—	—	—	—	1,300	—	75
		—	—	—	—	1,200	—	75
		—	—	—	—	1,100	—	75

[1]Hours/°C/percent relative humidity.

**TABLE 12-24   Mechanical and Electrical Properties of
Polytetrafluoroethylene-Coated Glass Fabric
(MIL-I-18746)**

THICKNESS (inch)	BREAKING STRENGTH MINIMUM lb/in OF WIDTH	DIELECTRIC STRENGTH MINIMUM AVERAGE OF 5 TEST VALUES volts/mil
0.002	20	800
0.003	40	800
0.005	70	700
0.007	100	600
0.010	150	500
0.012	175	500
0.015	200	500

after baking for 3–5 hours after the equipment reaches a temperature of 150°C. Nominal thickness is 0.014 inch. Nominal widths are $^3/_8$, $^1/_2$, $^3/_4$, $^5/_8$, and 1 inch.

NEMA/ICEA STANDARD FOR VARNISHED CLOTH INSULATED WIRE AND CABLE

The following standard is applicable:

• NEMA WC4/ICEA S-65, *Varnished Cloth Insulated Wire and Cable for the Transmission and Distribution of Electrical Energy.*

PROPERTIES

The minimum requirements for most widely used nonadhesive tapes and coated fabrics are described in the accompanying summaries of standards and specifications. In general:

• Oleoresinous varnishes on cotton fabric or kraft paper are limited to thermal Class 105 application. On glass fabric or dacron-glass fabric, varnishes may be formulated by addition of synthetic resins to meet thermal Class 130 operational requirements.

• Epoxy resin-coated glass fabric and polyester resin coated dacron-glass fabric qualify for thermal Class 155 with high chemical resistance.

• Polyurethane-coated glass cloth is rated thermal Class 155, and has outstanding abrasion resistance and toughness.

• High-temperature polyester and silicone resin-coated glass fabrics conform to thermal Class 180, and also have high resistance to moisture and solvents.

- Glass fabric-coated with silicone rubber carries thermal Class 200 rating with excellent flexibility and moisture resistance. Corona effect is minimal. Resistance to abrasion, aromatic solvents, and hot oils is low.

- Polytetrafluoroethylene resin-coated glass cloth has the highest thermal rating, and is designed for continuous use at 260°C (500°F). Moisture and chemical resistance are outstanding.

Dielectric strength for most varnished materials is in the range of 1,000–1,500 volts per mil for 10 mil specimens when tested under standard conditions. Breakdown strength for rubber tape and rubber-coated glass fabrics is in the range of 500–700 volts per mil for 30 mil specimens.

ELECTRICAL/ELECTRONIC USES

Nonadhesive tapes and coated fabrics are used primarily for coil and wire and cable wrapping, for phase and layer insulation in electrical and electronic equipment, and for slot-cell insulation (abrasion-resistant grades only).

DEVELOPMENT PROGRAMS

Even before EPA and OSHA standards and governmental regulations placed restrictions on solvent emissions, producers of varnish-coated fabrics have been trying to eliminate solvents from coatings. Although higher solids contents in varnishes have been achieved and there is much less use of toxic solvents, there is still no widespread use of solventless insulating varnishes for coating fabrics.

MARKET TRENDS

Usage of nonadhesive tapes and coated fabrics for electrical insulating applications should parallel production trends in the electrical equipment manufacturing industry.

TUBING AND SLEEVING

INSULATING TUBING

INTRODUCTION

There is some confusion as to the distinction between the terms "tubing" and "sleeving." Military specifications MIL-I-7444, MIL-I-23053C, and MIL-I-85080 use the term *sleeving* to describe extruded, unreinforced tubular products. These products are referred to as *tubing* in MIL-R-46846 and MIL-I-47049. *Tubing* is the term used in all applicable ASTM standards for these products. However, the International Electrotechnical Commission (IEC) Standard 684 uses *sleeving* to include extruded, unreinforced tubular products.

General usage in industry in the U.S. refers to extruded, unreinforced products as *tubing*, and to tubular, braided yarn products, coated or uncoated, as *sleeving*. Both terms are used to refer to tubular products made by spiral or convolute winding of tape or sheet materials.

Extruded, unreinforced insulating *tubing* covered in this section is made in all degrees of flexibility from very flexible to rigid (called *pipe* in larger sizes).

ASTM standards and Federal specifications, summarized herein, contain requirements and test procedures for widely used tubings and are an essential part of this section.

PRODUCERS OF EXTRUDED TUBING
UL APPROVED

Underwriters Laboratories (UL) Standard 224, *Extruded Insulating Tubing*, lists producers of UL-approved electrical grade and flame-retardant tubing showing company name, material, maximum voltage, and maximum operating temperature.

TECHNOLOGY

The term "flexible tubing" as used here includes tubing from extruded compounds, without reinforcement, whose characteristic constituents are thermoplastic, thermosetting, or elastomeric. Cross-sectional shapes are limited only by extrusion die configurations. All of the materials discussed in Chapter 5, *Extrusion Compounds*, can be extruded as insulating tubing. The more commonly used polymers and their useful temperature ranges include the following:

POLYMER	USEFUL TEMPERATURE RANGE °C (F°)
Butyl rubber	−46 to 150 (−51 to 300)
Neoprene	−70 to 120 (−94 to 250)
Polyolefin (ethylene and ethylene copolymers)	
High temperature	−55 to 135 (−67 to 275)
Low temperature	−75 to 125 (−103 to 257)
Polyvinyl chloride (PVC)	
High temperature	−10 to 105 (14 to 221)
Low temperature	−55 to 80 (−67 to 176)
General purpose	−30 to 90 (−22 to 194)
Silicone rubber	−75 to 200 (−103 to 392)
Fluoropolymers	
Fluorinated ethylene-propylene (FEP)	−55 to 200 (−67 to 392)
Perfluoroalkoxy (PFA)	−55 to 260 (−67 to 500)
Polychlorotrifluoroethylene (PCTFE)	−55 to 150 (−67 to 302)
Polytetrafluoroethylene (PTFE)	−55 to 260 (−67 to 500)
Polyvinylidene fluoride (PVDF)	−55 to 175 (−67 to 347)

Summaries of key ASTM standards and Federal specifications included in this section relate to the following tubing types:

- Chlorosulfonated polyethylene (MIL-I-85080/4)
- Neoprene (ASTM D 2903)
- Polychloroprene (MIL-I-23053/1, MIL-I-85080/5)
- Polyethylene terephthalate (MIL-I-23053/7)
- Polyolefin (ASTM D 3149, MIL-I-23053/4, /5, /6, /15)
- Polyurethane (MIL-I-85080/3)
- PVC (ASTM D 876, ASTM D 922, ASTM D 3150, MIL-I-7444, MIL-I-23053 /2, /3, MIL-I-85080/1)
- Silicone rubber (MIL-I-23053/9, /10, MIL-I-46846, MIL-I-47049, MIL-I-85080/2)
- Fluoroelastomer (MIL-I-23053/13)
- Fluoropolymers
  FEP (ASTM D 2902, MIL-I-23053/11)
  PFA (ASTM D 2902)
  PTFE (ASTM D 2902, MIL-I-23053/12, MIL-I-22129)
  PVDF (ASTM D 3144, MIL-I-23053/8)
  Ethylene-tetrafluoroethylene (MIL-I-23053/14)
- ASTM D 2671 covers testing methods for all types of heat-shrinkable tubing.

As can be noted in the accompanying summaries of standards and specifications, tubing is available both with fixed dimensions and shrinkable with application of heat, infrared radiation, or by evaporation of solvent. A shrinkable capability permits tubing to encapsulate or bind securely the parts, wires, or terminations over which it is shrunk. Commonly used shrink ratios are 3–1, 2–1, $1^3/_4$–1, $1^2/_3$–1, and $1^1/_3$–1.

Heat-shrinkable tubing on expansion acts as though its molecules are frozen in the expanded state. On heating under controlled conditions, the strain is released and the tubing tends to recover to the original extruded dimensions (elastic memory). Tubing which shrinks by evaporation of solvent is useful where application of heat is undesirable.

Heat-shrinkable tubing may or may not be formulated to crosslink readily. Chemical crosslinking with application of heat or radiation crosslinking makes polymers no longer thermoplastic and improves chemical resistance, thermal endurance, and hardness. It also makes expansion possible at temperatures above the crystalline melt point in the fully elastic state.

Standard grades of irradiated PVC, polyolefin, PVDF, neoprene, butyl rubber, and fluoroelastomer tubings shrink to about one-half their extruded diameter with little or no longitudinal change. Uncrosslinked tubings may shrink in length considerably on heat activation. Recommended temperatures for shrinking range from 135°C (275°F) for PVC to 350°C (662°F) for PTFE, as shown in the accompanying summaries of standards and specifications.

In the manufacturing process, tubing is first extruded into a cooling bath from a horizontal extruder. It is to this extruded diameter that the shrinkable tubing returns after being expanded to the desired ratio by one of several suitable devices involving heat and air pressure, and cooling in the expanded state to "freeze" the molecules. The crosslinking process, when used, occurs following the extrusion process and before the expansion process.

Several methods are employed, sometimes in combination, to induce shrinkage, including infrared radiation units, hot air blowers, contact tools, heat transfer fluids, ovens, open flame torches, and electric blanket heaters.

Spiral and convolute wound tubing, one of the oldest forms of electrical insulations, is made by wrapping layers of substrate to the required thickness on a steel rod or mandrel of specified size. Spiral tubes can be made in continuous lengths as tubes are drawn from mandrels and supported on long troughs. Substrate materials include kraft paper, aramid paper, vulcanized fibre, polyester film, polyimide film, and composites of these materials. Layers of substrate are bonded with suitable adhesive, usually proprietary. The tubing is cut to length and usually impregnated with an insulating varnish and baked to cure. Commonly used cross-sectional shapes commercially available include round, square, and rectangular.

GRADES

Grades for all types of tubing covered by standards and specifications summarized in this section are commercially available.

ASTM STANDARDS FOR ELECTRICAL INSULATING TUBING

The American Society for Testing and Materials has developed the following standards for electrical tubing:

SPECIFICATIONS

- Neoprene—D 2903
- Polyolefin—D 3149
- Polyvinyl chloride—D 922, 3150
- Fluoropolymers
  FEP—D 2902
  PFA—D 2902
  PTFE—D 2902
  PVDF—D 3144

TEST METHODS

- Extruded PVC tubing—D 876
- Heat-shrinkable tubing—D 2671

The following pages provide a more detailed examination of these standards.

- D 876, *Nonrigid Vinyl Chloride Polymer Tubing Used for Electrical Insulation.* This standard describes methods of testing general purpose (Grade A, low-temperature, Grade B) and high-temperature (Grade C) nonrigid vinyl chloride polymer tubing, or its copolymers with other materials, for use as electrical insulation. Nonrigid tubing shall have an initial elongation in excess of 100 percent at break. Table 13-1 lists the test methods.

Table 13-1   Property Tests for Nonrigid Vinyl Chloride Polymer Tubing (ASTM D 876)

PROPERTY MEASURED	ASTM METHOD
Brittleness temperature	D 746
Corrosion tests	D 876
Dielectric breakdown at high humidity	D 876
Dielectric breakdown test	D 149
Dimensional tests	D 876
Effect of elevated temperatures	D 876
Flammability test	D 876
Oil resistance test	D 876
Penetration test	D 876
Sampling	D 876
Strain relief test	D 876
Tension test	D 876
Test conditions	D 876
Volume resistivity	D 257

• D 922, *Nonrigid Vinyl Chloride Polymer Tubing*. This standard specification covers nonrigid tubing of vinyl chloride or its copolymers with other materials for use in electrical insulation. Three grades are designated:

Grade A—General purpose
Grade B—Low temperature
Grade C—High temperature

Standard sizes are shown in Table 13-2.

Table 13-2   ASTM Standard Sizes for Nonrigid Tubing (ASTM D 922)

| | INSIDE DIAMETER in. | | WALL THICKNESS | |
| | MAX | MIN | WALL THICKNESS in | TOLERANCES, PLUS OR MINUS, in |
SPECIFIED SIZE				
No. 24 (0.022 in)	0.027	0.020	0.012	0.002
No. 22 (0.027 in)	0.032	0.025	0.012	0.002
No. 20 (0.034 in)	0.039	0.032	0.016	0.003
No. 18 (0.042 in)	0.049	0.040	0.016	0.003
No. 16 (0.053 in)	0.061	0.051	0.016	0.003
No. 14 (0.066 in)	0.074	0.064	0.016	0.003
No. 12 (0.085 in)	0.091	0.081	0.016	0.003
No. 11 (0.095 in)	0.101	0.091	0.016	0.003
No. 10 (0.106 in)	0.112	0.102	0.016	0.003
No. 9 (0.118 in)	0.124	0.114	0.020	0.003
No. 8 (0.133 in)	0.141	0.129	0.020	0.003
No. 7 (0.148 in)	0.158	0.144	0.020	0.003
No. 6 (0.166 in)	0.178	0.162	0.020	0.003
No. 5 (0.186 in)	0.198	0.182	0.020	0.003
No. 4 (0.208 in)	0.224	0.204	0.020	0.003
No. 3 (0.234 in)	0.249	0.229	0.020	0.003
No. 2 (0.263 in)	0.278	0.258	0.020	0.003
No. 1 (0.294 in)	0.311	0.289	0.020	0.003
No. 0 (0.330 in)	0.347	0.325	0.020	0.003
$5/16$ in	0.334	0.312	0.025	0.003
$3/8$ in	0.399	0.375	0.025	0.003
$7/16$ in	0.462	0.438	0.025	0.003
$1/2$ in	0.524	0.500	0.025	0.003
$5/8$ in	0.655	0.625	0.030	0.003
$3/4$ in	0.786	0.750	0.035	0.005
$7/8$ in	0.911	0.875	0.035	0.005
1 in	1.036	1.000	0.035	0.005
$1 1/4$ in	1.290	1.250	0.040	0.005
$1 1/2$ in	1.550	1.500	0.045	0.006
$1 3/4$ in	1.812	1.750	0.055	0.008
2 in	2.070	2.000	0.060	0.010

^aMultiply inches by 25.4 to get millimeters.

(Reprinted with permission from the *Annual Book of ASTM Standards*. Copyright, ASTM, 1916 Race Street, Philadelphia, PA 19103.)

Requirements for this tubing are shown in Table 13-3.

Table 13-3   Requirements for Nonrigid Vinyl Polymer Tubing (ASTM D 922)

Tensile strength, min, psi (MPa)		
Grades A&C		2,000 (15)
Grade B		1,800 (13)
Brittleness temperature, max, °C (°F)		
Grade A		−30 (−22)
Grade B		−55 (−67)
Grade C		−10 (+14)
Volume resistivity, $\Omega \cdot$ cm		
Grade A		10^{12}
Grade B		10^{11}
Grade C		10^{12}

Dielectric breakdown, min, kvolts

Wall	thickness		
inches	mm	A & C	B
0.012	0.30	9.0	9.3
0.016	0.41	12.5	10.4
0.020	0.51	14.0	11.0
0.025	0.64	15.7	12.0
0.030	0.76	17.1	12.9
0.035	0.89	18.0	13.5
0.040	1.02	19.2	14.0
0.045	1.14	20.4	14.5
0.050	1.27	21.5	15.0
0.055	1.40	22.8	15.5
0.060	1.52	24.0	16.0

Breakdown at 96 percent humidity, percent of dry value	
Grade A	90
Grade B	75
Grade C	85

Note: Tests are conducted in accordance with ASTM D 876 procedures.

• D 2671, *Heat-Shrinkable Tubing for Electrical Use*. This standard describes methods of testing heat-shrinkable tubing used for electrical insulation. Materials used include poly(vinyl chloride), polyolefins, fluorocarbon polymers, silicone rubber, and other plastic or elastomeric compounds. Table 13-4 lists applicable test methods.

• D 2902, *Fluoropolymer Resin Heat-Shrinkable Tubing*. This standard specification covers flexible heat-shrinkable tubing made from tetrafluoroethylene resin, copolymer of tetrafluoroethylene and hexafluoropropylene, and from perfluoroalkoxy resin for use as electrical insulation. Three types of tubing are identified:

Type I—Tubing made from tetrafluoroethylene polymer (TFE) and capable of being heat shrunk at a temperature of 327°C (621°F).

Table 13–4   Property Tests for Heat-Shrinkable Tubing
(ASTM D 2671)

PROPERTY MEASURED	ASTM METHOD
Brittleness temperature	D 746
Color	D 1535
Color stability	D 1535
Conditioning	D 618
Corrosion testing	D 2671
Dielectric breakdown	D 149
Dimensions	D 2671
Flammability	D 876
Heat resistance	D 2671
Heat shock	D 2671
Low-temperature	D 2671
Restricted shrinkage	D 2671
Sampling	D 2671
Secant modulus	D 882
Shelf life	D 2671
Solvent resistance	D 2671
Specific gravity	D 792
Stress modulus	D 412
Tensile strength and ultimate elongation	D 2671
Thermal endurance	D 2671
Volume resistivity	D 257
Water absorption	D 570

Type II—Tubing made from a copolymer of tetrafluoroethylene and hexa-fluoropropylene (FEP) and capable of being heat shrunk at a temperature of 150°C (302°F).

Type III—Tubing made from perfluoroalkoxy resin (PFA) and capable of being heat shrunk at a temperature of 175°C (347F).

Dimensions for this tubing are shown in Table 13-5.

Tubing (all types) shall have low-temperature flexibility with no cracking at –55°C (–67°F). Dielectric breakdown after shrinkage shall not be less than the following values:

WALL THICKNESS (inches)	MINIMUM (kilovolts)
0.004–0.006	8
0.007–0.008	10
0.009	11.5
0.010–0.011	12.5
0.012–0.014	14.6
0.015	15
0.016–0.019	16.3
0.020 and larger	17

Note: Tests are conducted in accordance with ASTM D 2671 procedures.

Table 13-5   Dimension Ranges for Fluoropolymer Resin Heat-Shrinkable Tubing
(ASTM D 2902)

AS SUPPLIED	AFTER SHRINKAGE		
ID MINIMUM inches (mm)	ID MAXIMUM inches (mm)	WALL THICKNESS inches (mm)	STOCK LENGTHS feet
Type 1			
Heavy Wall 0.166–1.330 (4.22–33.8)	0.130–1.020 (3.30–25.9)	0.030–0.050 (0.76–1.27)	1, 3
Standard Wall 0.045–1.500 (1.14–38.1)	0.027–1.036 (0.69–26.3)	0.012–0.035 (0.30–0.76)	1, 2, 3
Thin Wall 0.034–1.125 (0.86–28.6)	0.015–0.786 (0.38–20.0)	0.009–0.030 (0.23–0.76)	1, 2, 3
Very Thin Wall 0.050–0.470 (1.27–11.94)	0.027–0.347 (0.69–8.81)	0.006–0.012 (0.15–0.30)	1, 2, 3

There are also special sizes for 2–1 shrink factor, 4–1 PTFE shrinkable tubing, and overexpanded tubing.

AS SUPPLIED	AFTER SHRINKAGE		
Types II and III			
1.33–1 and 1.20–1 Shrink Factor 0.031–12.1 (0.79–307.3)	0.027–10.6 (0.69–269.2)	0.008–0.20 (0.20–0.51)	3–10
1.67–1 Shrink Factor 0.093–2.00 (2.36–50.8)	0.056–1.20 (1.42–30.5)	0.008–0.030 (0.20–0.76)	3

Tubing shall shrink in accordance with the following schedule:
>   Type I—10 minutes at 350°C (662°F)
>   Type II—10 minutes at 175°C (347°F)
>   Type III—10 minutes at 204°C (400°F)

• D 2903, *Neoprene Heat-Shrinkable Tubing.* This standard specification covers electrical insulation grade flexible neoprene heat-shrinkable tubing that has been extruded, crosslinked, and then expanded to the required dimensions. Two types are identified:
>   Type I—Normal operating temperature (100°C (212°F))
>   Class 1—1.75:1 Shrink ratio
>   Class 2—2.0:1 Shrink ratio
>   Type II—Extended operating temperature (121°C (250°F))

Dimensions for this tubing are shown in Table 13-6.

• Type I tubing shall have low-temperature flexibility with no cracks at –55°C (–67°F). Type II tubing shall have low-temperature flexibility with no cracks at –70°C

Table 13-6   Dimension Ranges for Neoprene Heat-Shrinkable Tubing (ASTM D 2903)

AS SUPPLIED		AFTER SHRINKAGE	
NORMAL SIZE inches	ID MINIMUM inches (mm)	ID MAXIMUM inches (mm)	WALL THICKNESS inches (mm)
	Type I Class 1 and Type II		
¹/₄–4	0.250–4.00 (6.35–101.6)	0.143–2.280 (3.63–57.9)	0.035–0.140 (0.89–3.56)
	Type I Class 2		
	0.250–2.000 (6.4–50.8)	0.125–1.000 (3.2–25.4)	0.035–0.110 (0.89–2.8)

(–94°F). Dielectric breakdown for both types after shrinkage shall not be less than 300 volts/per mil for tubing sizes $1^3/_4$ inches ID and under, and 200 volts per mil for tubing sizes 2 inches and larger. Tests are conducted in accordance with ASTM D 2671 procedures. Tubing shall be shrunk at a temperature of 175°C (347°F) for 10 minutes.

• D 3144, *Poly(vinylidene Fluoride) Heat-Shrinkable Tubing for Electrical Insulation.* This standard specification covers semirigid, flame-retardant, crosslinked poly(vinylidene fluoride) heat-shrinkable tubing for electrical insulation applications. Dimensions for this tubing are shown in Table 13-7.

Tubing shall have low-temperature flexibility with no cracks at –55°C (–67°F). After 168 hours at 250°C (482°F), minimum elongation after shrinkage shall not be less than 50 percent. Dielectric breakdown after shrinkage shall not be less than 600 volts per mil. Tests are conducted in accordance with ASTM D 2671 procedures. Tubing shall be shrunk at a temperature of 392°F (200°C) for three minutes.

• D 3149, *Crosslinked Polyolefin Heat-Shrinkable Tubing for Electrical Insulation.* This standard specification covers crosslinked polyolefin heat-shrinkable tubing used for electrical insulating purposes. It is supplied in expanded form and will shrink to its extruded diameter when heated. Four types of tubing are identified:

> Type I—Flexible, flame-retardant opaque polyolefin tubing with a secant modulus at 2 percent strain less than 25,000 psi (172 MPa) (ASTM D 882).

Table 13-7   Dimension Ranges for Poly(vinylidene Fluoride) Heat-Shrinkable Tubing (ASTM D 3144)

AS SUPPLIED		AFTER SHRINKAGE	
NOMINAL SIZE inches	ID MINIMUM inches (mm)	ID MAXIMUM inches (mm)	WALL THICKNESS inches (mm)
³/₆₄–1	0.046–1.000 (1.16–25.4)	0.023–0.500 (0.59–12.7)	0.010–0.019 (0.25–0.48)

Type II—Flexible, nonflame-retardant, opaque or clear polyolefin tubing with a secant modulus at 2 percent strain less than 25,000 psi (172 MPa) (ASTM D 882).

Type III—Semi-rigid, flame-retardant, opaque polyolefin tubing with a secant modulus at 2 percent strain greater than 35,000 psi (241 MPa) (ASTM D 882).

Type IV—Semi-rigid, nonflame-retardant, opaque or clear polyolefin tubing with a secant modulus at 2 percent strain greater than 35,000 psi (241 MPa) (ASTM D 882).

Dimensions for this tubing are shown in Table 13-8.

Table 13-8   Dimension Ranges for Crosslinked Polyolefin Heat-Shrinkable Tubing (ASTM D 3149)

ALL TYPES AS SUPPLIED		AFTER HEAT SHRINKAGE		
		ALL TYPES	WALL THICKNESS	
NOMINAL SIZE inches	ID MINIMUM inches (mm)	ID MINIMUM inches (mm)	TYPES I, II inches (mm)	TYPES III, IV* inches (mm)
$^3/_{64}$–4	0.046–4.000 (1.16–101.6)	0.023–2.000 (0.59–50.1)	0.016–0.055 (0.41–1.40)	0.020–0.030 (0.56–0.76)

*Through nominal size $^1/_2$ in only.

Tubing shall have low-temperature flexibility with no cracks at −55°C (−67°F). After 168 hours at 175°C (347°F), minimum elongation shall not be less than 150 percent. Dielectric breakdown after shrinkage shall not be less than 500 volts per mil. Tests are conducted in accordance with ASTM D 2671 procedures. Tubing shall be shrunk at a temperature of 175°C (347°F) for 3 minutes.

• D 3150, *Crosslinked and Noncrosslinked Poly(vinylchloride) Heat-Shrinkable Tubing for Electrical Insulation*. This standard specification applies to extruded, flexible, crosslinked, and noncrosslinked poly(vinylchloride) heat-shrinkable tubing for electrical insulating purposes. It is supplied in an expanded form and will shrink to its extruded diameter when heated. Two types of tubing are identified:

Type I—Flexible, noncrosslinked poly(vinyl chloride) tubing capable of being shrunk at 135°C (275°F) in 15 minutes.

Type II—Flexible, crosslinked poly(vinyl chloride) tubing capable of being shrunk at 175°C (347°F) in 15 minutes.

Dimensions for this tubing are shown in Table 13-9.

Tubing shall have low-temperature flexibility with no cracks at −10°C (14°F) as received. After 168 hours at 136°C (277°F), minimum elongation shall not be less than 150 percent. Dielectric breakdown after shrinkage shall not be less than 400 volts per mil. Tests are conducted in accordance with ASTM D 2671 procedures.

Table 13-9   Dimension Ranges for Crosslinked and
Noncrosslinked Poly(vinyl Chloride) Heat-Shrinkable Tubing
(ASTM D 3150)

AS SUPPLIED	AFTER SHRINKAGE	
ID MINIMUM inches (mm)	ID MAXIMUM inches (mm)	WALL THICKNESS inches (mm)
	Type I	
0.063–6.000 (1.60–152.5)	0.037–3.600 (0.94–91.5)	0.014–0.034 (0.35–0.86)
	Types I and II	
0.046–2.000 (1.16–50.1)	0.023–1.000 (0.59–25.4)	0.020–0.050 (0.51–1.27)

FEDERAL SPECIFICATIONS FOR ELECTRICAL INSULATING TUBING

The following Federal specifications have been developed for electrical insulating tubing:

- Butyl rubber—MIL-R-46846
- Chlorosulfonated polyethylene—MIL-I-85080/4
- Neoprene—MIL-R-46846
- Polychloroprene—MIL-I-23053/1, MIL-I-85080/5
- Polyethylene terephthalate—MIL-I-23053/7
- Polyolefin—MIL-I-23053/4/5/6/15, MIL-R-46846
- Polyurethane—MIL-I-85080/3
- Polyvinyl chloride—MIL-I-7444, MIL-I-23053/2/3, MIL-I-85080/1
- Silicone rubber—MIL-I-23053/9/10, MIL-R-46846, MIL-I-47049, MIL-I-85080/2
- Fluoroelastomer—MIL-I-23053/13, MIL-R-46846
- Fluoropolymers
  FEP—MIL-I-23053/11
  PTFE—MIL-I-23053/12, MIL-I-22129
  Ethylene-tetrafluoroethylene—MIL-I-23053/14

The following pages provide a more detailed examination of these standards.

- MIL-I-7444, *Insulation Sleeving, Electrical, Flexible*. This specification is mandatory for use by all departments and agencies of the Department of Defense. It covers flexible, extruded, vinyl plastic tubing (sleeving). The tubing shall be identified as follows:

Type I—Colorless and transparent
Type II—no longer listed
Type III—Colored
  Class I (applicable to both types)—Cold brittle point at –90°F (–67°C)
and flame resistance of 45 seconds.
  Class 2 (applicable to both types)—Cold brittle point at –67°F (–55°C)
and flame resistance of 15 seconds.

The material shall be a vinyl chloride or vinyl chloride acetate copolymer. Nominal inside diameter shall be in the range of 0.022 inch (AWG No. 24)–2$^1/_2$ inches. Nominal wall thicknesses shall range from 0.012 inch for 0.022 inch ID tubing to 0.070 in for 2$^1/_2$ inch ID tubing. Tubing under 1 inch ID shall be wound on spools in lengths not less than 100 feet. Tubing 1 inch ID and over shall be wound on spools in lengths not less than 50 feet. Testing for brittle points is done both before aging and after aging for 168 hours at 158°F (70°C). Minimum dielectric strength of tubing shall be not less than 200 volts per mil when specimens are either dry or after immersion in water for 24 hours at 77°F (25°C).

• MIL-I-23053C, *Insulation Sleeving, Electrical, Heat Shrinkable, General Specification For*. This specification is approved for use by all departments and agencies of the Department of Defense. This is a general specification covering heat-shrinkable types listed in Table 13-10. The individual requirements for each type, including dimensions, are detailed in the specification sheet for that type.

Test procedures for all sleeving types are described with reference, where appropriate, to applicable ASTM standards. Property requirements after unrestricted shrinkage include the following:

Color
Color stability
Specific gravity
Dielectric constant
Odor
Fungus resistance
Tensile strength
Ultimate elongation
Dielectric strength
Volume resistivity
Low-temperature flexibility
Corrosion
Heat resistance
Water absorption
Flammability
Fluid resistance
Flow of inner wall

Table 13-10   Heat-Shrinkable Insulation Sleeving Types
Covered by MIL-I-23053C

SPECIFICATION SHEETS	
MIL-I-23053/1	Insulation Sleeving,* Electrical, Heat Shrinkable, Polychloroprene, Flexible
MIL-I-23053/2	Insulation Sleeving,* Electrical, Heat Shrinkable, Polyvinyl Chloride, Flexible, Crosslinked and Noncrosslinked
MIL-I-23053/3	Insulation Sleeving,* Electrical, Heat Shrinkable, Polyvinyl Chloride, Semi-Rigid, Crosslinked and Noncrosslinked
MIL-I-23053/4	Insulation Sleeving,* Electrical, Heat Shrinkable, Polyolefin, Dual-Wall, Rigid Outer Wall, Crosslinked
MIL-I-23053/5	Insulation Sleeving,* Electrical, Heat Shrinkable, Polyolefin, Flexible, Crosslinked
MIL-I-23053/6	Insulation Sleeving,* Electrical, Heat Shrinkable, Polyolefin, Semi-Rigid Crosslinked
MIL-I-23053/7	Insulation Sleeving,* Electrical, Heat Shrinkable, Polyethylene Tereph-thalate, Noncrosslinked
MIL-I-23053/8	Insulation Sleeving,* Electrical, Heat Shrinkable, Polyvinylidene Fluoride, Semi-Rigid, Crosslinked
MIL-I-23053/9	Insulation Sleeving,* Electrical, Heat Shrinkable, Fluorosilicone Rubber, Flexible (Cancelled)
MIL-I-23053/10	Insulation Sleeving,* Electrical, Heat Shrinkable, Silicone Rubber, Flexible
MIL-I-23053/11	Insulation Sleeving,* Electrical, Heat Shrinkable, Fluorinated Ethylene Propylene, Noncrosslinked
MIL-I-23053/12	Insulation Sleeving,* Electrical, Heat Shrinkable, Polytetrafluoroethylene
MIL-I-23053/13	Insulation Sleeving,* Electrical, Heat Shrinkable, Fluoroelastomer, Flexible
MIL-I-23053/14	Insulation Sleeving,* Electrical, Heat Shrinkable, Ethylene-Tetrafluoro-ethylene Fluoropolymer, Semi-Rigid
MIL-I-23053/15	Insulation Sleeving,* Electrical, Heat Shrinkable, Polyolefin, Heavy Wall, Coated, Flexible, Outer Wall Crosslinked

*Generally considered tubing.

• MIL-R-46846, *Rubber, Synthetic, Heat-Shrinkable*. This specification covers the requirements for flexible electrical insulating extruded tubing, molded parts, and extruded shapes whose dimensions will reduce to a predetermined size upon the application of heat. The following types and classes of material are identified:

Type I. Crosslinked, modified polychloroprene (Neoprene) tubing, molded parts, and extruded shapes, flexible general purpose cable-jacketing, cable harness, boots and transitions, service temperature range of −55 to 90°C.

Class 1. Tubing-cable-jacketing material, 1.75× expansion ratio

Class 2. Molded parts and extruded shapes, cable harness boots and transitions, 2.5× expansion ratio

Type II. Crosslinked modified flame-retarded silicone rubber tubing, molded parts and extruded shapes, highly flexible, ablative-resistant.

Class 1. Tubing, expansion ratio 1.75×, service temperature range of −70 to 180°C.

Class 2. Molded parts and extruded shapes, expansion ratio 2 to 4×, service temperature range −75 to 135°C.

Type III. Crosslinked, modified viton tubing, molded parts and extruded shapes, flexible for materials exposed to extended periods of high temperature and solvent attack.

Class 1. Tubing 2× expansion ratio, service temperature range −55 to 200°C

Class 2. Molded parts and extruded shapes, 3.5× expansion ratio, service temperature range of −55 to 200°C

Type IV. Crosslinked, modified, propellant-resistant butyl rubber tubing, molded parts and extruded shapes flexible, for protection of cable assemblies exposed to hypergolic fuels.

Class 1. Tubing, 2× expansion ratio, service temperature range of −55 to 125°C

Class 2. Molded parts and extruded shapes 2.5× expansion ratio, service temperature range of −55 to 125°C

Type V. Crosslinked, modified, flame-resistant elastomeric polyolefin tubing, highly flexible airborne harnessing or jacketing material where low weight, high flame resistance, and thin walls are required, low shrinkage temperature, 2–4× expansion ratios, service temperature range of −55 to 135°C.

Type VI. Crosslinked, modified, flame-resistant ethylene-propylene rubber tubing, molded parts, and extruded shapes, flexible, ozone resistant, mechanically tough.

Class 1. Tubing, 2× expansion ratio, service temperature of −70 to 115°C

Class 2. Molded parts and extruded shapes, 4× expansion ratio, service temperature range of −75 to 125°C

Type 1 synthetic rubber shall be fabricated from a crosslinked, stabilized, modified polychloroprene (Neoprene) composition. Type II synthetic rubber shall be fabricated from a crosslinked, stabilized, flame-retarded, modified silicone rubber composition. Type III synthetic rubber shall be fabricated from a crosslinked, stabilized, flame-resistant, modified Viton composition. Type IV synthetic rubber shall be fabricated from a crosslinked, stabilized, propellant-resistant, modified butyl rubber composition. Type V material shall be fabricated from a crosslinked, stabilized, flame-resistant, modified elastomeric polyolefin composition. Type VI synthetic rubber shall be fabricated from a crosslinked, stabilized, flame-resistant, modified ethylene-propylene rubber composition. These expanded synthetic rubbers shall shrink to predetermined dimensions upon the application of heat as follows:

Types I, IV, VI—Heat in excess of 135°C (275°F)

Types II, III—Heat in excess of 175°C (347°F)

Type V—Heat in excess of 90°C (194°F)

Types I–VI tubing shall have the sizes, tolerances, and properties specified in Tables 13-11 and 13-12.

Table 13-11 Heat-Shrinkable Synthetic Rubber Tubing Sizes (MIL-R-46846)

SIZE, inches	EXPANDED—AS SUPPLIED: MINIMUM INSIDE DIAMETER, inches						RECOVERED DIMENSIONS—AFTER HEATING: MAXIMUM INSIDE DIAMETER, inches						WALL THICKNESS, inch					
	TYPE I	TYPE II	TYPE III	TYPE IV	TYPE V	TYPE VI	TYPE I	TYPE II	TYPE III	TYPE IV	TYPE V	TYPE VI	TYPE I	TYPE II	TYPE III	TYPE IV	TYPE V	TYPE VI
3/64	—	—	—	—	.046	—	—	—	—	—	.023	—	—	—	—	—	.016 ± .003	—
1/16	—	—	—	—	.063	—	—	—	—	—	.031	—	—	—	—	—	.017 ± .003	—
3/32	—	—	—	—	.093	—	—	—	—	—	.046	—	—	—	—	—	.020 ± .003	—
1/8	—	—	—	—	.125	—	—	—	—	—	.062	—	—	—	—	—	.020 ± .003	—
3/16	—	—	—	—	.187	—	—	—	—	—	.093	—	—	—	—	—	.020 ± .003	—
1/4	.250	.250	.250	.250	.250	.250	.143	.143	.125	.125	.125	.125	.035 ± .010	.035 ± .010	.030 ± .010	.035 ± .010	.025 ± .003	.030 ± .010
3/8	.375	.375	.375	.375	.375	.375	.214	.214	.187	.187	.187	.187	.040 ± .010	.040 ± .010	.035 ± .010	.040 ± .010	.025 ± .003	.035 ± .010
1/2	.500	.500	.500	.500	.500	.500	.286	.286	.250	.250	.250	.250	.048 ± .010	.048 ± .010	.043 ± .010	.050 ± .015	.025 ± .003	.043 ± .010
5/8	.625	.625	.625	—	—	.625	.357	.357	.312	—	—	.312	.052 ± .015	.052 ± .015	.047 ± .012	—	—	.047 ± .012
3/4	.750	.750	.750	.750	.750	.750	.428	.428	.375	.375	.375	.375	.057 ± .015	.057 ± .015	.052 ± .015	.060 ± .015	.030 ± .003	.052 ± .015
7/8	.875	.875	.875	—	—	.875	.500	.500	.437	—	—	.437	.065 ± .015	.065 ± .015	.060 ± .015	—	—	.060 ± .015
1	1.00	1.00	1.00	1.00	1.00	1.00	.570	.570	.500	.500	.500	.500	.070 ± .015	.070 ± .015	.069 ± .015	.075 ± .020	.035 ± .005	.065 ± .015
1 1/4	1.25	1.25	1.25	1.00	—	1.25	.714	.714	.625	—	—	.625	.087 ± .020	.087 ± .020	.070 ± .020	—	—	.070 ± .020
1 1/2	1.50	1.50	1.50	1.50	1.50	1.50	.857	.857	.750	.750	.750	.750	.095 ± .020	.095 ± .020	.075 ± .020	.100 ± .020	.040 ± .006	.075 ± .020
1 3/4	1.75	1.75	—	—	—	—	1.00	1.00	—	—	—	—	.107 ± .020	.107 ± .020	—	—	—	—
2	2.00	2.00	2.00	2.00	2.00	—	1.14	1.14	1.00	1.00	1.00	—	.110 ± .020	.110 ± .020	.110 ± .020	.110 ± .020	.045 ± .007	—
2 1/2	—	—	—	2.50	—	—	—	—	1.00	1.25	—	—	—	—	—	.130 ± .020	—	—
3	3.00	—	—	3.00	3.00	—	1.71	—	—	1.50	1.50	—	.125 ± .020	—	—	.140 ± .020	.050 ± .008	—
4	4.00	—	—	4.00	4.00	—	2.28	—	—	2.00	2.00	—	.140 ± .020	—	—	.150 ± .020	.055 ± .009	—

Table 13-12 Requirements for Heat-Shrinkable Synthetic Rubber Tubing (MIL-R-46846)

PROPERTY	REQUIREMENT						METHOD
	TYPE I	TYPE II	TYPE III	TYPE IV	TYPE V	TYPE VI	
Longitudinal change, percent, maximum	10	10	20	15	5	15	MIL-R-46846
Tensile strength, psi, minimum	1,500	600	1,200	1,200	1,500	1,400	ASTM D 412, (See 4.5.1)
Tensile stress at 100 percent elongation, psi, maximum	1,500	500	1,400	1,200	1,000	1,000	ASTM D 412, (See 4.5.1)
Elongation, percent, minimum	200	200	250	250	200	200	ASTM D 412, (See 4.5.1)
Specific gravity, maximum	1.50	1.35	2.0	1.40	1.35	1.25	ASTM D 792
Hardness, shore A	80 ± 5	60 ± 5	82 ± 5	85 ± 5	85 ± 5	80 ± 5	ASTM D 2240
Low temperature	Pass or fail	Pass or fail	Pass or fail	Pass or fail	Pass or fail	Pass or fail	ASTM D 2671
Heat shock	No dripping, flowing, or cracking						MIL-R-46846
Heat aging, followed by tests for:							MIL-R-46846
Tensile strength, psi, minimum	1,200	500	1,200	1,000	1,500	1,200	ASTM D 412, (See 4.5.1)
Elongation, percent, minimum	150	120	200	100	200	100	ASTM D 412, (See 4.5.1)
Dielectric, strength volts/mil, minimum	200	300	150	100	500	220	ASTM D 876
Volume resistivity, $\Omega \cdot$ cm, minimum	1×10^{10}	1×10^{11}	1×10^{9}	1×10^{10}	1×10^{14}	1×10^{11}	ASTM D 257
Flammability, seconds, maximum	15; self-exting.*	60; self-exting.** (See Note 1)	15; self-exting.*	—	15; self-exting.*	15; self-exting.* Sec. 67–72	*ASTM D 876 **ASTM D 2671,
Fungus resistance	Rating of 0						ASTM D 1924
Water absorption, percent, maximum	1.0	1.0	0.5	0.5	0.5	1.0	ASTM D 570 Procedure A
Solvent resistance, followed by tests for:							MIL-R-46846
Tensile strength, psi, minimum	900	400	1,200	1,000	750	500	ASTM D 412, (See 4.5.1)
Elongation, percent, minimum	125	120	250	200	200	100	ASTM D 412, (See 4.5.1)
Eccentricity, percent, minimum	60	60	60	60	60	60	ASTM D 2671 (Sec. 6–11)

Note 1: Not more than 25 percent of indicator flag burned; no dripping or flowing.

• MIL-I-47049, *Insulation Tubing, Silicone Rubber, Heat Shrinkable*. This specification is approved for use by all departments and agencies of the Department of Defense. Covered are requirements for one type of silicone rubber heat-shrinkable tubing. The material shall be silicone rubber heat-shrinkable tubing supplied in an expanded form capable of being shrunk by the application of heat. The inside diameter and wall thickness of tubing shall be in accordance with Table 13-13.

Table 13-13   Inside Diameter and Wall Thickness of Heat-Shrinkable, Silicone Rubber Tubing (MIL-I-47049)

	EXPANDED AS SUPPLIED	RECOVERED DIMENSIONS AFTER HEATING	
SIZE NO.	INSIDE DIMATER (inches) MINIMUM	INSIDE DIAMETER (inches) MAXIMUM	WALL THICKNESS (inches) NOMINAL
1	0.125	0.067	0.020
2	0.250	0.125	0.035
3	0.375	0.187	0.040
4	0.500	0.250	0.048

Test procedures are described with reference, where appropriate, to applicable ASTM standards. Key requirements are shown in Table 13-14.

Table 13-14   Requirements for Heat-Shrinkable, Silicone Rubber Tubing (MIL-I-47049)

PROPERTY	REQUIREMENT
Dielectric strength, $1/16$ in thick wall	400 V/mil minimum
Dielectric constant	3.10 maximum
Volume resistivity	1×10^{14} minimum
Tensile strength	1200 psi minimum
Ultimate elongation	450% minimum
Specific gravity	1.20 maximum
Water absorption	0.5% maximum

Tubing shall be shrunk at 150° to 180°C (302°–356°F) for 2 minutes minimum.

• MIL-I-85080, *Insulation Sleeving,* * *Electrical, Shrinkable Without Heat, General Specification For*. This specification is approved for use by the Naval Air Systems Command, Department of the Navy, and is available for use by all departments and agencies of the Department of Defense. This specification establishes requirements for chemically expanded, flexible, electrical insulating sleeving that shrinks to a predetermined size upon solvent evaporation. This is a general specification covering shrinkable types listed in Table 13-15. The individual requirements for each type including dimensions are detailed in the specification sheet for that type.

Table 13-15   Insulation Sleeving Shrinkable Without
Heat Types Covered by MIL-I-85080

SPECIFICATION SHEETS

MIL-I-85080/1	Insulation Sleeving,* Electrical, Shrinkable Without Heat, Polyvinyl Chloride, Flexible, Noncrosslinked
MIL-I-85080/2	Insulation Sleeving,* Electrical, Shrinkable Without Heat, Silicone, Rubber, Flexible
MIL-I-85080/3	Insulation Sleeving,* Electrical, Shrinkable Without Heat, Polyurethane, Flexible
MIL-I-85080/4	Insulation Sleeving,* Electrical, Shrinkable Without Heat, Chlorosulfonated Polyethylene, Flexible
MIL-I-85080/5	Insulation Sleeving,* Electrical, Shrinkable Without Heat, Polychloroprene

*Generally considered tubing.

Test procedures for all sleeving types are described with reference, where appropriate, to applicable ASTM standards. Property requirements after unrestricted shrinkage include the following:

Water absorption

Dielectric strength

Volume resistivity

Fungus resistance

Color

Voltage withstand

Low-temperature flexibility

Heat shock

Heat resistance

Corrosion

Fluid resistance

Odor

Modulus

Tensile strength

Ultimate elongation

Flammability

Color stability

PROPERTIES

Properties of commonly used materials for extruded tubings are described in Chapter 5, *Extrusion Compounds*, and in standards and specifications summarized in this chapter. In general:

• Polyvinyl chloride and polyolefin tubings can be made with the widest range of flexibility and are low in cost.

• Butyl and neoprene tubings have good flexibility and have good resistance to fuels and oils (aerospace environments).

• Silicone rubber has high thermal rating (Class 200) and low-temperature brittleness point. Tubing is fungus resistant and exhibits excellent aging characteristics.

• Fluoropolymers are chemically inert and are highly moisture resistant. Polytetrafluoroethylene and perfluoroalkoxy have the best thermal rating (260°C (500°F)) of all polymers. Tubings from these polymers are the highest price.

ELECTRICAL/ELECTRONIC USES

The major electrical/electronic uses are to identify and bind wires together, to cover and insulate splices and terminations, as an outer jacket of covered wiring harnesses, and as protective packaging for electronic parts.

DEVELOPMENT PROGRAMS

Properties of existing tubings are adequate for virtually every tubing requirement. Therefore, development programs are aimed primarily at product refinements.

MARKET TRENDS

While usage of nonshrinkable tubing is tied to traditional applications in the electrical industry, shrinkable tubing should find new uses in the more rapidly expanding electronics industry.

INSULATING SLEEVING

INTRODUCTION

Insulating sleeving is similar to tubing in many applications. In construction, flexible tubing is made from unsupported extruded polymers and elastomers, while sleeving consists of coated or uncoated braided fabric.

ASTM and Federal specifications for sleeving, summarized herein, contain requirements and test procedures for widely used sleevings and are an essential part of this section.

PRODUCERS OF COATED SLEEVING
UL APPROVED

Underwriters Laboratories (UL) Standard 1441, *Coated Electrical Sleeving*, lists producers of UL-approved materials.

TECHNOLOGY

Coated sleeving consists of a tubular braided fiber product impregnated and/or coated with a resinous or elastomeric material. The various types of coated sleeving are described in the accompanying ASTM standards and Federal specifications. Commonly used products include the following, arranged by coating type (all on braided glass fiber sleeving):

- Acrylic thermal Class 155
- Oleoresinous varnish (ASTM D 372, MIL-I-3190C), thermal Class 105
- Polyimide (MIL-I-3190C), thermal Class 220
- Polytetrafluoroethylene (ASTM D 372), thermal Class 220
- Polyurethane thermal Class 130
- Polyvinyl chloride (ASTM D 372, MIL-I-3190, MIL-631, MIL-I-21557), F thermal Class 105
- Silicone resin (ASTM D 372, MIL-I-3190C), thermal Class 200
- Silicone rubber MIL-I-3190C, MIL-I-18057), thermal Class 200

Chemistry of the coatings is described in Chapter 8, pp. 244–257, *Insulating Coatings and Impregnants*. Substrates are made from braided fibers of glass (the predominant material), cotton, rayon, nylon, and dacron. E glass yarns used for most coated sleeving is useful up to 1,200°F (650°C). Special glass yarns retain life up to 1,600°F (870°C).

The traditional coating method consists of drawing lengths of sleeving onto wire mandrels, singeing off the hairs, dipping in varnish, draining, and baking. Dipping, draining, and baking steps are repeated until the desired number of varnish coats have been applied. The finished sleeving is then removed from the mandrels and trimmed to the specified length.

In another process, a steel ball is inserted in the untreated sleeving, which is then passed through a heated die constructed so the ball will not pass through. The sleeving is thus formed with circular cross section, after which it is wound under tension on a rectangular frame. The wound frame is dipped in varnish and baked the desired number of times, after which straight lengths of sleeving are cut from the frame.

In a process which produces continuous lengths, sleeving is coated in much the same manner as varnished cambric—by traveling through tanks of varnish and curing in vertical ovens. The tubing thus produced tends to be oval in cross section, caused by passing around rollers at the top of the oven. 2–12 return passes are required.

As well as being coated, sleeving may be used without any or with little impregnant, usually prior to subsequent varnish treatment and baking for a complete assembly. Sleeving provides better protection against mechanical abuse than extruded tubing, but it cannot be made to conform to wires and parts by shrinking action.

Polyvinyl chloride coatings are applied as plastisols, typically with 50 parts of plasticizer to 100 parts of resin. *Acrylics* are applied as water emulsions. *Silicone rubber* coatings are applied by extrusion directly over braided fabric or as solvent base elastomers. *Other resin* coatings are applied in organic solvents.

GRADES

Grades for all sleeving covered by standards and specifications summarized in this section are commercially available.

ASTM STANDARDS FOR ELECTRICAL INSULATING SLEEVING

The American Society for Testing and Materials has developed the following standards for electrical insulating sleeving. For a listing of sleeving types and related standards, see "Technology" (p. 394).

• D 350, *Flexible Treated Sleeving Used for Electrical Insulation*. This standard describes methods of testing flexible treated sleeving used as electrical insulation. Table 13-16 lists test methods:

• D 372, *Flexible Treated Sleeving Used for Electrical Insulation*. This standard covers six types of flexible treated sleeving as follows:

**Table 13-16   Property Tests for Flexible Treated Sleeving
(ASTM D 350)**

PROPERTY MEASURED	ASTM METHOD
Brittleness temperature	D 746
	D 876
Compatibility with magnet wire	D 350
	D 2307
Dielectric breakdown voltage	D 149
Dimensions	D 350
Flammability	D 350
Flexibility after heat aging	D 350
Oil resistance	D 350
Sampling	D 350
Test conditions	D 618
Thermal endurance	D 350

Type 1—A flexible treated sleeving made from organic-base materials such as cotton, rayon, nylon, or other organic fibers and impregnated or coated with an organic substance capable of operation at 105°C (221°F).

Type 2—A flexible treated sleeving made from inorganic-base materials such as fibrous glass and impregnated or coated with a thermosetting organic substance capable of operation at 130°C (266°F).

Type 3—A flexible treated sleeving made from inorganic-base materials such as fibrous glass and impregnated or coated with a poly(vinyl chloride) plastic compound capable of operation at 105°C (221°F).

Table 13-17   ASTM STANDARD SIZES FOR FLEXIBLE SLEEVING (ASTM D 372)

| SIZE | INSIDE DIAMETER, in (mm) | |
	MAX	MIN
1 in.	1.036 (26.3)	1.000 (25.4)
7/8 in.	0.911 (23.2)	0.875 (22.2)
3/4 in.	0.786 (20.0)	0.750 (19.1)
5/8 in.	0.655 (16.7)	0.625 (15.9)
1/2 in.	0.524 (13.3)	0.500 (12.7)
7/16 in.	0.462 (11.7)	0.438 (11.1)
3/8 in.	0.399 (10.1)	0.375 (9.5)
No. 0	0.347 (8.8)	0.325 (8.3)
No. 1	0.311 (7.9)	0.289 (7.3)
No. 2	0.278 (7.1)	0.258 (6.6)
No. 3	0.249 (6.3)	0.229 (5.8)
No. 4	0.224 (5.7)	0.204 (5.2)
No. 5	0.198 (5.0)	0.182 (4.6)
No. 6	0.178 (4.5)	0.162 (4.1)
No. 7	0.158 (4.0)	0.144 (3.7)
No. 8	0.141 (3.6)	0.129 (3.3)
No. 9	0.124 (3.2)	0.114 (2.9)
No. 10	0.112 (2.8)	0.102 (2.6)
No. 11	0.101 (2.6)	0.091 (2.31)
No. 12	0.091 (2.31)	0.081 (2.06)
No. 13	0.082 (2.08)	0.072 (1.83)
No. 14	0.074 (1.88)	0.064 (1.63)
No. 15	0.067 (1.70)	0.057 (1.45)
No. 16	0.061 (1.55)	0.051 (1.30)
No. 17	0.054 (1.37)	0.045 (1.14)
No. 18	0.049 (1.25)	0.040 (1.02)
No. 20	0.039 (0.99)	0.032 (0.81)
No. 22	0.032 (0.81)	0.025 (0.64)
No. 24	0.027 (0.69)	0.020 (0.51)

(Reprinted with permission from the *Annual Book of ASTM Standards*. Copyright, ASTM, 1916 Race Street, Philadelphia, PA 19103.)

Table 13-18   Grade Designations and Requirements for Flexible Treated Sleeving (ASTM D 372)

TYPE	GRADE	DIELECTRIC BREAKDOWN VOLTS C-48/23/50*		COMPATIBILITY WITH MAGNET WIRE DIELECTRIC BREAKDOWN % OF ORIGINAL AFTER EXPOSURE TO TEMPERATURE FOR 672 h		THERMAL ENDURANCE EXTRAPOLATED TEMPERATURE
		MINIMUM AVERAGE	MINIMUM INDIVIDUAL	°C	PERCENT	°C 20,000 h
1	A	7,000	5,000	130	50	105
	B	4,000	2,500	130	50	105
	C-1	2,500	1,500	130	50	105
	C-2	800	800	130	50	105
2	A	7,000	5,000	155	50	130
	B	4,000	2,500	155	50	130
	C-1	2,500	1,500	155	50	130
	C-2	1,500	800	155	50	130
3	A	8,000	6,000	155	50	105
	B	4,000	2,500	155	50	105
	C-1	2,500	1,500	155	50	105
	C-2	1,500	800	155	50	105
4	A	7,000	5,000	225	50	200
	B	4,000	2,500	225	50	200
	C-1	2,500	1,500	225	50	200
	C-2	1,500	800	225	50	200
5	A	8,000	6,000	225	50	200
	B	4,000	2,500	225	50	200
	C-1	2,500	1,500	225	50	200
	C-2	1,500	800	225	50	200
6	A	7,000	5,000	180	50	155
	B	4,000	2,500	180	50	155
	C-1	2,500	1,500	180	50	155
	C-2	1,500	800	180	50	155

*Conditioned for 48 h at 23°C and 50% relative humidity.

Note: Tests are conducted in accordance with ASTM D 350 procedures.

Type 4—A flexible treated sleeving made from inorganic-base materials such as fibrous glass and impregnated or coated with silicone resin or polytetrafluoroethylene materials capable of operation at 200°C (392°F).

Type 5—A flexible treated sleeving made from inorganic-base materials such as fibrous glass and impregnated or coated with silicone elastomer capable of operation at 200°C (392°F).

Type 6—A flexible treated sleeving made from inorganic-base materials such as fibrous glass and impregnated or coated with an insulating material such as some epoxies, polyesters, or acrylics capable of operation at 155°C (311°F).

Standard sizes for flexible sleeving are shown in Table 13-17.

Grade designations and required properties for flexible treated sleeving are shown in Table 13-18.

FEDERAL SPECIFICATION FOR ELECTRICAL INSULATING SLEEVING

The following Federal specification has been developed for sleeving. For a listing of sleeving types and related specifications, see "Technology" (p. 394).

• MIL-I-3190C, *Insulation Sleeving, Electrical, Flexible, Coated, General Specification For*. This is a general specification covering flexible coated sleeving in the classes, types, and categories shown below. The individual requirements for each type are detailed in the specification sheet for each type. The base material of the sleeving shall be glass fiber braid or other synthetic base braids as specified in the applicable specification sheet.

	THERMAL CLASS °(C)	
	130	
	155	
	200	
	220	
TYPE	**COATING**	
A	Nonchlorinated organic resins, singly or in admixture	
B	Polyvinyl chloride, homo- or copolymer	
C	Silicone resin	
D	Silicone rubber	
E	Polyimide	
CATEGORY	**HUMIDITY RESISTANCE**	**PERCENT RETENTION OF DIELECTRIC BREAKDOWN (MINIMUM) 96/23/96***
a	low	30
b	medium	50
c	high	80

*Hours/temperature °C/percent relative humidity.

Standard sizes for sleeving are shown in Table 13-19.

Nominal lengths of sleeving shall be 30-42 inches.

Table 13-19   Standard Sizes for Insulation Sleeving
(MIL-I-3190C)

| SIZE | INSIDE DIAMETER[1] | | | WALL THICKNESS |
	MAXIMUM inch	MINIMUM inch	NOMINAL inch	MAXIMUM inch
No. 24	0.027	0.020	0.022	0.030
No. 22	0.032	0.025	0.027	0.030
No. 20	0.039	0.032	0.034	0.030
No. 18	0.049	0.040	0.042	0.030
No. 17	0.054	0.045	0.047	0.030
No. 16	0.061	0.051	0.053	0.030
No. 15	0.067	0.057	0.059	0.030
No. 14	0.074	0.064	0.066	0.045
No. 13	0.082	0.072	0.076	0.045
No. 12	0.091	0.081	0.085	0.045
No. 11	0.101	0.091	0.095	0.045
No. 10	0.112	0.102	0.106	0.045
No. 9	0.124	0.114	0.118	0.045
No. 8	0.141	0.129	0.133	0.045
No. 7	0.158	0.144	0.148	0.045
No. 6	0.178	0.162	0.166	0.045
No. 5	0.198	0.182	0.186	0.045
No. 4	0.224	0.204	0.208	0.045
No. 3	0.249	0.229	0.234	0.045
No. 2	0.278	0.258	0.263	0.055
No. 1	0.311	0.289	0.294	0.055
No. 0	0.347	0.325	0.330	0.055
$3/8$ in	0.399	0.375	0.387	0.055
$7/16$ in	0.462	0.438	0.450	0.065
$1/2$ in	0.524	0.500	0.512	0.065
$5/8$ in	0.655	0.625	0.640	0.065
$3/4$ in	0.786	0.750	0.768	0.075
$7/8$ in	0.911	0.875	0.893	0.075
1 in	1.036	1.000	1.018	0.075

[1]The minimum inside diameters for sizes No. 20–0, inclusive, are the same as the diameters of copper wire for corresponding AWG or Brown and Sharpe gage sizes.

Test procedures for all sleeving types are described with reference, where appropriate, to applicable ASTM standards. Property requirements include the following:

Visual examination
Dimensional measurement
Dielectric breakdown
　Straight
　90° bend
Resistance to potting temperature
Cold brittleness
Flammability
Oil and solvent resistance

Hydrolytic stability
Push back after heat aging
Thermal endurance
Thermal stability
Fungus resistance

PROPERTIES

Property values for commonly used sleeving types are shown in the accompanying summary of ASTM Standard D 372, *Specification for Flexible Treated Sleeving Used for Electrical Insulation*.

In general:

• Oleoresinous varnish and polyvinylchloride coated sleevings are low cost, flexible, and resistant to attack by hot oil (105°C (221°F)), but limited to 105°C (221°F) use.

• Polyurethane-coated exhibits outstanding toughness and is unaffected by hot oil (105°C (221°F)). It is useful up to 130°C (266°F).

• Acrylic and modified acrylic (more flexible) coated sleevings have high resistance to acids, organic solvents, and water. Films are tough and compatible with most magnet wire enamels. These sleevings are useful up to 155°C (311°F).

• Silicone rubber coated sleeving is very flexible, has excellent aging characteristics, and is useful up to 200°C (392°F).

• Silicone resin and polytetrafluoroethylene resin coated sleevings are high priced and are for use at temperatures higher than those for which other sleevings are qualified.

ELECTRICAL/ELECTRONIC USES

Insulating sleevings are used primarily over wires where other insulation has been stripped away or where additional insulation or mechanical protection is needed.

DEVELOPMENT PROGRAMS

Since sleevings are available in all thermal Classes and degrees of flexibility, the requirements of design engineers can usually be met with existing products, and development is focused on product refinements.

MARKET TRENDS

Sleeving usage is directly tied to production trends in the electrical industry. There is growing usage of silicone and other high-temperature coatings at the expense of vinyl and traditional varnish coatings.

––––––– 14 –––––––

Mica and Mica Products

INTRODUCTION

Mica, a term derived from the Latin word *micare* (to shine or glitter), was mined in India for medicinal purposes as early as 2000 B.C. In medieval times, it was used for windows in houses because of its transparency, and later in stoves because of that property plus its resistance to fracture and heat. Mica has been used as a dielectric since the earliest days of the electrical industry. Principal manufacturers of mica products include Asheville-Schoonmaker, Cleveland Mica Insulating Materials, Macallen, Midwest Mica, Spruce Pine Mica, Tar Heel Mica, and U.S. Samica.

TECHNOLOGY AND GRADES

Micas are complex silicates of several types, of which two are used commercially as dielectrics:

• Muscovite mica, also referred to as India white, or ruby mica, mined primarily in India and Brazil, and in smaller amounts in Angola, Tanganyika, Argentina, and the United States

• Philogopite mica, also referred to as amber mica, mined in Madagascar, with small deposits in Canada and Mexico.

Table 14-1 compares the ranges of principal compounds in these mica types.

TABLE 14-1   Composition Ranges of Muscovite and Phlogopite Mica

	MUSCOVITE (%)	PHLOGOPITE (%)
SiO_2	45–47	38–43
Al_2O	30–38	12–17
K_2O	9–12	9–11
MgO	trace	23–29

There are also present traces of oxides of iron, calcium, lithium, sodium, titanium, and barium.

Mica is mined as blocks, plates, or slabs (7 mils and over thick), as films (0.8–4 mils thick), and as splittings (thin laminae split from blocks). After mining, blocks are trimmed to remove worthless layers and ragged edges. Full-trimmed mica is mica trimmed on all sides with all cracks, reeves (pockets), and cross grains removed. Mica has virtually perfect basal cleavage, permitting it to be split to exceptional thinness. Mica is often contaminated with impurities which render it unsuitable for dielectric applications.

The grading of muscovite mica, the most widely used as electrical insulation, is covered by ASTM Standard D 351, *Natural Muscovite Block Mica and Thins Based on Visual Quality*, and includes these categories:

V-1 Clear—Hard, of uniform color, flat, free from all stains and foreign inclusions, waves, cracks, buckles, and other similar defects.

V-2 Clear and Slightly Stained—Hard, of uniform color, nearly flat, free from all vegetable and mineral stains, cracks, buckles, and other similar defects, and foreign inclusions, except for a few tiny air inclusions in not more than one-fourth of the usable area. Slight crystallographic discoloration is permitted to a limited extent.

V-3 Fair Stained—Hard, of uniform color, free from all vegetable and mineral stains, cracks, buckles, and other similar defects, and foreign inclusions, except may be slightly wavy, and may contain slight air inclusions in not more than one-half of the usable area. These inclusions should not be concentrated. Slight crystallographic discoloration is permitted to a limited extent.

V-4 Good Stained—Hard, free from vegetable or mineral stains, cracks, buckles, and other similar defects, and foreign inclusions, except may be medium wavy but not rippled, and may contain medium air inclusions in not more than two-thirds of the usable area, but may not have heavily concentrated air inclusions in any of the usable area. Slight crystallographic discoloration is permitted to a limited extent.

V-5 Stained A Quality—Hard, free from cracks and other similar defects and foreign inclusions, except may be medium wavy and may contain slight vegetable stains, and the entire area may have air inclusions if not heavily concentrated. Crystallographic discoloration is permitted.

V-5.1 Stained A1 Quality—Hard, free from cracks and other similar defects and foreign inclusions, except may be wavy and may contain slight vegetable stains, slight mineral stains not exceeding two specks within the usable area, and the entire area may have air inclusions if not heavily concentrated over more than area equivalent to $1/4$ inch square (6.4 mm square) for grade 5 and up and over more than an area equivalent to $1/8$ inch square (3.2 mm square) for grade $5^1/_2$ and below. Crystallographic discoloration is permitted.

V-6 Stained B Quality—Hard, free from cracks, and other similar defects and foreign inclusions, except may be wavy and slightly buckled, and may contain heavy inclusions of air, medium vegetable, clay, and mineral stains. Crystallographic discoloration is permitted.

V-7 Heavy Stained—Hard, free from cracks and other similar defects and foreign inclusions, except may be wavy and buckled, and may contain heavy air inclusions, heavy vegetable and medium mineral stains. Crystallographic discoloration is permitted.

V-8 Densely Stained—May be soft and may contain heavy stains and inclusions, waves, cracks, buckles, and other defects. Crystallographic discoloration is permitted.

V-9 Black Dotted—Hard, free from cracks and other similar defects, but may be medium wavy, and may contain heavy air inclusions, vegetable stains, and dispersed black dots. Crystallographic discoloration is permitted.

V-10 Black Spotted—Hard, free from cracks and similar defects and foreign inclusions, except may be medium wavy and contain slight buckles and vegetable stains, black spotted or red dotted mineral stains, and heavy air inclusions. Crystallographic discoloration is permitted.

V-11 Black Stained—Hard, may contain medium waves, heavy air inclusions, smoky stains, black stains and red dots (mineral), green stains (vegetable type), and sand blast, medium black stains (mineral), slight red stains (mineral), and clay stains. Crystallographic discoloration is permitted.

V-12 Black/Red Stained—V-11 Quality, but may be soft and have black lines and/or short red bars or connected stains.

Federal Specification HH-I-536 also defines mica grades. Table 14-2 lists equivalent HH-I-536 and ASTM D 351 grades.

TABLE 14-2   Equivalent Grades of Mica

ASTM STANDARD D 351	FEDERAL SPECIFICATION HH-I-536
V-1	NMC_1
V-2	NMS_s
V-3	NMF_s
V-4	NMG_s
V-5	NMSA
V-6	NMSB
V-7	NMH_s
V-8	NMBD
V-9	$NMBS_p$
V-10	$NMBS_t$

**TABLE 14-3   Grade Sizes of Muscovite Uncut Mica
Block and Thins
(ASTM D 351)**

ASTM GRADE SIZES	AREA OF MINIMUM RECTANGLE		MINIMUM DIMENSION OF ONE SIDE	
	in^2	cm^2	in	mm
OOEE Special	100	650	4	100
OEE Special	80	520	4	100
EE Special	60	390	4	100
E Special	48	310	4	100
A-1 (Special)	36	235	$3^1/2$	89
No. 1	24	155	3	76
No. 2	15	97	2	51
No. 3	10	65	2	51
No. 4	6	40	$1^1/2$	38
No. 5	3	20	1	25
No. $5^1/2$	$2^1/4$	15	$7/8$	22
No. 6	1	6.5	$3/4$	19

(Reprinted with permission from the *Annual Book of ASTM Standards.* Copyright, ASTM, 1916 Race Street, Philadelphia, PA 19103.)

For critical dielectric applications, grades used are V-6 or better.

ASTM D 351 also establishes grade sizes based on the area of a rectangular surface of the mica containing the maximum usable surface within the specimen, and specifies the minimum dimension of one side for each grade as shown in Table 14-3.

Splittings may be loose or book packed. Loose packed splittings are irregular in shape and are randomly boxed, whereas book-packed splittings are consecutive laminae split from the same block and stacked as pages in a book. The size and thickness range of muscovite splittings are covered by ASTM D 2131, *Natural Muscovite Mica Splittings*, as shown in Table 14-4.

**TABLE 14-4   Requirements for Size and Average
Thickness of Muscovite Mica Splittings
(ASTM D 2131)**

ASTM GRADE	FORM	SIZE in² (cm²)	MINIMUM DIMENSION OF USABLE RECTANGLE in (mm)	THICKNESS OF TEN SPLITTINGS in (mm)
4	bookform	6–10 (39–65)	1¹/₂ (38)	0.006–0.009 (0.15–0.23)
5	bookform	3–6 (19–39)	1 (25)	0.006–0.009 (0.15–0.23)
5¹/₂	bookform	2–3 (13–19)	⁷/₈ (22)	0.006–0.009 (0.15–0.23)
6	bookform	1–2 (7–13)	¹/₄ (19)	0.006–0.10 (0.15–0.25)
4	loose with powder	6–10 (39–65)	1¹/₂ (39)	0.006–0.009 (0.15–0.23)
5	loose with powder	3–6 (19–39)	1 (25)	0.006–0.009 (0.15–0.23)
5¹/₂	loose with powder	1¹/₂–3 (10–19)	⁷/₈ (22)	0.007–0.010 (0.18–0.25)
6	loose with powder	1–1¹/₂ (6–10)	³/₄ (19)	0.007–0.010 (0.18–0.25)
6—1st	loose	1–1¹/₂ (70% min) (6–10) (70% min)		0.007–0.010 (0.018–0.25)
6 inter	loose	1–1¹/₂ (60% min) (6–10) (60% min) ³/₄–1 (25% min) (5–6) (25% min)		0.007–0.010 (0.18–0.25)
6—2nd	loose	1–1¹/₂ (50% min) (6–10) (50% min)		0.007–0.011 (0.18–0.28)
6—3rd	loose	³/₄ (65% min) (5) (65% min)		0.007–0.011 (0.18–0.28)
6—4th	loose	¹/₂ (30% min) (3.2) (30% min)		0.007–0.012 (0.18–0.30)

(Reprinted with permission from the *Annual Book of ASTM Standards*. Copyright, ASTM, 1916 Race Street, Philadelphia, PA 19103.)

SYNTHETIC MICA

Mica may also be grown electrothermally with a composition similar to natural mica, except without water content. A modified mica, fluorophlogopite, has excellent dielectric properties and higher thermal endurance than natural mica. With natural mica, when a temperature is reached where its water content is driven off (1,000°F for muscovite, 1,400°F for phlogopite), mica calcines, or crumbles, and is no longer useful unless bonded with high-temperature inorganic binders such as silicates, borates, and phosphates.

Synthetic mica splittings are used exclusively for glass-bonded mica insulations and are not available as splittings. (See "Glass-Bonded Mica," p. 412 in this chapter.)

PROPERTIES

The outstanding dielectric and thermal endurance, inertness, and nonflammable characteristics of mica separate it from virtually all other insulating materials. Mica is also highly resistant to the harmful effects of corona and radiation. However, without further fabrication into composites, splittings are usually limited to no more than 10 square inches in size.

Key property values for muscovite ruby, phlogopite amber, and fluorophlogopite synthetic micas are shown in Table 14-5.

TABLE 14-5   Key Property Values for Natural and Synthetic Micas (High Quality)

	MUSCOVITE RUBY NATURAL	PHLOGOPITE AMBER NATURAL	FLUORO-PHLOGOPITE SYNTHETIC
Specific gravity	2.8	2.8	2.9
Specific heat, cal/gm	0.207	0.207	0.194
Hardness (mohs)	2.8–3.1	2.5–3.0	3.0–3.5
Dielectric strength*			
3 mils thick, volts/mil	3,000	3,000	3,000
Dielectric constant*			
10^6 Hz	6.5–8.7	5–6	6.5
Dissipation factor*			
10^6 Hz	—	—	0.0002
Maximum use temperature, °F	1,000	1,400	1,800
(°C)	(540)	(760)	(980)

*At room temperature.

ELECTRICAL/ELECTRONIC USES

Block mica and mica films are used extensively as dielectrics in electronic capacitors (ASTM D 748). Mica provides a transparent path for microwaves, and is used in microwave windows. Other dielectric applications include stamped washers and other shapes for electrical/electronic equipment. There is still some use as spacers in the vacuum tubes which remain in the market.

BUILT-UP MICA

To make mica more useful, layers of splittings are bonded together with a variety of resins, including shellac, epoxy, alkyd, alkyd-vinyl, polyester, and silicone. The layering of splittings, originally done by hand, is now also done mechanically. This process consists of fluttering down mica flakes in a tower, much like snowflakes in a snowstorm. The mica flakes form a fairly uniform layer on a tray or moving belt. The desired amount of bonding agent is applied and the process is repeated, or when a belt is used, it carries the composite under another tower. Alternatively, powdered resin may be interspersed with the mica flakes in the tower. A still newer method involves a rectangular cowl fitting over a box of splittings. The mouth of the cowl is covered by a wire screen. Suction over the screen pulls splittings up against the screen. The mica box is drawn to one side and a perforated tray is substituted. The suction is released, and the layer of splittings is deposited on the tray. As this tray is withdrawn, it passes under a varnish sprinkler. The process is repeated to build the desired number of layers. There are many variations of these methods. After build-up, the composite is then cured in a multiple-opening, heated, hydraulic press. Milling of the cured mica plates brings them to the specified thickness.

Molding plate, although rigid at room temperature, softens when heated, and the splittings are able to slip relative to each other. The plate then becomes readily moldable. Binders include shellac, epoxy, silicone, and alkyd-vinyl. Binder content for this type of plate is relatively high, up to 25 percent for silicone binder. Uses include commutator V-rings, motor slot cells, bushings, washers, coil forms, sleeves, and transformer coil forms. Molding plate is available in thicknesses from 0.015 through 0.062 inch. Sheets are 36 × 36 inches.

Segment plate is made by laminating several layers of machine-laid mica splittings. Binders are the same as for molding plate, but binder content is less than 10 percent and the plate is fully cured, featuring maximum density combined with minimum compression when subjected to high temperatures and pressures. This material is used primarily in commutators for insulation between copper bars. Segment plate is available in thicknesses from 0.015 through 0.062 inch. Sheets are 36 × 36 inches.

Flexible plate contains binders of polyester, epoxy, and silicone. Binder contents is 10–15 percent. Sheets are cured to have optimum flexibility for up to six months or more. Flexible plate features high resistance to moisture, heat, and cut-through, high dielectric strength, and imperviousness to attack by oil. Uses include armature end-turn insulation, slot liners, field coil separators, and commutator core insulation. It is also used for transformer coil insulation, and insulating collars, channels, and rings. Flexible plate is available in thicknesses from 0.005 through 0.030 inch. Sheets are 36 × 36 inches and 36 × 72 inches.

Flexible composites use mica splittings backed and/or faced with reinforcing materials such as woven glass, glass mat, polyester film, polyester mat, polyimide film, polyamide paper, and electrical grade paper. Binder content is 10–20 percent with binders of alkyd, alkyd modified with pine tar, epoxy, polyester, and silicone. These composites are used where the high dielectric strength and high heat resistance of mica and the physical properties of the backing and facing materials are required. Applications include motor slot liners, coil and cell wrappers, and phase, layer, and core insulation. The governing military specification is MIL-I-3505, *Insulation Sheet and Tape, Electrical, Coil and Slot, High Temperature*. Flexible composites are available in thicknesses from 0.007 to 0.030 inch. Sheets are 36 × 36 inches.

Mica splittings tape is made by hand laying splittings selected for quality, large size, and uniformity in thickness and shape. Machine-laid tapes are also produced with B-stage epoxy binder. Because of limited shelf life, these tapes require cold storage if they are not to be used immediately. Woven glass cloth, typically 0.002 inch thick, is used as reinforcement on one or both sides of the mica, or on one side with 0.00025 inch polyester film on the other side to increase slipping property. Mica thickness is 0.002–0.008 inch. The product is available in rolls 36 inches wide by 36 or 50 yards long, or slit in widths from $^{1}/_{2}$ inch to 36 inches by 36 yards long.

MICA PAPER PRODUCTS

The manufacture of mica paper, commonly referred to as reconstituted mica, is described in Chapter 11, *Dielectric Papers and Boards*. In addition to flexible composites used as wrappers and tapes and covered in that chapter, these mica paper products are produced in forms similar to mica splittings products:

Mica paper molding plate contains up to 25 percent B-stage epoxy or silicone binder. Available in thicknesses from 0.010 through 0.062 inch.

Mica paper segment plate contains 7–14 percent fully cured epoxy or inorganic binder. Available in thicknesses from 0.010 through 0.062 inch.

Heater plate, a fully cured, rigid laminate of either mica paper or mica splittings, contains 5–12 percent silicone or inorganic binder. Available in thicknesses from 0.006 through 0.125 inch. Sheets are 36 × 36 inches. The principal use for heater plate is for the mechanical support and insulation of the heating elements of appliances such as toasters, hot combs, hair driers, irons, and space heaters where temperatures can reach 1,400°F (760°C).

Mica paper tubes, round, square, and rectangular, contain 16–22 percent fully cured epoxy or silicone binder. Minimum wall thickness is 0.020 inch, minimum ID is 0.125 inch, maximum length is 36 inches.

THERMAL CLASSIFICATION OF MICA PRODUCTS

In general, binders determine the thermal Class of mica products:

BINDER	THERMAL CLASS
Shellac	130
Epoxy	155
Alkyd-vinyl	155
Polyester	155
Silicone	180 and 200

ASTM STANDARD FOR BONDED MICA PRODUCTS

The American Society for Testing and Materials has developed the following standard for testing bonded mica products:

• D 352, *Pasted Mica Used in Electrical Insulation*. Covered are methods of testing bonded mica splittings and bonded mica paper. The following tests are described:
Compressive creep
Stability under heat and pressure
Mica or binder content
Molding test
Dielectric strength
Resistivity

FEDERAL SPECIFICATIONS FOR MICA PRODUCTS

The governing Federal specifications for mica products include the following:

• HH-I-538, *Insulation, Electrical, Pasted-Mica*. This specification was approved by the Commissioner, Federal Supply Services, General Services Administration, for the use of all Federal agencies. This specification covers pasted-mica electrical insulation for molded armature slot bases, commutator insulation, shells, heating elements, and other similar insulating purposes. Types and forms are designated as follows:
Type PMR—Pasted mica, rigid
Form S—Sheets (plates)
Type PMM—Pasted mica, moldable
Form R—Rings
Form S—Sheets (plates)

Form T—Tubes
Form X—Special shapes
Type PMF—Pasted mica, flexible
Form S—Sheets (plates)
Type PMH—Pasted mica, heat resistant
Form S—Sheets (plates)

Pasted-mica electrical insulation shall consist of, or contain, muscovite mica conforming to the requirements of ASTM D 2131 unless phlogopite mica is specifically required (which shall conform to the minimum standards for muscovite mica). The pasted mica shall consist of layers of overlapping mica splittings held together by a suitable binding material, which shall be chemically inert with respect to materials normally used in electrical equipment under conditions of rated operating temperature and the presence of humid salt atmosphere.

• MIL-I-3505, *Insulation Sheet and Tape; Electrical, Coil and Slot, High Temperature*. This specification has been approved by the Department of Defense and is mandatory for use by the Departments of the Army, the Navy, and the Air Force. This specification covers high-temperature mica-glass electrical insulation used in the coils and slots of rotating electrical machinery. The following classifications are employed:

Class B (130°C)—With organic varnish
Class H (180°C)—With silicone varnish
Form S—Sheet
Form T—Tape
Type Mg—Pasted mica laminated with one layer of glass fiber fabric
Type Pmg—Pasted mica laminated with one layer of paper on one side and
    one layer of glass fabric on the other side
Type Gmg—Pasted mica sandwiched between single layers of glass fabric
    insulation.

The binding material shall be high-quality insulating material used in the minimum amount needed to obtain specified properties, including tensile strength, flexibility, and dielectric strength.

• MIL-I-21070, *Insulating Sheet and Tape, Electrical, Reinforced Mica Paper*. This specification covers reinforced mica paper electrical insulation used in the coils and slots of electrical machinery. There are three Classes of this material supplied as flat sheets, rolls, and tapes:

Class 130—With organic resin binder and organic varnish coating (if used)
    on glass fiber woven fabric reinforcement
Class 155—With epoxy or polyester resin binder and an epoxy or polyester
    coating (if used) on glass fiber woven fabric reinforcement
Class 180—With silicone resin binder and silicone varnish coating (if used)
    on glass fiber woven fabric reinforcement

Constructions include mica paper backed and faced with woven glass or varnished woven glass fabric, or polyester film. Required performance characteristics for various constructions include dielectric strength, volume resistivity, tensile strength, and tear resistance.

GLASS-BONDED MICA

Finely ground mica, either natural or synthetic, may be bonded with a low-melting electrical grade glass to produce ceramic insulations that are both machinable and moldable. These materials are made by Mykroy/Mycalex Ceramics, with trade name Mykroy/Mycalex.

Glass-bonded mica products are covered by ASTM standard D 1039, *Standard Methods of Testing Glass-Bonded Mica Used as Electrical Insulation.*

The governing military specification for glass-bonded mica is MIL-I-10, *Insulating Materials, Ceramic.*

Commercially available Mykroy/Mycalex grades and their property values are shown in Table 14-6. Noteworthy characteristics include:

- Exceptionally high-dimensional stability
- Service temperatures up to 1,200°F (650°C)
- Good dielectric properties
- Nonflammable
- No arcing, tracking, or outgassing
- Impervious to moisture
- High mechanical properties
- Close tolerance molding and machining
- Thermal expansion coefficient close to that of steel used for molds

Typical electrical/electronic uses include circuit breakers, switchgear, arc barriers, connectors, relays, bushings, brush holders, coil forms, and printed circuit substrates.

DEVELOPMENT PROGRAMS

Because binders limit the maximum operating temperatures for mica composites, the search continues for higher thermal class bonding agents. Improved thermal properties would permit greater output from rotating electrical equipment using the same frame, an attribute of special importance for traction equipment.

Scientists at Bharat Heavy Electricals Limited, Bhopal, India, have developed a resin and mica composite system ascribed Class 275 and reported to be useful at temperatures up to 350°C (662°F).[1] The resin, comprised of linear chains, is polybenzimidazole (PBI or PBMI), formed by polymerizing benzimidiazole. Hydrogen bonding is pronounced in the polymer.

benzimidiazole

[1]M. V. Dalal and V. K. Ganpate, "A new 275 Class non-silicone micaceous insulation system," in *Proceedings of the 16th Electrical/Electronics Insulation Conference*, p. 32.

TABLE 14-6 Typical Property Values for Glass-Bonded Mica (Mykroy/Mycalex Grades)

			MACHINING/GRADES SHEET/POD			MOLDING GRADES			
GENERAL PROPERTIES	ASTM METHOD	UNITS	400	500	1100	410/751	555/761	620/1301	371
Specific gravity	C 373		2.5	2.7	2.8	3.8	3.7	3.9	4.8
Density		lb/in^3	0.09	0.10	0.10	0.14	0.13	0.14	0.17
Thermal conductivity		cal/s/cm^2/°C/cm	0.00100	0.00120	0.00140	0.00120	0.00130	0.00125	0.00136
		BTU/h/ft^2/ft/°F	0.24	0.29	0.39	0.29	0.31	0.30	0.31
Moisture absorption			Nil	Nil	Nil	Nil	Nil	Nil	Nil
Coefficient of thermal expansion ($\times 10^{-6}$)		in/in/°C	10.5	11.2	9.4	11.2	11.0	10.3	11.5
		in/in/°F	5.8	6.0	5.2	6.0	6.0	5.5	6.3
Specific heat		cal/gm/°C	0.12	0.12	0.11	0.24	0.24	0.23	0.25
Max continuous		°F	700	700	1100	750	750	1300	700
Operating temp.		°C	370	370	595	400	400	700	371
Flammability						Does not burn			
Mica filler			Natural	Synthetic	Synthetic	Natural	Synthetic	Synthetic	Natural
Radiation resistance (3×10^{10} rads-cobalt)			Good	Better	Excellent	Good	Better	Excellent	Good

ELECTRICAL PROPERTIES

			400	500	1100	410/751	555/761	620/1301	371
Dielectric strength	D 149	V/mil $^1/_8$ in thick	400	400	380	375	375	375	350
Arc resistance	D 495	s	300	300	325	325	325	350	300
Permittivity	D 150	1MHz	6.7	6.9	6.8	9.3/8.4	8.8/8.2	9.5/8.8	—
Dissipation factor	D 150	1MHz	0.0018	0.0013	0.0017	0.0012	0.0013	0.0015	0.006
Loss index	D 150	1MHz	0.012	0.009	0.012	0.0115	0.0110	0.0150	—
Surface resistivity	D 257	dry $\Omega \cdot$ cm (70°F)	10^{16}	10^{16}	10^{16}	10^{15}	10^{15}	10^{15}	10^{15}
Volume resistivity	D 257	$\Omega \cdot$ cm (70°F)	10^{12}	10^{13}	10^{14}	10^{13}	10^{14}	10^{14}	10^{14}
Surface resistivity	D 257	wet $\Omega \cdot$ cm (70°F)	10^{6}	10^{10}	10^{11}	10^{9}	10^{9}	10^{10}	10^{9}
Dielectric constant	D 150	1MHz	6.7	6.9	6.8	6.6	8.8	9.0	12.5

(continued)

TABLE 14-6 (*Continued*)

GENERAL PROPERTIES	ASTM METHOD	UNITS	MACHINING/GRADES SHEET/POD				MOLDING GRADES		
			400	500	1100	410/751	555/761	620/1301	371
MECHANICAL PROPERTIES									
Tensile strength	D 651	psi	6,000	6,000	5,000	6,500	6,000	6,000	6,500
Flexural strength	D 790	psi	15,000	15,000	12,000	12,000	10,000	10,000	11,000
Compressive strength	D 695	psi	45,000	40,000	32,000	33,000	34,000	30,000	35,000
Modulus of elasticity (in tension)		psi	11.0×10^6	12.0×10^6	10.6×10^6	7.0×10^6	7.2×10^6	9.0×10^6	—
Hardness—Rockwell H		H	90	90	90	90	90	90	90
Hardness—Brinell		H	56	56	56	56	56	56	56
Impact strength—IZOD (notched)		ft · lb/in	1.8	1.7	1.3	0.7	0.7	0.6	0.65

(Table courtesy of Mykroy/Mycalex.)

PBI is derived by the condensation of 3,3'-diaminobenzidien with dicarboxylic aromatic acids, such as a mixture of isophthalamide and terephthalamide. A solution of PBI resin is obtained by dissolving the resin in a complex solvent system, for example, dimethylformamide and dimethylacetamide in the presence of lithium chloride solution and triboronate catalyst. This solution is suitable for preparation of prepregs and flexible to rigid composites.

The composite consists of mica paper or splittings sandwiched between PBI saturated glass fabric layers and film. With mica paper, 35–40 percent by weight of PBI is required. Mica splittings require only 10–15 percent PBI. The amount of catalyst determines the degree of flexibility or stiffness after curing at 300°C.

MARKET TRENDS

Because of the Toxic Substances Control Act, inorganic bonded mica paper laminates are now widely used in high-temperature applications to replace asbestos products. It is forecast that this trend will continue.

—————————— 15 ——————————

Ceramic and Glass Insulations

INTRODUCTION

Ceramics and glasses are among the oldest dielectrics used in electrical equipment and in the transmission and distribution of power. A variety of ceramic and glass materials is available from several manufacturers, including the following:

MATERIALS	SUPPLIERS
Alumina	Alcoa, Kaiser, Ormet, Reynolds, Brush Wellman, Star
Aluminum nitride	Karamont
Aluminum silicate	Saxonburg
Beryllia	Brush Wellman
Boron nitride	General Electric (Borazon™), Carborundum (Combat™)
Cordierite	Saxonburg, Star, D. M. Stewart
Diamond	Norton
Magnesia	Dow Chemical
Porcelain	Southern Porcelain, Star, Coors
Sapphire	Crystal Systems, Tyco Laboratories, Union Carbide
Silica (fused)	Amersil, Corning Glass, General Electric, Raytheon, Thermal American, U.S. Fused Quartz
Steatite	D. M. Stewart, Saxonburg, Star
Zircon	Corning Glass, 3M

This chapter covers ceramics and glasses used for electrical insulation other than glass-bonded mica (see "Glass-Bonded Mica," p. 412 of Chapter 14, *Mica and Mica Products*) and substrates for printed circuits (see "Substrates for Printed Circuits," p. 285 of Chapter 9, *Unclad and Clad Structures*).

TECHNOLOGY

Ceramic structures are characterized by outstanding resistance to normal aging and the ability to withstand high operating temperatures, heat shock, and electrical degradation. However, it is to be noted that the structural strength of ceramics holds to much higher temperatures than do dielectric properties. The point at which there is a rapid increase in dielectric loss is referred to as the dielectric transition temperature. Above this temperature, dielectric breakdown occurs rapidly, typically accompanying localized heating and the establishment of high thermal gradients within the ceramic body, which may cause structural destruction.

CHEMISTRY

Early ceramics were composed of natural clays (hydrated aluminum silicates) with the generalized formula $Al_2O_3SiO_2 \cdot xH_2O$. Several materials used as dielectrics are now classed as ceramics. In general, they are crystalline structures bonded together by a smaller amount of a vitreous matrix formed during the firing process. The ratio of crystalline and amorphous phases significantly affects the dielectric and physical properties of the ceramic. The formation and rate of growth of the crystalline phase is determined by the temperature versus time profile of the firing operation. The firing temperature for most ceramics is in the range of 1,200° to 1,700°C (2,192°–3,092°F). By contrast, glass is technically an amorphous liquid of extremely high viscosity, composed of silica (75 percent), soda ash (20 percent), and lime (5 percent), often combined with metallic oxides such as those of calcium, lead, lithium, and cerium. The principal ceramics used in electrical insulations include:

- *Alumina*, or aluminum oxide (Al_2O_3), occurs naturally as the mineral corundum and in hydrated form in bauxite. Virtually all commercially produced alumina involves a caustic leach of the bauxite at elevated temperature and pressure. Lime is added to control the phosphorus content and to increase the solubility of alumina. The resulting slurry, 40–50 percent solids, is pumped to pressurized digesters, where high-pressure steam is used to raise the temperature. Alumina and some of the silica are dissolved during this step, soluble sodium aluminate is formed, and a complex sodium aluminum silicate is precipitated. The resulting slurry containing sodium aluminate in solution is cooled to atmospheric boiling temperature, and a coarse sand fraction is removed by gravity separators or wet cyclones. The liquor is clarified and cooled until it becomes supersaturated, and then is seeded with crystals of alumina trihydrate. Alumina is precipitated as the trihydrate. It is then filtered and washed, after which it is calcined (heated below the melting point) to drive off the water or hydration. High-grade ceramic shapes with melting point of 2,040°C (3,700°F) are made from alumina calcined at 1,480° to 1,590°C (2,700°–2,900°F).

- *Aluminum nitride* (AlN) powders are sintered at very high temperatures in nitrogen atmosphere to avoid its oxidation to alumina.

- *Aluminum silicate* ceramics contain varying amounts of alumina (Al_2O_3) and silica (SiO_2). They are made synthetically by heating aluminum fluoride at 1,000° to 1,200°C (1,830°–2,200°F) with silica and water vapor.

- *Beryllia*, or beryllium oxide (BeO), is derived from beryl ($Be_3Al_2Si_6O_{18}$), the ore of beryllium. the ore is wet-ground and screened to form a slurry that is transported to a processing plant, where it is leached with a 10 percent sulfuric acid solution. The insoluble residue is filtered and discarded. The remaining beryllium sulfate solution is treated by a solvent-extraction process using di-2-ethylhexyl phosphoric acid in kerosene. The beryllium in solution is subsequently converted to beryllium hydroxide, which is then calcined to form beryllia. Beryllium is the lightest structural metal known. The Occupational Safety and Health Administration (OSHA) has designated beryllium and its compounds highly toxic with a worker exposure limit of 0.002 milligram per cubic meter of air for an 8 hour time-weighted average.

- *Boron nitride* (BN) is formed at very high pressure (85,000 atmospheres) and temperatures (1,800°C, 3,270°F) from mixtures of boron and nitrogen, or from hexagonal boron nitride in the presence of catalysts such as lithium, calcium, magnesium, or their nitrides. Boron occurs in nature as sodium borate (borax), $Na_2B_4O_7 \cdot 10\ H_2O$. Boron is derived by several processes, including reduction by heating boric oxide with powdered metals (sodium, magnesium, aluminum); by vapor phase reduction of boron trichloride with hydrogen over hot filaments; and by electrolysis of fused salts. Boron dust ignites spontaneously in air, presenting a severe fire and explosion hazard.

• *Cordierite* is a naturally occurring magnesium aluminum silicate (2 MgO · 2 Al_2O_3 · 5 SiO_2).

• *Diamond* films are produced by mixing methane and hydrogen in a plasma environment and depositing pure diamond films on a substrate.

• *Magnesia*, or magnesium oxide (MgO), is produced by calcining magnesium carbonate or magnesium hydroxide; by treating magnesium chloride with lime and heating; and from sea water via the hydroxide. Two forms are produced. A light, fluffy material is prepared by a relatively low-temperature dehydration of the hydroxide. The other form is a dense material made by high-temperature furnacing of the oxide after it has been formed from the carbonate or hydroxide. Magnesia is toxic by inhalation of fumes. OSHA has placed a worker exposure limit of 15 milligrams per cubic meter of air for an 8 hour time-weighted average.

• *Porcelain*, or potassium aluminum silicate (4 K_2O · Al_2O_3 · 3 SiO_2), is formed by sintering mixtures of clays, quartz, and feldspar (potassium alumino-silicate). Ultra porcelain, the highest mechanical and dielectric grade, is composed of aluminum oxide, barium oxide, and clay.

• *Quartz*, or crystallized silicone dioxide (silica, SiO_2), is produced by crystal growth from molten silica under carefully regulated conditions. For a discussion of crystal growth, see "Semiconductor Silicon" (p. 286) in Chapter 9, *Unclad and Clad Structures*. Fused silica is obtained by the electric-arc fusion of high-purity quartz sand. Ingots thus formed are then crushed to silica grains for casting and kiln bonding.

• *Sapphire*, or corundum (Al_2O_3), is a crystalline variety of alumina. Purified, synthetic sapphire is made by crystal-growing techniques. For a discussion of these techniques, see "Semiconductor Silicon" (p. 286) in Chapter 9, *Unclad and Clad Structures*. In a proprietary EFG process, Saphikon (a Division of Tyco Laboratories) produces continuous lengths of sapphire crystal in almost any cross-sectional configuration from furnaces operating at almost 2,200°C (4,000°F).

• *Steatite* is a mixture of hydrous magnesium silicate, clay, and alkaline oxides.

• *Zircon*, or zirconium silicate ($ZrSiO_4$, or ZrO_2 · SiO_2), used in ceramics is a purified form of a naturally occurring mineral.

A new process for making ceramic multilayer circuit boards is based on co-firing alumina layers. Du Pont offers a system called *Green Tape*. Mixtures of alumina, glass, and other powders are combined with a plastic and a solvent, producing an uncured flexible tape. Tapes are cut, punched, and etched, and then holes are interconnected. The stack is pressed in a plastic press to make a multilayer laminate. The laminate is then fired at 800°C (1,472°F) which fuses the ceramic and burns out the plastic.

GRADES

Grades are available with varying degrees of basic mineral content with over 99.99 percent purity for single crystals such as quartz, and with up to 99.9 percent purity for other ceramic powders. Manufacturers of ceramic parts, including custom fabricators, often formulate compositions to meet their own special requirements by blending ceramic powders with binders and other additives.

PROCESSING

Ceramic powders are usually wet blended with enough water to wet the particles. Excess water causes the slurry to settle and should be avoided. It is important to use deionized water and lined tanks to minimize powder contamination. The slurry may then be vibro-energy-milled using ultrasonics, or ball-milled. This process breaks up agglomerates and produces a narrow size range of discrete particles. After milling, lubricants and organic binders, such as sodium lignum sulfonate, may be added during agitation. Moisture may then be completely removed by spray drying. The resulting particles flow freely to fill dies evenly with minimum voids and pits. The "green" parts thus obtained are then sintered, or fused, in high-temperature electric kilns with precise process controls. Organic components are burned out during this operation.

A wet process commonly used to manufacture large ceramic pieces is similar to that above, but omits the organic lubricants and binders. A portion of the water in the slurry is removed by filter press to make a putty or paste. The putty or paste is then formed into the desired shape, and subsequently dried to a sufficiently low moisture content so that it can be kiln fired without cracking.

In a simpler process, the ceramic powder is mixed with just enough water to make a stiff paste or putty, after which the desired shape is formed, dried, and fired at high temperature with exposure to air.

A "dry" process uses less water and produces parts with good dimensional properties. Mechanical and dielectric properties, however, are inferior to those obtained by the wet process.

During the kiln firing, crystalline growth proceeds at a rate which depends on the basic constituents, the temperature, and the duration of firing. Insulator porcelain is fired at 1,250°C (2,280°F), steatite at 1,400°C (2,550°F), and alumina and beryllia at 1,700°C (3,090°F).

ASTM STANDARDS FOR ELECTRICAL INSULATING CERAMICS

The American Society for Testing and Materials has developed the following standards for electrical insulating ceramics:

• D 116, *Vitrified Ceramic Materials for Electrical Appliances*. This standard outlines procedures for testing samples of vitrified ceramic materials that are to be used as electrical insulation. They are equally suited for both unglazed and glazed specimens. Test methods are described in Table 15-1.

• F 356, *Beryllia Ceramics for Electronic and Electrical Applications*. This specification covers fabricated beryllia parts suitable for electronic and electrical applications. Specified are limits and methods of test for electrical, mechanical, thermal, and

TABLE 15-1   Property Tests and Test Methods for
Vitrified Ceramic Materials
(ASTM D 116)

PROPERTY MEASURED	ASTM METHOD
Compressive strength	C 773
Dielectric constant and dissipation factor	D 2149, D 150, D 2520
Dielectric strength	D 149
Elastic properties	C 623
Electrical resistivity	D 257, D 1829
Flexural strength	C 674, F 417
Hardness	C 730, E 18
Porosity	C 373
Specific gravity	C 329, C 20
Thermal conductivity	C 408, C 177
Thermal expansion	F 228, C 539
Thermal shock resistance	D 116

general properties of the bodies used for these fabricated parts. Three types of materials are identified:

TYPE	BERYLLIA CONTENT WEIGHT PERCENT MINIMUM
I	95
II	98
III	99

Requirements for beryllia ceramics are shown in Table 15-2.

• D 2442, *Alumina Ceramics for Electrical and Electronic Applications*. This specification covers the requirements for fabricated alumina parts suitable for electronic and electrical applications, and ceramic-to-metal seals as used in electron devices. Specified are limits and methods of test for electrical, mechanical, thermal, and general properties of the bodies used for these fabricated parts. Four types of materials are identified:

TYPE	ALUMINA CONTENT WEIGHT PERCENT MINIMUM
I	82
II	93
III	97
IV	99

TABLE 15-2   Requirements for Beryllia Ceramics to be Used as Electrical Insulations (ASTM F 356)

	TYPE I	TYPE II	TYPE III	ASTM METHOD
Electrical				
Dielectric constant, max, 25°C				D 150
at 1 MHz	7.0	7.0	7.0	
at 10 GHz	6.8	6.8	6.9	
Dissipation factor, max, 25°C				D 150
at 1 MHz	0.0005	0.0005	0.0004	
at 10 GHz	0.001	0.001	0.001	
Volume resistivity, min, $\Omega \cdot$ cm				D 257
at 25°C	10^{14}	10^{15}	10^{15}	D 1829
at 300°C	10^{11}	10^{12}	10^{12}	
at 500°C	10^{10}	10^{11}	10^{11}	
at 700°C	10^{8}	10^{9}	10^{9}	
at 900°C	10^{6}	10^{8}	10^{8}	
Dielectric strength, min, 125 mils				D 149
volts/mil	300	300	300	
Mechanical				
Density, g/cm^3, min	2.80	2.80	2.85	F 77
Flexural strength, min, avg psi	20,000	24,000	25,000	F 417
Compressive strength, min, avg psi	200,000	222,000	240,000	C 773
Modulus of elasticity, min, psi	44×10^6	45×10^6	49×10^6	C 623
Thermal				
Mean coef. of linear thermal				
expansion, μm/m/°C				E 228
25°–1,000°C	9.0 min	8.7 min	9.4 min	
	10.0 max	9.4 max	9.7 max	
Thermal conductivity, (min)				C 408
cal/s/cm^2 °C/cm				
at 25°C	0.35	0.52	0.60	
at 100°C	0.30	0.40	0.41	
at 400°C	0.14	0.15	0.15	
at 800°C	0.05	0.05	0.06	
Specific heat at 100°C				C 351
cal/g/°C	0.30	0.30	0.31	
Thermal shock resistance*	pass	pass	pass	

*To be agreed upon between manufacturer and purchaser.

(Reprinted with permission from the *Annual Book of Standards*. Copyright, ASTM, 1916 Race Street, Philadelphia, PA 19103.)

**TABLE 15-3   Requirements for Alumina Ceramics to Be
Used as Electrical Insulation
(ASTM D 2442)**

	TYPE I	TYPE II	TYPE III	TYPE IV	ASTM METHOD
Electrical					
Dielectric constant, max, 25°C					D 150
at 1 MHz	8.8	9.6	9.8	10.1	
at 10 GHz	8.7	9.6	9.8	10.1	
Dissipation factor, max, 25°C					D 150
at 1 MHz	0.002	0.001	0.0005	0.0002	
at 10 GHz	0.002	0.001	0.0005	0.0002	
Volume resistivity, min, $\Omega \cdot$ cm					D 257
at 25°C	10^{14}	10^{14}	10^{14}	10^{14}	D 1829
at 300°C	1×10^{10}	1×10^{10}	1×10^{10}	7×10^{10}	
at 500°C	4×10^{7}	2×10^{7}	8×10^{7}	1×10^{8}	
at 700°C	4×10^{6}	2×10^{6}	6×10^{6}	1×10^{7}	
at 900°C	4×10^{5}	1×10^{5}	8×10^{5}	1×10^{6}	
Dielectric strength, min,					D 149
125 mils volts/mil	250	250	250	250	
Mechanical					
Density, g/cm³, min	3.37	3.57	3.72	3.78	F 77
Flexural strength, min, avg psi	35,000	40,000	40,000	40,000	F 417
Modulus of elasticity, min, psi	31×10^{6}	40×10^{6}	45×10^{6}	50×10^{6}	C 623
Thermal					
Mean coef. of linear thermal					E 228
expansion μm/m/°C					
25°–1,000°C	7.4 min	7.5 min	7.6 min	7.5 min	
	8.2 max	8.3 max	8.3 max	8.4 max	
Thermal conductivity,					C 408
cal/s/cm² °C/cm					
at 100°C	0.023 min	0.031 min	0.048 min	0.052 min	
	0.049 max	0.077 max	0.073 max	0.090 max	
at 400°C	0.015 min	0.014 min	0.022 min	0.023 min	
	0.022 max	0.036 max	0.033 max	0.047 max	
at 800°C	0.009 min	0.009 min	0.014 min	0.014 min	
	0.018 max	0.021 max	0.021 max	0.025 max	
Thermal shock resistance*	pass	pass	pass	pass	

*To be agreed upon between manufacturer and purchaser.

(Reprinted with permission from the *Annual Book of Standards*. Copyright, ASTM, 1916 Race Street, Philadelphia, PA 19103.)

    Requirements for alumina ceramics are shown in Table 15-3.

    • D 2757, *Impervious Steatite Ceramics for Electrical and Electronic Applications*. This specification covers the requirements for impervious steatite ceramics having low electrical loss characteristics. These ceramics are designated as Type I, Type II, Type III, and Type IV. There is no further description of these types other than the values appearing in the table of requirements Table 15-4.

TABLE 15-4   **Requirements for Impervious Steatite Ceramics to Be
Used for Electrical and Electronic Applications
(ASTM D 2757)**

	TYPE I	TYPE II	TYPE III	TYPE IV	ASTM METHOD
Physical					
Specific gravity, min	2.5	2.6	2.7	2.7	C 20
Flexural strength,					D 116
min avg psi	18,000	18,000	20,000	20,000	
Dye penetration	none	none	none	none	D 116
Thermal expansion,					D 116
25–700°C					
cm/cm · °C $\times 10^{-6}$	7–10	7–10	7–10	7–10	
Hardness, Rockwell 45N	45	45	50	50	E 18
Impact strength,					
min, in · lbf	4.0	4.5	5.0	5.5	D 116
Electrical					
Dielectric strength,					D 116
$^1/_4$ in thick					
volts/mil min	220	230	230	230	
Volume resistivity,					D 116
min Ω · cm					
at 25°C	10^{14}	10^{14}	10^{14}	10^{14}	
at 100°C	1×10^{12}	1×10^{13}	5×10^{13}	5×10^{13}	
at 300°C	1×10^{7}	1×10^{10}	5×10^{10}	5×10^{10}	
at 500°C	1×10^{5}	5×10^{7}	1×10^{8}	1×10^{9}	
at 700°C	1×10^{4}	1×10^{6}	1×10^{6}	1×10^{7}	
at 900°C	1×10^{3}	1×10^{5}	1×10^{5}	1×10^{6}	
Dielectric constant,					
at 25°C, max					
at 1 MHz	6.0	6.0	6.2	6.5	D 150
at 4 GHz	5.8	5.8	6.0	6.3	D 2520
Loss index, 25°C, max					
at 1 MHz	0.035	0.016	0.008	0.004	D 150
at 4 GHz	0.038	0.018	0.015	0.007	D 2520

(Reprinted with permission from the *Annual Book of Standards*. Copyright, ASTM, 1916 Race Street, Philadelphia, PA 19103.)

PROPERTIES

Thermal Properties

Thermal conductivity is the principal property that determines the dissipation of local heating. Of all electrical insulating ceramics, beryllia has by far the highest thermal conductivity. Where ceramics are used in conjunction with other materials, it is often important that rates of expansion be compatible. That is, the coefficients of linear thermal expansion should be as close as possible.

Thermal properties of selected ceramic materials are shown in Tables 15-5–15-8.

TABLE 15-5   Thermal Conductivities of Electrical Insulating Ceramics

| | TEMPERATURE | | |
	°F	°C	Btu/(h)(ft^2)(°F/ft)*
Beryllia	200	93	104
Magnesia	200	93	18
Alumina	212	100	16.4
Sapphire	200	93	15–16
Boron nitride	212	100	13.3–25.5
Porcelain	not stated		8.2
Aluminum silicate	77	25	4.8
Quartz	200	93	3.0–5.4
Zircon	200	93	2.4
Steatite	77	25	1.5
Cordierite	77	25	1.2
Silica (fused)	1,000	538	0.33

*To convert to cal/(s)(cm^2)(°C/cm), multiply by 0.00413.

Note: Because of variance in composition and processing methods, the actual value for a specific material in the above table may vary significantly from the value shown.

Sources: Handbook of Physics and Chemistry and others.

TABLE 15-6   Coefficient of Linear Expansion of Electrical Insulating Ceramics

	TEMPERATURE RANGE °C	inch/inch/°C $\times 10^6$*
Magnesia	25–400	12.8
Aluminum silicate	25–400	8.1
Quartz	0–80	8.0
Beryllia (99% BeO)	not stated	7–9
Alumina (99.9% Al$_2$O$_3$)	25–500	7.4
Porcelain	25–400	6.9
Steatite	25–400	6.6
Sapphire	25–400	4.5–7.0
Zircon	25–400	4.5
Cordierite	25–400	3.1
Boron nitride	25–350	0.18–3.1
Silica (fused)	0–800	0.55

*To convert to °F, divide by 1.8.

Note: Because of variance in composition and processing methods, the actual value for a specific material in the above table may vary significantly from the value shown.

Sources: Various.

TABLE 15-7 Maximum Operating Temperatures of Electrical Insulating Ceramics

	TEMPERATURE	
	°C	°F
Magnesia	2,200	4,000
Sapphire	2,000	3,630
Zircon	1,870	3,400
Alumina	1,700	3,100
Beryllia	1,700	3,100
Aluminum silicate	1,425	2,600
Silica (fused)	1,370	2,500
Steatite	1,315	2,400
Cordierite	1,300	2,380
Porcelain	1,200	2,200
Boron nitride (in oxidizing atmosphere)	985	1,805
Quartz	na	na

Note: Because of variance in composition and processing methods, the actual value for a specific material in the above table may vary significantly from the value shown.

na—not available.

Sources: Various.

TABLE 15-8 Thermal Shock Resistance of Electrical Insulating Ceramics

CERAMIC	RATING
Alumina	Good
Aluminum silicate	Excellent
Beryllia	Excellent
Boron nitride	Excellent
Cordierite	Good
Magnesia	Excellent
Porcelain	Poor
Quartz	Poor
Sapphire	na
Silica (fused)	Excellent
Steatite	Poor
Zircon	Good

Note: Because of variance in composition and processing methods, the actual value for a specific material in the above table may vary significantly from the value shown.

na—not available.

Sources: Various.

Physical Properties

Parts made of ceramics are characteristically hard, dense, and incombustible. Values for compressive strength are eight–ten times tensile strength values, indicating better performance under compression than under tension. Most ceramic bodies have low impact strength and tend to fracture when struck sharply with a hard object, although there are noteworthy exceptions. For example, beryllia can be fabricated into armor for both vehicles and personnel.

Ceramic parts are not affected by long exposure to the atmosphere, even at high operating temperatures. Densities and mechanical properties of electrical insulating ceramics are shown in Table 15-9.

TABLE 15-9   Densities and Mechanical Properties of Electrical Insulating Ceramics

	DENSITY g/cm³*	FLEXURAL STRENGTH psi (000)*	COMPRESSIVE STRENGTH psi (000)*	TENSILE STRENGTH psi (000)*
Alumina (99.9% Al₂O₃)	4.0	41	370	30
Aluminum silicate	1.9	3	10	1.5
Beryllia (99% BeO)	2.9	34	240 min	21
Boron nitride	2.1	na	34–45	na
Cordierite	2.1–2.9	15	50	5.5
Magnesia	2.3–2.5	2–10	2–10	na
Porcelain	3.0	12.5	75	8.5
Quartz	2.7	na	na	na
Sapphire	4.0	na	450	60
Silica (fused)	2.0	1.5	5	na
Steatite	2.9	21	90	9
Zircon	3.6	4.6	27	2

*At room temperature

Note: Because of variance in composition and processing methods, the actual value for a specific material in the above table may vary significantly from the value shown.

na—not available.

Sources: Various.

Dielectric Properties

Dielectric strengths of most electrical grade ceramics are in the range of 100–400 volts per mil for 125 mil specimens. Dielectric constants are in the 4–10 range, with alumina near the high end of the range and fused silica at the low end. Dielectric constant values are often sensitive to temperature, and may rise, fall, or remain unaffected by temperature changes, depending on the type and density of the ceramic. The dissipation, or loss factor, also sensitive to temperature, is lowest for beryllia, followed by alumina.

Moisture accumulating on the surface or penetrating into the body of a ceramic significantly degrades all dielectric properties, and may also degrade mechanical properties via surface tension phenomena which can enhance crack propagation in highly stressed ceramics. Dust, or other deposits (as on transmission line insulators), worsens dielectric degradation and becomes conducting with high moisture content. Glazes reduce the ability of dust or other deposits to adhere to ceramic bodies and make their surfaces more washable, but they may incur significant power losses. Dielectric greases have also been used to protect ceramic surfaces in contaminated environments, but greases often require periodic cleaning and replacement.

Values for dielectric properties of ceramics are shown in Table 15-10.

TABLE 15-10   Dielectric Properties of Electrical Insulating Ceramics

	DIELECTRIC STRENGTH 125 mils volts/mil* (ASTM D 149)	DIELECTRIC CONSTANT 1 MHz* (ASTM D 150)	DISSIPATION FACTOR 1 MHz* (ASTM D 150)
Alumina (99.9% Al_2O_3)	340	10.1	0.0002
Aluminum silicate	150	4.1	0.0027
Beryllia (99% BeO)	350	6.4	0.0001
Boron nitride	950	4.2	0.00034
Cordierite	200	4.8	0.0050
Magnesia	na	5.4	<0.0003
Porcelain	na	8.5	0.005
Quartz	na	3.8	0.0038
Sapphire	na	9.3–11.5	0.00003–0.000086
Silica (fused)	na	3.2	(0.0045 @ 10 GHz)
Steatite	230–390	5.5–7.2	0.0010
Zircon	na	5.0	0.0023

*At 25°C (77°F).

Note: Because of variance in composition and processing methods, the actual value for a specific material in the above table may vary significantly from the value shown.

na—not available.

Sources: Various.

SUMMARY OF PROPERTIES AND USES

Specific property values are shown in the preceding tables. The following are note-worthy features of the principal ceramics used for electrical insulation.

• *Alumina* has the highest mechanical properties throughout a wide useful temperature range. A dissipation factor of 0.0002 at 1 megahertz is among the lowest of all ceramics, although other dielectric properties are not outstanding. Thermal conductivity is relatively high and dimensional stability is excellent. Overall, alumina is one of the best ceramic insulations available. It is used to insulate spark plugs, for vacuum-tight seals to metal, and for a variety of electrical/electronic components.

• *Aluminum nitride* substrates and packages appeared in the late 1980s as an alternative to beryllium. This ceramic has high thermal conductivity, thermal expansion close to silicon, and excellent dielectric properties.

• *Aluminum silicate* is noted for its heat shock resistance, and also has one of the lowest dielectric constants (4.1 at 1 megahertz) of all ceramics. Mechanical properties are poor. It is used for specialized applications.

• *Beryllia* has the highest thermal conductivity and specific heat of any oxide ceramic. Machined parts can be held to a tolerance of ±0.0002 inch. Exceptionally high impact strength can be achieved by special processing. Dielectric strength is high and a dissipation factor of 0.0001 is even lower than alumina. Beryllia ceramics are used extensively for semiconductor packages, and for heat sinks, ceramic-to-metal seals, waveguide windows, radomes, and electrical/electronic components.

• *Boron nitride* is less brittle than most ceramics, has high thermal shock resistance, high thermal conductivity, and a very low rate of thermal expansion. Dielectric strength is outstanding and the dielectric constant is among the lowest for all ceramics. Mechanical strength and moisture resistance are low unless compounded with 40–60 percent silica. Uses include additives to other ceramics to increase thermal conductivity and reduce thermal expansion, transistor and rectifier mounting wafers, and nonabrading electronic parts.

• *Cordierite* has excellent heat shock resistance and fair dielectric properties. Mechanical strength is low. It is suitable for applications where heat shock is severe.

• *Diamond film* is a material of exceptionally good properties for electronics applications. It has a thermal conductivity five times higher than copper, making it a unique heat sink to reduce chip temperatures. Dielectric properties are among the best for most ceramics.

• *Magnesia* has the highest coefficient of thermal expansion of all ceramics. Dielectric properties are good, but mechanical properties are weak. Its main application is as insulation in sheathed heating units and thermocouples. A semiconductor grade is available.

• *Porcelain* is chemically inert and has satisfactory dielectric properties for applications at power frequencies, where it is widely used. Mechanical properties and thermal shock resistance are low. Glazed porcelain is used extensively for insulation on electrical transmission lines, circuit breakers, cutouts, and spark plugs. Low voltage uses include switches, bushings, fuse blocks, and light sockets.

• *Quartz* and fused silica possess the lowest dielectric loss properties of any inorganic material. Fused quartz is virtually immune to thermal shock. Thermal expansion is very low. The dissipation factor is among the lowest for ceramics. Physical properties are generally very low. Uses include radomes, nose cones, wave filters, and other electronic components.

• *Sapphire* has a single crystal surface desirable for integrated circuit and other electronic systems. A dielectric constant of 11.5 is the highest of all ceramics. High-temperature stability is excellent. Uses include electron and microwave tubes, and substrates for integrated circuits.

• *Steatite* has good dielectric properties, average mechanical properties, and poor thermal shock resistance. It can be fabricated to close tolerances. Die wear is low. It is suitable for making electronic components where mechanical properties are not demanding.

• *Zircon* ceramics offer improved mechanical and dielectric properties over other porcelains. Thermal shock resistance is also better. The material is abrasive and die wear is high. Uses include stand-off insulators, pole lightning arrestors, and applications requiring high arc resistance.

DEVELOPMENT PROGRAMS

Property values of ceramics are significantly affected by changes in the ratios of basic materials in the composition, sometimes with synergistic effects. Of promise is a blend of silicon carbide and beryllia which is being evaluated as an improved material for integrated circuits.

Development continues on polymer compounds as replacements for ceramics in general use as high-voltage transmission and distribution line insulators. In addition to being mechanically suitable, materials must have high resistance to forming carbonizing paths and to erosion under outdoor high-voltage service conditions. Surface degradation of polymer insulator is a combination of surface erosion and interaction with contaminants. Surface contamination is caused by industrial pollution, wind-blown agricultural chemicals, and salt fog along coastal regions. Contaminated surfaces tend to wet and erode readily. Eventual failure of the insulator is caused by flashover and/or tracking encouraged by wet, contaminated surfaces, as well as by imperfections incurred in the manufacture of the insulator.

Silicone rubber, one of the polymeric candidates for replacing ceramics, has inherently low arc resistance which, however, is improved to the acceptable level by addition of alumina trihydrate (ATH) fillers. Ethylene-propylene-diene copolymer (EPDM rubber) with ATH fillers is also being used to replace ceramic high-voltage power line insulators. ATH fillers may also be added to room-temperature-vulcanizing (RTV) silicone rubber dispersions as surface coatings for ceramic, fiberglass reinforced resin, ethylene-propylene rubber, and epoxy insulators to provide enduring water repellency. These coatings also have high resistance to corona and arcing, and are more permanent than greases which require reapplication.

Thus, the selection of a suitable polymer in preference to porcelain includes several criteria involving the environment to which the insulator will be exposed:

- Temperature range
- Prevailing humidity and rainfall
- Salt fog and mist
- Industrial pollutants and agricultural chemicals
- Corona attack
- Arcing

MARKET TRENDS

For electrical applications, ceramics will lose market share and volume as the acceptance and use of polymeric materials becomes widespread. Ceramics used in electronic devices and packages continue to keep pace with the dynamic growth of high-technology applications. Ceramics are especially important in meeting the military's needs for materials with long-term environmental stability. Japanese ceramics for electronic uses already dominate world markets, and will undoubtedly continue to provide formidable competition to United States manufactured products.

Dielectric Gases

INTRODUCTION

The first gases used in the insulation of electrical equipment were air and its major constituent, nitrogen. These gases, however, possessed inherent limitations, and the need for gases with improved characteristics grew as the equipment requiring gaseous insulation became more complex and demanding. Answering this need, other gases were developed or adapted for use as dielectrics. Dielectric gases may be classified as follows:

- Simple gases—air, nitrogen, hydrogen, helium, oxygen
- Oxide gases—carbon dioxide, sulfur dioxide
- Hydrocarbon gases—methane, ethane, propane, butane, hexane
- Electronegative gases (all but SF_6 used mainly as refrigerants)
    - Sulfur hexafluoride (SF_6)
    - Dichlorodifluoromethane (CCl_2F_2, Freon 12, Genetron 12)
    - Chlorotrifluoromethane ($CClF_3$, Freon 13)
    - Carbon tetrafluoride (CF_4, Freon 14)
    - Chlorodifluoromethane ($CHClF_2$, Freon 22)
    - Chloropentafluoroethane (CF_3CF_2Cl, Freon 115)
    - Hexafluoroethane (CF_3CF_3, Freon 116)

Electronegative gases are gases with the ability to remove free electrons by negative ion formation, thereby increasing dielectric strength over gases which do not have this property, such as nitrogen and air. The principal electronegative dielectric gases are halides of sulfur and carbon. This chapter is primarily concerned with SF_6, the most widely used electronegative gaseous dielectric. Because of harmful effects on the ozone layer, chlorofluorocarbons (CFCs) will not be produced after 1995. Principal suppliers of electronegative gases include:

- AlliedSignal Corporation, SF_6 and halogenated hydrocarbons (Genetron™)
- Ashland Chemical, SF_6
- Du Pont, halogenated hydrocarbons (Freons™)

Conversion factors and formulas for frequently used units are shown in the Appendix.

TECHNOLOGY

Breakdown in Gases

In a dielectric gas subjected to low electric stress, there can be no current flowing between electrodes unless there are free electrons or ions present in significant numbers. However, a small quantity of charged particles caused by environmental radioactivity and cosmic radiation is always present, and as the electric field strength is increased, more collisions occur between electrons and gas molecules, creating additional charged particles. As this process proceeds, the gas becomes conductive at some point. It is characteristic of molecules of electronegative gases to attach free electrons before they cause conduction, which makes them the most effective dielectric gases.

An electron "avalanche" can occur if the field strength is high enough and if the free electrons are not captured by attachment. A single electron ionizes a gas molecule by collision, producing two free electrons and a positive ion. The two free electrons may then cause further ionization and more free electrons. This cumulative process eventually produces a swarm of electrons and ions, and the gas becomes conductive. Secondary avalanches may be produced at the cathode by the impact of positive ions or photons from an ongoing avalanche. These in turn initiate more avalanches, creating a self-sustaining discharge known as the Townsend theory of breakdown.

Another theory, the *streamer theory*, was developed to explain the breakdown of a gas where secondary processes are not a factor. According to this theory, there are two requirements for initiation and propagation of a streamer.

• The space charge produced in the avalanche causes sufficient distortion of the electric field that those free electrons move toward the avalanche head, generating further, cumulative avalanches.

• Photons from the avalanche head produce free electrons by photoionization.

This theory applies when the number of electrons in the average avalanche reaches a critical value, generally taken as 10^8.

The sudden occurrence of a voltage greater than the breakdown voltage of the system does not immediately result in dielectric failure. This time lag is caused by the interval between the application of voltage and the event of the first free electrons plus the time required for the avalanche to build from the initiating electrons to the point where dielectric breakdown occurs.

In essence, then, free electrons or negative ions from any source colliding with electronegative gas molecules can cause a dielectric breakdown when the rate at which electrons are produced exceeds the rate at which they are absorbed.

Paschen's Law

The dielectric strength of most electrical insulating gases follows Paschen's Law up to a pressure of about 150 psi, which includes pressures used in most applications. Factors that may cause significant deviations from the law are:

• Shape, surface condition, and material of electrodes
• Gap between electrodes
• Degree of purity of the gas
• Uniformity of the electrical field
• Extremes of temperatures and pressures

In brief, Paschen's Law states that over a certain pressure and temperature range in a uniform field, using parallel plate electrodes, the sparking potential of a gas depends on the product of gas pressure and electrode spacing:

$$V_s = f(pd)$$

where

V_s is the sparking potential,

p is the gas pressure at a given temperature,

d is the gap spacing, and

f is a function of pd only.

Paschen's Law is also sometimes stated as

$$V_s = f(Nd)$$

where N is the gas number density in molecules per cm^3.

The latter form of the law is useful in considering a scaled system such as a gas insulated cable or substation bus where pressure increases with temperature, but N remains constant, so the breakdown strength remains unchanged. Likewise, if the pressure is doubled (at constant temperature) and the gap spacing is halved, the breakdown voltage remains unchanged.

Above the pressure range where Paschen's Law applies, the breakdown voltage rises more slowly than the pressure. The law also does not apply at extremely low gas pressures approaching a perfect vacuum, where the reduced number of molecules severely limits ionization by collision, resulting in high dielectric strength. However, the engineering problems of maintaining a high vacuum preclude its use for most electrical equipment applications, except for electronic capacitors.

Because of the increased dielectric strength values obtainable, it is common practice to place a gas under pressure when it functions as a dielectric. However, a system should be designed so that liquefaction is not possible under foreseeable operating conditions plus a safety factor, since liquefaction acts to lower dielectric strength values and to promote instability and corrosion.

Electrode Condition and Configuration

The breakdown strength of most strongly electronegative gases is especially sensitive to conductor surface condition. Grooved, pitted, or roughened surfaces create nonuniform electrical fields and appreciably decrease dielectric breakdown strengths of gases below that for a uniform field gap. With practical electrode finishes, the dielectric strength of SF_6 is reduced by at least 50 percent of its theoretical value at 4 atmospheres pressure. Breakdown strength is also affected by dust, moisture, gas decomposition products, and other contaminants on the surfaces of electrodes.

The effect of electrode configuration is demonstrated for SF_6 by rapidly declining breakdown voltage with increasing pressure as the radius of the rod is decreased in a rod-plane electrode-type tester with gap and testing conditions remaining constant. In one study,[1] the breakdown voltage was over three times higher using a 5 mm radius with 20 mm gap versus using a 0.5 mm radius at 4 atmospheres pressure, although at 1 atmosphere pressure, breakdown values were not significantly different. Thus, different electrode configurations have different effects on space charge for-

[1] S. Sangkasaad, Doctor of Technical Sciences Dissertation, Swiss Federal Institute of Technology Diss. ETH No. 5738, Zurich, 1976.

mation and corona stabilization processes. While the physics of this phenomenon is not fully understood, it involves the ionic space charges produced when corona discharges occur from sharp points. This type of breakdown is usually referred to as *corona stabilized breakdown*.

Controlled insulating coatings on the electrodes can significantly increase breakdown strength by 25–50 percent, so that Paschen's Law will hold to higher pressures and field strengths than for bare electrodes.

The approved industry method for determining dielectric breakdown voltage is described in ASTM Standard D 2477, *Test Method for Dielectric Breakdown Voltage and Dielectric Strength of Insulating Gases at Commercial Power Frequencies*. This method employs vertically mounted sphere and plane electrodes. The sphere shall be a precision steel ball bearing 0.75 inch (19.1 mm in diameter). The plane electrode shall be of brass 1.50 inches (38.1 mm) in diameter with rounded edges. The gap setting shall be 0.100 ± 0.001 inch (2.54 ± 0.025 mm). This procedure covers testing only under standard conditions of 760 torr and 25°C (77°F). There is no provision for testing under higher or lower pressures or temperatures. A formula is given for converting ambient to standard conditions:

Multiply ambient conditions by F = (273 + T)/0.392P

where

T = ambient temperature, degrees C

P = gas pressure, torr.

Test results using this method are stated to be reproducible within acceptable limits for samples of the same gas tested in different laboratories under similar test conditions.

However, because of the considerable effect on dielectric breakdown of electrode test gaps, shapes, and surface irregularities, it is generally accepted practice to compare breakdown strengths of gases and gas mixtures with the breakdown strengths of nitrogen or SF_6 determined in the same test equipment under similar test conditions. Using this approach, mean breakdown strengths of selected dielectric gases and gas mixtures are shown in "Gas Mixtures" (p. 449) in this chapter. Three pure gases from this table have as high or higher dielectric strengths as SF_6, but their higher boiling points would prevent them from serving the full useful temperature range covered by SF_6.

	BREAKDOWN STRENGTH	BOILING POINT	
	%	°C	°F
Chloropentafluoroethane, CF_3CF_2Cl	114	–38.7	–37.7
Sulfur dioxide, SO_2	102	–10.0	+14.0
Dichlorodifluoromethane, CCl_2F_2	100	–29.8	–21.6
Sulfur hexafluoride, SF_6	100	–63.9	–83.0

Conducting Particles

The presence of free conducting particles in a gaseous insulating system may lead to reduction in dielectric strength and failure of the system. For example, the dielectric strength of SF_6 may be significantly reduced at higher gas pressures when conducting particles are present.[2] Conducting particles can be introduced during assembly or be produced by abrasion between components during assembly. These particles can be lifted by the electric field, and migrate to the conductor or insulators where they can initiate breakdown at voltages well below those of particle-free systems. It is reported that particle traps incorporating adhesive material in trap cavities offer effective protection against particle contamination, thus increasing reliability of the system.

Moisture

It is important to keep moisture in a system below a critical level determined by the minimal utilization temperature in order to avoid condensation on insulators.

CHEMISTRY

SF_6 is made by directly combining sulfur vapor with fluorine. Byproducts are removed by treatment with caustic, followed by heating to 400°C to break down large molecules of sulfur and fluorine to SF_4 and SF_6. A second scrubbing with caustic removes SF_4, but not SF_6, which does not react with caustic. SF_6 is then dried in a sulfuric acid tower to remove all traces of moisture, liquefied, and distilled to its high degree of purity.

Dichlorodifluoromethane is made by two methods:

• Reaction of carbon tetrachloride and anhydrous hydrogen fluoride in the presence of an antimony halide catalyst

• High-temperature chlorination of vinylidene fluoride (made by addition of hydrogen fluoride to acetylene).

GRADES

Only grades with the lowest moisture content are suitable for use as gaseous dielectrics. Commercially available SF_6 is supplied to these specifications:

Dew point	–60°C (–76°F)
Air as nitrogen, weight percent	0.04 maximum
Acidity as HF, weight percent	0.00003 maximum

[2]Steinar, J. Dale, and Melvin D. Hopkins, "Methods of particle control in SF_6 insulated CGIT systems," paper presented at the IEEE PES 1981 Transmission and Distribution Conference and Exposition, Minneapolis, MN, Sept. 20–25, 1981.

CF_4, weight percent	0.04 maximum
Oil, weight percent	0.0005 maximum
Toxicity	Nontoxic
Odor	None
Assay weight percent	99.9 minimum

ASTM STANDARDS FOR DIELECTRIC GASES

The American Society for Testing and Materials has developed the following standards for dielectric gases.

• D 1933, *Nitrogen Gas as an Electrical Insulating Material.* This specification covers three types of nitrogen used as an electrical insulating material in electrical equipment:

Type I, obtained from air by liquefaction processes and dried

Type II, obtained from air by liquefaction processes, deoxidized with hydrogen over a platinum catalyst and dried

Type III, obtained from air by liquefaction processes and, if necessary, deoxidized by suitable means.

The required characteristics for these types are shown in Table 16-1.

• D 2472, *Sulfur Hexafluoride.* Sulfur hexafluoride for use as an electrical insulation material shall conform to the requirements in Table 16-2.

In addition to the above requirements, the manufacturer shall be able to certify that the material is at least as nontoxic as Group VI of the Underwriters' Laboratories classification.

• D 3283, *Air as an Electrical Insulating Material.* Air used as an electrical insulation material shall conform to the requirements in Table 16-3.

TABLE 16-1   Composition and Properties for Nitrogen Dielectric Gas (ASTM D 1933)

	TYPE I	TYPE II	TYPE III
Nitrogen and rare gases, minimum volume %	99.8	98.998	99.993
Hydrogen, maximum volume %	0.0	1.0	0.005
Oxygen, maximum volume %	0.2	0.002	0.002
Dew point, maximum °F (°C)	−67	−67	−75
	(−55)	(−55)	(−60)

*Corresponds to a water content of 8.9 ppm by weight or 71 ppm by volume.

(Reprinted with permission from the *Annual Book of Standards.* Copyright, ASTM, 1916 Race Street, Philadelphia, PA 19103).

TABLE 16-2   Requirements for Sulfur Hexafluoride Dielectric Gas (ASTM D 2472)

	VALUE
Water content, maximum dew point, °F (°C)	−50* (−45)
Hydrolyzable fluorides, expressed as HF acidity maximum ppm by weight	0.3
Air, expressed as N_2, maximum weight %	0.05
Carbon tetrafluoride, maximum weight %	0.05
Molecular weight	$146 \pm 2\%$
Assay, minimum weight %	99.8

*Corresponds to a water content of 8.9 ppm by weight or 71 ppm by volume.

(Reprinted with permission from the *Annual Book of Standards*. Copyright, ASTM, 1916 Race Street, Philadelphia, PA 19103.)

TABLE 16-3   Composition and Properties for Air Dielectric Gas (ASTM D 3283)

	VALUE
Oxygen, volume %	19 to 23
Carbon monoxide, maximum volume %	0.0020
Carbon dioxide, maximum volume %	0.10
Nitrogen	balance
Dewpoint, maximum °F (°C)	−72 (−58)

(Reprinted with permission from the *Annual Book of Standards*. Copyright, ASTM, 1916 Race Street, Philadelphia, PA 19103.)

PROPERTIES

Important factors to be considered in evaluating gaseous dielectrics are:

- Degree to which the gas is electronegative as measured by electron affinity and electron capture
- Toxicity of a gas and its decomposition byproducts
- Formation of carbon deposits under arcing conditions
- Nonflammability
- Inertness
- Chemical stability
- Thermal stability
- Critical temperature
- Critical pressure

- Dielectric properties
- Heat transfer rate
- Compatibility with other materials used in electrical equipment
- Availability
- Price

Toxicity

Gases that are toxic to humans and that are hazardous to the environment are virtually ruled out from further consideration as dielectrics, however attractive their other properties. Table 16-4 lists the limits of worker exposure to selected gases as determined by NIOSH (see Chapter 19, *Government Activities*).

Thus, sulfur dioxide is high in toxicity, while SF_6 and dichlorodifluoromethane are much safer to use in work areas, although precautions must be taken even with these gases. The major problem of toxicity, however, occurs when gases are decom-

TABLE 16-4   Air Contaminants Limits of Worker Exposure (8 h Time-Weighted Averages)

GAS	PARTS/MILLION (volt) @ 25°C 760 mm MERCURY PRESSURE
Carbon dioxide	5,000
Sulfur hexafluoride	1,000
Freons 12, 13, 14, 22, 115, 116	1,000
Propane	1,000
n-Hexane	500
Sulfur dioxide	5

Source: NIOSH.

TABLE 16-5   Threshold-Limiting Values* of Potential Gaseous Decomposition Products of Sulfur Hexafluoride

GAS	PARTS/MILLION (vol) @ 25°C 760 mm MERCURY PRESSURE
Sulfuryl fluoride (SO_2F_2)	5
Hydrofluoric acid (HF)	3
Sulfur dioxide (SO_2)	2
Thionyl fluoride (SOF_2)	1.6
Silicon tetrafluoride (SiF_4)	0.6
Sulfur monofluoride (S_2F_2)	0.5
Thionyl tetrafluoride (SOF_4)	0.5
Sulfur tetrafluoride (SF_4)	0.1
Tungsten hexafluoride (WF_6)	0.1
Disulfur decafluoride (S_2F_{10})	0.01

*Concentration to which a worker can be safely exposed during a 40-hour week.

Source: American Conference of Governmental Industrial Hygienists.

posed by high temperatures or an electric arc. While simple gases, such as nitrogen, present no problem, more complex gases may form toxic byproducts when decomposed (as solid insulations may also do).

The most severe application for a dielectric gas is in circuit breakers. Scrubbers, incorporating soda lime (CaO and NaOH), or activated alumina (Al_2O_3), installed in breakers absorb small amounts of decomposition products, but when arcing or corona discharge cause equipment failure, large amounts of these products are formed. For example, SF_6 may decompose into a variety of toxic byproducts, usually through reaction with oxygen or water. The most common byproducts are sulfur dioxide, thionyl fluoride, and sulfuryl fluoride. The threshold limiting value (TLV) for each possible decomposition product is shown in Table 16-5.

Arcing

SF_6 is more stable in arcs than most other electronegative gases and, in fact, arcing does not lessen the dielectric strength of the gas. All halogenated hydrocarbons have carbon, a conductor, as a potential breakdown byproduct. Free carbon deposits from breakdown in fluorocarbon gases are decreased somewhat by the addition of SF_6 in concentrations of 20–30 percent, usually with improved dielectric strength.

Chemical and Thermal Stability

Along with low toxicity, dielectric gases should be inert, nonflammable, and be chemically and thermally stable over a wide useful temperature range. All the dielectric gases covered here, except the hydrocarbon gases, are considered unreactive, nonflammable, and chemically stable at room temperature and atmospheric pressure. Hydrocarbon gases are inherently flammable. For example, the closed cup flash points of propane and n-heptane are –104°C (–155°F) and –21°C (–5.8°F), respectively. Liquid propane, in fact, is widely used as a fuel.

All chemical compounds have temperature limits above which they are no longer chemically stable. The decomposition rates of dry SF_6 in contact with metals used in electrical equipment are shown in Table 16-6.

**TABLE 16-6   Stability of Sulfur Hexafluoride
in Selected Materials of Construction**

	DECOMPOSITION %/year	
MATERIAL	@ 200°C	@ 250°C
Aluminum	—	0.006
Copper	0.18	1.4
Silicon steel	0.005	0.01
Mild steel	0.2	2

Source: Chemical Sector, AlliedSignal.

Decomposition rates increase significantly with higher concentrations of moisture and other contaminants.[3] A feature of nitrogen is its chemical stability at all operating temperatures to which electrical apparatus could be exposed. Tests by Du Pont indicate that under certain conditions, Freon 116 (C_2F_6) has somewhat higher thermal stability than SF_6. However, both gases exhibit significantly increased fluoride ion formation as moisture content rises, and this is the paramount factor in determining the thermal and electrical stability of both gases. There are difficulties in measuring moisture content, and different equipment can produce different results. Decomposition products can also affect the instrument probe sensitivity.

Thermal Properties

The cooling capacity of a dielectric gas in electrical apparatus depends on its specific heat, its specific thermal conductivity, and its convection characteristics. Specific heat at a given temperature and pressure is the amount of heat required to raise a unit weight of substance 1 degree of temperature at either constant pressure or constant volume. Since gases have extremely low densities compared with liquids and solids, specific heat values are usually not high enough to affect cooling capacity significantly.

Thermal conductivity is the quantity of heat passing in a unit time through a unit thickness of a substance across a unit area for a unit difference in temperature. The specific thermal conductivity of a gas rises with temperature and pressure, indicating improved ability to transfer heat as this occurs. Table 16-7 shows specific thermal conductivities for selected dielectric gases.

Convection characteristics, measured by viscosity, are the most important cooling mechanism for gases. The coefficient of viscosity of a material is defined as the tangential force per unit area of either of two horizontal planes at unit distance apart, one of which is fixed, while the other moves with a unit velocity, the space between being filled with the material to be tested. Thus, the higher the coefficient of viscosity, the poorer is the convection and dissipation of heat. Coefficients for gases increase with temperature (they decrease for liquids). Values for selected gases are shown in Table 16-8.

Critical Constants

Each gas has its own critical temperature above which it cannot be liquefied by pressure alone. The critical pressure of a gas is the minimum pressure under which a substance may exist in gas/liquid equilibrium at the critical temperature. Below the critical temperature, less pressure is required to maintain this equilibrium. Table 16-9 shows the critical constants for certain dielectric gases.

[3]R. J. Van Brunt, "Production rates for oxyfluorides, SOF_2, SO_2F_2, and SOF_4 in SF_6 corona discharges." *Journal of Research of the National Bureau of Standards*, vol. 90, no. 3, 1985.

TABLE 16-7   Specific Thermal Conductivities for
Selected Dielectric Gases
(At Normal Atmospheric Pressure)

	TEMPERATURE		Calories*/s/cm/°C
	°C	°F	for 1 cm^2
Helium (He)	26.7	80	360.4×10^{-6}
Oxygen (O$_2$)	26.7	80	63.6×10^{-6}
Nitrogen (N$_2$)	26.7	80	62.4×10^{-6}
Air	26.7	80	62.2×10^{-6}
Carbon dioxide (CO$_2$)	26.7	80	39.7×10^{-6}
Hexafluoroethane (CF$_3$CF$_3$)	25	77	36.5×10^{-6}
Sulfur hexafluoride (SF$_6$)	30	86	33.6×10^{-6}
Chloropentafluoroethane, Freon 115 (CF$_3$CF$_2$Cl)	25	77	33.1×10^{-6}
Chlorodifluoromethane, Freon 22 (CHClF$_2$)	26.7	80	28.1
Dichlorodifluoromethane, Freon 12, Genetron 12 (CCl$_2$F$_2$)	26.7	80	22.7
Sulfur dioxide (SO$_2$)	0	32	15.0

*One calorie is the amount of heat required at a pressure of 1 atm to raise the temperature of 1 g of water from 4°C to 5°C (where water has its greatest density).

Source: Handbook of Physics and Chemistry.

TABLE 16-8   Coefficients of Viscosity for Selected
Dielectric Gases
(At Normal Atmospheric Pressure)

	TEMPERATURE		
	°C	°F	POISES*
Sulfur dioxide (SO$_2$)	20	68	125×10^{-6}
Dichlorodifluoromethane, Freon 12 (Cl$_2$F$_2$)	25	77	130×10^{-6}
Chlorodifluoromethane, Freon 22 (CHClF$_2$)	25	77	130×10^{-6}
Chloropentafluoroethane, Freon 115 (CF$_3$CF$_2$Cl)	25	77	130×10^{-6}
Carbon dioxide (CO$_2$)	20	68	160×10^{-6}
Sulfur hexafluoride (SF$_6$)	25	77	161×10^{-6}
Air	20	68	181×10^{-6}
Nitrogen (N$_2$)	20	68	184×10^{-6}
Helium (He)	20	68	200×10^{-6}
Oxygen (O$_2$)	20	68	206×10^{-6}

*One poise is equal to 1 dyn · s/cm^2.

Source: Handbook of Physics and Chemistry, Lange's Handbook of Chemistry.

TABLE 16-9   Critical Constants for Selected Dielectric Gases

	CRITICAL TEMPERATURE °C (°F)	CRITICAL PRESSURE (atmospheres)
Helium (He)	−268.0 (−450.4)	2.26
Nitrogen (N2)	−147.1 (−232.3)	33.5
Air	−140.7 (−221.3)	32.2
Oxygen (O_2)	−118.4 (−202.5)	50.1
Carbon tetrafluoride (CF_4)	−45.7 (−50.3)	37.0
Hexafluoroethane (CF_3CF_3)	19.7 (67.5)	29.4
Chlorotrifluoromethane ($CClF_3$)	28.9 (84.0)	38.2
Carbon dioxide (CO_2)	31.1 (88.0)	73.0
Sulfur hexafluoride (SF_6)	45.6 (114.1)	37.2
Chloropentafluoroethane (CF_3CF_2Cl)	80.0 (176.0)	30.8
Chlorodifluoromethane ($CHClF_2$)	96.0 (204.8)	49.1
Dichlorodifluoromethane (CCl_2F_2)	112.0 (233.6)	40.6
Sulfur dioxide (SO_2)	157.5 (315.5)	77.9

Source: Handbook of Physics and Chemistry.

The estimated pressures required to liquefy SF_6 and hexafluoroethane at various temperatures are shown in Tables 16-10 and 16-11. The working pressure for many practical SF_6 systems is in the range of 3–5 atmospheres.

Boiling Temperatures

The boiling or sublimation temperature of a dielectric gas at normal atmospheric pressure should be well below the minimum temperature to which the equipment it insulates could be exposed. Otherwise, placing the gas under pressure would cause

TABLE 16-10   Pressures at Selected Temperatures Required to Liquefy SF_6

TEMPERATURE			
°F	°C	psi	atmospheres
−83	−63.9	14.7	1
−40	−40.0	33	2.2
−20	−28.9	57	3.9
0	−17.8	90	6.1
20	−6.7	130	8.8
40	4.4	180	12.2
60	15.6	260	17.7
80	26.7	330	22.4
100	37.8	430	29.3
114	45.6 (critical)	546	37.2

Source: Estimated from the Vapor Pressure Versus Temperature Chart, Chemical Sector, AlliedSignal, Technical Bulletin 524-003 AccuDri SF_6.

TABLE 16-11   Pressures at Selected Temperatures Required to Liquefy Hexafluoroethane

TEMPERATURE			
°F	°C	psi	atmospheres
–40	–40.0	80	5.4
–20	–28.9	125	8.5
0	–17.8	175	11.9
20	–6.7	240	16.3
40	4.4	300	20.4
60	15.6	400	27.2
67.4	19.7 (critical)	432	29.4

Source: Estimated from the Vapor Pressure Versus Temperature Chart, "Freon" Products Division, Du Pont, Product Information Bulletin EL-15, "Freon" 116 Dielectric Gas.

liquefaction. Table 16-12 lists the boiling or sublimation temperatures for selected gases with values of –10°C (14°F) or lower.

Most gas-insulated equipment is designed to allow a minimum temperature of –40°C (–40°F) without auxiliary heaters. Heaters may be required in circuit breakers for pressures above 5 atmospheres.

Dielectric Constant

The dielectric constant of all dielectric gases is generally accepted as 1.0, although values are actually slightly higher. The dielectric constant of SF_6 gas, for example, is reported as 1.002; for nitrogen and carbon tetrafluoride, the dielectric constant is 1.0006.

TABLE 16-12   Boiling or Sublimation Temperatures for Selected Dielectric Gases (At Normal Atmosphere Pressure)

	°C	°F
Helium (He)	–269.8	–453.8
Nitrogen (N_2)	–195.8	–320.4
Oxygen (O_2)	–183.0	–297.4
Carbon tetrafluoride (CF_4)	–128.0	–198.4
Chlorotrifluoromethane ($CClF_3$)	–81.4	–114.5
Carbon dioxide (sublimes) (CO_2)	–78.5	–109.3
Hexafluoroethane (CF_3CF_3)	–78.2	–108.8
Sulfur hexafluoride (sublimes) (SF_6)	–63.9	–83.0
Chlorodifluoromethane ($CHClF_2$)	–40.75	–41.35
Chloropentafluoroethane (CF_3CF_2Cl)	–38.7	–37.7
Dichlorodifluoromethane (CCl_2F_2)	–29.8	–21.6
Sulfur dioxide (SO_2)	–10.0	+14.0

Source: Handbook of Physics and Chemistry.

Since SF_6 is nonpolar, its dielectric constant does not vary with frequency. Pressure changes do, however, affect its dielectric constant. Over a pressure increase of 22 atmospheres, the increase in dielectric constant is about 7 percent.

Compatibility

Compatibility of components in electrical equipment is of paramount importance to its performance and longevity. Of at least equal consequences is the freedom of the system from moisture and impurities. Thus, performance characteristics of individual components do not always predict accurately the behavior of equipment in actual service. Therefore, it is prudent for a manufacturer to ensure a high degree of compatibility among components, and also the capability to assemble them into equipment free from moisture and other contaminants before settling on final specifications. This is usually not a problem for pure SF_6, since the gas is chemically inert at normal operating temperatures. Problems of compatibility arise only when the gas is decomposed by electrical discharge and forms reactive byproducts such as hydrogen fluoride and free fluorine.

Availability

SF_6, halogenated hydrocarbons, and nitrogen, the most commonly used dielectric gases, are readily available from two or more sources each. Although the price of SF_6 is relatively high compared with most other dielectric gases, its cost in equipment is usually 10 percent or less. Technologically it is still unsurpassed for most applications.

PROPERTIES OF SF_6

The basic properties of SF_6 which make it the preeminent dielectric gas may be summarized as follows:

- It suppresses carbonization.
- It is useful over the normal range of utility operating conditions.
- It has the best obtainable arc-interrupting properties.
- It has excellent dielectric properties.
- It is nontoxic and easy to handle.
- It is chemically inert.
- It has good heat transfer properties.
- It is highly electronegative.

Comparative properties of SF_6 with other gases are shown in the preceding tables. Table 16-13 shows specific properties of SF_6, the principal dielectric gas.

TABLE 16-13   **Properties of Sulfur Hexafluoride (SF$_6$)**

	VALUE
Physical	
Molecular weight	146.05
Sublimation temperatures at 1 atm °C	−63.9
°F	−83
Melting point at 32.5 psia °C	−50.8
°F	−59.4
Density at 21.1°C, 1 atm, g/L	6.139
at 70°F, 1 atm, lb/ft^3	0.382
Surface tension at −°20C, dyn/cm	8.02
Viscosity, centipoises	
liquid at 13.52°C (56.34°F)	0.305
gas at 31.16°C (88.08°F)	0.0157
Index of refraction at 0°C (32°F), 1 atm, n_D	1.000738
Critical temperature, °K	318.70
Critical pressure, bars	37.71
atmospheres	37.22
Critical volume, cm^3/g	1.356
Solubility at 1 atm	
in transformer oil at 27°C (80.6°F), mL SF$_6$/mL oil	0.408
in water at 24.85°C (76.73°F), cm^3 SF$_6$/cm^3 H$_2$O	0.0055
Solubility of water in SF$_6$, percent by weight	0.0097
Electrical	
Dielectric strength relative to nitrogen = 1 at 60 Hz–1.2 MHz	2.3–2.5
Dielectric constant at 25°C (77°F)	1.002
Loss tangent (tan δ) at 1 atm	$<2 \times 10^{-7}$
Thermodynamic	
Heat of sublimation, cal/gram · mol	5,640
Heat of fusion, cal/g · mol	1,200
Heat of vaporization at 70°F, Btu/lb	28.380
at 25°C, cal/g	15.767
Heat of formation at 25°C, kcal/g · mol	−291.77
Thermal conductivity at 30°C, cal/s · cm · °C	3.36×10^{-5}

Source: AlliedSignal Technical Bulletin *Sulfur Hexafluoride.*

GAS MIXTURES

Because of the relatively high cost of SF$_6$, its higher than desirable boiling point, and its sensitivity to surface imperfections on electrodes, there is a continuing search for dielectric gases that would provide satisfactory alternatives. Since no single gas appears to be an adequate replacement, the principal focus is on gas mixtures.[4]

Since SF$_6$ liquefies more readily than nitrogen, dispensing mixtures of these gases in correct proportions requires special precautions.

[4]EPRI Report EL.-2620, "Gases superior to SF$_6$ for insulation and interruption," 1982. 5 EPRI Report EL-2620 "Gases ect."

Table 16-14 shows breakdown strengths for selected gases in mixtures with SF_6.

TABLE 16-14 Mean Breakdown Strengths of Selected Dielectric Gases and Binary Gas Mixtures with SF_6 ($SF_6 = 100$)

GAS	FORMULA	UNMIXED	PERCENT OF GAS MIXED WITH SF_6		
			75	50	25
Chloropentafluoroethane	CF_3CF_2Cl	114	111	108	104
Sulfur dioxide	SO_2	102	111	116	108
Dichlorodifluoromethane	CCl_2F_2	100	108	107	106
Hexafluoroethane	CF_3CF_3	81	88	90	95
Chlorotrifluoromethane	$CClF_3$	58	78	88	95
Chlorodifluoromethane	$CHClF_2$	43	84	92	97
Carbon tetrafluoride	CF_4	42	63	78	89
Air	$N_2 + O_2$	37	78	85	94
Nitrogen	N_2	37	77	88	95
Carbon dioxide	CO_2	32	65	80	91

Source: Measurements are made in uniform fields using an aluminum sphere with radius 49 mm and a high-voltage plane electrode 26 mm in diameter. Gap is 5 mm and pressure is 0.15 MPa with ultraviolet radiation.

ELECTRICAL/ELECTRONIC USES

The principal application for SF_6 is in circuit breakers. These devices are used for current interruption in high-voltage equipment. Circuit breakers for 765 kV operation and with continuous ratings of up to 3,000 amperes are in service. As even higher voltages come into use, the advantages of SF_6 become more important. There is some replacement of oil or air circuit breakers with "puffer" breakers using SF_6. Typically, this type of breaker includes three cylindrical aluminum-interrupting modules containing SF_6 at a pressure of 75 psig. Within each module is mounted an epoxy filament-wound tube. A hydraulic or spring-driven mechanism opens and closes the breaker contacts and compresses the spring which provides the energy for opening the breaker. When the breaker is tripped, the arcing contacts part and the compressed gas flows along the arc until interruption occurs at ac current zero.

SF_6 is also the gaseous insulation for high-voltage coaxial power transmission lines with voltage ratings up to 765 kV for use above ground, underground, and underwater where their high cost is justified. These lines can be laid close together, requiring $1/10$–$1/20$ the space for conventional overhead air-insulated power lines.

Waveguides with SF_6 insulation can transport seven–ten times more microwave power than with air or nitrogen at the same pressure, or the same power as with air or nitrogen at lower pressure. Similarly, the ratings of both Van de Graaff and linear accelerators can be raised by using SF_6. The arc-extinguishing ability of SF_6 is also important in these applications.

SF_6 is used to insulate transmitter sections on large high-power radar installations such as AWACS.

Nitrogen is used primarily in liquid-filled transformers to prevent oxidation of the oils.

DEVELOPMENT PROGRAMS

There is growing interest in minisubstations, some of which are in operation in Europe and Japan. Because of its inherent properties, SF_6 insulated circuit breakers, transformers, cables, disconnect switches, and other equipment can be made more compact with these advantages:

• Much less land area would be needed for minisubstations

• Conventional substations are dangerous, unsightly, and in the open, whereas minisubstations can be installed inside structures, protected from the weather, pollutants, and vandalism.

Compact capacitor designs using SF_6 are under investigation.

A new application for SF_6 could be in high-voltage dc transmission systems.

Evaluation of gas mixtures continues with the aim of determining a mixture that would be technologically satisfactory and lower cost than pure SF_6. An 80% SF_6/20% N_2 mixture has nearly the same breakdown strength as pure SF_6 due to synergistic effects. There is interest in ternary mixtures composed of 40–60 percent nitrogen, 20–30 percent SF_6, and as much halocarbon gas as permissible without condensation under operating pressures and low temperatures and without formation of carbon under arcing conditions. Such a mixture, if technologically acceptable to industry, would significantly decrease gas cost versus 100 percent SF_6. For example, 50% N_2 + 40% SF_6 + 10% CCl_2F_2 has 85 percent the dielectric strength of SF_6 alone at approximately 40 percent of the cost of SF_6 alone. However, as noted earlier, the cost of gas is usually a small fraction of the cost of the equipment it insulates, and the use of chlorofluorocarbons is likely to be strictly regulated because of the effect they have on depleting ozone in the upper atmosphere.

MARKET TRENDS

It is likely that SF_6 will continue to be the predominant gaseous dielectric for the foreseeable future.

REFERENCES FOR FURTHER READING

• EPRI EL-2620, "Gases Superior to SF_6 for Insulation and Interruption," obtainable from Research Reports Center, Box 50490, Palo Alto, CA 94303 (contains numerous references on all aspects of subject), 1982.

- EPRI EL-1007, *Investigation of High-Voltage Particle-Initiated Breakdown in Gas-Insulated Systems* (obtainable as above), 1979.

- L. G. Christophorou et al., *Recent Advances in Gaseous Dielectrics*, from Conference Record of the 1984 IEEE International Symposium on Electrical Insulations, June 11–13, Montreal (obtainable from author, Oak Ridge National Laboratory, Oak Ridge, TN 37831).

- J. M. Meek and J. D. Craggs, *Electrical Breakdown of Gases*, John Wiley & Sons, ISBN 0-471-995533, 1979.

- E. Nasser, *Fundamentals of Gaseous Ionization & Plasma Electronics*, Wiley Interscience, ISBN 0-471-63056-X, 1971.

- R. J. Van Brunt and J. T. Herron, *Fundamental Process of SF_6 Decomposition in Glow and Corona Discharges*, IEEE Transactions on Dielectrics and Electrical Insulation, vol. 25, pp. 75–94, 1990.

———————————— 17 ————————————

Dielectric Liquids

INTRODUCTION

Dielectric liquids have long served as electrical insulation for transformers, capacitors, cables, and switchgear. Some of the formerly widely used fluids, polychlorinated biphenyls (PCBs, askarels), are no longer available because of their toxicity and detrimental environmental effects (see "Toxic Substances Control Act," p. 516, in Chapter 19, *Government Activities*). Tetrachloroethylene (perchloroethylene) has also been banned by the Naval Facilities Engineering Command for use in transformers because of its toxicity. Consequently, fluids have been developed that do not have these problem properties, but which provide functionality for their specific uses. Thus, there is no single fluid type suitable for all applications, but rather, there is an assortment of fluids, each of which is tailored to a certain end use. Liquids now available include mineral (petroleum) oils, high-molecular-weight paraffinic oils, silicones, fluoropolymers, and other synthetic oils. Major producers include:

COMPANY	PRODUCT
Allco Chemicals	Dixylylethane (DXE)
Dow Corning	Silicones
Exxon	Mineral Oil (Univolt^{TM})
General Electric	Silicones
Mobil Oil	Mineral Oil (Mobilect^{TM})
RTE Corporation	High-Molecular-Weight Oils (RTEmp^{TM}), (Envirotemp^{TM+})
Texaco	Mineral Oil
Union Carbide	Silicones

TECHNOLOGY

The fundamental processes leading to breakdown in dielectric liquids are generally stated to be similar to breakdown in dielectric gases (see discussion in Chapter 16, *Dielectric Gases*). Briefly, breakdown is the cumulative process in which charged particles accelerated by an electric field produce additional charged particles through collision with neutral liquid molecules or atoms, resulting in a cascade multiplication of ions sufficient to conduct a current.

Because dielectric liquids come into contact with a variety of other insulating and metallic materials, as well as moisture and foreign particles introduced in the course of storage and equipment manufacture, the breakdown strength of a liquid in an installation is almost always significantly lower than that of a laboratory-tested pure sample. Particles may jump between electrodes. The charge that a particle acquires near one electrode causes it to be repelled toward the opposing electrode where it gives up its charge and is repelled back to the other electrode. In liquids contaminated with moisture or gases, operating temperatures can cause formation of bubbles around submicron sized particles, thus lowering dielectric properties.

It is, therefore, of utmost importance that liquids and the equipment in which they are to be used be as free as possible from moisture and particle contamination for optimum performance.

For further reading on breakdown in liquids, see *Electrical Insulation*, A. Brad-well, Ed., Peter Peregrinus Ltd., London, UK, 1983.

CHEMISTRY

Historically, mineral insulating oils have been naphthenic-base products derived from petroleum. More recently, paraffinic-base oils of high molecular weight have become widely used, a development encouraged by the unavailability of PCBs. The exact com-position of these oils is, however, highly complex and proprietary, depending on the petroleum source, degree and method of refining, additives, and many other factors. Oils containing principally saturated hydrocarbons are oxidized readily, but this may be controlled by addition of oxidation inhibitors. To reduce the volatile content and increase the fire point of mineral oils, special hydrogenation and stripping processes are employed, with a resultant increase in viscosity. The oxidation rate of conventional mineral oils may be slowed significantly by refining methods that leave in the oil some naturally occurring aromatic compounds while removing other aromatic compounds which promote oxidation and the formation of sludge.

Silicone dielectric liquids are polydimethylsiloxanes. These linear polymers have the general structure:

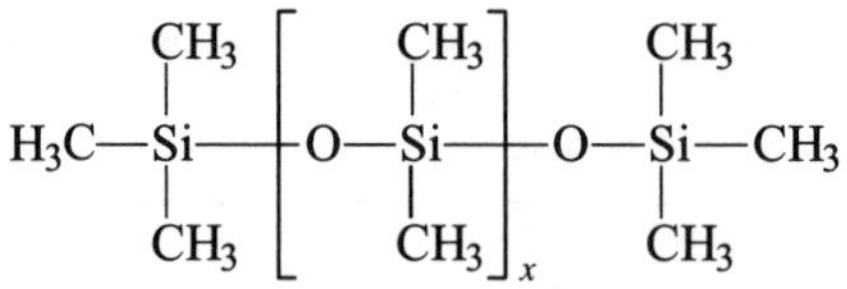

$$H_3C-\underset{\underset{CH_3}{|}}{\overset{\overset{CH_3}{|}}{Si}}-\left[O-\underset{\underset{CH_3}{|}}{\overset{\overset{CH_3}{|}}{Si}}\right]_x-O-\underset{\underset{CH_3}{|}}{\overset{\overset{CH_3}{|}}{Si}}-CH_3$$

These liquids are characterized by narrower molecular weight distribution than mineral oils or askarels. They have volatile content of less than 0.5 percent, which per-mits operation at higher temperatures with a corresponding lower vapor pressure.

DXE (1,1-di(ortho-xylyl)ethane, dixylylethane) is a thermally stable aromatic alkane derived from petroleum with the structural formula:

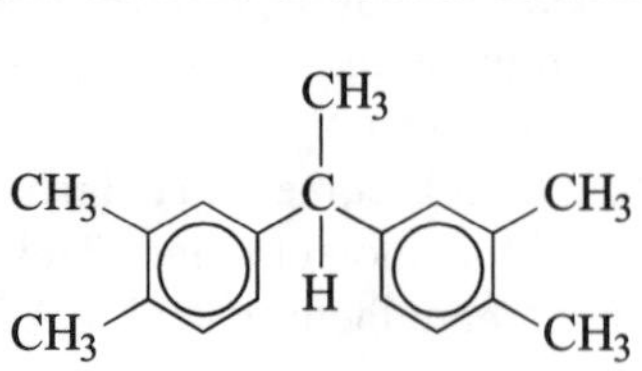

GRADES

Mineral oils and silicone liquids with a wide range of viscosities are available. The most widely used new mineral oil products have viscosities at 40°C (104°F) of approx-imately 10 centistokes, and for silicone liquids, viscosities are approximately 40 cen-tistokes (nominal 50 centistokes). The viscosity of high-molecular-weight mineral oil is much higher, about 130 centistokes. Conventional new mineral oils are available with normal and greater oxidation resistance. DXE is available with a viscosity of 12.5 centistokes at 38°C (100°F).

PROCESSING

Too much emphasis cannot be placed on the importance of removing from dielectric liquids particulate contamination, water, and dissolved air. Cartridge filters with ratings of 5 microns (the largest size particle that can be passed) or less are generally sufficient for removing solids. Water, including free water, is more difficult to remove by filtration, as filter elements become saturated and the water is then dispersed in the liquid. This is evidenced by the liquid becoming milky. It is then necessary to allow the water to settle in a cool, dry environment. This water may then be removed by syphoning or draining. Remaining dissolved water is removed by filtration or vacuum degasification until dielectric measurements indicate the appropriate degree of purity has been reached. Dissolved gases are also removed by this process, which requires specialized equipment. The best procedures are based on exposure to vacuum of a thin layer of liquid over a large area. Of course, it is necessary to take equal precautions to decontaminate other components of the equipment in which it is installed, since upon final assembly, equilibrium of moisture and gas will occur in all materials.

SIGNIFICANCE OF FUNCTIONAL PROPERTIES OF MINERAL INSULATING OIL

The following properties specifically refer to mineral insulating oils, but apply in general to all dielectric oils.

Physical Properties

• Aniline Point—The lowest temperature at which equal volumes of aniline and the test liquid are miscible.

• Color—An increase in the color number during service is an indicator of deterioration of the oil.

• Flash Point—The safe operation of the apparatus requires an adequately high flash point.

• Interfacial Tension—A high value indicates the absence of undesirable polar contaminants.

• Pour Point—The lowest temperature at which the oil will just flow.

• Specific Gravity—Influences the heat transfer rate.

• Viscosity—Influences the heat transfer rate and, consequently, the temperature rise of apparatus. High viscosity affects the speed of moving parts, a major concern in cold climates.

• Visual Examination—Indicates the absence or presence of undesirable contaminants.

Dielectric Properties

• Dielectric Breakdown Voltage, 60Hz—Indicates the ability of the oil to resist breakdown at power frequencies in electrical apparatus.
• Dielectric Breakdown, Disk Electrodes—Used for assessing the quality of oil as received in tank cars, tank trucks, or drums.
• Dielectric Breakdown, VDE Electrodes—Verband Deutscher Elektrotechniker (VDE) Specification 0370 electrodes are spherically capped electrodes used to determine if a processed oil meets the minimum acceptable breakdown strength of new oils.
• Dielectric Breakdown Voltage, Impulse—Indicates the ability of an oil to resist electrical breakdown under transient voltage stresses (lightning and switching surges).
• Power Factor (Dissipation Factor)—Measures the dielectric losses in an oil. A low value indicates a low level of soluble contaminants.

Chemical Properties

• Oxidation Inhibitor Content—Oxidation inhibitor added to mineral oil retards the formation of oil sludge and acidity under oxidative conditions.
• Corrosive Sulfur—The absence of elemental sulfur and thermally unstable sulfur-bearing compounds is necessary to prevent the corrosion of metals such as copper and silver in contact with the oil.
• Water Content—A low water content of the oil is necessary to achieve adequate dielectric strength and low dielectric loss characteristics, to maximize the insulation system life, and to minimize metal corrosion.
• Neutralization Number—A low total acid content of oil is necessary to minimize electrical conduction and metal corrosion and to maximize the useful life of the insulation system.
• Oxidation Stability—The development of oil sludge and acidity resulting from oxidation during storage, processing, and long service life should be held to a minimum in order to minimize electrical conduction and metal corrosion, to maximize insulation system life and electrical breakdown strength, and to ensure satisfactory heat transfer.
• Gassing—The gassing tendency of oil is a measure of the rate of absorption or desorption of hydrogen into or out of the oil under prescribed conditions.

ASTM STANDARDS FOR DIELECTRIC LIQUIDS

The American Society for Testing and Materials has developed the following standards for dielectric liquids:

Specifications
- Silicone fluids—D 2225
- Polybutene oil for capacitors—D 2296
- Mineral oils for electrical apparatus—D 3487
- Chlorinated aromatic hydrocarbons (Askarels) for capacitors—D 2233
- Chlorinated aromatic hydrocarbons (Askarels) for transformers—D 2283

Test Methods
- Petroleum-based oils—D 117
- Synthetic fluids for capacitors—D 3809
- D 117, *Electrical Insulating Oils of Petroleum Origin*. This standard describes

methods of testing and specifications for oils of petroleum origin intended for use as dielectrics or heat transformer media in electrical cables, transformers, oil circuit breakers, capacitors, and other electrical apparatus. Table 17-1 lists the test methods.

TABLE 17-1   Property Tests and Test Methods for Dielectric Oils of Petroleum Origin (ASTM D 117)

PROPERTY MEASURED	ASTM METHOD
PHYSICAL	
Aniline point	D 611
Coefficient of thermal expansion	D 1903
Color	D 1500
Examination	D 1500
	D 1524
Flash and fire point	D 92
Interfacial tension	D 971
	D 2285
Molecular weight	D 2224
Pour point	D 97
Refractive index	D 1218
	D 1807
Sampling	D 923
	D 2759
	D 3613
Specific gravity	D 287
	D 1298
	D 1481
Steam emulsion	D 1935
Viscosity	D 88
	D 445
	D 2161
ELECTRICAL	
Dielectric breakdown voltage	D 877
	D 1816
	D 3300

(continued)

TABLE 17-1   (*Continued*)

PROPERTY MEASURED	ASTM METHOD
Gas evolution	D 2298
	D 2300
Power factor	D 294
Resistivity	D 1169
CHEMICAL	
Acidity, approximate	D 1534
	D 1902
Carbon-type composition	D 2140
Characteristic groups	D 2007
Compatibility with construction materials	D 3455
Copper content	D 2608
	D 2675
Gas analysis	D 3612
Gas content	D 831
	D 1827
	D 2945
Inorganic chlorides and sulfates	D 878
Neutralization number	D 664
	D 974
Oxidation inhibitor content, phenolic	D 2668
Oxidation stability	D 1313
	D 1934
	D 2112
	D 2440
Peroxide number	D 1563
Saponification number	D 94
Sediment and soluble sludge	D 1698
Sulfur, corrosive	D 1275
Water content	D 1533
SPECIFICATIONS	
Transformer and circuit breaker oils	D 3487

• D 2225, *Silicone Fluids Used for Electrical Insulation.* This standard covers the testing of silicone fluids for use in transformers, capacitors, and electronic assemblies as an insulating or cooling medium. Table 17-2 lists the test methods.

• D 2296, *Continuity of Quality of Electrical Insulating Polybutene Oil for Capacitors.* This specification provides the limits within which the properties of Type D27-3000-1 electrical insulating polybutene oil for capacitors shall conform in order to ensure an unvarying continuity in the quality and grade of oil supplied. Requirements for this type oil are shown in Table 17-3. ASTM test methods are listed in D 117.

• D 3487, *Mineral Insulating Oil Used in Electrical Apparatus.* This specification applies only to new insulating oil as received prior to any processing. It covers mineral oil of petroleum origin for use as an insulating and cooling medium in new and existing power and distribution electrical apparatus, such as transformers, regulators, reactors, circuit breakers, switch-gear, and attendant equipment. This oil should be functionally interchangeable and miscible with existing oils. Two types of this oil are designated:

TABLE 17-2   Property Tests and Test Methods for Silicone Dielectric Fluids (ASTM D 2225)

PROPERTY MEASURED	ASTM METHOD
PHYSICAL	
Color	D 2129
Flash point	D 92
Fire point	D 92
Pour point	D 97
Refractive index	D 1807
Specific gravity	D 1298
Volatility	
Viscosity	D 445
CHEMICAL	
Neutralization number	D 974
Water content	
ELECTRICAL	
Dielectric constant	D 924
Dielectric breakdown voltage	D 877
Power factor	D 924
Specific resistance	D 1169
Compatibility	D 3455

TABLE 17-3   Requirements for Type D-27-3000-1 Polybutene Oil (ASTM D 2296)

PROPERTY	REQUIREMENTS
PHYSICAL	
Specific gravity, maximum, 15.6/15.6°C (60/60°F)	0.890
Pour point, maximum	10°C (50°F)
Flash point, minimum	218°C (425°F)
Viscosity: kinematic, centistokes at 100°C (212°F)	600–771
Color, maximum	1
Water content as shipped, ppm	40
CHEMICAL	
Chlorine, maximum, ppm	40
Inorganic chlorides and sulfates	none distinguishable
Neutralization number, maximum, mg KOH/g	0.04
Sulfur, corrosive	noncorrosive
ELECTRICAL	
Dielectric breakdown, min. kV at 25°C (77°F), 0.10 in gap	35 (ASTM D 877)
Power factor, 60 Hz, maximum at 100°C (212°F)	0.0003
Resistivity, minimum, $10^{12}\ \Omega \cdot$ cm at 100°C (212°F)	50
AGING TESTS (AFTER HEATING 96 h at 115°C (239°F)	
Color, maximum	1.5
Neutralization number, maximum, mg KOH/g	0.04
Power factor, 60 Hz, maximum at 100°C (212°F)	0.0005
Resistivity, minimum, $10^{12}\ \Omega \cdot$ cm at 100°C (212°F)	10

(Reprinted with permission from the *Annual Book of Standards.* Copyright, ASTM, 1916 Race Street, Philadelphia, PA 19103).

Type I Mineral Oil—an oil for apparatus where normal oxidation resistance is required.

Type II Mineral Oil—an oil for apparatus where greater oxidation resistance is required.

Requirements for these oils are shown in Table 17-4.

Table 17-4   Requirements for Type I and Type II Mineral Oils (ASTM D 3487)

PROPERTY	LIMIT TYPE I	TYPE II	ASTM METHOD
PHYSICAL			
Aniline point °C	63–84	63–84	D 611
Color, maximum	0.5	0.5	D 1500
Flash point, minimum °C	145	145	D 92
Interfacial tension at 25°C, minimum, dyn/cm	40	40	D 971
Pour point, maximum, °C	−40	−40	D 97
Specific gravity, 15/15°C maximum	0.91	0.91	D 1298
Viscosity, maximum, centistokes			D 445 or
at 100°C	3.0	3.0	D 88
at 40°C	12.0	12.0	
at 0°C	76.0	76.0	
Visual examination	clear, bright	clear, bright	D 1524
ELECTRICAL			
Dielectric breakdown voltage at 60 Hz, 25°C			
disk electrodes, minimum kV, 0.10 in gap	30	30	D 877
VDE electrodes, minimum kV			D 1816
0.040 in (1.02 mm) gap	28	28	
0.080 in (2.03 mm) gap	56	56	
Dielectric breakdown, impulse conditions 25°C, minimum, kV, needle negative to sphere grounded, 1 in (25.4 mm gap)	145	145	D 3300
Power factor, 60 Hz, maximum percent			D 924
at 25°C	0.05	0.05	
at 100°C	0.30	0.30	
CHEMICAL			
Oxidation stability (acid-sludge test) 72 h			D 2440
percent sludge, maximum, by mass	0.3	0.2	
total acid number, maximum, mg KOH/g	0.6	0.4	
Oxidation stability (rotating bomb test) minimum, minutes	—	195	D 2112
Oxidation inhibitor content, maximum, percent by mass	0.08	0.3	D 1473
Sulfur, corrosive	noncorrosive	noncorrosive	D 1275
Water, maximum, ppm	35	35	D 1533
Neutralization number, total acid number, maximum, mg KOH/g	0.03	0.03	D 974

(Reprinted with permission from the *Annual Book of Standards.* Copyright, ASTM, 1916 Race Street, Philadelphia, PA 19103).

• D 3809, *Synthetic Dielectric Fluids for Capacitors*. This standard covers testing synthetic dielectric fluids in use for capacitors. Table 17-5 lists the test methods.

• D 2233, *Chlorinated Aromatic Hydrocarbons (Askarels) for Capacitors*. This specification covers synthetic nonflammable liquids of the chlorinated aromatic type used as a dielectric and cooling medium in capacitors. Included are liquids that were previously available and are still in use. Four types are identified:

Type A—biphenyl with chlorine content of 42 percent by weight

Type B—biphenyl with chlorine content of 54 percent by weight

Type C—a mixture of approximately 70 percent Type B and 25 percent trichlorobenzene

Type D—The same as Type A, except with higher boiling homologs removed to a maximum of 0.4 percent.

Detailed requirements for these askarels are listed in this standard.

• D 2283, *Chlorinated Aromatic Hydrocarbons (Askarels) for Transformers*. This specification covers synthetic nonflammable liquids of the chlorinated aromatic type used as a dielectric and cooling medium in liquid-filled transformers. Eight types are identified:

Type A—50 percent hexachlorobiphenyl chlorinated to a chlorine content of 60 percent by weight, 40 percent trichlorobenzene isomers, plus 0.18–0.22 percent phenoxypropene oxide

Type B—45 percent hexachlorobiphenyl chlorinated to a chlorine content of 60 percent by weight, 55 percent blend of trichlorobenzene and tetrachlorobenzene, plus 0.115–0.135 percent diepoxide-type compound

Type C—80 percent trichlorobiphenyl chlorinated to a chlorine content of 42 percent by weight, 20 percent blend of trichlorobenzene and tetrachlorobenzene, plus 0.115–0.135 percent diepoxide-type compound

TABLE 17-5   Property Tests and Test Methods for Synthetic Dielectric Fluids for Capacitors (ASTM D 3809)

PROPERTY MEASURED	ASTM METHOD
PHYSICAL	
Coefficient of thermal expansion	D 1903
Flash point	D 92
Pour point	D 97
Refractive index	D 1218
Specific gravity	D 1298
Viscosity	D 445
CHEMICAL	
Acid number	D 664
Water content	D 1533
ELECTRICAL	
Dielectric constant	D 924
Dielectric strength	D 877
Power factor	D 924

Type D—70 percent pentachlorobiphenyl chlorinated to a chlorine content of 45 percent by weight, 30 percent trichlorobenzene isomers, plus 0.18–0.22 percent oxide

Type E—100 percent trichlorobiphenyl plus 0.18–0.22 percent phenoxypropene oxide by weight

Type F—45 percent pentachlorobiphenyl chlorinated to a chlorine content of 45 percent by weight, 55 percent blend of isomers of trichlorobenzene and tetrachlorobenzene, plus 0.115–0.135 diepoxide-type compound

Type G—60 percent pentachlorobiphenyl chlorinated to a chlorine content of 45 percent by weight, 40 percent trichlorobenzene isomers, plus 0.115–0.135 percent diepoxide-type compound

Type H—100 percent blend of isomers of trichlorobenzene and tetrachlorobenzene, plus 0.10–0.135 percent diepoxide compound. This type does not contain biphenyls.

The diepoxide-type compound in the above formulations is 3,4-epoxycyclohexylmethyl-3,4-epoxycyclohexane carboxylate.

Detailed requirements for these askarels are listed in this standard.

Current governmental regulations prohibit the manufacture and sale of askarels except Type H. The information in D 2233 and D 2283 is being maintained as a reference for those units still in service. (The EPA document regulating PCBs in effect in 1994 is *40 CFR Part 761* (July 1, 1991 Edition).)

PROPERTIES

There are several factors to be considered which vary in importance with each application:

- Toxicity
- Nonflammability (or high flash and fire points)
- Purity
- Inertness
- Chemical stability
- Thermal stability
- Dielectric characteristics
- Heat transfer rate
- Compatibility with other materials in electrical equipment
- Availability
- Price

Toxicity

With the exception of PCBs and tetrachloroethylene, widely used dielectric liquids are essentially nontoxic. Decomposition products, however, may have varying amounts of toxicity, but this is generally not a deterrent to their use. Some proposed chlorinated substitutes for PCBs, such as tetrachloroethylene, are, however, considered toxic by the National Institute of Occupational Safety and Health and the Clean Water Act, and this has slowed their adoption. Trichlorotrifluoroethane is thought to be potentially detrimental to the atmospheric ozone layer.

Flammability

Flammable liquids are defined by the National Fire Protection Association and the Department of Transportation (DOT) as those having a flash point below 100°F (37.7°C) (closed cup) and a vapor pressure of not over 40 psia at 100°F. The term "noncombustible" (often confused with "nonflammable") refers to any substance that will not burn, such as highly halogenated liquids. Even these liquids, however, when decomposed by arcing, may form highly flammable products. The applicable standard approved by Underwriters Laboratories is UL 340, *Tests for Comparative Flammability of Liquids.*

Except in low-risk, isolated areas, flammability is a major concern in selecting a dielectric liquid. Silicone oils and highly refined high-molecular-weight paraffinic oils, with fire points of 600°F (315°C) or over, have low flammability. The 1981 National Electric Code, Article 450-23 states: "Transformers insulated with less-flammable liquids shall be permitted to be installed without a vault in noncombustible occupancy areas of noncombustible buildings, provided there is a liquid confinement area and the liquid is listed as having a fire point of not less than 300°C (572°F)."

Purity, Inertness, and Chemical and Thermal Stability

A high degree of purity, inertness, and chemical and thermal stability is necessary if a dielectric liquid is to withstand long service without deterioration. Commercially available products of all types have a high degree of purity when shipped as mandated by ASTM standards. Of the most widely used liquids, silicones are the most stable. They begin to break down thermally at temperatures over 450°F (230°C), with formation of more volatile cyclic silicone compounds. Silicone transformer liquid will begin to oxidize slowly with oxygen present at temperatures over 175°C (347°F). As it oxidizes, it gradually polymerizes, increasing in viscosity until gelation occurs. No harmful acids or sludges are formed, and dielectric properties remain stable. In service, no degradation takes place over the useful service life of a 65°C (149°F) rise or a 110°C (230°F) rise transformer.

Mineral oils without inhibitor oxidize readily in the presence of oxygen with the formation of sludge and acidity. This tendency may be slowed by addition of an oxidation inhibitor, such as 2,6-ditertiary-butyl paracresol. High-molecular-weight paraffinic fluids oxidize at a much slower rate than conventional mineral oils. Equipment may also be designed to minimize the effects of oil oxidation. Transformers can be designed to limit the supply of air in contact with the oil. Another effective measure is to blanket the liquid with an inert gas, such as nitrogen.

The oxidation rate of all mineral oils is increased with increasing temperature. An often used rule of thumb is that the rate of oxidation is doubled for each increase of

10°C (18°F). In equipment containing mineral oil, commonly used metals, such as copper, iron, and lead, act as catalysts to further increase the oxidation rate of the oil. The effect is different for each metal and its concentration, and depends on the amount of available oxygen, being insignificant where the supply of oxygen is severely limited.

Arcing

Both silicones and mineral oils evolve flammable gases under high current arcing, and can produce a flash or fireball under catastrophic failure conditions. Arcing gases from both liquids include hydrogen (the major gas), nitrogen, methane, oxygen–argon, carbon monoxide, with traces of ethylene, acetylene, and carbon dioxide. When decomposed by arcing, halogenated compounds form toxic and corrosive products which attack cellulosics. The energies involved in arcing determine how the evolution of gases develops.

Water Content

Liquids may be exposed to water during shipment and storage, as well as during the manufacture of electrical equipment when liquids may come into contact with materials containing relatively high moisture content. However, it is essential that the moisture content of dielectric liquids be kept to a minimum. Otherwise, moisture will react with other contaminants, accelerating chemical and dielectric degradation of the liquid. When excess moisture is present, water may separate from the liquid at low temperatures with formation of emulsions and significant reduction in dielectric values. Table 17-6 lists water content specification maximum limits for widely used dielectric liquids. The delivered liquids are usually much lower in water content.

Dielectric Properties

The dielectric breakdown voltage of a liquid at commercial power frequencies is influenced by the following factors:

- Shape, area, surface condition, and material of the electrodes
- Spacing between electrodes

TABLE 17-6   Water Content Maximum Limits in Specifications for Typical Dielectric Liquids (Determined by ASTM D 1533)

	ppm WATER
New mineral oil	25
High molecular weight mineral oil	15
Silicone fluid	50
DXE	300 max

Sources: Manufacturers' data sheets.

- Length of time for which the liquid is under electrical stress
- Temperature of the liquid
- Degree of purity
- Concentration of dissolved gases and gassing tendency.

In designing apparatus containing liquid dielectrics, ample allowance should be made for the cumulative and, in some cases, interacting effects of the above factors. The insulation systems of transformers, cables, and capacitors often include paper and other cellulosic and absorbant materials with greater affinity for water than have dielectric liquids. Thus, dielectric systems must establish moisture equilibrium before dielectric measurements can be meaningful.

Because dielectric measurements among manufacturers are often made under different test conditions in key matters such as test methods employed, electrode types and spacings, frequencies used, moisture content, degree of purity, etc., it is difficult to compare dielectric values among insulating liquids. The most meaningful comparisons are those made on complete systems, although as a practical matter, end use conditions often dictate the liquid to be used. For widely used products, dielectric values for dry, filtered, and degassed transformer liquids fall within the ranges shown in Table 17-7.

A noteworthy feature of all liquid dielectrics is their self-healing properties after dielectric breakdown, as the device containing the liquid is immediately reinsulated by liquid flow.

Heat Transfer Rate

The suitability of a dielectric liquid as a cooling medium is determined by its heat transfer properties, namely:

- Viscosity (resistance to flow)
- Density (weight per unit volume)
- Thermal conductivity (the quantity of heat passing in a unit time through a unit thickness of a substance across a unit area for a unit difference in temperature)
- Specific heat (the ratio of the quantity of heat required to raise the temperature of a body 1 degree to that required to raise the temperature of an equal mass of water 1 degree)

TABLE 17-7   Dielectric Properties of Typical Dielectric Liquids for Transformers

	ASTM METHOD	TEMPERATURE	RANGE
Dielectric strength, 0.080 gap, kV	D 1816	25°C (77°F)	56–60
Dielectric constant, power frequencies	D 924	25°C (77°F)	2.2–2.7
Volume resistivity, $\Omega \cdot cm$	D 1169	25°C (77°F)	1×10^{13} 1×10^{15}

- Coefficient of thermal expansion (the ratio of the increase in volume of a liquid for a unit rise in temperature to the original volume).

For optimum heat transfer, the viscosity of a liquid should be as low as possible, and other property values should be as high as possible. The thermal properties of widely used dielectric liquids are shown in Tables 17-8 to 17-13.

**TABLE 17-8   Viscosities and Pour Points of Typical
Transformer Dielectric Liquids
(ASTM D 445)**

	CENTISTOKES			
	0°C (32°F)	40°C (104°F)	100°C (212°F)	POUR POINT
New mineral oil	55	9	2.3	−45°C (−49°F)
High-molecular-weight mineral oil	—	130	13	−24°C (−11°F)
Silicone	95	38	16	−55°C (−67°F)
DXE	—	13	—	−34°C (−30°F)

Sources: Manufacturers' data sheets.

**TABLE 17-9   Densities of Typical Transformer Dielectric
Liquids
(ASTM D 1298)**

	TEMPERATURE	g/cm³
New mineral oil	15°C (59°F)	0.91
High-molecular-weight mineral oil	25°C (77°F)	0.87
Silicone fluid	25°C (77°F)	0.96
Trichlorotrifluoroethane	25°C (77°F)	1.42
Tetrachloroethylene	25°C (77°F)	1.63
DXE	16°C (60°F)	0.98

Sources: Manufacturers' data sheets.

**TABLE 17-10   Thermal Conductivities of Typical
Transformer Dielectric Liquids**

	TEMPERATURE	cal/s cm2 °C/cm
New mineral oil	25°C (77°F)	3.5×10^{-4}
High-molecular-weight mineral oil	25°C (77°F)	3.1×10^{-4}
Silicone fluid	25°C (77°F)	3.6×10^{-4}
DXE	—	1.2×10^{-4}

Sources: Manufacturers' data sheets.

TABLE 17-11   Specific Heats of Typical Transformer Dielectric Liquids

	TEMPERATURE	cal/g/°C
New mineral oil	25°C (77°F)	0.53
High-molecular-weight mineral oil	25°C (77°F)	0.46
Silicone fluid	25°C (77°F)	0.36
DXE	260°C (500°F)	0.63

Sources: Manufacturers' data sheets.

**TABLE 17-12   Coefficients of Thermal Expansion of
Typical Transformer Dielectric Liquids
(ASTM D 1903)**

	TEMPERATURE	$cm^3/cm^3/°C$
New mineral oil	na	na
High-molecular-weight mineral oil	25°C (77°F)	0.0007
Silicone fluid	25°C (77°F)	0.0010
DXE	25°C (77°F)	0.00063

Sources: Manufacturers' data sheets.

**TABLE 17-13   Flash and Fire Points of Typical
Transformer Dielectric Liquids
(ASTM D 92)**

	FLASH POINT °C (°F)	FIRE POINT °C (°F)
New mineral oil	145 (293)	na
High-molecular-weight mineral oil	284 (543)	312 (594)
Silicone fluid	300 (572)	315 (600)
DXE	163 (325)	193 (380)

Sources: Manufacturers' data sheets.

Compatibility

Liquids that have achieved wide usage as dielectrics are, of necessity, compatible with most other materials used in the construction of electrical equipment, for example, cellulose and polyamide papers, films, wire enamels, varnishes, mica products, wood, and metals. However, care is needed in selecting seal and gasket materials, and testing is required to ensure compatibility. Contact of silicone liquids with silicone rubber gaskets, silicone rubber wire coverings, and other silicone materials should be avoided. Silicone rubber and silicone transformer liquid are chemically similar, and the liquid is readily absorbed into the rubber, causing it to swell and lose physical properties. Silicone liquids also tend to leach plasticizers from natural and synthetic rubber compounds, causing an increase in hardness and loss of effectiveness as gasket materials. Where replacement of

PCB liquids is required, only liquids designated as suitable should be considered, for example, high-molecular-weight paraffinic oils or certain silicone oils.

Availability and Price

Dielectric liquids of the types discussed in this chapter are available from several sources as noted. Where specifications and circumstances permit consideration of alternative liquids, price is the predominant factor. Conventional mineral oil is by far the lowest in price, and silicone liquids are the most expensive. High-molecular-weight paraffinic oils, although not as expensive as silicone liquids, are much higher in price than conventional mineral oils. It is important, however, to consider the total cost of an installation. In some cases, when there are alternatives, the lower cost of conventional mineral oil may be more than offset by requirements for vaults and fire protection equipment not necessary when less-flammable, higher-price liquids are used.

ELECTRICAL/ELECTRONIC USES

Mineral oil is used extensively in electrical equipment that is often nitrogen blanketed, such as distribution and substation transformers. It is also used in low- and high-pressure cable systems, and in capacitors, regulators, circuit breakers, and switchgear in isolated areas or where its flammability does not present a hazard. It is interchangeable and miscible with other dielectric oils.

High-molecular-weight paraffinic oils are used primarily in transformers and load break switchgear. They are also suitable as make-up fluids replacing askarels. They are less flammable than conventional mineral oils, and can be used with fewer precautions than are required of conventional oils.

Silicone liquids are used in small and medium transformers, capacitors, and other apparatus primarily because of their high-temperature properties, low volatility, and their resistance to oxidation and sludge formation.

Polybutene liquids are of generally high purity and have excellent dielectric characteristics, making them suitable for use in capacitors. They are also used as paper impregnants and as pipe oils.

Fluoropolymers are used as convective or evaporative dielectric coolants, where they are up to 100 times as effective as mineral oil in free convection. They are also used in applications where nonflammability is required.

DXE is used as a heat transfer and dielectric liquid in transformers.

NEW CAPACITOR DIELECTRIC LIQUIDS

There have been intensive efforts to evaluate alternatives to traditional capacitor dielectric fluids (askarels, mineral oils, silicones). It has been found[1] that highly aromatic fluids give the best high-voltage results. Two of the most promising products are:

[1]R. L. Miller, L. Mandelcorn, and G. E. Mercier, *Evaluation of Capacitor Impregnants*, paper presented at the Conference on the Properties and Applications of Dielectric Materials, Xian, China, June 24–28, 1985. Reprint, Westinghouse R&D Center, Pittsburgh, PA 15235.

meta   para

isopropylbiphenyl (Westinghouse's "Wemcol™")

(isomer mixture)

phenylxylyl ethane (Nippon Oil's "PXE™")

The aliphatic groups in these products contribute low-temperature fluidity and low-vapor pressure. However, since aliphatic groups tend to be susceptible to high-voltage effects, the optimum percentage of aliphatic carbons appears to be in the 15–25 percent range. Key properties of Wemcol and PXE are shown in Tables 17-14 and 17-15.[2]

[2]Ibid.

TABLE 17-14   Physical Properties of Two Newer Capacitor Fluids

| | | CENTISTOKES | | | |
	DENSITY	−50°C (−55°F)	25°C (77°F)	100°C (212°F)	POUR POINT
Wemcol	0.99	11,000	7.0	1.4	−50°C (−58°F)
PXE	0.98	23,000	6.5	1.6	−45°C (−49F)

TABLE 17-15   Dielectric Properties of Two Newer Capacitor Fluids

| | DIELECTRIC CONSTANT | | DISSIPATION FACTOR | | BREAKDOWN VOLTAGE, kV (ASTM D 877) |
	25°C (77°F)	100°C (212°F)	25°C (77°F)	100°C (212°F)	
Wemcol	2.6	2.5	0.02	0.02	60
PXE	2.6	2.5	0.02	0.03	60

DEVELOPMENT PROGRAMS

Better dielectric mineral oils are under development through improvements in refining method, degree of refining, and antioxidants, as well as in evaluation of crudes for the best feedstock.

Although improvements can undoubtedly be made in silicone dielectric liquids, industry seems to have settled on polydimethylsiloxanes as satisfying most of the needs silicones are expected to fulfill.

MARKET TRENDS

While much effort has been made to develop alternatives to askarels, it should be noted that the applications served by these materials in reality represent less than 10 percent of the total market for dielectric liquids. This market is directly linked to the growth of the electrical industry.

18

Industry Activities

INTRODUCTION

The electrical/electronic industries are fortunate to be served by a number of long-established, progressive organizations described in this chapter. Of these, ANSI serves solely as a clearinghouse for consensus voluntary standards. UL and ASTM primarily are dedicated to testing and standardization of materials, products, processes, and devices for all the nation's industries, among which the electrical/electronic industries are prominent. The SAE is broad in scope within the automotive and aerospace industries, but directs part of its efforts to electrical/electronic interests. All other organizations are concerned only with the electrical/electronic industries.

Important industry developments are presented at conferences/symposia, listed in this chapter, sponsored by industry and government groups.

Also included in this chapter are titles of encyclopedias and periodicals containing information on electrical/electronic insulating materials and their applications.

AMERICAN NATIONAL STANDARDS INSTITUTE (ANSI)

11 West 42nd Street, New York, NY 10036

ANSI was founded in 1918 as American Engineering Standards Committee. Its 1250 members are served by a staff of 107. The principal purpose is to serve as a clearinghouse for nationally coordinated voluntary safety, engineering, and industrial standards. ANSI gives status to standards developed by all groups concerned in such areas as definitions, terminology, symbols, and abbreviations; materials performance and characteristics; procedures and methods of rating; methods of testing and analysis; size, weight, volume; safety and health; and building construction. ANSI gathers and provides information on foreign standards, and represents the United States' interests in international standardization work.

Standards developed by any organization can be submitted to ANSI in one of the following ways:

1. A committee may be set up by the organization within ANSI; this committee is a balanced group of representatives from all areas interested in the standard.

2. The organization may submit a list of qualified parties throughout the industry whom the organization has designated to review the standard.

3. An organization can have their standards procedures approved by ANSI. ASTM, IEEE, and SAE are ANSI accredited. In each case, ANSI reviews and evaluates the proposed standard.

The end result of the work of ANSI is the publication of consensus standards developed by representatives from all groups that will be affected by the use of the standard, including manufacturers, consumers, government, and industry.

Publications include:

- *Standards Action*, biweekly
- *Catalogue of American National Standards*, annual, listing about 8,000 standards

ANSI holds an annual convention.

AMERICAN SOCIETY FOR TESTING AND MATERIALS (ASTM)

1916 Race Street, Philadelphia, PA 19103

Formed in 1898 by 70 members of the American section of the International Association for Testing Methods, headquartered in Europe, ASTM grew to over 33,000 members by 1993, comprising engineers, scientists, managers, professionals, academicians, consumers, and technicians holding membership as individuals or representatives of manufacturers, educational institutions, laboratories, and government agencies. In 1993, ASTM employed a staff of 200.

The mission of the society is the development of consensus standards on characteristics and performance of products, materials, systems, and services, and for the promotion of related knowledge. Standards include definitions, classifications, specifications, test methods, and recommended practices. Work is carried out by 134 technical committees. Anyone technically qualified may be invited to participate in the work of a committee. Committees of interest to electrical/electronic industries include *Electrical Conductors*, *Electrical Insulating Materials*, *Materials for Electronics*, and *Plastics*.

Publications include:

- *Annual Book of ASTM Standards*, containing all current standards
- *Standardization News*, monthly, includes new standards
- *Journal of Testing and Evaluation*, bimonthly
- *Composites Technology Review*, quarterly
- *Geotechnical Testing Journal*, quarterly
- Numerous technical papers and reports

ASSOCIATION OF EDISON ILLUMINATING COMPANIES (AEIC)

600 North 18th Street, P.O. Box 2641, Birmingham, AL 35291

One of the oldest associations in the electrical industry, AEIC was formed in 1885 by investor-owned public utilities. In 1993, membership was 72 companies. Activities center around these committees: Electric Power Apparatus, Meter and Services, Power Distribution, Power Generation, Research.

Publications include reports, for example, on crosslinked polyethylene insulated cable and oil-impregnated paper-insulated cables.

AEIC convention/meetings are held each fall.

EDISON ELECTRIC INSTITUTE (EEI)

701 Pennsylvania Avenue NW, Washington, DC 20004

EEI was formed in 1933 by U.S. investor-owned electrical utility companies. The mission of the Institute is to advance the art of producing and transmitting electricity, and to provide a source of statistics related to the electrical industry. EEI has 7 divisions and 60 committees which cover every area of interest or concern to the industry, and which conduct and sponsor studies, surveys, and reports.

The Institute serves as a forum in which member companies exchange ideas, experiences, and information. EEI participates with other industries in joint activities of mutual interest.

Publications include a *Publications Catalogue* with a comprehensive listing of publications covering every aspect of the electric utility industry:

- Engineering
- Alternative fuels
- The environment
- Economics, finance, statistics

EEI holds an annual convention/meeting in June.

ELECTRIC POWER RESEARCH INSTITUTE (EPRI)

3412 Hillview Avenue, Palo Alto, CA 94303

EPRI was formed in 1972 by electric power utilities to do research on long-term improvements in conventional electric energy production and on development of alternative energy sources. Activities are supported by U.S. electric utilities representing 70 percent of generating capacity, and the focus has shifted to shorter-term projects such as causes of and remedies for network failures, improved materials, and lower cost, higher efficiency transmission methods.

EPRI is organized to work in these areas:

- Electrical Systems
- Environment
- Exploratory and Applied Research
- Generation and Storage
- Nuclear Power
- Technical Operations

Projects are typically performed by commercial firms with EPRI in the role of project planning and contract management. EPRI publishes a monthly journal and *Technology Applications*, a series of one-page case histories prepared by the first utility to use an EPRI product or research result.

ELECTRICAL APPARATUS SERVICE ASSOCIATION (EASA)

1331 Baur Boulevard, St. Louis, MO 63132

EASA represents a service network of over 2,700 member companies throughout the world engaged in the repair, sale, and maintenance of electrical equipment, including motors, generators, transformers, controls, and power tools. A full-time engineering staff at headquarters is available to members for consultation. There are also management and training programs, and a convention and exhibition held each summer. EASA also maintains a library of training films and technical instructions as a service to members.

Publications include:

• *Currents*, monthly

ELECTRICAL MANUFACTURING AND COIL WINDING ASSOCIATION (EMCWA)

P.O. Box 278, Imperial Beach, CA 91933

EMCWA was organized in 1977 as the International Coil Winding Association to serve as the authoritative source of coil winding technology. In 1993, the name was changed to the Electrical Manufacturing and Coil Winding Association, with the mission to provide an array of educational opportunities which will enhance the general knowledge and use of electrical manufacturing technologies, including a scholarship program and an annual industry conference. In 1994, membership consisted of 120 company and 430 individual members. Publications include a *Proceedings* of the annual conference.

ELECTRONIC INDUSTRIES ASSOCIATION (EIA)

2001 Pacific Avenue NW, Washington, DC 20006

Started in 1924, EIA has since absorbed several electronic industry associations, and in 1993, membership totaled over 350 manufacturers of electronic parts; solid-state components; radio, TV, and video systems; governmental electronic systems; and industrial and communications electronic products. Divisions include Communications, Consumer Electronics Group, Distributor Products, Government, Industrial Electronics, Parts, Solid State Products, and Tubes.

EIA services all facets of the electronics industry with such activities as compiling statistics, monitoring legislative actions affecting the industry, representing the industry in government agencies by presenting consensus views and recommendations, and testifying at Congressional public hearings involving the industry. EIA develops industry standards and promotes their use. Most standards are submitted to ANSI.

Publications include:

- *Electronic Market Trends*, monthly
- *Electronic Market Data Book*, annual, provides information on industry sales and production
- *Publications Index*, listing publications covering all aspects of the Association's activities
- *Government Market Planning Information Service*, includes information on budget submissions and policy statements of Federal Departments
- *Consumer Electronics Annual Review*
- *Trade Directory*, annual
- *Electronics Foreign Trade*, features data on imports and exports
- *EIA Standards and Engineering Publications*, annual

EIA sponsors several events, including winter and summer consumer electronics shows, international telecommunications seminars and exhibits, and annual Design Engineers Electronic Components Conference.

INSTITUT DE RESERCHE D'HYDRO-QUÉBEC (IREQ)

1800, Montée Sainte-Julie, Varennes, Québec, Canada J3X 1S1

IREQ was created in 1967 to solve the technical problems that arise in the operation of Hydro-Québec's extensive power generation, transmission, and distribution system, and to study future needs, including alternative energy sources. It is widely recognized as one of the world's foremost research and development centers devoted to the technology, equipment, and materials of the electrical and electronics industries. Solid, liquid, and gaseous insulations are under continuous investigation.

IREQ works under contract for clients around the world wanting to carry out research or to develop new products. It also offers training courses and welcomes visits by technicians and scientists from other organizations.

Available from the Institute are conference papers, technical reports, addresses, and other publications on investigations undertaken by IREQ.

IREQ employs a staff of about 600 in permanent positions.

THE INSTITUTE FOR INTERCONNECTING AND PACKAGING ELECTRONIC CIRCUITS (IPC)

7380 North Lincoln Avenue, Lincolnwood, IL 60646

Formerly the Institute of Printed Circuits (organized in 1957), IPC is international in scope, serving manufacturers and users of electronic interconnecting devices, suppliers to the industry, and representatives from over 60 government agencies. The present name was adopted in 1977. Membership passed 1800 in 1993, with over 100 committees and working groups developing programs regarding interconnection technology, including printed wiring boards, flexible circuits, flat cable, hybrid circuits, connectors, and discrete wiring devices. IPC interfaces with government agencies, other industry organizations, and related professional groups. IPC employs a staff of 42.

IPC has developed a comprehensive set of *Standards and Specifications* covering all aspects of industry technology. Other publications include:

- *IPC Guidelines and Technical Reports*
- *The IPC Design Guide*
- *IPC Assembly/Joining Handbook*
- *IPC Test Manual*
- *IPC Gold Handbook*
- *Technical Reports*
- *The Technical Review*, monthly bulletin

IPC sponsors two major meetings annually plus numerous workshops and technical courses.

INSTITUTE OF ELECTRICAL AND ELECTRONICS ENGINEERS (IEEE)

345 East 47th Street, New York, NY 10017

IEEE was formed in 1963 by the merger of the American Institute of Electrical Engineers (founded in 1884) and the Institute of Radio Engineers (founded in 1912). By 1993, membership grew to approximately 320,000 in 145 countries, making it the largest professional engineering organization in the world. IEEE comprises 31 societies, including the *Dielectrics & Electrical Insulation Society*, each of which publishes its own periodicals and sponsors its own conferences. There are 530 local groups which meet periodically to hear lectures on and discuss topics of current engineering and scientific interest. IEEE has a staff of 500.

IEEE fosters the advancement of the theory and practice of electrical and electronic engineering and related arts and sciences through awards, conferences, publications, and the development of standards. The Institute supports the *Engineering Societies Library* in New York City, perhaps the largest electrical engineering library in the world.

Publications, with broad scope available to all IEEE members are:

• *IEEE Transactions on Dielectrics and Electrical Insulation*, monthly, publishes original papers edited to encourage deeper understanding and greater effectiveness in observing, classifying, interpreting, and reporting facts and theories germane to dielectric behavior and properties of electrical insulating materials and systems. It also lists scheduled national and international meetings and conferences in fields related to dielectric phenomena and measurements.

• *IEEE Electrical Insulation Magazine*, bimonthly, publishes articles dealing with materials, systems, and applications. Also included are columns dealing with tutorial and educational subjects, as well as columns expressing opinions and viewpoints. Other features are announcements of new products, calendar of events, and reports of matters of interest.

• *IEEE Spectrum*, monthly, contains articles on advanced technology, new product applications, industry news, information on IEEE activities, and employment opportunities.

• IEEE Standards covering the electrical/electronic industries

• *Proceedings of the IEEE*, monthly, contains fundamental papers of broad significance and long range interest.

• *IEEE Standard Dictionary of Electrical and Electronics Terms*

• Periodicals from each of its 31 societies

• *IEEE Potentials*, a magazine for students covering career issues, technical topics, and subjects of general interest to engineering students

IEEE is one of the sponsors of the *Electrical Insulation Conference*, held biennially in odd-numbered years, at which papers and exhibits are presented on all aspects of electrical insulations.

INSULATED CABLE ENGINEERS ASSOCIATION (ICEA)

P.O. Box 440, South Yarmouth, MA 02664

The predecessor to ICEA, the Insulated Power Cable Engineers Association, was formed in 1925. In 1993, there were 100 members, including the top engineers from most wire and cable manufacturers. The purpose of ICEA is to promote the safety, reliability, and economy of insulated conductors for the transmission and distribution of electrical energy. The principal activity of ICEA is to develop jointly with the Wire and Cable section of NEMA standards which are submitted to ANSI. ICEA also coordinates its activities on standards, where appropriate, with IEEE, UL, and ASTM, which organizations publish the standards.

There are five sections in ICEA:

• Extruded Dielectric Power Cable Section, including all cables with extruded insulations used from the transmission and distribution of electrical energy to equipment fixed in place.

- Laminar Dielectric Power Cable Section, including all cables with laminar insulations used for the transmission and distribution of electrical energy to equipment fixed in place.
- Communications Cable Section, including all insulated cables used to transmit voice, video, and data signals.
- Control and Instrumentation Cable Section, including all insulated cables used to convey electrical signals used for controlling or monitoring processes or electrical systems.
- Portable Cable Section, including all insulated electrical cables for equipment portable or movable in nature or used in mines or other similar applications.

Publications include standards, guides, and books on cable subjects. There is an annual conference.

INTERNATIONAL CONFERENCE ON LARGE HIGH VOLTAGE ELECTRICAL SYSTEMS (CIGRÉ)

112 Boulevard Hausmann, F–75008, Paris, France

CIGRÉ (Conference Internationale des Grands Reseaux Electricques à Haute Tension) was established in 1921 by the International Electrotechnical Commission to facilitate the interchange of technical knowledge and data between all countries in the general field of electricity generation and transmission at high voltages. All together, there are over 700 Collective Members and 3,000 Individual Members.

CIGRÉ Headquarters are in Paris, France. Conferences are held there and elsewhere biennially for the presentation of technical papers and for the free discussion thereof. Continuity of work between conferences is carried on by 15 special groups known as International Study Committees. Reports of their discussions and progress are published in the journal *ELECTRA*.

Publications include:

- *ELECTRA* (in English and French), bimonthly
- *Proceedings of the Session*, biennial

THE INTERNATIONAL ELECTROTECHNICAL COMMISSION (IEC)

United States National Committee of IEC
A Committee of the American National Standards Institute
11 West 42nd Street, New York, NY 10036

The IEC was founded in 1906, following a resolution passed in 1904 by the Chamber of Government Delegates, consisting of eminent engineers and scientists, at the International Electrical Congress in St. Louis, MO, U.S.A. It is the body responsible for preparing and publishing international standards for the electrical and electronics fields. The IEC is a nongovernmental organization composed of National Committees in 43 countries that together represent over 80 percent of the world's population, and produce over 95 percent of the world's electric energy.

The principal organ of the IEC is the Council, consisting of representatives from all National Committees, with annual meetings. The management of the technical work of the IEC is delegated to the Committee of Action, comprising members from 12 National Committees elected by the Council. Technical committees develop standards, kept under constant review, covering the entire electrical and electronic fields. These standards are used as the basis of national rules and international trade. Committees include Insulating Materials and Systems, and Insulation Coordination in a total of 77 technical committees.

A general meeting of IEC takes place annually, comprising sessions of the Council, the Committee of Action, and of some 30–40 technical committees and subcommittees. Other meetings are held as required.

A complete list of publications can be obtained from the National Committee or from the IEC Central Office in Geneva, Switzerland. Participants in the IEC Registered Subscribers Scheme receive all or selected groups of IEC publications and draft standards.

Selected publications include the following:

- 371: *Specification for insulating materials based on mica*
- 394: *Varnished fabrics for electrical purposes*
- 454: *Specification for pressure-sensitive adhesive tapes for electrical purposes*
- 455: *Specification for solventless polymerizable resinous compounds used for electrical insulation*
- 464: *Specification for insulating varnishes containing solvent*
- 554: *Specification for cellulosic papers for electrical purposes*
- 626: *Specification for combined flexible materials for electrical insulation*
- 641: *Specification for pressboard and presspaper for electrical purposes*
- 667: *Specification for vulcanized fibre for electrical purposes*
- 672: *Specification for ceramic and glass insulating materials*
- 674: *Specification for plastic films for electrical purposes*
- 763: *Specification for laminated pressboard*

NATIONAL ELECTRICAL MANUFACTURERS ASSOCIATION (NEMA)

2101 L Street NW, Washington, DC 20037

NEMA was formed in 1926 by the merger of the Associated Manufacturers of Electrical Supplies and the Electric Power Club. Membership in 1993 included more than 600 companies engaged in the manufacture of equipment for the generation, transmission, distribution, and utilization of electric power. The association, with a staff of 100, is the largest trade organization of electrical product companies in the United States.

Principal objectives of NEMA are to develop and promote the use of standards for the industries it serves, to provide an exchange of information on industry and product statistics and engineering safety, and to keep members informed on governmental activities affecting electrical/electronic industries. Many NEMA standards are submitted to ANSI.

NEMA publications include newsletters, standards, and manuals. Standards of special relevance are listed in the Appendix.

NEMA is one of the sponsors of the *Electrical Insulation Conference*, held biennially in odd-numbered years in the fall, at which papers and exhibits are presented on all aspects of insulating materials.

SOCIETY OF AUTOMOTIVE ENGINEERS (SAE)

400 Commonwealth Drive, Warrendale, PA 15096

The origins of SAE date back to 1905. Although primarily comprised of automotive engineers, SAE enrolls in a special affiliation engineers and scientists in fields of self-propelled sea, air, and space vehicles, and engineering students. The mission of SAE is to advance the arts, sciences, standards, and engineering practices related to self-propelled mechanisms. Research on utilization of fuels and lubricants is carried out with the American Petroleum Institute through the Coordinating Research Council.

A principal activity of SAE is the development of standards and specifications with emphasis on motor vehicle and aerospace industries. General interest standards are submitted to ANSI. The society conducts research on a contract basis for both industry and government agencies.

Publications include:

• *SAE Handbook*, an annual compilation of standards, recommended practices, and information reports

• *Automotive Engineering*, monthly, covers current topics in the automotive industry

• *Advances in Engineering Series*

• *Aerospace Standards*

• *Aerospace Material Specifications*

An annual convention is held in Detroit, MI, with attendance over 30,000. In excess of 400 papers are presented at its technical sessions.

SOCIETY OF PLASTICS ENGINEERS (SPE)

14 Fairfield Drive, Brookfield, CT 06804

Founded in 1942, SPE, with 1993 membership of 37,000, is a professional society of plastics scientists, engineers, educators, students, and others interested in the design, development, production, and utilization of plastics materials, products, and equipment. The society maintains an awards program in recognition of fundamental contributions to the technology of polymer science and engineering, and for achievements in engineering and technology, education, business management, research, and unique plastic products in both industrial and consumer fields. Divisions include *Advanced Polymer Composites, Blow Molding, Color and Appearance, Electrical and Electronics, Engineering Properties and Structure, Extrusion, Injection Molding, Molding, Moldmaking and Mold Design, Plastics Analysis, Plastics in Automotive, Thermoplastics Materials and Foams, Thermosetting Molding,* and *Vinyl Plastics.*

Publications include:

- *Polymer Engineering and Science*, 24 semi-monthly issues per year
- *Plastics Engineering*, monthly
- *Journal of Vinyl Technology*, quarterly
- *Polymer Composites*, bimonthly
- *Preprint Volumes*, as required
- *Plastics Engineering Series*, books

SPE holds an annual convention in May.

SOCIETY OF THE PLASTICS INDUSTRY (SPI)

1275 K Street NW, Washington, DC 20005

Founded in 1937, SPI membership in 1983 comprised over 2,000 manufacturers and processors of molded, extruded, fabricated, laminated, calendered, and reinforced plastics; manufacturers of raw materials, machinery, tools, dies, and molds; and research, development, and testing laboratories. The society supports the preparation of voluntary product standards and commercial standards in cooperation with other agencies. SPI maintains the Plastics Hall of Fame and staffs a speakers bureau.

Publications include:

- *Facts and Figures*, annual, contains statistics on plastics markets and production
- *Proceedings of Conferences*, annual
- *Membership Directory and Buyers Guide*, annual

SPI holds an annual meeting in Washington, DC, an annual APC/I conference, and a triennial National Plastics Exposition.

UNDERWRITERS LABORATORIES (UL)

333 Pfingsten Road, Northbrook, IL 60062

Organized in 1894, primarily to test products for electrical and fire hazards for insurance companies, by 1993 the scope of UL had expanded to include the investigation and testing of materials, devices, products, equipment, constructions, methods, and systems with respect to hazards to life and property with a staff of 3,900 in laboratories in Northbrook, IL; Melville, NY; Santa Clara, CA; and Research Triangle Park, NC.

There are seven departments in UL: *Burglary Protection*; *Casualty and Chemical Hazards*; *Electrical*; *Fire Protection*; *Heating, Air-Conditioning and Refrigeration*; and *Marine*. The Electrical Department evaluates appliances, rebuilt motors and generators, and electrical construction materials and wiring used to distribute power inside buildings. Approved products are given a UL label or mark.

In conjunction with testing work, UL has developed and published hundreds of standards. Most of these are submitted to ANSI, where they are generally accepted.

UL publishes annual lists of approved materials, devices, products, equipment, constructions, methods, and systems. Other publications include pamphlets on safety and fire protection, research bulletins with test procedures, and the quarterly *Lab Data* covering product testing and safety.

Specifically related to insulating materials are the following publications:

- UL *746A Polymeric Materials—Short-Term Property Evaluations*
- UL *746B Polymeric Materials—Long-Term Property Evaluations*
- UL *746C Polymeric Materials—Use in Electrical Equipment*
- UL *1446 Systems of Insulating Materials—General*
- UL *94—Tests for Flammability of Plastic Materials for Parts in Devices and Appliances*.

CONFERENCES/SYMPOSIUMS

Dielectric materials are discussed and displayed at many meetings, conferences, and symposiums, including the following:

- *BEMA International Electrical Insulation Conference*, held periodically. *Proceedings* are available from:
> BEMA Limited
> Leicester House, 8 Leicester Street
> London WC2H 7BN, England

- *Conference on Electrical Insulation and Dielectric Phenomena*, held annually in the fall. Sponsor is the Dielectrics and Electrical Insulation Society of the Institute of Electrical and Electronics Engineers. *Proceedings* are available from:
> Institute of Electrical and Electronics Engineers
> Service Center
> 445 Hoes Lane
> Piscataway, NJ 08855

- *Electrical/Electronics Insulation Conference*, held biennially in odd-numbered years in the fall. Sponsors are the Dielectrics and Electrical Insulation Society of the Institute of Electrical and Electronics Engineers, the Electrical Manufacturing and Coil Winding Association, and the National Electrical Manufacturers Association. *Proceedings* are available from:

Institute of Electrical and Electronics Engineers
Service Center
445 Hoes Lane
Piscataway, NJ 08855

• *IEEE International Symposium on Electrical Insulation*, held periodically since 1976. *Proceedings* are available from:

Institute of Electrical and Electronics Engineers
Service Center
445 Hoes Lane
Piscataway, NJ 08855

• *International Conference on Dielectric Materials, Measurements and Applications*. Principal organizer is the Science, Education and Technology Division of the Institution of Electrical Engineers. *Proceedings* are available from:

The Institution of Electrical Engineers—LS(C)
P.O. Box 26, Hitchin
Herfordshire SG5 1SA, England

• *International Wire and Cable Symposium*, held annually in the fall. Sponsor is the International Wire and Cable Symposium, Inc. and the U.S. Army Electronics Command. *Proceedings* are available from:

International Wire and Cable Symposium Inc.
174 Main Street
Eatontown, NJ 07724

• *Printed Circuit World Expo*, held periodically. Sponsors are *ELECTRI*ONICS* magazine, *Microelectronic Manufacturing and Testing* magazine, and Worldwide Convention Management Company, a Division of Lake Publishing Corporation. *Proceedings* are available from:

Worldwide Convention Management Company
Division of Lake Publishing Corporation
17730 West Peterson Road
Box 159
Libertyville, IL 60048

PUBLICATIONS

There are several encyclopedias, handbooks, and periodicals covering electrical/electronic insulating materials:

ENCYCLOPEDIAS AND HANDBOOKS

Encyclopedia of Chemical Technology, Kirk—Othner
John Wiley & Sons, Inc.
605 3rd Avenue
New York, NY 10158

Concise Encyclopedia of Polymer Science and Technology
John Wiley & Sons, Inc.

Encyclopedia of Science and Technology
McGraw-Hill, Inc.
P.O. Box 602
Hightstown, NJ 08520

Modern Plastics Encyclopedia
McGraw-Hill, Inc.

Plastics 1980
The International Plastics Selector, Inc.
2251 San Diego Avenue, Suite A216
San Diego, CA 92110

Plastics Technology Manufacturing Handbook and Buyers Guide
633 Third Avenue
New York, NY 10017

PERIODICALS

Circuit News
125 Jericho Turnpike
Jericho, NY 11753

Circuits Assembly
13760 Noel Road, Ste. 500
Dallas, TX 75240
circulation 32,543

EPRI Journal
P.O. Box 10412, 12X
Palo Alto, CA 94303

Electrical Engineering in Japan
605 3rd Avenue
New York, NY 10158
circulation 400

Electrical Review—England
Perrymount Road/Haywards Heath
West Sussex RH16 3BR
England

Electrical World
11 West 19th Street
New York, NY 10011

Electronic Engineering Times
111 East Shore Road
Manhasset, NY 11030

Electronic News
7 East 12th Street
New York, NY 10003
circulation 72,000

Electronic Packaging and Production
1350 East Touhy Avenue
Des Plaines, IL 60018

Electronics
611 Route 46 West
Hasbrouck Heights, NJ 07604

Electronics Today International
Ryrie House
15 Boundary Street
Rushcutter's Bay, NSW 2011
Australia
circulation 33,500

IEEE Electrical Insulation
IEEE Publishing Services Department
445 Hoes Lane
Piscataway, NJ 08855

IEEE Spectrum
IEEE Publishing Services Department
445 Hoes Lane
Piscataway, NJ 08855
circulation 320,000

Journal of Elastomers & Plastics
851 New Holland Avenue
P.O. Box 3535
Lancaster, PA 17604
circulation 350

Journal of Vinyl Technology
Society of Plastics Engineers
14 Fairfield Drive
Brookfield, CT 06804

Modern Plastics
1221 Avenue of the Americas
New York, NY 10020
circulation 50,000

Plastics Engineering
Society of Plastics Engineers

14 Fairfield Drive
Brookfield, CT 06804
circulation 36,113

Plastics World
Society of Plastics Engineers
14 Fairfield Drive
Brookfield, CT 06804
circulation 1,380

Polymer Composites
Society of Plastics Engineers
14 Fairfield Drive
Brookfield, CT 06804

Polymer Engineering & Science
Society of Plastics Engineers
14 Fairfield Drive
Brookfield, CT 06804

Polymer Plastics Technology & Engineering
270 Madison Avenue
New York, NY 10016
circulation 800

19

Government Activities

INTRODUCTION

In recent years, the federal government has become increasingly active in developing and administering regulations and standards which affect the manufacture and use of dielectrics and their raw materials. As a result, these groups are faced with problems of compliance with a multitude of federal, state, regional, and local statutes enacted to protect the health and safety of workers, and to ensure a clean, safe environment.

This chapter discusses the principal governmental agencies involved in these activities and lists targeted materials. Also included is information on agencies developing and distributing standards and specifications, and an example of a recent military specification on polyamide-imide materials.

WORKER SAFETY AND HEALTH

In launching the program to improve worker safety and health, Congress passed the Williams–Steiger Occupational Safety and Health Act on December 29, 1970, to become effective April 28, 1971. The stated purpose of the act is

> To assure safe and healthful working conditions for working men and women: by authorizing enforcement of the standards developed under the act; by assisting and encouraging the States in their efforts to assure safe and healthful working conditions; by providing for research, information, education, and training in the field of occupational safety and health and for other purposes.

Two government organizations were established to administer the act:

• *The Occupational Safety and Health Administration (OSHA)* functions within the Department of Labor, setting and enforcing standards, encouraging states to develop their own safety and health programs, and establishing certain record-keeping requirements

• *The National Institute for Occupational Safety and Health (NIOSH)*, a research activity of the Department of Health and Human Services, which conducts research and recommends criteria to OSHA on health standards, undertakes education and training programs, and performs work in the area of safety engineering.

Current occupational safety and health standards are published in *Code of Federal Regulations, Title 29, Part 1910*, effective June 30, 1993, obtainable from the Department of Labor or OSHA regional or field offices. This document lists approximately 600 materials as air contaminants with limits of permissible concentrations in air to which workers are exposed.

An employee's exposure to any material in Table 19-1 in any 8-hour work shift of a 40-hour work week shall not exceed the values for that material in the table.

TABLE 19-1   Air Contaminants Limits of Worker Exposure
(8 hour Time-Weighted Averages)

SUBSTANCE	Parts/Million (vol) @ 25°C, 760 mm Mercury Pressure	mg/m³ of Air
Acetaldehyde	200	360
Acetic acid	10	25
Acetic anhydride	5	20
Acetone	1,000	2,400
Acetonitrile	40	70
Acetylene dichloride, see 1,2-Dichloroethylene		
Acetylene tetrabromide	1	14
Acrolein	0.1	0.25
Acrylamide—Skin		0.3
Aldrin—Skin		0.25
Allyl alcohol—Skin	2	5
Allyl chloride	1	3
Allyl propyl disulfide	2	12
2-Aminoethanol, see Ethanolamine		
2-Aminopyridine	0.5	2
Ammonia	50	35
Ammonium sulfamate (Ammate)		15
n-Amyl acetate	100	525
sec-Amyl acetate	125	650
Aniline—Skin	5	19
Anisidine (o, p-isomers)—Skin		0.5
Antimony and compounds (as Sb)		0.5
ANTU (alpha naphthyl thiourea)		0.3
Arsenic organic compounds (as As)		0.5
Arsine	0.05	0.2
Azinphos-methyl—Skin		0.2
Barium (soluble compounds)		0.5
p-Benzoquinone, see Quinone		
Benzoyl peroxide		5
Benzyl chloride	1	5
Biphenyl, see Diphenyl		
Boron oxide		15
Bromine	0.1	0.7
Bromoform—Skin	0.5	5
Butadiene (1,3-butadiene)	1,000	2,200
Butanethiol, see Butyl mercaptan		
2-Butanone	200	590
2-Butoxy ethanol (Butyl Cellosolve)—Skin	50	240
Butyl acetate (n-butyl acetate)	150	710
sec-Butyl acetate	200	950
tert-Butyl acetate	200	950
Butyl alcohol	100	300
sec-Butyl alcohol	150	450
tert-Butyl alcohol	100	300
n-Butyl glycidyl ether (BGE)	50	270
Butyl mercaptan	10	35
p-tert-Butyltoluene	10	60
Calcium oxide		5
Camphor		2

(*continued*)

TABLE 19-1 (*Continued*)

SUBSTANCE	Parts/Million (vol) @ 25°C, 760 mm Mercury Pressure	mg/m³ of Air
Carbaryl (Sevin®)		5
Carbon black		3.5
Carbon dioxide	5,000	9,000
Carbon monoxide	50	55
Chlordane—Skin		0.5
Chlorinated camphene—Skin		0.5
Chlorinated diphenyl oxide		0.5
Chlorine dioxide	0.1	0.3
a-Chloroacetophenone (phenacychloride)	0.05	0.3
Chlorobenzene (monochlorobenzene)	75	350
o-Chlorobenzylidene malononitrile (OCBM)	0.05	0.4
Chlorobromomethane	200	1,050
2-Chloro-1,3-butadiene, see Chloroprene		
Chlorodiphenyl (42 percent Chlorine)—Skin		1
Chlorodiphenyl (54 percent Chlorine)—Skin		0.5
1-Chloro, 2,3-epoxypropane, see Epichlorhydrin		
2-Chloroethanol, see Ethylene chlorohydrin		
Chloroethylene, see Vinyl chloride		
1-Chloro-1-nitropropane	20	100
Chloropicrin	0.1	0.7
Chloroprene (2-chloro-1,3-butadiene)—Skin	25	90
Chromium, sol. chromic, chromous salts as Cr		0.5
Metal and insol. salts		1
Coal tar pitch volatiles (benzene soluble fraction) anthracene, BaP, phenanthrene, acridine, chrysene, pyrene		0.2
Cobalt, metal fume and dust		0.1
Copper fume		0.1
Dusts and mists		1
Cotton dust (raw)		11
Crag® herbicide		15
Cresol (all isomers)—Skin	5	22
Crotonaldehyde	2	6
Cumene—Skin	50	245
Cyanide (as CN)—Skin		5
Cyclohexane	300	1,050
Cyclohexanol	50	200
Cyclohexanone	50	200
Cyclohexene	300	1,015
Cyclopentadiene	75	200
2,4-D		10
DDT—Skin		1
DDVP—Skin		1
Decaborane—Skin	0.05	0.3
Demeton®—Skin		0.1
Diacetone alcohol (4-hydroxyl-4-methyl-2-pentanone)	50	240
1,2-Diaminoethane, see Ethylene-diamine		
Diazomethane	0.2	0.4
Diborane	0.1	0.1

(*continued*)

TABLE 19-1   (*Continued*)

SUBSTANCE	Parts/Million (vol) @ 25°C, 760 mm Mercury Pressure	mg/m^3 of Air
Dibutyl phosphate	1	5
Dibutylphthalate		5
p-Dichlorobenzene	75	450
Dichlorodifluoromethane	1,000	4,950
1,3-Dichloro-5,5-dimethyl hydantoin		0.2
1,1-Dichloroethane	100	400
1,2-Dichloroethylene	200	790
Dichloromethane, see Methylenechloride		
Dichloromonofluoromethane	1,000	4,200
1,2-Dichloropropane, see Propylenedichloride		
Dichlorotetrafluoroethane	1,000	7,000
Dieldrin—Skin		0.25
Diethylamine	25	75
Diethylamino ethanol—Skin	10	50
Diethylether, see Ethyl ether		
Difluorodibromomethane	100	860
Dihydroxybenzene, see Hydroquinone		
Diisobutyl ketone	50	290
Diisopropylamine—Skin	5	20
Dimethoxymethane, see Methylal		
Dimethyl acetamide—Skin	10	35
Dimethylamine	10	18
Dimethylaminobenzene, see Xylidene		
Dimethylaniline (N-dimethyl-aniline)—Skin	5	25
Dimethylbenzene, see Xylene		
Dimethyl 1,2-dibromo-2,2-dichloroethyl phosphate, (Dibrom)		3
Dimethylformamide—Skin	10	30
2,6-Dimethylheptanone, see Diisobutyl ketone		
1,1-Dimethylhydrazine—Skin	0.5	1
Dimethylphthalate		5
Dimethylsulfate—Skin	1	5
Dinitrobenzene (all isomers)—Skin		1
Dinitro-o-cresol—Skin		0.2
Dinitrotoluene—Skin		1.5
Dioxane (Diethylene dioxide)—Skin	100	360
Diphenyl	0.2	1
Diphenylmethane diisocyanate, see Methylene bisphenyl isocyanate (MDI)		
Dipropylene glycol methyl ether—Skin	100	600
Di-sec, octyl phthalate (Di-2-ethylhexylphthalate)		5
Endrin—Skin		0.1
Epichlorohydrin—Skin	5	19
EPN—Skin		0.5
1,2-Epoxypropane, see Propyleneoxide		
2,3-Epoxy-1-propanol, see Glycidol		
Ethanethiol, see Ethylmercaptan		
Ethanolamine	3	6
2-Ethoxyethanol—Skin	200	740
2-Ethoxyethylacetate (Cello-solve acetate)—Skin	100	540
Ethyl acetate	400	1,400

TABLE 19-1   (*Continued*)

SUBSTANCE	Parts/Million (vol) @ 25°C, 760 mm Mercury Pressure	mg/m³ of Air
Ethyl acrylate—Skin	25	100
Ethyl alcohol (ethanol)	1,000	1,900
Ethylamine	10	18
Ethyl sec-amyl ketone (5-methyl-3-heptanone)	25	130
Ethyl benzene	100	435
Ethyl bromide	200	890
Ethyl butyl ketone (3-Heptanone)	50	230
Ethyl chloride	1,000	2,600
Ethyl ether	400	1,200
Ethyl formate	100	300
C Ethyl mercaptan	10	25
Ethyl silicate	100	850
Ethylene chlorohydrin—Skin	5	16
Ethylenediamine	10	25
Ethylene glycol monomethyl ether acetate, see Methyl cellosolve acetate		
Ethylene imine—Skin	0.5	1
Ethylene oxide	50	90
Ethylidine chloride, see 1,1-Dichloroethane		
N-Ethylmorpholine—Skin	20	94
Ferbam		15
Ferrovanadium dust		1
Fluoride (as F)		2.5
Fluorine	0.1	0.2
Fluorotrichloromethane	1,000	5,600
Formic acid	5	9
Furfural—Skin	5	20
Furfuryl alcohol	50	200
Glycidol (2,3-Epoxy-1-propanol)	50	150
Glycol monoethyl ether, see 2-Ethoxyethanol		
Guthion®, see Azinphosmethyl		
Hafnium		0.5
Heptachlor—Skin		0.5
Heptane (n-heptane)	500	2,000
Hexachloroethane—Skin	1	10
Hexachloronaphthalene—Skin		0.2
Hexane (n-hexane)	500	1,800
2-Hexanone	100	410
Hexone (Methyl isobutyl ketone)	100	410
sec-Hexyl acetate	50	300
Hydrazine—Skin	1	1.3
Hydrogen bromide	3	10
Hydrogen cyanide—Skin	10	11
Hydrogen peroxide (90%)	1	1.4
Hydrogen selenide	0.05	0.2
Hydroquinone		2
Iron oxide fume		10
Isoamyl acetate	100	525
Isoamyl alcohol	100	360
Isobutyl acetate	150	700

(*continued*)

TABLE 19-1   (*Continued*)

SUBSTANCE	Parts/Million (vol) @ 25°C, 760 mm Mercury Pressure	mg/m^3 of Air
Isobutyl alcohol	100	300
Isophorone	25	140
Isopropyl acetate	250	950
Isopropyl alcohol	400	980
Isopropylamine	5	12
Isopropylether	500	2,100
Isopropyl glycidyl ether (IGE)	50	240
Ketene	0.5	0.9
Lindane—Skin		0.5
Lithium hydride		0.025
L.P.G. (liquefied petroleum gas)	1,000	1,800
Magnesium oxide fume		15
Malathion—Skin		15
Maleic anhydride	0.25	1
Mesityl oxide	25	100
Methanethiol, see Methyl mercaptan		
Methoxychlor		15
2-Methoxyethanol, see Methyl cellosolve		
Methyl acetate	200	610
Methyl acetylene (propyne)	1,000	1,650
Methyl acetylene-propadiene mixture (MAPP)	1,000	1,800
Methyl acrylate—Skin	10	35
Methylal (dimethoxymethane)	1,000	3,100
Methyl alcohol (methanol)	200	260
Methylamine	10	12
Methyl amyl alcohol, see Methyl isobutyl carbinol		
Methyl (n-amyl) ketone (2-Heptanone)	100	465
Methyl butyl ketone, see 2-Hexanone		
Methyl cellosolve—Skin	25	80
Methyl cellosolve acetate—Skin	25	120
Methyl chloroform	350	1,900
Methylcyclohexane	500	2,000
Methylcyclohexanol	100	470
O-Methylcyclohexanone—Skin	100	460
Methyl ethyl ketone (MEK), see 2-Butanone		
Methyl formate	100	250
Methyl iodide—Skin	5	28
Methyl isobutyl carbinol—Skin	25	100
Methyl isobutyl ketone, see Hexone		
Methyl isocyanate—Skin	0.02	0.05
Methyl methacrylate	100	410
Methyl propyl ketone, see 2-Pentanone		
Molybdenum:		
Soluble compounds		5
Insoluble compounds		15
Monomethyl aniline—Skin	2	9
Morpholine—Skin	20	70
Naphtha (coaltar)	100	400
Naphthalene	10	50
Nickel carbonyl	0.001	0.007
Nickel, metal and soluble cmpds, as Ni		1

TABLE 19-1   (*Continued*)

SUBSTANCE	Parts/Million (vol) @ 25°C, 760 mm Mercury Pressure	mg/m^3 of Air
Nicotine—Skin		0.5
Nitric acid	2	5
Nitric oxide	25	30
p-Nitroaniline–Skin	1	6
Nitrobenzene—Skin	1	5
p-Nitrochlorobenzene—Skin		1
Nitroethane	100	310
Nitrogen trifluoride	10	29
Nitromethane	100	250
1-Nitropropane	25	90
2-Nitropropane	25	90
Nitrotoluene—Skin	5	30
Nitrotrichloromethane, see Chloropicrin		
Octachloronaphthalene—Skin		0.1
Octane	500	2,350
Oil mist, mineral		5
Oxmium tetroxide		0.002
Oxalic acid		1
Oxygen difluoride	0.05	0.1
Ozone	0.1	0.2
Paraquat—Skin		0.5
Parathion—Skin		0.1
Pentaborane	0.005	0.01
Pentachloronaphthalene—Skin		0.5
Pentachlorophenol—Skin		0.5
Pentane	1,000	2,950
2-Pentanone	200	700
Perchloromethyl mercaptan	0.1	0.8
Perchloryl fluoride	3	13.5
Petroleum distillates (naphtha)	500	2,000
Phenol—Skin	5	19
p-Phenylene diamine—Skin		0.1
Phenyl ether (vapor)	1	7
Phenyl ether-biphenyl mixture (vapor)	1	7
Phenylethylene, see Styrene		
Phenyl glycidyl ether (PGE)	10	60
Phenylhydrazine—Skin	5	22
Phosdrin (Mevinphos®)—Skin		0.1
Phosgene (carbonyl chloride)	0.1	0.4
Phosphine	0.3	0.4
Phosphoric acid		1
Phosphorus (yellow)		0.1
Phosphorus pentachloride		1
Phosphorus pentasulfide		1
Phosphorus trichloride	0.5	3
Phthalic anhydride	2	12
Picric acid—Skin		0.1
Pival® (2-Pivalyl-1,3-indandione)		0.1
Platinum (Soluble salts) as Pt		0.002
Propane	1,000	1,800

(*continued*)

TABLE 19-1   (*Continued*)

SUBSTANCE	Parts/Million (vol) @ 25°C, 760 mm Mercury Pressure	mg/m³ of Air
n-Propyl acetate	200	840
Propyl alcohol	200	500
n-Propyl nitrate	25	110
Propylene dichloride	75	350
Propylene imine—Skin	2	5
Propylene oxide	100	240
Propyne, see Methylacetylene		
Pyrethrum		5
Pyridine	5	15
Quinone	0.1	0.4
Rhodium, Metal fume and dusts, as Rh		0.1
Soluble salts		0.001
Ronnel		15
Rotenone (commercial)		5
Selenium compounds (as Se)		0.2
Selenium hexafluoride	0.05	0.4
Silver, metal and soluble compounds		0.01
Sodium fluoroacetate (1080)—Skin		0.05
Sodium hydroxide		2
Stibine	0.1	0.5
Stoddard solvent	500	2,900
Strychnine		0.15
Sulfur dioxide	5	13
Sulfur hexafluoride	1,000	6,000
Sulfuric acid		1
Sulfur monochloride	1	6
Sulfur pentafluoride	0.025	0.25
Sulfuryl fluoride	5	20
Systox, see Demeton®		
2,4,5T		10
Tantalum		5
TEDP—Skin		0.2
Tellurium		0.1
Tellurium hexafluoride	0.02	0.2
TEPP—Skin		0.05
1,1,1,2-Tetrachloro-2,2-difluorethane	500	4,170
1,1,2,2-Tetrachloro-1,2-difluorethane	500	4,170
1,1,2,2-Tetrachloroethane—Skin	5	35
Tetrachloromethane, see Carbon tetrachloride		
Tetrachloronaphthalene—Skin		2
Tetraethyl lead (as Pb)—Skin		0.075
Tetrahydrofuran	200	590
Tetramethyl lead (as Pb)—Skin		0.075
Tetramethyl succinonitrile—Skin	0.5	3
Tetranitromethane	1	8
Tetryl (2,4,6-trinitrophenyl methyl-nitramine)—Skin		1.5
Thallium (soluble compounds)—Skin as T1		0.1
Thiram		5
Tin (inorganic cmpds, except oxides)		2
Tin (organic cmpds)		0.1
Titaniumdioxide		15

TABLE 19-1 (*Continued*)

SUBSTANCE	Parts/Million (vol) @ 25°C, 760 mm Mercury Pressure	mg/m^3 of Air
o-Toluidine—Skin	5	22
Toxaphene, see Chlorinated camphene		
Tributyl phosphate		5
1,1,1-Trichloroethane, see Methyl chloroform		
1,1,2-Trichloroethane—Skin	10	45
Trichloromethane, see Chloroform		
Trichloronaphthalene—Skin		5
1,2,3-Trichloropropane	50	300
1,1,2-Trichloro 1,2,2-trifluoroethane	1,000	7,600
Triethylamine	25	100
Trifluoromonobromomethane	1,000	6,100
2,4,6-Trinitrophenol, see Picric acid		
2,4,6-Trinitrophenylmethyl-nitramine, see Tetryl		
Trinitrotoluene—Skin		1.5
Triorthocresyl phosphate		0.1
Triphenyl phosphate		3
Turpentine	100	560
Uranium (soluble compounds)		0.05
Uranium (insoluble compounds)		0.25
Vinyl benzene, see Styrene		
Vinylcyanide, see Acrylonitrile		
Vinyl toluene	100	480
Warfarin		0.1
Xylene (xylol)	100	435
Xylidine—Skin	5	25
Yttrium		1
Zinc chloride fume		1
Zinc oxide fume		5
Zirconium compounds (as Zr)		5

Source: Occupational Safety and Health Administration, *Toxic and Hazardous Substances 29 CFR Part 1910.*

The cumulative exposure for an 8-hour work shift is computed as follows:

$$E = (C_a T_a + C_b T_b + \cdots C^n T^n) \div 8$$

where

E is the equivalent exposure for the working shift

C is the concentration during any period of time T where the concentration remains constant.

For compliance, the value of E should not exceed the 8-hour time-weighted average limit in the table.

An employee's exposure to any material in Table 19-2 shall not exceed the ceiling value given for that material in the table.

TABLE 19-2   Air Contaminants Limits of Worker Exposure (Ceiling Limits)

SUBSTANCE	Parts/Million (vol) @ 25°C, 760 mm Mercury Pressure	mg/m^3 of Air
Allylglycidyl ether (AGE)	10	45
Boron trifluoride	1	3
Butylamine—Skin	5	15
Tert-Butyl chromate (CrO$_3$) Skin	—	0.1
Chlorine	1	3
Chlorine trifluoride	0.1	0.4
Chloroacetaldehyde	1	3
Chloroform (trichloromethane)	50	240
o-Dichlorobenzene	50	300
Dichloroethyl ether—Skin	15	90
1,1-Dichloro-1-nitroethane	10	60
Diglycidyl ether (DGE)	0.5	2.8
Ethyl mercaptan	10	25
Ethylene glycol dinitrate—Skin	0.2	1
Hydrogen chloride	5	7
Iodine	0.1	1
Manganese	—	5
Methyl bromide—Skin	20	80
Methyl mercaptan	10	20
Methyl styrene	100	480
Methylene bisphenyl isocyanate (MDI)	0.02	0.2
Monomethyl hydrazine—Skin	0.2	0.35
Nitrogen dioxide	5	9
Nitroglycerin—Skin	0.2	2
Terphenyls	1	9
Soluene-2,4-diisocyanate (TDI)	0.02	0.14
Vanadium		
V$_2$O$_5$ dust	—	0.5
V$_2$O$_5$ fume	—	0.1

TABLE 19-3   Air Contaminants Limits of Worker Exposure
(8 h Time-Weighted Averages, Ceiling Limit, and Acceptable Maximum Peak Above Ceiling Limit)

MATERIAL	8 h TIME-WEIGHTED AVERAGE	ACCEPTABLE CEILING CONCENTRATION	ACCEPTABLE MAXIMUM PEAK ABOVE THE ACCEPTANCE CEILING CONCENTRATION FOR AN 8 h SHIFT	
			CONCENTRATION	MAXIMUM DURATION
Benzene	10 p.p.m.	25 p.p.m.	50 p.p.m.	10 min
Beryllium and beryllium compounds	2 μg/M^3	5 μg/M^3	25 μg/M^3	30 min
Cadmium fume	0.1 mg/M^3	0.3 mg/M^3		
Cadmium dust	0.2 mg/M^3	0.6 mg/M^3		
Carbon disulfide	20 p.p.m.	30 p.p.m.	100 p.p.m.	30 min
Carbon tetrachloride	10 p.p.m.	25 p.p.m.	200 p.p.m.	5 min in any 4 h
Chromic acid and chromates		1 mg/10M^3		
Ethylene dibromide	20 p.p.m.	30 p.p.m.	50 p.p.m.	5 min
Ethylene dichloride	50 p.p.m.	100 p.p.m.	200 p.p.m.	5 min in any 3 h
Formaldehyde	3 p.p.m.	5 p.p.m.	10 p.p.m.	30 min
Hydrogen fluoride	3 p.p.m.			
Hydrogen sulfide		20 p.p.m.	50 p.p.m.	10 min once only if no other measurable exposure occurs
Fluoride as dust	2.5 mg/M^3			
Mercury		1 mg/10M^3		
Methyl chloride	100 p.p.m.	200 p.p.m.	300 p.p.m.	5 min in any 3 h
Methylene chloride	500 p.p.m.	1,000 p.p.m.	2,000 p.p.m.	5 min in any 2 h
Organo (alkyl) mercury	0.01 mg/M^3	0.04 mg/M^3		
Styrene	100 p.p.m.	200 p.p.m.	600 p.p.m.	5 min in any 3 h
Tetrachloroethylene	do	do	300 p.p.m.	5 min in any 3 h
Toluene	200 p.p.m.	300 p.p.m.	500 p.p.m.	10 min
Trichloroethylene	100 p.p.m.	200 p.p.m.	300 p.p.m.	5 min in any 2 h

An employee's exposure to any material listed in Table 19-3, in any 8-hour work shift of a 40-hour work week, shall not exceed the 8-hour time-weighted average limit given for that material in the table, or its ceiling concentration or maximum peak above ceiling limits, except as indicated in the table.

Vinyl chloride permissible exposure limits are as follows:

• No employee may be exposed to vinyl chloride at concentrations greater than one part per million of atmosphere averaged over any 8-hour period.

• No employee may be exposed to vinyl chloride at concentrations greater than five parts per million of atmosphere averaged over any period not exceeding 15 minutes.

• No employee may be exposed to vinyl chloride by direct contact with liquid vinyl chloride.

Acrylonitrile permissible exposure limits are as follows:

• No employee may be exposed to acrylonitrile at concentrations greater than two parts per million of atmosphere averaged over any 8-hour period.

• No employee may be exposed to acrylonitrile at concentrations greater than ten parts per million of atmosphere as averaged over any 15-minute period during the work day.

• No employee may be exposed to skin or eye contact with acrylonitrile.

HAZARD COMMUNICATION STANDARD

On November 25, 1983, OSHA issued a regulation referred to as the *Hazard Communication Standard*, requiring manufacturers, importers, and distributors to establish an information and training program for employees exposed to hazardous chemicals listed in Table 19-1 earlier in this chapter.

This requires employers to establish a comprehensive hazard communication program which includes the mandated container labeling and material safety data sheets (MDS), elements of which are shown in Exhibit 19-1. Also required is a training program containing information on how the employer plans to meet the criteria of the standards governing each hazardous chemical present in each work area. A list of these materials must be posted for workers to see.

Labels are required on all containers of hazardous chemicals leaving work areas, with the appropriate identity hazard warnings, and the name and address of the manufacturer. A sign may be posted if there are a number of stationary containers within a workplace instead of individually labeling each piece of equipment. The standard also requires that employers ensure that labels on incoming containers are not removed or defaced unless they are immediately replaced. Any container leaving a facility must be appropriately labeled.

Chemical manufacturers and importers are required to develop material safety data sheets for each hazardous chemical they produce or import. Employers are

ANATOMY OF A MATERIAL SAFETY DATA SHEET
(OSHA FORM 174)

SECTION I, CHEMICAL IDENTITY

Who makes the product
Their emergency phone number(s)
Date the MSDS was prepared

The chemical should be listed by its chemical and common name.

SECTION II, HAZARDOUS INGREDIENTS

If it is a tested mixture, the information will include the substance's hazardous components, which includes all the chemicals in the mixture which are hazardous.

If it is not a tested mixture, when all components of the chemical have been mixed, then the chemical and common name of all chemicals included in the mixture that are hazardous in concentrations of 1% or more must be listed.

The chemical and common name of each hazardous chemical that are carcinogens and are in concentrations of 0.1% or more must also be listed.

SECTION III, PHYSICAL AND CHEMICAL CHARACTERISTICS

Boiling Point
Vapor Pressure
Vapor Density
Specific Gravity
Evaporation Rate
Melting Point
Water Solubility
Appearance and Odor (Under Normal Conditions)

SECTION IV, PHYSICAL HAZARD DATA

Flash Point
Flammability Limits
Fire Extinguishing Media(s)
Special Firefighting Instructions
Fire/Explosion Hazards

(continued)

Exhibit 19-1.

SECTION V, HEALTH HAZARD DATA

Any known allowable/permissible exposure limits
Signs and symptoms of exposure and overexposure
Primary routes of entry
Cancer data
Applicable precautions for safe use and handling

SECTION VI, REACTIVITY DATA

Stability of material
Conditions to avoid
Hazardous decomposition products
Occurrence of hazardous polymerization
Incompatibilities

SECTION VII, SPILL AND DISPOSAL DATA

Steps to take if the material is released
How to dispose of the material

SECTION VIII, SPECIAL PROTECTION INFORMATION

Proper respiratory protection
Proper type of ventilation
Other types of protective equipment: gloves, glasses, etc.

SECTION IX, SPECIAL HANDLING AND STORAGE

Any special precautions for handling and storage

SECTION X, MISCELLANEOUS INFORMATION

Dept. of Transportation shipping classes
Shipping information/methods
Any other useful information

Exhibit 19-1 (*cont.*)

required to obtain or develop an MSDS for each hazardous chemical used in their workplace. These sheets must be accessible to all employees in the work area where the chemicals exist. MSDS must include the identity of the chemical, ingredients physically present within the chemical, and precautionary measures, emergency and first aid procedures, and identification of the person responsible for the sheet. New information on a chemical must be added to the MSDS within three months. Information is permitted to be withheld from MSDS if certain conditions are met:

- If claimed that the information withheld is a trade secret;
- If information concerning properties and effects of the hazardous chemical is disclosed as required on the appropriate MSDS;
- If the MSDS indicate that a specific chemical identity is being withheld as a trade secret;
- If the specific chemical identity is made available to health professionals under certain specified situations.

In the case of a medical emergency, the specific chemical identity of a hazardous chemical must be made available to a treating physician or nurse when the information is needed for proper emergency or first aid treatment.

Training is to be provided to the employee at the time of initial assignment and when a new hazardous chemical is introduced in the work area. Training programs must inform employees where in their work area hazardous chemicals are present; where written communication regarding chemicals are kept; how to detect them (i.e., odor of chemicals); methods to be taken by employees to protect themselves from the hazards; and specific procedures implemented by the employer to provide protection.

A sample of the OSHA Hazard Communication Program is shown in Exhibit 19-2.

COMPANY NAME
SAMPLE
HAZARD COMMUNICATION PROGRAM

I. GENERAL

To comply with the Federal Hazard Communication Standards (29 CFR 1910.1200 and 1926.59) as required by the Occupational Safety and Health Administration, the following Hazard Communication Program has been established. The following pages document the actions we have taken regarding our chemical information list, material safety data sheets, labels, and employee information and training. This program will be available in each department supervisor's office for review by all employees at any time and on all shifts.

(continued)

Exhibit 19-2.

II. EMPLOYEE TRAINING AND INFORMATION

Before starting work, each new employee will attend a safety class which will educate and train them as to:

    a. The Company's Hazard Communication Program.

    b. Processes used within their department and how they can be a hazard.

    c. Product inventory lists for their department and how to use them.

    d. Material Safety Data Sheets (MSDSs).

    e. Product labeling requirements.

    f. How exposure to hazardous products/chemicals can be controlled by such means as work practices, and personal protective equipment both during normal use and foreseeable emergencies.

    g. What the Company has done to lessen or prevent workers' exposure to products/chemicals.

    h. Procedures to follow if exposed to products/chemicals.

After attending the class, each employee will sign a form (sample attached) stating that they have received the training outlined above.

Before any new hazardous product is introduced into any department, each employee will be given information in the same manner as when newly hired. The department supervisor will be responsible for seeing that MSDSs on the new product(s) are available.

Monthly safety meetings will be held within each department, and hazardous products/chemicals used in the department will be discussed. Attendance is mandatory for all employees.

Notices will be posted on the employee bulletin boards that provide an explanation of our container labeling system and the location of the written hazard communication program.

III. CONTAINER LABELING

The supervisor will verify that all product/chemical containers received for use by the department are clearly labeled with:

    a. the trade name of the product;

    b. the name and address of the manufacturer; and

    c. the appropriate hazard warnings (target organs and type of protective equipment required).

No containers will be released for use until the above data are verified. Material in unlabeled piping will be addressed in the same manner.

If hazardous products/chemicals are transferred from the original shipping containers to other containers, the department supervisor is responsible for the labeling with the appropriate from a. & c. above.

Exhibit 19-2 (*cont.*)

IV. MATERIAL SAFETY DATA SHEETS (MSDSs)

It is our company policy not to use a hazardous chemical for which no MSDS has been received. We therefore require all suppliers of hazardous chemicals to provide an appropriate MSDS. Our company will not accept any new hazardous materials without an accompanying MSDS for the product.

Copies of MSDSs for all of the hazardous products to which department employees may be exposed will be kept in the department supervisor's office (with this program) and in the plant office.

MSDSs will be available to all employees during each workshift. Copies of MSDSs will be made available to any employee upon request to the department supervisor.

It is important to know what a MSDS is and what it is designed to do; they help you identify health and physical hazards of the chemicals you work with.

Please note that the information on an MSDS may not be in the exact same order about to be addressed, but will be very similar. While MSDSs may take different shapes, they must all contain the information identified.

V. ADDITIONAL PROVISIONS

A. Informing Contractors

It is the responsibility of the facility manager to coordinate with the department supervisor(s) to ensure that contractors and their employees are provided with the following information prior to entering the worksite:

1. Hazardous products to which they may be exposed while on the job site;
2. Measures the employees may take to lessen the possibility of exposure;
3. Steps the company has taken to lessen the risks;
4. MSDSs for all hazardous products are on file in the plant office, and the department supervisor has a copy of each one; and
5. Procedures to follow if they are exposed.

By the same token, contractors must notify the facility manager of the hazardous materials which they will be using. The facility manager must then notify the plant employees of this information.

B. Hazardous Nonroutine Tasks

It is the Company's policy that no employee will begin any "nonroutine" task without first receiving a safety briefing. Each "nonroutine" task must be described in detail and the following information discussed:

1. Specific chemical hazards;
2. Protective equipment and safety measures the employee(s) are to use; and

(continued)

Exhibit 19-2 (*cont.*)

3. Measures the company has taken to lessen the hazards (engineering controls, protective equipment, additional employees, and emergency procedures).

VI. INVENTORY OF HAZARDOUS MATERIALS

Attached is an inventory of hazardous products used in each department. The list is arranged in alphabetical order by common name, and either the chemical name or common name is the identity shown on the MSDS and container labels. Further information on each hazardous product listed can be obtained by reviewing the product's MSDS in the supervisor's office. (Use the corresponding MSDS # for quick cross-reference.)

PRODUCT TRADE NAME	MANUFACTURER/ ADDRESS	PHONE #	MSDS NO

_____

Company President

EMPLOYEE ACKNOWLEDGMENT
OF
HAZARD COMMUNICATION TRAINING

I, _____, have been trained in the company's hazard communication program. The materials/processes in my work area have been explained, and I am aware of the material safety data sheets (MSDSs) which apply, as well as their location.

_____   _____

Employee's Signature   Date

_____   _____

Supervisor's Signature   Date

Exhibit 19-2 (*cont.*)

OSHA REGIONAL OFFICES

OSHA regional offices are shown in Table 19-4.

TABLE 19-4   OSHA Regional and Area Offices

REGION 1
(CT, MA, ME, NH, RI, VT)
133 PORTLAND STREET
BOSTON, MA 02114

Bridgeport, CT 06604
  One Lafayette Square

Hartford, CT 06103
  Federal Office Bldg.
  450 Main Street

Augusta, ME 04330
  U.S. Federal Bldg.
  40 Western Avenue

Concord, NH 03301
  279 Pleasant Street

Springfield, MA 01103-1493
  1145 Main Street

Braintree, MA 02184
  639 Granite Street

Methuen, MA 01844
  Valley Office Park
  13 Branch Street

Providence, RI 02903
  380 Westminster Hall

REGION 2
(NJ, NY, PR, VI)
201 VARICK STREET
NEW YORK, NY 10014

Avenel, NJ 07001
  Plaza 35
  1030 Saint Georges Avenue

Hasbrouck Heights, NJ 07604
  Teterboro Airport
  Professional Bldg.
  500 Route 17 South

Marlton, NJ 08053
  Marlton Executive Park
  701 Route 73 South

Parsippany, NJ 07054
  299 Cherry Hill Road

Hato Rey, PR 00918
  U.S. Courthouse & FOB
  Carlos Chardon Street

Albany, NY 12205-3826
  401 New Karner Road

Bowmansville, NY 14026
  5360 Genesee Street

Bayside, NY 11361
  42-40 Bell Blvd.

New York, NY 10007
  90 Church Street

Syracuse, NY 13260
  100 South Clinton Street

Tarrytown, NY 10591
  660 White Plains Road

Westbury, NY 11590
  990 Westbury Road

(continued)

TABLE 19-4   (*Continued*)

REGION 3
(DC, DE, MD, PA, VA, WV)
GATEWAY BLDG.
3535 MARKET STREET
PHILADELPHIA, PA 19104

Baltimore, MD 21201
Federal Bldg.
40 Western Avenue

Allentown, PA 18102
850 North 5th Street

Erie, PA 18506
3939 West Ridge Road

Harrisburg, PA 17109
Progress Plaza
49 North Progress Street

Philadelphia, PA 19106
U.S. Custom House
Second and Chestnut Streets

Pittsburgh, PA 15222
Federal Bldg.
1000 Liberty Avenue

Wilkes-Barre, PA 18701
Penn Place
20 North Pennsylvania Ave.

Norfolk, VA 23510
Federal Office Bldg.
200 Granby Mall

Charlestown, WV 25301
550 Egan Street

REGION 4
(AL, FL, GA, KY, MS, NC, SC, TN)
1375 PEACHTREE STREET, N.E.
ATLANTA, GA 30367

Birmingham, AL 35216
2047 Canyon Road—Todd Mall

Mobile, AL 36693
3737 Government Blvd.

Savannah, GA 31401
1600 Dryton Street

Smyrna, GA 30080
Herodian Way

Tucker, GA 30084
Bldg. 7
La Vista Perimeter Office Park

Jackson, MS 39269
Federal Bldg.
100 West Capitol Street

Columbia, SC 29201
1835 Assembly Street

Nashville, TN 37215
2002 Richard Jones Road

Fort Lauderdale, FL 33324
Jacaranda Executive Court
8040 Peters Road

Jacksonville, FL 32216
3100 University Blvd. South

Tampa, FL 33602
700 Twiggs Street

Frankfort, KY 40601
John C. Watts Federal Bldg.
330 West Broadway

Raleigh, NC 27601
Century Station
300 Fayetteville Street Mall

TABLE 19-4 (*Continued*)

REGION 5
(IL, IN, MI, MN, OH, WI)
230 SOUTH DEARBORN STREET
CHICAGO, IL 60604

Calumet City, IL 60409
1600 167th Street

Des Plaines, IL 60018
2360 East Devon Avenue

North Aurora, IL 60524
344 Smoke Tree Business Park

Peoria, IL 61614-1223
2001 West Willow Knolls Road

Appleton, WI 54915
2618 North Ballard Road

Madison, WI 53713
2934 Fish Hatchery Road

Milwaukee, WI 53203
310 West Wisconsin Avenue

Indianapolis, IN 46204
46 East Ohio Street

Minneapolis, MN 55401
110 South 4th Street

Lansing, MI 48917
801 South Waverly Road

Cincinnati, OH 45246
36 Triangle Park Drive

Cleveland, OH 44199
Federal Office Bldg.
1240 East 9th Street

Columbus, OH 43215
Federal Office Bldg.
200 North High Street

Toledo, OH 43604
Federal Office Bldg.
234 North Summit Street

REGION 6
(AR, LA, NM, OK, TX)

Little Rock, AR 72201
Savers Bldg.
320 West Capitol Avenue

Baton Rouge, LA 70806
2156 Wooddale Blvd.
Hoover Annex

Albuquerque, NM 87102
320 Central Avenue, S.W.

Oklahoma City, OK 73102
420 West Main Place

Austin, TX 78701
611 East 6th Street

Corpus Christi, TX 78401
Government Plaza
400 Mann Street

Dallas, TX 75228
8344 East R.L. Freeway

Fort Worth, TX 76180-7604
North Star 2 Bldg.

Houston, TX 77004
2320 La Branch Street

Houston, TX 77060
350 N. Sam Houston Pkwy.

Lubbock, TX 79401
Federal Bldg.
1205 Texas Avenue

(*continued*)

TABLE 19-4 (*Continued*)

REGION 7
(IA, KS, MO, NE)
911 WALNUT STREET
KANSAS CITY, MO 64106

Des Moines, IA 50309
    210 Walnut Street

Kansas City, MO 64106
    6200 Connecticut Avenue

St. Louis, MO 63120
    4300 Goodfellow Blvd.

Wichita, KS 67202
    216 North Waco

Omaha, NE 68106
    Overland-Wolf Bldg.
    6910 Pacific Street

REGION 8
(CO, MT, ND, SD, UT, WY)
FEDERAL BLDG.
1961 STOUT STREET
DENVER, CO 80294

Denver, CO 80204
    1244 Speer Blvd.
    Colonade Center

Englewood, CO 80111
    7935 E. Prentice Avenue

Bismarck, ND 58501
    Federal Bldg.

Billings, MT 59101
    19 North 25th Street

Salt Lake City, UT 84165-0200
    1781 South 300 West

REGION 9
(AMERICAN SAMOA, AZ, CA, GUAM, HI, NV,
TRUST TERRITORIES OF THE PACIFIC)
71 STEVENSON STREET
SAN FRANCISCO, CA 94105

Phoenix, AZ 85016
    3221 North 16th Street

Honolulu, HI 96850
    300 Ala Moana Blvd.

San Francisco, CA 94105
    71 Stevenson Street

Carson City, NV 68106
    1050 East Williams Street

TABLE 19-4   (*Continued*)

REGION 10
(AK, ID, OR, WA)
1111 THIRD AVENUE
SEATTLE, WA 98101-3212

Anchorage, AK 99513-7571
    Federal Bldg.
    222 West 7th Avenue

Portland, OR 97204
    1220 S.W. Third Avenue

Boise, ID 83702
    3050 North Lake Harbor Lane

Bellevue, WA 98004
    121 107th Avenue, N.E.

ENVIRONMENTAL PROTECTION AGENCY (EPA)

The EPA is an independent federal agency, created in 1970 to protect the country's environment. The EPA administers the Toxic Substances Control Act (TSCA), the Federal Air Pollution Program, the Federal Water Pollution Program, and the Resource Conservation and Recovery Act (RCRA). All federal agencies are required to file with the EPA a written analysis of the environmental impact of a proposed action, including alternatives and any expected adverse effects. The EPA must comment on these impact statements before final action by the Council on Environmental Quality (CEQ), which coordinates Federal environmental activities.

The Office of Research and Development of EPA has established the Technology Information Transfer and Support Program to disseminate information on successful technology to control air, water, and solid waste pollution. Seminars, held on a regional basis, are used as the method of communication backed by special publications, films, photos, and exhibits. The publications include *Technical Capsule Reports*, a series on proven pollution control techniques. The information presented at seminars is published in the *Seminar Publication Series*. Basic to the seminars are process design manuals which present information on all of the usable and practical technology needs of the areas in which they are held.

The EPA also publishes semi-annually a *Regulatory Agenda* which summarizes the status of regulations under development, revision, and review. The purpose is to keep interested parties informed of the progress of these regulations. Persons wishing to be on the mailing list for Regulatory Agenda may contact:

Regulatory Development Branch (PM-223Y)
Environmental Protection Agency
401 M Street SW
Washington, DC 20460

TOXIC SUBSTANCES CONTROL ACT (TSCA)

The mission of TSCA is to prevent risk of injury from harmful chemicals. The EPA is authorized to regulate against such risk at any stage in the life cycle of a chemical, including manufacture, processing, distribution, use, and disposal. The EPA may require manufacturers and processors of a designated chemical to test for certain adverse health or environmental effects and to keep appropriate records of results. EPA has the authority to require testing after finding that:

1. a chemical substance may present an unreasonable risk of injury to human health or the environment, or the chemical is produced in substantial quantities which could result in significant or substantial human or environmental exposure,
2. available data to evaluate the chemical are inadequate,
3. testing is needed to develop the necessary data.

The EPA's Office of Pollution Prevention and Toxics administers the Existing Chemical Program which screens, establishes testing requirements for, assesses, and develops strategies for managing risks posed by approximately 14,000 chemicals (other than polymers) that are produced in quantities of more than 10,000 pounds per year. Excluded are eight product categories: pesticides, tobacco, nuclear material, firearms and ammunitions, food, food additives, drugs, and cosmetics. An important component of this program is the Master Testing List of 320 specific chemicals (1992 List). This list may be obtained from:

Office of Pollution Prevention and Toxics
U.S. Environmental Protection Agency
Washington, DC 20460

Actions taken on asbestos, dioxin, and polychlorinated biphenyls (PCBs) have been widely publicized. Asbestos is being phased out of industrial uses, including electrical and thermal insulation.

Dioxin, 2,3,7,8-tetrachlorodibenzo-p-dioxin (TCDD) is a highly toxic contaminant in the manufacture of a herbicide. The disposal of dioxin in wastes is now closely monitored and controlled by EPA.

The rule governing use of PCBs as liquid dielectric in electrical equipment became effective August 18, 1982. This rule allows the use of certain electrical equipment containing PCBs to continue under specified conditions. The rule prohibits use of PCBs in transformers and electromagnets (with concentration of 500 parts per million or greater), posing an exposure to food or feed, after October 1, 1985. A weekly inspection of this equipment is required until that date. Other provisions include:

• A requirement for a quarterly inspection of other PCB transformers for their remaining useful life
• A requirement to maintain records of inspection/maintenance for at least three years after disposing of PCB transformers
• An authorization for the use of large PCB capacitors located in contained and restricted access electrical substations for the remainder of their useful lives

• A prohibition on the use of all other large PCB capacitors after October 1, 1988.

Formaldehyde has been investigated in a TSCA study conducted by the National Cancer Institute involving 26,561 workers at ten facilities that produce either formaldehyde or formaldehyde-containing products. The study showed mortality rates no different among these workers than in the general population. The study concluded that formaldehyde can be used safely in the workplace and in consumer products.

FEDERAL AIR POLLUTION PROGRAM

The major law, administered by the EPA, governing air pollution is the *Clean Air Act*, as amended. This act is designed to enhance the quality of air resources, and thereby concerns manufacturers and users of organic coatings.

The EPA prepares a Guideline Series to assist regional, state, and local jurisdictions in the development of air pollution regulations for volatile organic compounds (VOC) which contribute to the formation of photochemical oxidants.

The official definition of a volatile organic compound is stated in the *Federal Register*, Vol. 57, No. 22/Monday, February 3, 1992, /Rules and Regulations, p. 3945. As generally understood, a volatile organic compound is any organic compound that, when released into the atmosphere, can remain long enough to participate in a photochemical reaction. Known exceptions, which have been determined to have negligible photochemical activity, include methane, ethane, methyl chloride, 1,1,1,-trichloroethane (methyl chloroform), and certain fluorocarbons.

Table 19-5 lists hazardous air pollutants from Section 112 of the Clean Air Act, as amended and effective in 1993.

TABLE 19-5   EPA Title III—Hazardous Air Pollutants

Acetaldehyde	Biphenyl
Acetamide	Bis(2-ethylhexyl)phthalate (DEHP)
Acetonitrile	Bis(chloromethyl)ether
Acetophenone	Bromoform
2-Acetylaminofluorene	1,3-Butadiene
Acrolein	Calcium cyanamide
Acrylamide	Caprolactam
Acrylic acid	Captan
Acrylonitrile	Carbaryl
Allyl chloride	Carbon disulfide
4-Aminobiphenyl	Carbon tetrachloride
Aniline	Carbonyl sulfide
o-Anisidine	Catechol
Asbestos	Chloramben
Benzene (including benzene from gasoline)	Chlordane
Benzidine	Chlorine
Benzotrichloride	Chloroacetic acid
Benzyl chloride	2-Chloroacetophenone

(continued)

TABLE 19-5 (*Continued*)

Chlorobenzene
Chlorobenzilate
Chloroform
Chloromethyl methyl ether
Chloroprene
Cresols/Cresylic acid (isomers and mixture)
o-Cresol
m-Cresol
p-Cresol
Cumene
2,4-D, salts and esters
DDE
Diazomethane
Dibenzofurans
1,2-Dibromo-3-chloropropane
Dibutylphthalate
1,4-Dichlobenzene(p)
3,3,-Dichlorobenzidene
Dichloroethyl ether (Bis(2-chloroethyl)ether)
1,3-Dichloropropene
Dichlorvos
Diethanolamine
N,N-Diethyl aniline (N,N-Dimethylaniline)
Diethyl sulfate
3,3-Dimethoxybenzidine
Dimethyl aminoazobenzene
3,3-Dimethyl benzidine
Dimethyl carbamoyl chloride
Dimethyl formamide
1,1-Dimethyl hydrazine
Dimethyl phthalate
Dimethyl sulfate
4,6-Dinitro-o-cresol, and salts
2,4-Dinitrophenol
2,4-Dinitrotoluene
1,4-Dioxane (1,4-Diethyleneoxide)
1,2-Diphenylhydrazine
Epichlorohydrin (1-Chloro-2,3-epoxypropane)
1,2-Epoxybutane
Ethyl acrylate
Ethyl benzene
Ethyl carbamate (Urethane)
Ethyl chloride (Chloroethane)
Ethylene dibromide (Dibromoethane)
Ethylene glycol
Ethylene imine (Aziridine)
Ethylene oxide
Ethylene thiourea
Ethylidene dichloride (1,1-Dichloroethane)
Formaldehyde
Heptachlor
Hexachlorobenzene

Hexachlorobutadiene
Hexachlorocyclopentadiene
Hexachloroethane
Hexamethylene-1,6-diisocyanate
Hexamethylphosphoramide
Hexane
Hydrazine
Hydrochloric acid
Hydrogen fluoride (Hydrofluoric acid)
Hydrogen sulfide
Hydroquinone
Isophorone
Lindane (all isomers)
Maleic anhydride
Methanol
Methoxychlor
Methyl bromide (Bromomethane)
Methyl chloride (Chloromethane)
Methyl chloroform (1,1,1-Trichloroethane)
Methyl ethyl ketone (2-Butanone)
Methyl hydrazine
Methyl iodide (Iodomethane)
Methyl isobutyl ketone (Hexone)
Methyl isocyanate
Methyl methacrylate
Methyl tert butyl ether
4,4-Methylene bis(2-chloroaniline)
Methylene chloride (Dichloromethane)
Methylene diphenyl diisocyanate (MDI)
4,4-Methylenedianiline
Naphthalene
Nitrobenzene
4-Nitrobiphenyl
4-Nitrophenol
2-Nitropropane
N-Nitroso-N-methylurea
N-Nitrosodimethylamine
N-Nitrosomorpholine
Parathion
Pentachloronitrobenzene (Quintobenzene)
Pentachlorophenol
Phenol
p-Phenylenediamine
Phosgene
Phosphine
Phosphorus
Phthalic anhydride
Polychlorinated biphenyls (Aroclors)
1,3-Propane sultone
beta-Propiolactone
Propionaldehyde
Propoxur (Baygon)

TABLE 19-5   (*Continued*)

Propylene dichloride (1,2-Dichloropropane)	Vinyl bromide
Propylene oxide	Vinyl chloride
1,2-Propylenimine (2-Methyl aziridine)	Vinylidene chloride (1,1-Dichloroethylene)
Quinoline	Xylenes (isomers and mixture)
Quinone	o-Xylenes
Styrene	m-Xylenes
Styrene oxide	p-Xylenes
2,3,7,8-Tetrachlorodibenzo-p-dioxin	INORGANIC MATERIALS
1,1,2,2,-Tetrachloroethane	Antimony Compounds
Tetrachloroethylene (Perchloroethylene)	Arsenic Compounds (inorganic including arsine)
Titanium tetrachloride	Beryllium Compounds
Toluene	Cadmium Compounds
2,4-Toluene diamine	Chromium Compounds
2,4-Toluene diisocyanate	Cobalt Compounds
o-Toluidine	Coke Oven Emissions
Toxaphene (chlorinated camphene)	Cyanide Compounds *1
1,2,4-Trichlorobenzene	Glycol ethers *2
1,1,2-Trichloroethane	Lead Compounds
Trichloroethylene	Manganese Compounds
2,4,5-Trichlorophenol	Mercury Compounds
2,4,6-Trichlorophenol	Fine mineral fibers *3
Triethylamine	Nickel Compounds
Trifluralin	Polycyclic Organic Matter *4
2,2,4-Trimethylpentane	Radionuclides (including radon) *5
Vinyl acetate	Selenium Compounds

ASTM D 2369, *Standard Method of Test for Volatile Content of Coatings*, is recommended by EPA for determining compliance with standards. It is recognized that this method may not be suitable for determining the total VOC content of all coatings cured at elevated temperatures where photochemically reactive compounds may be formed and released to the atmosphere. Additional test methods may be required as more scientific information on this subject becomes available.

The document which is most directly related to insulation coatings is EPA-450/2-77-033 OAQPS No. 1.2-087, *Control of Volatile Emissions from Existing Stationary Sources, Volume IV: Coating for Insulation of Magnet Wire*. A wire enameling plant with only a few uncontrolled ovens could easily exceed 91 megagrams per year (100 tons per year) of VOC emissions. The most common control technique used for wire coating ovens is incineration. Essentially, all solvent emissions from the ovens can be directed to an incinerator with a combustion efficiency of at least 90 percent. This efficiency is reasonable to attain. Thermal incinerators have an efficiency range from 90 to 99 percent. Catalytic oxidizers have an efficiency range of 90–95 percent, if not fouled.

Low-polluting coatings are beginning to be used in the wire coating industry. It is reasonable to exempt an oven from the incineration requirement if the coatings used contain less than the recommended limitation for low-solvent coatings of 0.20 kilograms of solvent per liter of coating, (1.7 pounds of solvent per gallon of coating), minus any water. This emission limit can be met with high-solids coatings having greater than 77 percent solids by volume. Powder coatings and hot melt coatings will both achieve this. This emission limit can also be met with a waterborne coating which contains 29 volume percent solids, 8 volume percent organic solvent, and 63 volume percent water. A waterborne emulsion with no organic solvent would, of course, meet the recommended limit.

Approximately the same amount of solvent will be emitted from a low-solvent coating meeting the above limitation as from an equal volume of solids applied as a conventional coating with 90 percent incineration of solvent emissions from the conventional coating.

The solvent content of solvent-type wire enamels typically ranges from 67 to 85 percent. Cresylic acid and various cresols, xylene, and mixtures of C_8 to C_{12} aromatics are widely used. Other solvents include:

- alcohols
- high-flash naphtha (known as hi-flash naphtha)
- methyl ethyl ketone
- phenol
- N-methyl pyrrolidine

The relevant document for control technologies is EPA/625/6-91/014, June 1991: *Handbook—Control Technologies for Hazardous Air Pollutants*.

FEDERAL WATER POLLUTION PROGRAM

The principal law governing pollution of the Nation's waterways is the Federal Water Pollution Control Act, or Clean Water Act. Originally enacted in 1948, it was totally revised by amendments in 1972 that gave the Act its current (1993) shape. The 1972 legislation spelled out ambitious programs for water quality improvement that are still being implemented by industries and municipalities. Congress made certain fine-tuning amendments in 1977, revised portions of the law in 1981, and enacted further amendments in 1987.

The Clean Water Act has as its objective the restoration and maintenance of the "chemical, physical, and biological integrity of the Nation's waters." Two goals also were established in the 1972 legislation: zero discharge of pollutants by 1985 and, where possible, water quality that is both fishable and swimmable by mid-1983. While those dates have passed without full implementation, the goals remain, and efforts to attain them continue.

The EPA issues regulations containing the best practicable technology (BPT) and best available technology (BAT) effluent standards applicable to various categories of industrial sources, including organic chemical manufacturing and related industries. This Act embodies a philosophy of Federal–State partnership in which the Federal government sets the agenda and standards for pollution abatement, while states carry out day-to-day activities of implementation and enforcement.

The EPA has prepared these publications relevant to companies in and serving the electrical and electronics industries. To inquire about their availability, write to:

U.S. Environmental Protection Agency
Office of Water Resource Center
RC-4100
401 M Street, SW
Washington, DC 20460

• *Development Document for Effluent Limitations Guidelines and New Source Performance Standards for the Organic Chemicals, Plastics, and Synthetic Fibers Point Source Category*
   Volume I EPA# 440/1-87-009a
   Volume II EPA# 440/1-87-009b

• *Supplement to the Development Document for Effluent Limitations Guidelines and New Source Performance Standards for the Organic Chemicals, Plastics, and Synthetic Fibers Point Source Category—Final*
   EPA# 821-R-93-007

• *Re-Evaluation of the Economic Impact Analysis of Effluent Limitations Guidelines and Standards for the Organic Chemicals, Plastics, and Synthetic Fibers Industry*
   EPA# 821/R-93-008

• *Development Document for Effluent Limitations Guidelines and New Source Performance Standards for the Plastics Molding and Forming Point Source Category—Final*
   EPA# 440/1-84-069

• *Economic Impact Analysis of Effluent Limitations Guidelines and Standards for the Plastics Molding and Forming Industry*
   EPA# 440/2-84-025

A list of 65 toxic pollutants, shown in Table 19-6, has been developed by a multidisciplinary task force of scientists, forming a basis for the development of effluent limitations from point sources. The EPA may, when appropriate, revise the list to add or remove pollutants, taking into account their toxicity, persistence, degradability, and the extent of their effect on the usual or potential organisms found in any waters.

TABLE 19-6 EPA Toxic Pollutant List for Effluents to Waterways

 1. Acenaphthene
 2. Aerolein
 3. Acrylonitrile
 4. Aldrin/Dieldrin[1]
 5. Antimony and compounds[2]
 6. Arsenic and compounds
 7. Asbestos
 8. Benzene
 9. Benzidine[1]
10. Beryllium and compounds
11. Cadmium and compounds

(*continued*)

TABLE 19-6 (*Continued*)

12. Carbon tetrachloride
13. Chlordane (technical mixture and metabolites)
14. Chlorinated benzenes (other than dichlorobenzenes)
15. Chlorinated ethanes (including 1,2-dichloroethane, 1,1,1-trichloroethane, and hexachloroethane)
16. Chloroalkyl ethers (chloromethyl, chloroethyl, and mixed ethers)
17. Chlorinated naphthalene
18. Chlorinated phenols (other than those listed elsewhere; includes trichlorophenols and chlorinated cresols)
19. Chloroform
20. 2-chlorophenol
21. Chromium and compounds
22. Copper and compounds
23. Cyanides
24. DDT and metabolites[1]
25. Dichlorobenzenes (1,2-, 1,3-, and 1,4-dichlorobenzenes)
26. Dichlorobenzidine
27. Dichloroethylenes (1,1-, and 1,2-dichloroethylene)
28. 2,4-dichlorophenol
29. Dichloropropane and dichloropropene
30. 2,4-dimethylphenol
31. Dinitrotoluene
32. Diphenylhydrazine
33. Endosulfan and metabolites
34. Endrin and metabolites[1]
35. Ethylbenzene
36. Fluoranthene
37. Haloethers (other than those listed elsewhere; includes chlorophenylphenyl ethers, bromophenylphenyl ether, bis(dichloroisopropyl) ether, bis-((chloroethoxy) methane and polychlorinated diphenyl ethers)
38. Halomethanes (other than those listed elsewhere; includes methylene chloride, methylchloride, methylbromide, bromoform, dichlorobromomethane, trichlorofluoromethane, dichlorodifluoromethane)
39. Heptachlor and metabolites
40. Hexachlorobutadiene
41. Hexachlorocyclohexane (all isomers)
42. Hexachlorocyclopentadiene
43. Isophorone
44. Lead and compounds
45. Mercury and compounds
46. Naphthalene
47. Nickel and compounds
48. Nitrobenzene
49. Nitrophenols (including 2,4-dinitrophenol, dinitrocresol)
50. Nitrosamines
51. Pentachlorophenol
52. Phenol
53. Phthalate esters
54. Polychlorinated biphenyls (PCBs)[1]
55. Polynuclear aromatic hydrocarbons (including benzanthracenes, benzopyrenes, benzofluoranthene, chrysenes, dibenzanthracenes, and indenopyrenes)

TABLE 19-6   (*Continued*)

56. Selenium and compounds
57. Silver and compounds
58. 2,3,7,8-tetrachlorodibenzo-p-dioxin (TCDD)
59. Tetrachloroethylene
60. Thallium and compounds
61. Toluene
62. Toxaphene[1]
63. Trichloroethylene
64. Vinyl chloride
65. Zinc and compounds

[1]Effluent standard promulgated (40 CFR Part 129).

[2]The term "compounds" shall include organic and inorganic compounds.

EPA REGIONAL OFFICES

EPA regional offices are shown in Table 19-7.

TABLE 19-7   EPA Regional Offices

REGION	STATES COVERED
JFK Federal Building Boston, MA 02203	Connecticut, Maine, Massachusetts, New Hampshire, Rhode Island, Vermont
26 Federal Plaza New York, NY 10278	New Jersey, New York, Virgin Islands, Puerto Rico
Curtis Building 6th and Walnut Streets Philadelphia, PA 19106	Delaware, Maryland, Pennsylvania, Virginia, West Virginia, District of Columbia
345 Courtland Street, NE Atlanta, GA 30365	Alabama, Florida, Georgia, Kentucky, Mississippi, North Carolina, South Carolina, Tennessee
230 South Dearborn Street Chicago, IL 60604	Indiana, Illinois, Michigan, Minnesota, Ohio, Wisconsin
1201 Elm Street Dallas, TX 75270	Arkansas, Louisiana, New Mexico, Oklahoma, Texas
324 East 11th Street Kansas City, MO 64106	Iowa, Kansas, Missouri, Nebraska
1860 Lincoln Street Denver, CO 80295	Colorado, Montana, North Dakota, South Dakota, Utah, Wyoming
215 Fremont Street San Francisco, CA 94105	Arizona, California, Hawaii, Nevada, American Samoa, Guam, Trust Territories of the Pacific, Wake Island
1200 Sixth Avenue Seattle, WA 98101	Alaska, Idaho, Oregon, Washington

NATIONAL INSTITUTE OF STANDARDS AND TECHNOLOGY (NIST)

U.S. Department of Commerce
Gaithersburg, MD 20899-001

NIST's primary mission is to promote U.S. economic growth by working with industry to develop and apply technology, measurements, and standards. It administers a portfolio of four major programs:

- A rigorously competitive *Advanced Technology Program* providing cost-sharing grants to industry for development of high-risk technologies with significant commercial potential;
- A grassroots *Manufacturing Extension Partnership* helping small and medium-sized companies adopt new technologies;
- A strong laboratory effort planned and implemented in cooperation with industry and focused on measurements, standards, evaluated data, and test methods;
- A highly visible quality outreach program associated with the Malcolm Baldridge National Quality Award.

In 1993, the staff comprised 3,200 scientists, engineers, technicians, and support personnel, plus some 1,200 visiting researchers. The Institute operates more than 40 different research facilities that are accessible to outside researchers for collaborative or independent research. Some are one-of-a-kind facilities not available elsewhere in the United States.

Among the more than 350 publications of NIST are the following:

- *Journal of Research of the National Institute of Standards and Technology*, bimonthly reports on research and development results in physics, chemistry, engineering, mathematics, and computer science.
- *Technology at a Glance*, a free newsletter providing updates on NIST research, grants, and other program activities. Issued quarterly.
- *Guide to NIST*, provides an overview of the activities of NIST with brief descriptions of more than 250 research projects, grants, and industry outreach programs, services and facilities, followed by contact names and phone numbers.

Noteworthy programs include the following:

- *Electrical and Optical Properties of Polymers*
- *Gaseous Electronics*
- *Polymer Blends*
- *Polymer Combustion Research*
- *Polymer Mechanics and Structure*
- *Processing Science for Polymer Composites*
- *Semiconductor Electronics*
- *Testing Electronic Systems*

NATIONAL CENTER FOR STANDARDS AND
CERTIFICATION INFORMATION (NCSCI)

Established in 1965 as part of NIST, The National Center for Standards and Certification Information provides information on U.S., foreign, and international voluntary standards, government regulations, and rules of certification for nonagricultural products. The Center serves as a referral service and focal point for information about standards and standards-related information. The NCSCI reference collection of standards-related documents includes:

- Microfilm files of military and federal specifications, U.S. industry and national standards, and international and selected foreign national standards;
- Reference books, including directories, technical and scientific dictionaries, encyclopedias, and handbooks;
    - Articles, pamphlets, reports, and handbooks on standardization and certification;
    - Standards-related periodicals and newsletters.

These documents are available in the Center for review only. NCSCI does not provide copies of its documents. This facility is open to visitors Monday–Friday, 8:30 AM–5:00 PM at Technology Building, Room B157, NIST, Gaithersburg, MD (about 25 miles northwest of Washington, DC).

RURAL ELECTRIFICATION
ADMINISTRATION (REA)

U.S. Department of Agriculture, Washington, DC 20250

REA, an agency of the U.S. Department of Agriculture, was created by Executive Order in 1935 to finance electric and telephone facilities in rural areas. Loans for these activities reached a peak of over $7 billion in 1980, dropping to $1.2 billion in 1992, when agency staff headquartered in Washington was 417.

Electric Loan Program

REA makes loans to nonprofit and cooperative associations, public bodies, and other electric utilities. The loans finance the construction and operation of generating plants, transmission and distribution lines, or systems to provide initial and continued adequate electric service to persons in rural areas. The agency also provides engineering and management assistance to its borrowers. Loans are repaid from the operating revenues of the systems REA finances. Publications are available covering all aspects of this program.

Telephone Loan Program

Telephone loans may be made to telephone companies, to public bodies, and to cooperative nonprofit, limited-dividend or mutual associations. The agency also provides engineering and management assistance to its borrowers. Loans are repaid from the operating revenues of the systems REA finances. Publications are available covering all aspects of this program.

U.S. GOVERNMENT SPECIFICATIONS

There are several thousand U.S. Government specifications relating to electrical/ electronic insulating materials. Information about both government and industry specifications and standards is available from many government sources. Major sources are listed below. (The end of this chapter contains Exhibit 19-4, Military Specification MIL-P-46179 (MR), *Plastic Molding and Extrusion Material, Polyamide-imide*, which is typical for a plastic material.) Note references to ASTM standards and UL tests.

U.S. GOVERNMENT PRINTING OFFICE

Washington, DC 20402

Several key publications may be obtained from the Superintendent of Documents, U.S. Government Printing Office:

• *Department of Defense Index of Specifications and Standards (DODISS)*, D7.14:981. Subscription service consists of two parts. Part I, Alphabetical Index; Part II, Numerical Index. This is a consolidated edition of the indexes of Military Specifications and Standards used by the Department of the Air Force, Department of the Army, and the Department of the Navy, which were formerly issued separately.

• *Index of Federal Specifications, Standards and Commercial Item Descriptions*, GS2.8/2:981. Subscription service provides alphabetical, numerical, and Federal Supply Classification listings of Federal and Interim Federal Standards, Federal Handbooks, and Qualified Products Lists in general use throughout the Federal Government.

• *Plastics for Aerospace Vehicles, Part I—Reinforced Plastics (Handbook)*, MIL-HDBK-17A

• *Plastics (Handbook)*, MIL-HDBK-700

• *Adhesives—A Guide to Their Properties and Uses as Described by Federal and Military Specifications (Handbook)*, MIL-HDBK-725

GENERAL SERVICES ADMINISTRATION (GSA)

Federal specifications are available from the GSA Business Service Centers at the addresses shown in Table 19-8.

TABLE 19-8   GSA Business Service Centers

ADDRESS AND TELEPHONE NUMBER	SERVICE AREA
10 Causeway Street Boston, MA 02222 (617) 565-8100	*Region 1*—Connecticut, Massachusetts, New Hampshire, Maine, Rhode Island, and Vermont
26 Federal Plaza New York, NY 10278 (212) 264-1234	*Region 2*—New Jersey, New York, Puerto Rico, and Virgin Islands
Seventh and D Streets, SW Washington, DC 20407 (202) 708-5804	*Region 3 (National Capital Region)*—District of Columbia, nearby Maryland and Virginia
100 Penn Square East Philadelphia, PA 19107 (215) 656-5525	*Region 3*—Delaware, Pennsylvania, Maryland and Virginia (other than above), and West Virginia
401 West Peachtree Street Atlanta, GA 30365 (404) 331-5103	*Region 4*—Alabama, Florida, Georgia, Kentucky, Mississippi, North Carolina, South Carolina, and Tennessee
230 South Dearborn Street Chicago, IL 60604 (312) 353-5383	*Region 5*—Illinois, Indiana, Michigan, Minnesota, Ohio, and Wisconsin
1500 East Bannister Road Kansas City, MO (816) 926-7203	*Region 6*—Iowa, Kansas, Missouri, and Nebraska
819 Taylor Street Fort Worth, TX 76102 (817) 334-3284	*Region 7*—Arkansas, Louisiana, New Mexico, Oklahoma, and Texas
Denver Federal Center, Building 41 Denver, CO 80225 (303) 236-7408	*Region 8*—Colorado, Montana, North Dakota, South Dakota, Utah, and Wyoming
525 Market Street San Francisco, CA 94105 (415) 744-5050	*Region 9*—Northern California, Hawaii, and all Nevada except Clark County
300 North Los Angeles Street Los Angeles, CA 90012 (213) 894-3210	*Region 9*—Los Angeles, Southern California, Arizona, and Clark County, Nevada
400 15th Street, SW Auburn, WA 98001 (206) 931-7956	*Region 10*—Alaska, Idaho, Oregon, and Washington

THE NAVAL PUBLICATIONS AND FORMS CENTER

5801 Tabor Avenue, Philadelphia, PA 19120

This facility has responsibility for the operation of Defense Single Stock Point for Standardization Documents included in the DoD Index of Military Specifications and Standards.

Publications available include:

• *Department of Defense Single Stock Point (DoD-SSP) for Specifications and Standards, A Guide for Private Industry*

• *Army, Navy, and Air Force Specifications, including MIL and JAN Specifications*

• *Plastics (Handbook)*, MIL-HDBK-700A

• *Adhesives—A Guide to Their Properties and Uses as Described by Federal and Military Specifications* (Handbook), MIL-HDBK-725

Exhibit 19-3 shows the order form for Military Specifications and Standards.

DEPARTMENT OF THE NAVY
NAVAL PUBLICATIONS AND FORMS CENTER
5801 Tabor Ave
Philadelphia, Pa. 19120
OFFICIAL BUSINESS
PENALTY FOR PRIVATE USE $300

POSTAGE AND FEES PAID
DEPARTMENT OF THE NAVY
DOD - 316

Please self address the above label

SPECIFICATIONS AND STANDARDS REQUISITION

Send the number of copies of documents listed below which are in the DoD Index of Specifications and Standards. Limit 5 line items per request. Requests submitted on this form will speed service. Reorder forms will be enclosed with each shipment. *FORWARD REQUEST TO:* Naval Publications and Forms Center, Code 3015, 5801 Tabor Avenue, Philadelphia, PA 19120.

1. QUANTITY	2. STANDARDIZATION DOCUMENT SYMBOL	3. TITLE *(From DoD Index of Specifications and Standards)*	4. NONAVAILABILITY CODE * *(NPFC use only)*

*NONAVAILABILITY CODE EXPLANATION

B — Item temporarily not in stock. Resubmit request in 30 days.
E — Item not identifiable as an active document listed in the DODISS. Direct questions concerning this action to NPFC, Attn: Code 1052 or call (215) 697-3321.
I — For Official Use Only. Submit request via cognizant DoD Inspection Office or Contract Administrator for certification of "need to know."
L — Industry standardization documents will not be furnished by NPFC to commercial concerns. Copies may be purchased from the appropriate Industry Association.

5. SIGNATURE OF REQUESTOR	6. DATE PREPARED *(YY, MM, DD)*	7. CLOSING DATE *(YY, MM, DD)* *(IFB, RFQ, or RFP)*

DD FORM 1425
82 MAY

Previous editions of this form are obsolete

S/N 0102-LF-001-4252

Exhibit 19-3   (a) Order Form DD 1425 for Military Specifications and Standards.

INSTRUCTIONS FOR COMPLETING DD FORM 1425

Item   1. **Quantity:** Enter the number of publications requested.

   *2. **Standardization Document Symbol:** Indicate the appropriate Standardization Document Symbol for each document requested.

    3. **Title:** Enter the title of the document requested.

   *4. **Nonavailability Code:** (For NPFC use only): The Naval Publications and Forms Center will enter the appropriate code to indicate the reason the requested document is unavailable.

    5. **Signature of Requestor:** Self-explanatory

   *6. **Date Prepared:** Enter the date prepared in the sequence of year, month, day (e.g., July 1, 1980 would be entered 80/07/01).

   *7. **Closing Date:** Enter the closing date in the following format: year, month, day.

"General Note for Personnel Processing This Report:

Items marked with an asterisk (*) have been registered in the DoD Data Element Program."

Exhibit 19-3   (b) DD FORM 1425 (*Back*).

POLYAMIDE-IMIDE MILITARY SPECIFICATION MIL-P-46179

The following (Exhibit 19-4) is an example of a military specification for a plastic insulating material.

MIL-P-46179(MR)
3 December 1979

MILITARY SPECIFICATION

PLASTIC MOLDING AND EXTRUSION MATERIAL, POLYAMIDE-IMIDE

This specification is approved for use by the Army Materials and Mechanics Research Center, Department of the Army, and is available for use by all Departments and Agencies of the Department of Defense.

1. SCOPE AND CLASSIFICATION

1.1 *Scope*. This specification covers two grades of polyamide-imide resins.

1.2 *Classification*.

1.2.1 *Grades*. The polyamide-imide resins shall be of the following grades, as specified (see 6.2).

   Grade GP—General purpose
   Grade E—Electrical

2. APPLICABLE DOCUMENTS

2.1 *Issues of documents*. The following documents, of the issues in effect on date of invitation for bids or request for proposal, form a part of this specification to the extent specified herein.

SPECIFICATIONS
 FEDERAL
   PPP-D-723—Drums, Fiber
   PPP-D-729—Drums, Metal, 55-Gallon (for Shipment of Noncorrosive Material)
STANDARDS
 MILITARY
   MIL-STD-105—Sampling Procedures and Tables for Inspection by Attributes

Exhibit 19-4.

MIL-STD-129—Marking for Shipment and Storage
MIL-STD-1188—Commercial Packaging of Supplies and Equipment

(Copies of specifications, standards, drawings, and publications required by suppliers in connection with specific procurement functions should be obtained from the procuring activity or as directed by the contracting officer.)

2.2 *Other publications*. The following documents form a part of this specification to the extent specified herein. Unless otherwise indicated, the issue in effect on date of invitation for bids or request for proposal shall apply.

AMERICAN SOCIETY FOR TESTING AND MATERIALS (ASTM) STANDARDS
D 150—AC Loss Characteristics and Dielectric Constant (Permittivity) of Solid Electrical Insulating Materials
D 257—D-C Resistance or Conductance of Insulating Materials
D 618—Conditioning Plastics and Electrical Insulating Materials
D 648—Deflection Temperature of Plastics Under Load
D 785—Rockwell Hardness of Plastics and Electrical Insulating Materials
D 792—Specific Gravity and Density of Plastics by Displacement
D 1708—Tensile Properties of Plastics by Use of Microtensile Specimens
(Application for copies should be addressed to the American Society for Testing and Materials, 1916 Race Street, Philadelphia, PA 19103).

UNDERWRITERS' LABORATORIES (UL), INC.
UL 94—Tests for Flammability of Plastic Materials for Parts in Devices and Appliances, dated December 1973.
(Application for copies should be made to the Underwriters' Laboratories, Inc., 207 E. Ohio Street, Chicago, IL 60611).

3. REQUIREMENTS

3.1 *Material*. The material shall consist of polyamide-imide resins with a basic polymer structure consisting of aromatic diamines linked to trimellitic anhydrides. The resins may contain up to 10 percent pigments, 5 percent lubricants and no fibrous reinforcement. Addition of reground material shall be permitted to the extent that the resultant material meets all property requirements specified herein.

3.2 *General property values (applicable to grades GP and E)*. The material shall conform to the property values specified in table I, when tested as specified in the applicable procedure of 4.3.

Table I. General property values

Property	Value
Specific gravity	1.36 to 1.44
Tensile strength MPa (psi), min.	172 (25,000)
Tensile elongation at break, percent, min.	10
Deflection temperature °C (°F), min.	268 (515)
Hardness, Rockwell scale E, min.	E 77
Flammability	94 V-0[1]

[1] 94 V-0 by Underwriters' Laboratories test specified in 4.3.7 correlation with flammability under actual use conditions is not necessarily implied.

Exhibit 19-4 (*cont.*)

3.3 *Electrical property values (applicable to grade E only).*

Table II. Electrical property values

Property	Value
Dielectric constant at 10^3 Hz, min.	3.65
Dissipation factor at 10^3 Hz, max.	0.022
Volume resistivity, ohms-cm (ohm-in), min.	0.8×10^{16} (2×10^{16})
Surface resistivity, ohms, min.	5×10^{16}

3.4 *Color.* Unless otherwise specified, the color shall be opaque greenish brown (see 6.2).

3.5 *Form.* The form shall be pellets. When specified by the procuring activity, the form shall be in the size and shape specified (see 6.2).

3.6 *Workmanship.* The material shall be free of contamination and color and form shall be uniform from container to container within each lot (see 4.2.2.1.1).

4. QUALITY ASSURANCE PROVISIONS

4.1 *Responsibility for inspection.* Unless otherwise specified in the contract, the contractor is responsible for the performance of all inspection requirements as specified herein. Except as otherwise specified in the contract, the contractor may use his own or any other facilities suitable for the performance of the inspection requirements specified herein, unless disapproved by the Government. The Government reserves the right to perform any of the inspections set forth in the specification where such inspections are deemed necessary to assure supplies and services conform to prescribed requirements.

4.2 *Sampling for inspection and acceptance.* Sampling for inspection shall be performed in accordance with the provisions set forth in MIL-STD-105 except where otherwise indicated. For purposes of sampling, an inspection lot for examination and tests shall consist of all material submitted for delivery at one time.

4.2.1 *Inspection of materials and components.* In accordance with 4.1, the supplier is responsible for insuring that materials and components used were manufactured, tested and inspected in accordance with the requirements of referenced subsidiary specifications and standards to the extent specified. In event of conflict, this specification shall govern. A supplier's certificate of compliance with 3.1 shall be furnished.

4.2.2 *Inspection of material.*

4.2.2.1 *Examination of the material.* Examination of the material shall be made in accordance with the classification of defects, inspection levels and acceptable quality levels (AQLs) set forth below. The lot size, for purpose of determining the sample size in accordance with MIL-STD-105, shall be expressed in units of 100 pounds for examination in 4.2.2.1.1, and in units of shipping containers for examination in 4.2.2.1.2.

4.2.2.1.1 *Examination of the material for defects in appearance and workmanship.* The sample unit for this examination specified in table III shall be approximately one pound.

Exhibit 19-4 (*cont.*)

Table III. Examination of the material for defects in appearance and workmanship

Examine	Defect
Appearance and workmanship	When required, form size not as specified
	Not uniform in form from container to container
	Not uniform in color from container to container
	Not clean, presence of foreign material

4.2.2.1.2 *Examination of the preparation for delivery.* An examination shall be made in accordance with table IV to determine that packing and marking comply with section 5 requirements. The sample unit for this examination shall be one shipping container fully packed, selected just prior to the closing operation. Shipping containers fully prepared for delivery shall be examined for closure defects.

Table IV. Examination of preparation for delivery

Examine	Defects
Packing	Not level specified; not in accordance with contract requirements.
	Any nonconforming component, component missing, damaged or otherwise defective affecting serviceability.
	Inadequate application of components such as: incomplete closures of case liners; container flaps, loose or inadequate strappings, bulged or distorted container.
Quantity of material	Less than specified or indicated quantity.
Weight	Gross weight exceeds specified requirements.
Markings	Interior or exterior markings omitted, illegible, incorrect, incomplete, of improper size, location, sequence, method of application, or not in accordance with contract requirements.

4.2.2.1.3 *Inspection levels and acceptable quality levels (AQL's) for examinations.* The inspection levels for determining the sample size and the acceptable quality level (AQL) expressed as defects per 100 units shall be as follows:

Examination paragraph	Inspection level	AQL
4.2.2.1.1	II	2.5
4.2.2.1.2	S-2	2.5

4.2.3 *Testing.* The molding material shall be tested for the characteristics listed in table I and table II, as applicable. Testing shall be in accordance with the test methods specified herein for each lot submitted for inspection. The lot size, for the purpose of determining the sample size for testing shall be expressed in units of 100 pounds of material. The

Exhibit 19-4 (*cont.*)

sample unit of product shall consist of sufficient material to prepare all required specimens. The inspection level shall be S-1, with an acceptance number of 0. The results for
each test shall be the averaged results of the specimens, unless only one specimen is
required for testing.

4.2.3.1 *Classification of tests.* All tests shall be classified, as follows:

    a. Lot acceptance tests (see 4.2.3.2).

    b. Periodic lot check tests (see 4.2.3.3).

4.2.3.2 *Lot acceptance tests.* Lot acceptance tests shall be made on each lot of material
and shall be the basis for acceptance or rejection of the lot, except when periodic lot check
tests are required. Lot acceptance tests shall consist of testing for tensile strength, tensile
elongation at break, and hardness. For grade E, lot acceptance tests shall include the
additional tests specified in table II.

4.2.3.3 *Periodic lot check tests.* Periodic lot check tests shall be made on the first lot
of material furnished under this specification, and on any subsequent lot specified by the
procuring agency (see 6.2). If not specified by the procuring agency, periodic lot check
tests shall be repeated at least once every year. Periodic lot check tests shall consist of all
tests specified in table I and table II, as applicable. When periodic lot check tests are
made, they shall be included in the basis for acceptance or rejection of the lot.

4.3 *Test methods.*

4.3.1 *Preparation of specimens.* Unless otherwise specified, test specimens shall be
prepared by injection molding. Specimens shall be post cured for 24 ± 0.1 hours at
166° ± 5°C (330° ± 9°F), followed by 24 ± 0.1 hours at 243 ± 3°C (470° ± 5°F), followed by 24 ± 0.1 hours at 257 ± 3°C (495° ± 5°F).

4.3.2 *Test conditions.* Unless otherwise specified, all tests shall be performed at
23° ± 2°C (73.4° ± 3.6°F) and 50 ± 5 percent relative humidity. Specimens shall be
conditioned in accordance with procedure A of ASTM D 618. Prior to conditioning,
specimens shall be dried for 24 ± 0.1 hours at 166° ± 5°C (330° ± 9°F).

4.3.3 *Specific gravity.* One specimen shall be tested in accordance with method A-1 or
A-2 of ASTM D 792 using the test specimen size specified in this method.

4.3.4 *Tensile strength and elongation at break.* Five specimens, each 3.2 ± 0.1 mm
(0.125 ± 0.005 inches) thick, shall be tested in accordance with ASTM D 1708.

4.3.5 *Deflection temperature under load.* Three specimens shall be tested in accordance with ASTM D 648 using a fiber stress of 1820 KPa (264 psi). Length and width of
test specimens shall be as specified in D 648 and thickness shall be 3.2 ± 0.3 mm
(0.125 ± 0.015 inch).

4.3.6 *Hardness.* Five measurements shall be made in accordance with ASTM D 785.
The determinations shall be made using procedure A and scale E as specified in ASTM
D 785.

4.3.7 *Flammability.* Specimens shall be tested in accordance with Underwriters' Laboratories UL 94 using 1.6 mm (1/16 inch) thick specimens. The number of specimens to be
tested and dimensions for length and width shall be as specified in Underwriters' Laboratories UL 94.

4.3.8 *Dielectric constant and dissipation factor.* Three specimens shall be tested in
accordance with ASTM D 150, using any electrode system specified in this test method.

4.3.9 *Volume and surface resistivity.* Three specimens, each 3.2 ± 0.3 mm
(0.125 ± 0.0125 inches) thick shall be tested in accordance with ASTM D 257. Other
specimen dimensions shall be as allowed in ASTM D 257.

5. PACKAGING

Exhibit 19-4 (*cont.*)

Application. The requirements of section 5 apply only to purchase by or direct shipment to the Government.

5.1 *Packing.* Packing shall be level A, B, or Commercial as specified (see 6.2).

5.1.1 *Level A.* Unless otherwise specified, the material shall be packed in one of the following types of containers:

    a. Fiber drums conforming to PPP-D-723, type II, grade A, or type III, grade A in quantities of 200 pounds, maximum.

    b. Metal drums conforming to PPP-D-729, type III or type IV, in quantities of 400 pounds, maximum.

Insofar as practical, drums shall be of uniform shape and size, with minimum cube and tare consistent with the protection required. Drums shall contain identical quantities and shall be closed in accordance with the applicable container specification. Fiber drums shall be furnished with a 0.004 inch thick polyethylene liner property heat sealed.

5.1.2 *Level B.* Unless otherwise specified, the material shall be packed in one of the following types of containers:

    a. Fiber drums conforming to PPP-D-723, type I, grade A in quantities of 200 pounds maximum.

    b. Metal drums conforming to PPP-D-729, type III or type IV, in quantities of 400 pounds, maximum.

Insofar as practical, drums shall be of uniform shape and size with minimum cube and tare consistent with the protection required. Drums shall contain identical quantities and shall be closed in accordance with the applicable container specification. Fiber drums shall be furnished with a 0.004 inch thick polyethylene liner properly heat sealed.

5.1.3 *Commercial.* The material shall be packed in accordance with MIL-STD-1188.

5.2 *Marking.* In addition to any special marking required by the contract or purchase order, shipping containers shall be marked in accordance with MIL-STD-129, with the exception that commercial marking in accordance with MIL-STD-1188 applies for commercial packaging only.

6. NOTES

6.1 *Intended use.* The thermoplastic material is intended for use in applications requiring good mechanical and electrical properties at 250°C (500°F). Strength is retained at cryogenic temperatures making possible applications requiring a wide temperature range. The material is resistant to acids, hydrocarbon oils and most common solvents. Stability in vacuum is good and resistance to gamma radiation is excellent.

6.2 *Ordering data.* Procurement documents should specify the following:

    a. Title, number and date of this specification.

    b. Grade required (see 1.2.1).

    c. Color, if different (see 3.4).

    d. Specific pellet size, if required (see 3.5).

    e. Periodic lot check tests, if required (see 4.2.3.3).

    f. Level of packing required (see 5.1).

Custodian: Preparing activity:

   Army—MR   Army—MR

                                               Project No. 9330-A860

Exhibit 19-4 (*cont.*)

Appendix

CONVERSION FACTORS FOR FREQUENTLY USED UNITS

Measures and Weights

CLASSIFICATION UNIT	EQUIVALENTS IN OTHER UNITS OF SAME SYSTEM	METRIC EQUIVALENT
Weight (*avoirdupois*)		
Grain	0.036 dram 0.002285 ounce	0.0648 gram
Dram	27.343 grains 0.0625 ounce	1.771 grams
Ounce	16 drams 437.5 grains	28.349 grams
Pound	16 ounces 7,000 grains	0.453 kilogram
Short ton	2,000 pounds	0.907 metric ton
Long ton	2,240 pounds	1.016 metric ton
(*troy*)		
Grain	0.042 pennyweight 0.002083 ounce	0.0648 gram
Pennyweight	24 grains 0.05 ounce	1.555 grams
Ounce	20 pennyweight 480 grains	31.103 grams
Pound	12 ounces 240 pennyweight 5,760 grams	0.373 kilogram
(*apothecaries'*)		
Grain	0.05 scruple 0.002083 ounce 0.0166 dram	0.0648 gram
Scruple	20 grains 0.333 dram	1.295 grams
Dram	3 scruples 60 grains	3.887 grams
Ounce	8 drams 480 grains	31.103 grams
Pound	12 ounces 5,760 grains	0.373 kilogram
Volume		
Cubic inch	0.00058 cubic foot 0.000021 cubic yard	16.378 cubic centimeters
Cubic foot	1,728 cubic inches 0.0370 cubic yard	0.028 cubic meter
Cubic yard	27 cubic feet 46,656 cubic inches	0.765 cubic meter

(*continued*)

Measures and Weights (*continued*)

CLASSIFICATION UNIT	EQUIVALENTS IN OTHER UNITS OF SAME SYSTEM	METRIC EQUIVALENT
U.S. liquid measure		
Minim	1/60 fluidram	0.061610 milliliters
	0.003759	
Fluidram	60 minims	3.696 milliliters
	0.225 cubic inch	
Fluidounce	8 fluidrams	29.573 milliliters
	1.804 cubic inches	
Gill	4 fluidounces	118.291 milliliters
	7.218 cubic inches	
Pint	4 gills	0.473 liter
	28.875 cubic inches	
Quart	2 pints	0.946 liter
	57.75 cubic inches	
Gallon	4 quarts	3.785 liters
	231 cubic inches	
U.S. dry measure		
Pint	33.600 cubic inches	0.550 liter
Quart	2 pints	1.101 liters
	67.200 cubic inches	
Peck	8 quarts	8.809 liters
	537.605 cubic inches	
Bushel	4 pecks	35.238 liters
	2150.42 cubic inches	
Length		
Inch	0.083 foot	2.540 centimeters
	0.027 yard	
Foot	12 inches	30.48 centimeters
	0.333 yard	
Yard	3 feet	0.914 meter
	36 inches	
Rod	5.50 yards	5.029 meters
	16.5 feet	
Mile	5,280 feet	1.609 kilometers
	1,760 yards	
Area		
Square inch	0.007 square foot	6.451 square centimeters
	0.00077 square yard	
Square foot	144 square inches	0.093 square meter
	0.111 square yard	
Square rod	30.25 square yards	25.293 square meters
	0.006 acre	
Acre	4,840 square yards	0.405 hectare
	43,560 square feet	4,047 square meters
Square mile	640 acres	2,590 square kilometers
	102,400 square rods	259.2 hectares

Pressure

atm		psia		in Hg*		mm Hg*		bar		kg/cm²		dyn/cm²		N/m² (pascal)
1	=	14.6960	=	29.921	=	760.0	=	1.01325	=	1.03323	=	1,013,250	=	101,325

Volume

in³		ft³		gal		L		m³		cm³
1	=	5.787×10^{-4}	=	4.329×10^{-3}	=	0.0163871	=	1.63871×10^{-5}	=	16.3871

Density

lb/ft³		lb/gal		g/cm³		kg/m³ = g/L
1	=	0.133680	=	0.016018	=	16.018463

*Mercury.

Temperature Formulas

To convert from	to	
Degrees Celsius	Kelvin	$T_K = T_C + 273.15$
Degrees Fahrenheit	Kelvin	$T_K = (T_F + 459.67)/1.8$
Degrees Rankine	Kelvin	$T_K = T_R/1.8$
Degrees Fahrenheit	Celsius	$T_C = (T_F - 32)/1.8$
Degrees Kelvin	Celsius	$T_C = T_K - 273.15$

Temperature Equivalents

°F	READING IN °F OR °C TO BE CONVERTED	°C	°F	READING IN °F OR °C TO BE CONVERTED	°C
—	−458	−272.22	—	−422	−252.22
—	−456	−271.11	—	−420	−251.11
—	−454	−270.00	—	−418	−250.00
—	−452	−268.89	—	−416	−248.89
—	−450	−267.78	—	−414	−247.78
—	−448	−266.67	—	−412	−246.67
—	−446	−265.56	—	−410	−245.56
—	−444	−264.44	—	−408	−244.44
—	−442	−263.33	—	−406	−243.33
—	−440	−262.22	—	−404	−242.22
—	−438	−261.11	—	−402	−241.11
—	−436	−260.00	—	−400	−240.00
—	−434	−258.89	—	−398	−238.89
—	−432	−257.78	—	−396	−237.78
—	−430	−256.67	—	−394	−236.67
—	−428	−255.56	—	−392	−235.56
—	−426	−254.44	—	−390	−234.44
—	−424	−253.33	—	−388	−233.33

Temperature Equivalents (*continued*)

°F	READING IN °F OR °C TO BE CONVERTED	°C	°F	READING IN °F OR °C TO BE CONVERTED	°C
—	−386	−232.22	—	−292	−180.00
—	−384	−231.11	—	−290	−178.89
—	−382	−230.00	—	−288	−177.78
—	−380	−228.89	—	−286	−176.67
—	−378	−227.78	—	−284	−175.56
—	−376	−226.67	—	−282	−174.44
—	−374	−225.56	—	−280	−173.33
—	−372	−224.44	—	−278	−172.22
—	−370	−223.33	—	−276	−171.11
—	−368	−222.22	—	−274	−170.00
—	−366	−221.11	−457.6	−272	−168.89
—	−364	−220.00	−454.0	−270	−167.78
—	−362	−218.89	−450.4	−268	−166.67
—	−360	−217.78	−446.8	−266	−165.56
—	−358	−216.67	−443.2	−264	−164.44
—	−356	−215.56	−439.2	−262	−163.33
—	−354	−214.44	−436.0	−260	−162.22
—	−352	−213.33	−432.4	−258	−161.11
—	−350	−212.22	−428.8	−256	−160.00
—	−348	−211.11	−425.2	−254	−158.89
—	−346	−210.00	−421.6	−252	−157.78
—	−344	−208.89	−418.0	−250	−156.67
—	−342	−207.78	−414.4	−248	−155.56
—	−340	−206.67	−410.8	−246	−154.44
—	−338	−205.56	−407.2	−244	−153.33
—	−336	−204.44	−403.6	−242	−152.22
—	−334	−203.33	−400.0	−240	−151.11
—	−332	−202.22	−396.4	−238	−150.00
—	−330	−201.11	−392.8	−236	−148.89
—	−328	−200.00	−389.2	−234	−147.78
—	−326	−198.89	−385.6	−232	−146.67
—	−324	−197.78	−382.0	−230	−145.56
—	−322	−196.67	−378.4	−228	−144.44
—	−320	−195.56	−374.8	−226	−143.33
—	−318	−194.44	−371.2	−224	−142.22
—	−316	−193.33	−367.6	−222	−141.11
—	−314	−192.22	−364.0	−220	−140.00
—	−312	−191.11	−360.4	−218	−138.89
—	−310	−190.00	−356.8	−216	−137.78
—	−308	−188.89	−353.2	−214	−136.67
—	−306	−187.78	−349.6	−212	−135.56
—	−304	−186.67	−346.0	−210	−134.44
—	−302	−185.56	−342.4	−208	−133.33
—	−300	−184.44	−338.8	−206	−132.22
—	−298	−183.33	−335.2	−204	−131.11
—	−296	−182.22	−331.6	−202	−130.00
—	−294	−181.11	−328.0	−200	−128.89

Temperature Equivalents (*continued*)

°F	READING IN °F OR °C TO BE CONVERTED	°C	°F	READING IN °F OR °C TO BE CONVERTED	°C
−324.4	−198	−127.78	−155.2	−104	−75.56
−320.8	−196	−126.67	−151.6	−102	−74.44
−317.2	−194	−125.56	−148.0	−100	−73.33
−313.6	−192	−124.44	−144.4	−98	−72.22
−310.0	−190	−123.33	−140.8	−96	−71.11
−306.4	−188	−122.22	−137.2	−94	−70.00
−302.8	−186	−121.11	−133.6	−92	−68.89
−299.2	−184	−120.00	−130.0	−90	−67.78
−295.6	−182	−118.89	−126.4	−88	−66.67
−292.0	−180	−117.78	−122.8	−86	−65.56
−288.4	−178	−116.67	−119.2	−84	−64.44
−284.8	−176	−115.56	−115.6	−82	−63.33
−281.2	−174	−114.44	−112.0	−80	−62.22
−277.6	−172	−113.33	−108.4	−78	−61.11
−274.0	−170	−112.22	−104.8	−76	−60.00
−270.4	−168	−111.11	−101.2	−74	−58.89
−266.8	−166	−110.00	−97.6	−72	−57.78
−263.2	−164	−108.89	−94.0	−70	−56.67
−259.6	−162	−107.78	−90.4	−68	−55.56
−256.0	−160	−106.67	−86.8	−66	−54.44
−252.4	−158	−105.56	−83.2	−64	−53.33
−248.8	−156	−104.44	−79.6	−62	−52.22
−245.2	−154	−103.33	−76.0	−60	−51.11
−241.6	−152	−102.22	−72.4	−58	−50.00
−238.0	−150	−101.11	−68.8	−56	−48.89
−234.4	−148	−100.00	−65.2	−54	−47.78
−230.8	−146	−98.89	−61.6	−52	−46.67
−227.2	−144	−97.78	−58.0	−50	−45.56
−223.6	−142	−96.67	−54.4	−48	−44.44
−220.0	−140	−95.56	−50.8	−46	−43.33
−216.4	−138	−94.44	−47.2	−44	−42.22
−212.8	−136	−93.33	−43.6	−42	−41.11
−209.2	−134	−92.22	−40.0	−40	−40.00
−205.6	−132	−91.11	−36.4	−38	−38.89
−202.0	−130	−90.00	−32.8	−36	−37.78
−198.4	−128	−88.89	−29.2	−34	−36.67
−194.8	−126	−87.78	−25.6	−32	−35.56
−191.2	−124	−86.67	−22.0	−30	−34.44
−187.6	−122	−85.56	−18.4	−28	−33.33
−184.0	−120	−84.44	−14.8	−26	−32.22
−180.4	−118	−83.33	−11.2	−24	−31.11
−176.8	−116	−82.22	−7.6	−22	−30.00
−173.2	−114	−81.11	−4.0	−20	−28.89
−169.6	−112	−80.00	−0.4	−18	−27.78
−166.0	−110	−78.89	+3.2	−16	−26.67
−162.4	−108	−77.78	+6.8	−14	−25.56
−158.8	−106	−76.67	+10.4	−12	−24.44

Temperature Equivalents (*continued*)

°F	READING IN °F OR °C TO BE CONVERTED	°C	°F	READING IN °F OR °C TO BE CONVERTED	°C
+14.0	−10	−23.33	+100.4	+38	+3.33
+17.6	−8	−22.22	+102.2	+39	+3.89
+19.4	−7	−21.67	+104.0	+40	+4.44
+21.2	−6	−21.11	+105.8	+41	+5.00
+23.0	−5	−20.56	+107.6	+42	+5.56
+24.8	−4	−20.00	+109.4	+43	+6.11
+26.6	−3	−19.44	+111.2	+44	+6.67
+28.4	−2	−18.89	+113.0	+45	+7.22
+30.2	−1	−18.33	+114.8	+46	+7.78
+32.0	±0	−17.78	+116.6	+47	+8.33
+33.8	+1	−17.22	+118.4	+48	+8.89
+35.6	+2	−16.67	+120.2	+49	+9.44
+37.4	+3	−16.11	+122.0	+50	+10.00
+39.2	+4	−15.56	+123.8	+51	+10.56
+41.0	+5	−15.00	+125.6	+52	+11.11
+42.8	+6	−14.44	+127.4	+53	+11.67
+44.6	+7	−13.89	+129.2	+54	+12.22
+46.4	+8	−13.33	+131.0	+55	+12.78
+48.2	+9	−12.78	+132.8	+56	+13.33
+50.0	+10	−12.22	+134.6	+57	+13.89
+51.8	+11	−11.67	+136.4	+58	+14.44
+53.6	+12	−11.11	+138.2	+59	+15.00
+55.4	+13	−10.56	+140.0	+60	+15.56
+57.2	+14	−10.00	+141.8	+61	+16.11
+59.0	+15	−9.44	+143.6	+62	+16.67
+60.8	+16	−8.89	+145.4	+63	+17.22
+62.6	+17	−8.33	+147.2	+64	+17.78
+64.4	+18	−7.78	+149.0	+65	+18.33
+66.2	+19	−7.22	+150.8	+66	+18.89
+68.0	+20	−6.67	+152.6	+67	+19.44
+69.8	+21	−6.11	+154.4	+68	+20.00
+71.6	+22	−5.56	+156.2	+69	+20.56
+73.4	+23	−5.00	+158.0	+70	+21.11
+75.2	+24	−4.44	+159.8	+71	+21.67
+77.0	+25	−3.89	+161.6	+72	+22.22
+78.8	+26	−3.33	+163.4	+73	+22.78
+80.6	+27	−2.78	+165.2	+74	+23.33
+82.4	+28	−2.22	+167.0	+75	+23.89
+84.2	+29	−1.67	+168.8	+76	+24.44
+86.0	+30	−1.11	+170.6	+77	+25.00
+87.8	+31	−0.56	+172.4	+78	+25.56
+89.6	+32	±0.00	+174.2	+79	+26.11
+91.4	+33	+0.56	+176.0	+80	+26.67
+93.2	+34	+1.11	+177.8	+81	+27.22
+95.0	+35	+1.67	+179.6	+82	+27.78
+96.8	+36	+2.22	+181.4	+83	+28.33
+98.6	+37	+2.78	+183.2	+84	+28.89

Temperature Equivalents (*continued*)

°F	READING IN °F OR °C TO BE CONVERTED	°C	°F	READING IN °F OR °C TO BE CONVERTED	°C
+185.0	+85	+29.44	+269.6	+132	+55.56
+186.8	+86	+30.00	+271.4	+133	+56.11
+188.6	+87	+30.56	+273.2	+134	+56.67
+190.4	+88	+31.11	+275.0	+135	+57.22
+192.2	+89	+31.67	+276.8	+136	+57.78
+194.0	+90	+32.22	+278.6	+137	+58.33
+195.8	+91	+32.78	+280.4	+138	+58.89
+197.6	+92	+33.33	+282.2	+139	+59.44
+199.4	+93	+33.89	+284.0	+140	+60.00
+201.2	+94	+34.44	+285.8	+141	+60.56
+203.0	+95	+35.00	+287.6	+142	+61.11
+204.8	+96	+35.56	+289.4	+143	+61.67
+206.6	+97	+36.11	+291.2	+144	+62.22
+208.4	+98	+36.67	+293.0	+145	+62.78
+210.2	+99	+37.22	+294.8	+146	+63.33
+212.0	+100	+37.78	+296.6	+147	+63.89
+213.8	+101	+38.33	+298.4	+148	+64.44
+215.6	+102	+38.89	+300.2	+149	+65.00
+217.4	+103	+39.44	+302.0	+150	+65.56
+219.2	+104	+40.00	+303.8	+151	+66.11
+221.0	+105	+40.56	+305.6	+152	+66.67
+222.8	+106	+41.11	+307.4	+153	+67.22
+224.6	+107	+41.67	+309.2	+154	+67.78
+226.4	+108	+42.22	+311.0	+155	+68.33
+228.2	+109	+42.78	+312.8	+156	+68.89
+230.0	+110	+43.33	+314.6	+157	+69.44
+231.8	+111	+43.89	+316.4	+158	+70.00
+233.6	+112	+44.44	+318.2	+159	+70.56
+235.4	+113	+45.00	+320.0	+160	+71.11
+237.2	+114	+45.56	+321.8	+161	+71.67
+239.0	+115	+46.11	+323.6	+162	+72.22
+240.8	+116	+46.67	+325.4	+163	+72.78
+242.6	+117	+47.22	+327.2	+164	+73.33
+244.4	+118	+47.78	+329.0	+165	+73.89
+246.2	+119	+48.33	+330.8	+166	+74.44
+248.0	+120	+48.89	+332.6	+167	+75.00
+249.8	+121	+49.44	+334.4	+168	+75.56
+251.6	+122	+50.00	+336.2	+169	+76.11
+253.4	+123	+50.56	+338.0	+170	+76.67
+255.2	+124	+51.11	+339.8	+171	+77.22
+257.0	+125	+51.67	+341.6	+172	+77.78
+258.8	+126	+52.22	+343.4	+173	+78.33
+260.6	+127	+52.78	+345.2	+174	+78.89
+262.4	+128	+53.33	+347.0	+175	+79.44
+264.2	+129	+53.89	+348.8	+176	+80.00
+266.0	+130	+54.44	+350.6	+177	+80.56
+267.8	+131	+55.00	+352.4	+178	+81.11

Temperature Equivalents (*continued*)

°F	READING IN °F OR °C TO BE CONVERTED	°C	°F	READING IN °F OR °C TO BE CONVERTED	°C
+354.2	+179	+81.67	+449.6	+232	+111.11
+356.0	+180	+82.22	+453.2	+234	+112.22
+357.8	+181	+82.78	+456.8	+236	+113.33
+359.6	+182	+83.33	+460.4	+238	+114.44
+361.4	+183	+83.89	+464.0	+240	+115.56
+363.2	+184	+84.44	+467.6	+242	+116.67
+365.0	+185	+85.00	+471.2	+244	+117.78
+366.8	+186	+85.56	+474.8	+246	+118.89
+368.6	+187	+86.11	+478.4	+248	+120.00
+370.4	+188	+86.67	+482.0	+250	+121.11
+372.2	+189	+87.22	+485.6	+252	+122.22
+374.0	+190	+87.78	+489.2	+254	+123.33
+375.8	+191	+88.33	+492.8	+256	+124.44
+377.6	+192	+88.89	+496.4	+258	+125.56
+379.4	+193	+89.44	+500.0	+260	+126.67
+381.2	+194	+90.00	+503.6	+262	+127.78
+383.0	+195	+90.56	+507.2	+264	+128.89
+384.8	+196	+91.11	+510.8	+266	+130.00
+386.6	+197	+91.67	+514.4	+268	+131.11
+388.4	+198	+92.22	+518.0	+270	+132.22
+390.2	+199	+92.78	+521.6	+272	+133.33
+392.0	+200	+93.33	+525.2	+274	+134.44
+393.8	+201	+93.89	+528.8	+276	+135.56
+395.6	+202	+94.44	+532.4	+278	+136.67
+397.4	+203	+95.00	+536.0	+280	+137.78
+399.2	+204	+95.56	+539.6	+282	+138.89
+401.0	+205	+96.11	+543.2	+284	+140.00
+402.8	+206	+96.67	+546.8	+286	+141.11
+404.6	+207	+97.22	+550.4	+288	+142.22
+406.4	+208	+97.78	+554.0	+290	+143.33
+408.2	+209	+98.33	+557.6	+292	+144.44
+410.0	+210	+98.89	+561.2	+294	+145.56
+411.8	+211	+99.44	+564.8	+296	+146.67
+413.6	+212	+100.00	+568.4	+298	+147.78
+415.4	+213	+100.56	+572.0	+300	+148.89
+417.2	+214	+101.11	+575.6	+302	+150.00
+419.0	+215	+101.67	+579.2	+304	+151.11
+420.8	+216	+102.22	+582.8	+306	+152.22
+422.6	+217	+102.78	+586.4	+308	+153.33
+424.4	+218	+103.33	+590.0	+310	+154.44
+426.2	+219	+103.89	+593.6	+312	+155.56
+428.0	+220	+104.44	+597.2	+314	+156.67
+431.6	+222	+105.56	+600.8	+316	+157.78
+435.2	+224	+106.67	+604.4	+318	+158.89
+438.8	+226	+107.78	+608.0	+320	+160.00
+442.4	+228	+108.89	+611.6	+322	+161.11
+446.0	+230	+110.00	+615.2	+324	+162.22

Temperature Equivalents (*continued*)

°F	READING IN °F OR °C TO BE CONVERTED	°C	°F	READING IN °F OR °C TO BE CONVERTED	°C
+618.8	+326	+163.33	+791.6	+422	+216.67
+622.4	+328	+164.44	+795.2	+424	+217.78
+626.0	+330	+165.56	+798.8	+426	+218.89
+629.6	+332	+166.67	+802.4	+428	+220.00
+633.2	+334	+167.78	+806.0	+430	+221.11
+636.8	+336	+168.89	+809.6	+432	+222.22
+640.4	+338	+170.00	+813.2	+434	+223.33
+644.0	+340	+171.11	+816.8	+436	+224.44
+647.6	+342	+172.22	+820.4	+438	+225.56
+651.2	+344	+173.33	+824.0	+440	+226.67
+654.8	+346	+174.44	+827.6	+442	+227.78
+658.4	+348	+175.56	+831.2	+444	+228.89
+662.0	+350	+176.67	+834.8	+446	+230.00
+665.6	+352	+177.78	+838.4	+448	+231.11
+669.2	+354	+178.89	+842.0	+450	+232.22
+672.8	+356	+180.00	+845.6	+452	+233.33
+676.4	+358	+181.11	+849.2	+454	+234.44
+680.0	+360	+182.22	+852.8	+456	+235.56
+683.6	+362	+183.33	+856.4	+458	+236.67
+687.2	+364	+184.44	+860.0	+460	+237.78
+690.8	+366	+185.56	+863.6	+462	+238.89
+694.4	+368	+186.67	+867.2	+464	+240.00
+698.0	+370	+187.78	+870.8	+466	+241.11
+701.6	+372	+188.89	+874.4	+468	+242.22
+705.2	+374	+190.00	+878.0	+470	+243.33
+708.8	+376	+191.11	+881.6	+472	+244.44
+712.4	+378	+192.22	+885.2	+474	+245.56
+716.0	+380	+193.33	+888.8	+476	+246.67
+719.6	+382	+194.44	+892.4	+478	+247.78
+723.2	+384	+195.56	+896.0	+480	+248.89
+726.8	+386	+196.67	+899.6	+482	+250.00
+730.4	+388	+197.78	+903.2	+484	+251.11
+734.0	+390	+198.89	+906.8	+486	+252.22
+737.6	+392	+200.00	+910.4	+488	+253.33
+741.2	+394	+201.11	+914.0	+490	+254.44
+744.8	+396	+202.22	+917.6	+492	+255.56
+748.4	+398	+203.33	+921.2	+494	+256.67
+752.0	+400	+204.44	+924.8	+496	+257.78
+755.6	+402	+205.56	+928.4	+498	+258.89
+759.2	+404	+206.67	+932.0	+500	+260.00
+762.8	+406	+207.78	+935.6	+502	+261.11
+766.4	+408	+208.89	+939.2	+504	+262.22
+770.0	+410	+210.00	+942.8	+506	+263.33
+773.6	+412	+211.11	+946.4	+508	+264.44
+777.2	+414	+212.22	+950.0	+510	+265.56
+780.8	+416	+213.33	+953.6	+512	+266.67
+784.4	+418	+214.44	+957.2	+514	+267.78
+788.0	+420	+215.56	+960.8	+516	+268.89

Temperature Equivalents (*continued*)

°F	READING IN °F OR °C TO BE CONVERTED	°C	°F	READING IN °F OR °C TO BE CONVERTED	°C
+964.4	+518	+270.00	+1133.6	+612	+322.22
+968.0	+520	+271.11	+1137.2	+614	+323.33
+971.6	+522	+272.22	+1140.8	+616	+324.44
+975.2	+524	+273.33	+1144.4	+618	+325.56
+978.8	+526	+274.44	+1148.0	+620	+326.67
+982.4	+528	+275.56	+1151.6	+622	+327.78
+986.0	+530	+276.67	+1155.2	+624	+328.89
+989.6	+532	+277.78	+1158.8	+626	+330.00
+993.2	+534	+278.89	+1162.4	+628	+331.11
+996.8	+536	+280.00	+1166.0	+630	+332.22
1000.4	+538	+281.11	+1169.6	+632	+333.33
+1004.0	+540	+282.22	+1173.2	+634	+334.44
+1007.6	+542	+283.33	+1176.8	+636	+335.56
+1011.2	+544	+284.44	+1180.4	+638	+336.67
+1014.8	+546	+285.56	+1184.0	+640	+337.78
+1018.4	+548	+286.67	+1187.6	+642	+338.89
+1022.0	+550	+287.78	+1191.2	+644	+340.00
+1025.6	+552	+288.89	+1194.8	+646	+341.11
+1029.2	+554	+290.00	+1198.4	+648	+342.22
+1032.8	+556	+291.11	+1202.0	+650	+343.33
+1036.4	+558	+292.22	+1205.6	+652	+344.44
+1040.0	+560	+293.33	+1209.2	+654	+345.56
+1043.6	+562	+294.44	+1212.8	+656	+346.67
+1047.2	+564	+295.56	+1216.4	+658	+347.78
+1050.8	+566	+296.67	+1220.0	+660	+348.89
+1054.4	+568	+297.78	+1223.6	+662	+350.00
+1058.0	+570	+298.89	+1227.2	+664	+351.11
+1061.6	+572	+300.00	+1230.8	+666	+352.22
+1065.2	+574	+301.11	+1234.4	+668	+353.33
+1068.8	+576	+302.22	+1238.0	+670	+354.44
+1072.4	+578	+303.33	+1241.6	+672	+355.56
+1076.0	+580	+304.44	+1245.2	+674	+356.67
+1079.6	+582	+305.56	+1248.8	+676	+357.78
+1083.2	+584	+306.67	+1252.4	+678	+358.89
+1086.8	+586	+307.78	+1256.0	+680	+360.00
+1090.4	+588	+308.89	+1259.6	+682	+361.11
+1094.0	+590	+310.00	+1263.2	+684	+362.22
+1097.6	+592	+311.11	+1266.8	+686	+363.33
+1101.2	+594	+312.22	+1270.4	+688	+364.44
+1104.8	+596	+313.33	+1274.0	+690	+365.56
+1108.4	+598	+314.44	+1277.6	+692	+366.67
+1112.0	+600	+315.56	+1281.2	+694	+367.78
+1115.6	+602	+316.67	+1284.8	+696	+368.89
+1119.2	+604	+317.78	+1288.4	+698	+370.00
+1122.8	+606	+318.89	+1292.0	+700	+371.11
+1126.4	+608	+320.00	+1295.6	+702	+372.22
+1130.0	+610	+321.11	+1299.2	+704	+373.33

Temperature Equivalents (*continued*)

°F	READING IN °F OR °C TO BE CONVERTED	°C	°F	READING IN °F OR °C TO BE CONVERTED	°C
+1302.8	+706	+374.44	+1468.4	+798	+425.56
+1306.4	+708	+375.56	+1472.0	+800	+426.67
+1310.0	+710	+376.67	+1475.6	+802	+427.78
+1313.6	+712	+377.78	+1479.2	+804	+428.89
+1317.2	+714	+378.89	+1482.8	+806	+430.00
+1320.8	+716	+380.00	+1486.4	+808	+431.11
+1324.4	+718	+381.11	+1490.0	+810	+432.22
+1328.0	+718	+381.11	+1493.6	+812	+433.33
+1328.0	+720	+382.22	+1497.2	+814	+434.44
+1331.6	+722	+383.33	+1500.8	+816	+435.56
+1335.2	+724	+384.44	+1504.4	+818	+436.67
+1338.8	+726	+385.56	+1508.0	+820	+437.78
+1342.4	+728	+386.67	+1511.6	+822	+438.89
+1346.0	+730	+387.78	+1515.2	+824	+440.00
+1349.6	+732	+388.89	+1518.8	+826	+441.11
+1353.2	+734	+390.00	+1522.4	+828	+442.22
+1356.8	+736	+391.11	+1526.0	+830	+443.33
+1360.4	+738	+392.22	+1529.6	+832	+444.44
+1364.0	+740	+393.33	+1533.2	+834	+445.56
+1367.6	+742	+394.44	+1536.8	+836	+446.67
+1371.2	+744	+395.56	+1540.4	+838	+447.78
+1374.8	+746	+396.67	+1544.0	+840	+448.89
+1378.4	+748	+397.78	+1547.6	+842	+450.00
+1382.0	+750	+398.89	+1551.2	+844	+451.11
+1385.6	+752	+400.00	+1554.8	+846	+452.22
+1389.2	+754	+401.11	+1558.4	+848	+453.33
+1392.8	+756	+402.44	+1562.0	+850	+454.44
+1396.4	+758	+403.33	+1565.6	+852	+455.56
+1400.0	+760	+404.44	+1569.2	+854	+456.67
+1403.6	+762	+405.56	+1572.8	+856	+457.78
+1407.2	+764	+406.67	+1576.4	+858	+458.89
+1410.8	+766	+407.78	+1580.0	+860	+460.00
+1414.4	+768	+408.89	+1583.6	+862	+461.11
+1418.0	+770	+410.00	+1587.2	+864	+462.22
+1421.6	+772	+411.11	+1590.8	+866	+463.33
+1425.2	+774	+412.22	+1594.4	+868	+464.44
+1428.8	+776	+413.33	+1598.0	+870	+465.56
+1432.4	+778	+414.44	+1601.6	+872	+466.67
+1436.0	+780	+415.56	+1605.2	+874	+467.78
+1439.6	+782	+416.67	+1608.8	+876	+468.89
+1443.2	+784	+417.78	+1612.4	+878	+470.00
+1446.8	+786	+418.89	+1616.0	+880	+471.11
+1450.4	+788	+420.00	+1619.6	+882	+472.22
+1454.0	+790	+421.11	+1623.2	+884	+473.33
+1457.6	+792	+422.22	+1636.8	+886	+474.44
+1461.2	+794	+423.33	+1630.4	+888	+475.56
+1464.8	+796	+424.44	+1634.0	+890	+476.67

Temperature Equivalents (*continued*)

°F	READING IN °F OR °C TO BE CONVERTED	°C	°F	READING IN °F OR °C TO BE CONVERTED	°C
+1637.6	+892	+477.78	+1806.8	+986	+530.00
+1641.2	+894	+478.89	+1810.4	+988	+531.11
+1644.8	+896	+480.00	+1814.0	+990	+532.22
+1648.4	+898	+481.11	+1817.6	+992	+533.33
+1652.0	+900	+482.22	+1821.2	+994	+534.44
+1655.6	+902	+483.33	+1824.8	+996	+535.56
+1659.2	+904	+484.44	+1828.4	+998	+536.67
+1662.8	+906	+485.56	+1832.0	+1000	+537.78
+1666.4	+908	+486.67	+1850.0	+1010	+543.33
+1670.0	+910	+487.78	+1868.0	+1020	+548.89
+1673.6	+912	+488.89	+1886.0	+1030	+554.44
+1677.2	+914	+490.00	+1904.0	+1040	+560.00
+1680.8	+916	+491.11	+1922.0	+1050	+565.56
+1684.4	+918	+492.22	+1940.0	+1060	+571.11
+1688.0	+920	+493.33	+1958.0	+1070	+576.67
+1691.6	+922	+494.44	+1976.0	+1080	+582.22
+1695.2	+924	+495.56	+1994.0	+1090	+587.78
+1698.8	+926	+496.67	+2012.0	+1100	+593.33
+1702.4	+928	+497.78	+2030.0	+1110	+598.89
+1706.0	+930	+498.89	+2048.0	+1120	+604.44
+1709.6	+932	+500.00	+2066.0	+1130	+610.00
+1713.2	+934	+501.11	+2084.0	+1140	+615.56
+1716.8	+936	+502.22	+2102.0	+1150	+621.11
+1720.4	+938	+503.33	+2120.0	+1160	+626.67
+1724.0	+940	+504.44	+2138.0	+1170	+632.22
+1727.6	+942	+505.56	+2156.0	+1180	+637.78
+1731.2	+944	+506.67	+2174.0	+1190	+643.33
+1734.8	+946	+507.78	+2192.0	+1200	+648.89
+1738.4	+948	+508.89	+2210.0	+1210	+654.44
+1742.0	+950	+510.00	+2228.0	+1220	+660.00
+1745.6	+952	+511.11	+2246.0	+1230	+665.56
+1749.2	+954	+512.22	+2264.0	+1240	+671.11
+1752.8	+956	+513.33	+2282.0	+1250	+676.67
+1756.4	+958	+514.44	+2300.0	+1260	+682.22
+1760.0	+960	+515.56	+2318.0	+1270	+687.78
+1763.6	+962	+516.67	+2336.0	+1280	+693.33
+1767.2	+964	+517.78	+2354.0	+1290	+698.89
+1770.8	+966	+518.89	+2372.0	+1300	+704.44
+1774.4	+968	+520.00	+2390.0	+1310	+710.00
+1778.0	+970	+521.11	+2408.0	+1320	+715.56
+1781.6	+972	+522.22	+2426.0	+1330	+721.11
+1785.2	+974	+523.33	+2444.0	+1340	+726.67
+1788.8	+976	+524.44	+2462.0	+1350	+732.22
+1792.4	+978	+525.56	+2480.0	+1360	+737.78
+1796.0	+980	+526.67	+2498.0	+1370	+743.33
+1799.6	+982	+527.78	+2516.0	+1380	+748.89
+1803.2	+984	+528.89	+2534.0	+1390	+754.44

Temperature Equivalents (*continued*)

°F	READING IN °F OR °C TO BE CONVERTED	°C	°F	READING IN °F OR °C TO BE CONVERTED	°C
+2552.0	+1400	+760.00	+3398.0	+1870	+1021.1
+2570.0	+1410	+765.56	+3416.0	+1880	+1026.7
+2588.0	+1420	+771.11	+3434.0	+1890	+1032.2
+2606.0	+1430	+776.67	+3452.0	+1900	+1037.8
+2624.0	+1440	+782.22	+3470.0	+1910	+1043.3
+2642.0	+1450	+787.78	+3488.0	+1920	+1048.9
+2660.0	+1460	+793.33	+3506.0	+1930	+1054.4
+2678.0	+1470	+798.89	+3524.0	+1940	+1060.0
+2696.0	+1480	+804.44	+3542.0	+1950	+1065.6
+2714.0	+1490	+810.00	+3560.0	+1960	+1071.1
+2732.0	+1500	+815.56	+3578.0	+1970	+1076.7
+2750.0	+1510	+821.11	+3596.0	+1980	+1082.2
+2768.0	+1520	+826.67	+3614.0	+1990	+1087.8
+2786.0	+1530	+832.22	+3632.0	+2000	+1093.3
+2804.0	+1540	+837.78	+3650.0	+2010	+1098.9
+2822.0	+1550	+843.33	+3668.0	+2020	+1104.4
+2840.0	+1560	+848.89	+3686.0	+2030	+1110.0
+2858.0	+1570	+854.44	+3704.0	+2040	+1115.6
+2876.0	+1580	+860.00	+3722.0	+2050	+1121.1
+2894.0	+1590	+865.56	+3740.0	+2060	+1126.7
+2912.0	+1600	+871.11	+3758.0	+2070	+1132.2
+2930.0	+1610	+876.67	+3776.0	+2080	+1137.8
+2948.0	+1620	+882.22	+3794.0	+2090	+1143.3
+2966.0	+1630	+887.78	+3812.0	+2100	+1148.9
+2984.0	+1640	+893.33	+3830.0	+2110	+1154.4
+3002.0	+1650	+898.89	+3848.0	+2120	+1160.0
+3020.0	+1660	+904.44	+3866.0	+2130	+1165.6
+3038.0	+1670	+910.00	+3884.0	+2140	+1171.1
+3056.0	+1680	+915.56	+3902.0	+2150	+1176.7
+3074.0	+1690	+921.11	+3920.0	+2160	+1182.2
+3092.0	+1700	+926.67	+3938.0	+2170	+1187.8
+3110.0	+1710	+932.22	+3956.0	+2180	+1193.3
+3128.0	+1720	+937.78	+3974.0	+2190	+1198.9
+3146.0	+1730	+943.33	+3992.0	+2200	+1204.4
+3164.0	+1740	+948.89	+4010.0	+2210	+1210.0
+3182.0	+1750	+954.44	+4028.0	+2220	+1215.6
+3200.0	+1760	+960.00	+4046.0	+2230	+1221.1
+3218.0	+1770	+965.56	+4064.0	+2240	+1226.7
+3236.0	+1780	+971.11	+4082.0	+2250	+1232.2
+3254.0	+1790	+976.67	+4100.0	+2260	+1237.8
+3272.0	+1800	+982.22	+4118.0	+2270	+1243.3
+3290.0	+1810	+987.78	+4136.0	+2280	+1248.9
+3308.0	+1820	+993.33	+4154.0	+2290	+1254.4
+3326.0	+1830	+998.89	+4172.0	+2300	+1260.0
+3344.0	+1840	+1004.4	+4190.0	+2310	+1265.6
+3362.0	+1850	+1010.0	+4208.0	+2320	+1271.1
+3380.0	+1860	+1015.6	+4226.0	+2330	+1276.7

Temperature Equivalents (*continued*)

°F	READING IN °F OR °C TO BE CONVERTED	°C	°F	READING IN °F OR °C TO BE CONVERTED	°C
+4244.0	+2340	+1282.2	+4946.0	+2730	+1498.9
+4262.0	+2350	+1287.8	+4964.0	+2740	+1504.4
+4280.0	+2360	+1293.3	+4982.0	+2750	+1510.0
+4298.0	+2370	+1298.9	+5000.0	+2760	+1515.6
+4316.0	+2380	+1304.4	+5018.0	+2770	+1521.1
+4334.0	+2390	+1310.0	+5036.0	+2780	+1526.7
+4352.0	+2400	+1315.6	+5054.0	+2790	+1532.2
+4370.0	+2410	+1321.1	+5072.0	+2800	+1537.8
+4388.0	+2420	+1326.7	+5090.0	+2810	+1543.3
+4406.0	+2430	+1332.2	+5108.0	+2820	+1548.9
+4424.0	+2440	+1337.8	+5126.0	+2830	+1554.4
+4442.0	+2450	+1343.3	+5144.0	+2840	+1560.0
+4460.0	+2460	+1348.9	+5162.0	+2850	+1565.6
+4478.0	+2470	+1354.4	+5180.0	+2860	+1571.1
+4496.0	+2480	+1360.0	+5198.0	+2870	+1576.7
+4514.0	+2490	+1365.6	+5216.0	+2880	+1582.2
+4532.0	+2500	+1371.1	+5234.0	+2890	+1587.8
+4550.0	+2510	+1376.7	+5252.0	+2900	+1593.3
+4568.0	+2520	+1382.2	+5270.0	+2910	+1598.9
+4586.0	+2530	+1387.8	+5288.0	+2920	+1604.4
+4604.0	+2540	+1393.3	+5306.0	+2930	+1610.0
+4622.0	+2550	+1398.9	+5324.0	+2940	+1615.6
+4640.0	+2560	+1404.4	+5342.0	+2950	+1621.1
+4658.0	+2570	+1410.0	+5360.0	+2960	+1626.7
+4676.0	+2580	+1415.6	+5378.0	+2970	+1632.2
+4694.0	+2590	+1421.1	+5396.0	+2980	+1637.8
+4712.0	+2600	+1426.7	+5414.0	+2990	+1643.3
+4730.0	+2610	+1432.2	+5432.0	+3000	+1648.9
+4748.0	+2620	+1437.8	+5450.0	+3010	+1654.4
+4766.0	+2630	+1443.3	+5468.0	+3020	+1660.0
+4784.0	+2640	+1448.9	+5486.0	+3030	+1665.6
+4802.0	+2650	+1454.4	+5504.0	+3040	+1671.1
+4820.0	+2660	+1460.0	+5522.0	+3050	+1676.7
+4838.0	+2670	+1465.6	+5540.0	+3060	+1682.2
+4856.0	+2680	+1471.1	+5558.0	+3070	+1687.8
+4874.0	+2690	+1476.7	+5576.0	+3080	+1693.3
+4892.0	+2700	+1482.2	+5594.0	+3090	+1698.9
+4910.0	+2710	+1487.8	+5612.0	+3100	+1704.4
+4928.0	+2720	+1493.3			

INDUSTRY AND FEDERAL STANDARDS AND SPECIFICATIONS

Following is a sequential listing of key industry and Federal standards and specifications referenced in this handbook.

American Society for Testing and Materials (ASTM) Test Methods (cont.) Page

American Society for Testing and Materials (ASTM) Test Methods (cont.)	Page
D 3288	Magnet Wire Enamels	203
D 3300	Dielectric Breakdown Voltage of Insulating Oils of Petroleum Origin Under Impulse Conditions	459
D 3312	Percent Relative Monomer in Solventless Varnishes	238
D 3377	Weight Loss for Solventless Varnishes	238
D 3394	Methods of Sampling and Testing Electrical Insulating Board	323
D 3420	Dynamic Ball Burst (Pendulum) Impact Resistance of Plastic Film	314
D 3455	Compatibility of Construction Materials with Electrical Insulating Oils of Petroleum Origin	460
D 3612	Analysis of Gases Dissolved in Electrical Insulating Oil by Gas Chromatography	460
D 3613	Sampling Electrical Insulating Oils for Gas Analysis and Determination of Water Content	459
D 3809	Synthetic Dielectric Fluids for Capacitors	463
D 4217	Gel Time of Thermosetting Coating Powder	238
D 4325	Nonmetallic Conducting and Electrically Insulating Rubber Tapes	357, 364
D 4733	Solventless Electrical Insulating Varnishes	238
D 4880	Salt Water Proofness of Insulating Varnishes Over Enamelled Magnet Wire	202
D 4881	Thermal Endurance of Varnished Fibrous or Film Wrapped Magnet Wire	202
D 4882	Bond Strength of Electrical Insulating Varnishes	202
D 5109	Copper-Clad Thermosetting Laminates for Printed Wiring Boards	293
D 5214	Polyimide Resin Film for Electrical and Dielectric Applications	309
E 18	Rockwell Hardness and Rockwell Superficial Hardness of Metallic Materials	423
E 96	Water Vapor Transmission of Materials in Sheet Form	314
E 228	Linear Thermal Expansion of Rigid Solids with a Vitreous Silica Dilatometer	423
F 77	Apparent Density of Ceramics for Electron Device and Semiconductor Application	424
F 356	Beryllia Ceramics for Electronic and Electrical Applications	422, 424
F 417	Flexural Strength (Modulus of Rupture) of Electronic Grade Ceramics	424

Association of Edison Illuminating Companies (AEIC) Standard	Page
CS5	Thermoplastic and Crosslinked Polyethylene (PE) Insulated Shielded Power Cables Rated 5,000 to 69,000 volts	158

Index